Meteorology at the Millennium

This is volume 83 in the
INTERNATIONAL GEOPHYSICS SERIES
A series of monographs and textbooks
Edited by RENATA DMOWSKA, JAMES R. HOLTON and H. THOMAS ROSSBY
A complete list of books in this series appears at the end of this volume.

Meteorology at the Millennium

Edited by

R. P. Pearce

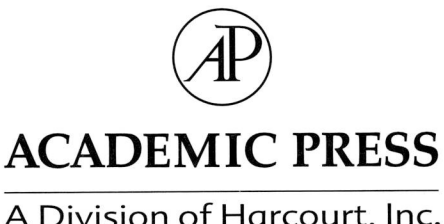

ACADEMIC PRESS
A Division of Harcourt, Inc.

San Diego   San Francisco   New York   Boston   London   Sydney   Tokyo

This book is printed on acid-free paper.

Copyright © 2002 by ACADEMIC PRESS
Except Chapters 13, 14, 16 and 21

Chapters 13, 14, 16 and 21
Copyright © British Crown Copyright 2002

All Rights Reserved.
No part of this publication may be reproduced or transmitted in any form or by any means, electronic or mechanical, including photocopying, recording, or any information storage and retrieval system, without permission in writing from the publisher.

Academic Press
A division of Harcourt, Inc.
Harcourt Place, 32 Jamestown Road, London NW1 7BY, UK
http://www.academicpress.com

Academic Press
A division of Harcourt, Inc.
525 B Street, Suite 1900, San Diego, California 92101–4495, USA
http://www.academicpress.com

ISBN 0-12-548035-0

Library of Congress Catalog Number: 2001089149

A catalogue record for this book is available from the British Library

Typeset by Fakenham Photosetting Limited, Fakenham, Norfolk
Printed and bound in Italy by LEGO, Vicenza

02 03 04 05 06 07 LE 9 8 7 6 5 4 3 2 1

Contents

Foreword xi
Preface xiii
Contributors xv

Part 1. Weather Prediction at the Millennium

Predicting Uncertainty in Numerical Weather Forecasts (T. N. Palmer)

A. INTRODUCTION 3
B. THE LIOUVILLE EQUATION 3
C. THE PROBABILITY DENSITY FUNCTION OF MODEL ERROR 5
D. PROBABILITY FORECASTING BY ENSEMBLE PREDICTION 7
E. VERIFYING PROBABILITY FORECASTS 10
F. THE ECONOMIC VALUE OF PROBABILITY FORECASTS 11
G. CONCLUDING REMARKS 12
REFERENCES 12

Extratropical Cyclones
An Historical Perspective (A. J. Thorpe)

A. INTRODUCTION 14
B. THE THERMAL THEORY 14
C. THE BERGEN SCHOOL 15
D. BAROCLINIC INSTABILITY MODEL 17
E. POTENTIAL VORTICITY IDEAS 18

F. PREDICTABILITY, FORECASTING AND OBSERVATIONS 20
G. THE FUTURE 21
REFERENCES 22

Numerical Methods for Atmospheric General Circulation Models
Refinements or Fundamental Advances? (D. L. Williamson)

A. NUMERICAL WEATHER PREDICTION 23
B. CLIMATE MODELS 24
C. NWP VERSUS CLIMATE MODELLING 25
D. FUTURE EVOLUTION OF NUMERICS 26
REFERENCES 27

Mesoscale Mountains and the Larger-scale Atmospheric Dynamics A Review (C. Schär)

A. INTRODUCTION 29
B. REGIME DIAGRAM FOR FLOW PAST TOPOGRAPHY 30
1. Balanced Solutions 30
2. Wake Formation and Transition into the Dissipative Regime 30
3. Flow Regimes for Major Topographic Obstacles 33
C. INTERACTIONS WITH THE BALANCED LARGER-SCALE DYNAMICS 34
1. Surface Potential Temperature Anomalies 34
2. Potential Vorticity Anomalies 35

D. NUMERICAL SIMULATIONS OF ALPINE WAKES   37

E. OUTLOOK   39

REFERENCES   41

The Upscale Turbulent Cascade *Shear Layers, Cyclones and Gas Giant Bands* (W. R. Peltier and G. R. Stuhne)

A. INTRODUCTION   43

B. PAIRING IN UNSTRATIFIED SHEAR FLOW   44

C. PAIRING IN ROTATING STRATIFIED FLOW: SUBSYNOPTIC BAROCLINIC INSTABILITY   48

D. PAIRING IN VORTEX DYNAMICS ON THE SPHERE: ROSSBY-WAVE ARREST OF THE UPSCALE CASCADE AND THE "PARITY" OF GAS GIANT BANDS   51

E. CONCLUSIONS   59

REFERENCES   60

The Role of Water in Atmospheric Dynamics and Climate (K. A. Emanuel)

A. HISTORICAL OVERVIEW   62

B. WATER IN ATMOSPHERIC DYNAMICS   64

1. Zonally Symmetrical Equilibrium States   64
2. Baroclinic Instability   65
3. The Tropics   66

C. WATER IN THE CLIMATE SYSTEM   69

D. SUMMARY   70

REFERENCES   71

New Observational Technologies *Scientific and Societal Impacts* (F. Fabry and I. Zawadzki)

A. INTRODUCTION   72

B. REMOTE SENSING OF THE ATMOSPHERE   72

C. REMOTE SENSORS AND THEIR SCIENTIFIC IMPACTS   73

1. Air Temperature and Moisture   73
2. Clouds and Precipitation   74
3. Wind   75
4. Others   78
5. Related Scientific Considerations   79

D. SOCIETAL IMPACTS   80

E. CONCLUSIONS   81

REFERENCES   82

Turning Dynamical Ideas into Forecast Practice *A Proposal for a Renewed Graphical Summary of the Synoptic Scale Situation* (A. Joly and P. Santurette)

A. INTRODUCTION   83

B. THE EXISTING GRAPHICAL SUMMARY MAP: ITS IMPORTANCE AND ITS PROBLEMS   84

C. HISTORICAL PERSPECTIVE   86

1. The Original Intent of the Bergen Method   86
2. Evolution of the Form of the Graphical Summary   87

D. RATIONALE OF A DYNAMICALLY GROUNDED SYNOPTIC SUMMARY   88

E. EXAMPLES OF USE OF THE NEW GRAPHICAL FRAMEWORK   90

1. A Large Transoceanic Cloud Band   90
2. Comma-Cloud–Baroclinic-Zone Interaction   96
3. Blocking   96
4. A Summer Situation   96

F. POTENTIAL FOR UPGRADING   98

1. Towards a Dual Representation of Meteorological Structures: Graphical Objects and Field Anomalies   98
2. Towards a Subjective Forecast that Looks Like a Numerical Model Output   100

G. CONCLUSION   101

APPENDIX: THE GRAPHICAL CODE OF ANASYG AND PRESYG   102

REFERENCES   104

Weather Forecasting *From Woolly Art to Solid Science* (P. Lynch)

A. THE PREHISTORY OF SCIENTIFIC FORECASTING   106

1. Vilhelm Bjerknes   106

2. Lewis Fry Richardson 108
3. Richardson's Forecast 109

B. THE BEGINNING OF MODERN NUMERICAL WEATHER PREDICTION 110

1. John von Neumann and the Meteorology Project 110
2. The ENIAC Integrations 111
3. The Barotropic Model 112
4. Primitive Equation Models 113

C. NUMERICAL WEATHER PREDICTION TODAY 113

1. ECMWF 114
2. HIRLAM 116

D. CONCLUSIONS 116

REFERENCES 118

Part 2. Climate Variability and Change

Problems in Quantifying Natural and Anthropogenic Perturbations to the Earth's Energy Balance (K. P. Shine and E. J. Highwood)

A. INTRODUCTION 123

B. ESTIMATES OF THE GLOBAL-MEAN RADIATIVE FORCING 124

C. TIME VARIATION OF RADIATIVE FORCING 126

D. DEVELOPMENTS IN STUDIES OF THE INDIRECT AEROSOL EFFECT 129

E. CONCLUDING REMARKS 130

REFERENCES 131

Changes in Climate and Variability over the Last 1000 Years (P. D. Jones)

A. INTRODUCTION 133

B. THE INSTRUMENTAL PERIOD (1850–) 133

C. EARLY EUROPEAN INSTRUMENTAL RECORDS 137

D. THE PRE-INSTRUMENTAL PERIOD (BACK TO 1000) 137

1. Palaeoclimatic Sources 137
2. Multiproxy Averages from High-Frequency Sources 138
3. Low-Frequency Sources 139

E. DISCUSSION AND CONCLUSIONS 140

1. The Instrumental Period 140
2. The Last 1000 Years 140

REFERENCES 141

Climatic Variability over the North Atlantic (J. Hurrell, M. P. Hoerling and C. K. Folland)

A. INTRODUCTION 143

B. WHAT IS THE NORTH ATLANTIC OSCILLATION AND HOW DOES IT IMPACT REGIONAL CLIMATE? 143

C. WHAT ARE THE MECHANISMS THAT GOVERN NORTH ATLANTIC OSCILLATION VARIABILITY? 144

1. Atmospheric Processes 144
2. Ocean Forcing of the Atmosphere 146

D. CONCLUDING COMMENTS ON THE OTHER ASPECTS OF NORTH ATLANTIC CLIMATE VARIABILITY 149

REFERENCES 150

Prediction and Detection of Anthropogenic Climate Change (J. F. B. Mitchell)

A. INTRODUCTION 152

B. AN EARLY GENERAL CIRCULATION MODEL SIMULATION OF THE CLIMATIC EFFECTS OF INCREASES IN GREENHOUSE GASES 152

C. A RECENT SIMULATION OF THE EFFECTS OF INCREASES IN GREENHOUSE GASES 153

D. DETECTION AND ATTRIBUTION OF ANTHROPOGENIC CLIMATE CHANGE 158

E. SOME CONCLUDING REMARKS 162

REFERENCES 163

Absorption of Solar Radiation in the Atmosphere *Reconciling Models with Measurements* (A. Slingo)

A. INTRODUCTION 165

B. THE GLOBAL RADIATION BALANCE 165

C. SATELLITE MEASUREMENTS 166

D. ABSORPTION BY THE SURFACE 168

E. ABSORPTION BY THE ATMOSPHERE 168

F. DISCUSSION 172

REFERENCES 172

The Impact of Satellite Observations of the Earth's Spectrum on Climate Research (J. E. Harries)

A. THE CLIMATE CHANGE DETECTION AND ATTRIBUTION PROBLEM 174

B. SATELLITE SPECTRAL SIGNATURES OF CLIMATE CHANGE 175

C. RECENT WORK ON SPECTRAL SIGNATURES 176

D. IRIS AND IMG: CHANGE OF OLR(v) BETWEEN 1970 AND 1997 177

1. Previous Broad Band Measurements 177
2. Spectrally Resolved Data 177

E. ANALYSIS OF IMG AND IRIS DATA 178

1. Accuracy 178
2. Validation 178
3. Results 178
4. Cloudy Scenes 181

F. SUMMARY 181

REFERENCES 182

Part 3. The Atmosphere and Oceans

Atlantic Air–Sea Interaction Revisited (M. J. Rodwell)

A. INTRODUCTION 185

B. DATA AND MODELS 186

C. THE ANALYSIS METHOD 187

D. ATMOSPHERIC FORCING OF NORTH ATLANTIC SEA SURFACE TEMPERATURES 188

E. NORTH ATLANTIC SEA SURFACE TEMPERATURE FORCING OF THE ATMOSPHERE 190

1. Observational Evidence 190
2. Model Results 193

F. POTENTIAL SEASONAL PREDICTABILITY BASED ON THE ATMOSPHERE GENERAL CIRCULATION MODEL 194

G. CONCLUSIONS AND DISCUSSION 195

REFERENCES 197

The Monsoon as a Self-regulating Coupled Ocean–Atmosphere System (P. J. Webster, C. Clark, G. Cherikova, J. Fasullo, W. Han, J. Loschnigg and K. Sahami)

A. INTRODUCTION 198

B. REGULATION OF THE MONSOON ANNUAL CYCLE 201

1. The Climatological Annual Cycle 201
2. Processes Determining the Annual Cycle of the Monsoon 203
3. Role of Ocean Dynamics in the Annual Heat Balance of the Indian Ocean 204
4. Regulation of the Annual Cycle of the Monsoon: an Ocean–Atmosphere Feedback System 205

C. INTERANNUAL VARIABILITY OF THE MONSOON 206

1. Modes of Interannual Variability in the Monsoon 207
2. Interannual Modes in Ocean Heat Transport 212
3. Interannual Regulation of the Monsoon 212

D. GENERAL THEORY OF REGULATION OF THE COUPLED OCEAN–ATMOSPHERIC MONSOON SYSTEM 214

E. CONCLUSIONS 214

REFERENCES 217

Laminar Flow in the Ocean Ekman Layer (J. T. H. Woods)

A. INTRODUCTION 220

B. THE EFFECT OF A STABLE DENSITY GRADIENT 220

C. THE FATAL FLAW 221

D. FLOW VISUALIZATION 222

E. THE DISCOVERY OF LAMINAR FLOW 222

F. FINE STRUCTURE 222

G. WAVE-INDUCED SHEAR INSTABILITY 222

H. BILLOW TURBULENCE 223

I. REVERSE TRANSITION 224

J. REVISED PARADIGM 224

K. ONE-DIMENSIONAL MODELLING OF THE UPPER OCEAN 224

L. DIURNAL VARIATION 225

M. BUOYANT CONVECTION 225

N. BILLOW TURBULENCE IN THE DIURNAL THERMOCLINE 225

O. CONSEQUENCES FOR THE EKMAN CURRENT PROFILE 225

P. SOLAR RADIATION 226

Q. APPLICATIONS 226
1. Slippery Seas of Acapulco 226
2. Pollution 226
3. Afternoon Effect in Sonar 230
4. Patchiness 230
5. Fisheries 230
6. Climate 230

R. DISCUSSION 231

S. CONCLUSION 231

REFERENCES 231

Ocean Observations and the Climate Forecast Problem (C. Wunsch)

A. INTRODUCTION 233

B. SOME HISTORY: OBSERVATIONAL AND THEORETICAL 233

C. THE PERCEPTION 234

D. THE REALITY 238

E. CONCLUSIONS BASED UPON MISAPPREHENSIONS 242

F. WHERE DO WE GO FROM HERE? 243

REFERENCES 245

Part 4. The Biogeochemical System

Biochemical Connections Between the Atmosphere and the Ocean (P. S. Liss)

A. INTRODUCTION 249

B. THE CHEMICAL COMPOSITION OF THE EARTH'S ATMOSPHERE 250

C. AIR–SEA EXCHANGE OF GASES OF IMPORTANCE 250
1. Ozone 250
2. Manmade Carbon Dioxide 250
3. Dimethyl Sulphide 251
4. Dimethyl Selenide 252
5. Ammonia 253

C. IMPACT OF ATMOSPHERIC DUST ON OCEAN BIOCHEMISTRY 253

D. GLOBAL PERSPECTIVES ON BIOGEOCHEMICAL FLUXES ACROSS THE AIR–SEA INTERFACE 256
1. DMS and the CLAW Hypothesis 256
2. Iron 257

REFERENCES 257

Modelling Vegetation and the Carbon Cycle as Interactive Elements of the Climate system (P. M. Cox, R. A. Betts, C. D. Jones, S. A. Spall and I. J. Totterdell)

A. INTRODUCTION 259

B. MODEL DESCRIPTION 260
1. Ocean–Atmosphere GCM (HadCM3L) 260
2. The Hadley Centre Ocean Carbon Cycle Model (HadOCC) 260

3. The Dynamic Global Vegetation Model (TRIFFID) 262

C. PRE-INDUSTRIAL STATE 264

1. Spin-up Methodology 264
2. The Mean Pre-industrial State 265

D. A FIRST TRANSIENT CLIMATE–CARBON CYCLE SIMULATION 268

1. 1860–2000 271
2. 2000–2100 271

E. DISCUSSION 272

1. Sink-to-source Transitions in the Terrestrial Carbon Cycle 272

F. CONCLUSIONS 276

REFERENCES 277

Part 5. Middle Atmosphere and Solar Physics: Other Planets and Epochs

Some Fundamental Aspects of Atmospheric Dynamics, with a Solar Spinoff (M. E. McIntyre)

A. INTRODUCTION 283

B. THERMAL AND CHEMICAL EVIDENCE 284

C. WAVE-INDUCED MOMENTUM TRANSPORT: SOME SIMPLE EXAMPLES 287

D. THE QUASI-BIENNIAL OSCILLATION (QBO) 289

E. THE MICHELSON–MORLEY PRINCIPLE 291

F. THE NONACCELERATION CONSTRAINT 291

G. EXTRATROPICAL LATITUDES 291

H. GYROSCOPIC PUMPING 292

I. ROSSBY- AND GRAVITY-WAVE BREAKING 294

J. THE JIGSAW PUZZLE: BAROTROPIC MODELS 295

K. GENERALIZATION TO REALISTIC STRATIFIED FLOW 299

L. THE SUN'S RADIATIVE INTERIOR 299

1. Inevitability of a Magnetic Field 299
2. The Tachopause and the Lithium Problem 302

REFERENCES 303

Atmospheric Dynamics of the Outer Planets (A. P. Ingersoll)

A. INTRODUCTION 306

B. VOYAGER OBSERVATIONS OF WINDS AND TEMPERATURES 306

C. MODELS OF VORTICES IN SHEAR 308

D. GALILEO PROBE AND VERTICAL STRUCTURE 309

E. GALILEO ORBITER OBSERVATIONS OF WATER AND LIGHTNING 312

REFERENCES 315

Palaeoclimate Studies at the Millennium *The Role of the Coupled System* (P. J. Valdes)

A. INTRODUCTION 316

B. PRE-QUATERNARY PALAEOCLIMATES 316

C. QUATERNARY PALAEOCLIMATES 318

1. Observational Evidence 318
2. Modelling Studies 319

D. SUBORBITAL TIMESCALE VARIABILITY 323

E. CONCLUSIONS 323

REFERENCES 324

Index 327

Foreword

"No man is an island" wrote John Donne, and the same truism applies with equal force to science. The specialism and technicality of science means that a scientist can become parochial, with a gaze concentrated inwards, toward his or her own discipline. Yet the great paradigm shifts and imaginative leaps in a science are so often the result of cross-fertilization by ideas from an apparently unrelated field.

With this background, in 1995, we began to think that the 150th anniversary of the founding of the Royal Meteorological Society in the year 2000 ought to be the occasion of a major conference. Our Royal Charter charges us with advancing meteorology and related subjects. It seemed both timely and natural that the theme should be the interfaces between atmospheric sciences and other fields, and the ways in which ideas from the atmospheric sciences have enriched other fields and vice versa. The title eventually was chosen as "Meteorology at the Millennium: its relationship to other sciences and technology and to society". Looking around for possible venues, it was apparent that St Johns College Cambridge provided just the right environment. The conference took place there from 10 to 14 July 2000, and was attended by some 250 people.

The conference was organized around seven themes ranging through meteorological understanding, climate, weather prediction, observational techniques, atmosphere–ocean, other planets and epochs, and biogeochemical cycles. There were no parallel sessions, as we wanted to make the breadth a positive feature. There was also a determined attempt to make the impact of the younger element of the participants felt. In general both these aspects worked well. There were invited speakers and contributed oral and poster papers. The general standard was remarkably high.

We did not start out with the idea of making a book from the conference. However, the range and quality of the speakers was such that it was clearly an attractive idea, provided that we did not frighten them away! The major task then was to find an editor of suitable scientific stature and organizational ability. We were very glad that our former Head of Department Bob Pearce agreed. As a result, this volume has been produced and is now a permanent reminder of the occasion, one that can be shared with a much wider audience. We hope that this book plays its part in the "advancement of meteorology and related subjects" that will be reflected in the 200th anniversary of the Royal Meteorological Society!

Brian Hoskins (President of the Royal Meteorological Society)
Ian James (Chair of the Scientific Programme Committee for the Conference)

Preface

The background to the conference from which this volume has arisen has been described in the Foreword. My task has been to encourage the keynote speakers to prepare articles covering the material they presented and to assemble these articles into a coherent overview of the main concerns of meteorologists and their associates at the start of the new millennium.

The conference programme was broken into seven themes. Five of these could clearly be taken as part titles and the papers submitted under the remaining two, concerned respectively with fundamental theory and observational techniques, could be comfortably fitted into the parts on weather prediction and climate. Consequently these parts (1 and 2) are longer than the remaining three. The contribution by Professor McIntyre on processes in the middle atmosphere and the sun's interior, placed in Part 5, has considerable relevance also to the themes of Parts 1–3. However, as the only contribution to solar physics, it seemed most appropriate to place it in the much wider context of our solar system and its evolution in Part 5—an area of study likely to play an increasing role in the new millennium.

It has been the aim of the Society to publish this millennium volume as soon as possible following the July 2000 150th Anniversary Conference at Cambridge. The wholehearted cooperation of the staff at RMS headquarters in processing the large amount of e-mail involved in dealing with the authors of the 24 contributions has been critical to expediting the preparation of the final manuscript for the publisher to meet this objective. Thanks are due especially to Richard Pettifer, the Executive Secretary and his assistant Mrs Margaret McGraw. They could not have been more helpful.

Mrs Gioia Ghezzi and later Mr Simon Crump have had the responsibility of overseeing the publication for Academic Press and I am most grateful to them both for their most helpful and valuable advice.

Robert P. Pearce, Editor

Contributors

R. A. Betts Hadley Centre for Climate Prediction and Research, UK Meteorological Office, London Road, Bracknell RG12 2SZ, UK

G. Cherikova Program in Atmospheric and Oceanic Science, University of Colorado, Boulder Campus, Box 311, Boulder, CO 80304, USA

C. Clark Program in Atmospheric and Oceanic Science, University of Colorado, Boulder Campus, Box 311, Boulder, CO 80304, USA

P. M. Cox Hadley Centre for Climate Prediction and Research, UK Meteorological Office, London Road, Bracknell RG12 2SZ, UK

K. A. Emanuel Program in Atmospheres, Oceans and Climate, Room 54–1620, Massachusetts Institute of Technology, 77 Massachusetts Avenue, Cambridge, MA 02139, USA

F. Fabry Department of Atmospheric and Oceanic Sciences, McGill University, 805 Sherbrooke St. West, Montreal, QCH3A 2K6, Canada

J. Fasullo Program in Atmospheric and Oceanic Science, University of Colorado, Boulder Campus, Box 311, Boulder, CO 80304, USA

C. K. Folland Hadley Centre for Climate Prediction and Research, UK Meteorological Office, London Road, Bracknell RG12 2SZ, UK

W. Han Program in Atmospheric and Oceanic Science, University of Colorado, Boulder Campus, Box 311, Boulder, CO 80304, USA

J. E. Harries Blackett Laboratory, Department of Physics, Imperial College, Prince Consort Road, London SW7 2BZ, UK

E. J. Highwood Department of Meteorology, University of Reading, Earley Gate, PO Box 243, Reading, RG6 6BB, UK

M. P. Hoerling Climate Diagnostics Center, NOAA/ERL, Boulder, CO 80307, USA

B. J. Hoskins Royal Meteorological Society, University of Reading, 104 Oxford Road, Reading RG1 7LL, UK

J. Hurrell National Center for Atmospheric Research, Climate Analysis Section, PO Box 3000, Boulder, CO 80307–3000, USA

A. P. Ingersoll Division of Geological and Planetary Sciences, California Institute of Technology, Mail code 150–21, 1200 E. California Blvd., Pasadena, CA 91125, USA

I. James Royal Meteorological Society, University of Reading, 104 Oxford Road, Reading RG1 7LL, UK

A. Joly Météo-France, Centre National de Recherches Météorologiques, CNRM/GMME/RECYF, 42 av. G Coriolis, F-31057 Toulouse Cedex 1, France

C. D. Jones Hadley Centre for Climate Prediction and Research, UK Meteorological Office, London Road, Bracknell RG12 2SZ, UK

P. D. Jones Climatic Research Unit, University of East Anglia, Norwich NR4 7TJ, UK

P. S. Liss School of Environmental Sciences, University of East Anglia, Norwich NR4 7TJ, UK

J. Loschnigg International Pacific Research Center, University of Hawaii, Honolulu, Hawaii, USA

P. Lynch Met Éireann, Glasnevin Hill, Dublin 9, Ireland

M. E. McIntyre Centre for Atmospheric Science at the Department of Applied Mathematics and Theoretical Physics, Silver St, Cambridge CB3 9EW, UK

J. F. B. Mitchell Hadley Centre for Climate Prediction and Research, UK Meteorological Office, Bracknell, RG12 2SZ, UK

T. N. Palmer ECMWF, Shinfield Park, Reading, RG2 9AX, UK

R. P. Pearce Department of Meteorology, University of Reading, Earley Gate, PO Box 243, Reading RG6 6BB, UK.

W. R. Peltier Department of Physics, University of Toronto, Toronto, Ontario, Canada M5S 1A7

M. J. Rodwell Hadley Centre for Climate Prediction and Research, UK Meteorological Office, Bracknell, RG12 2SZ, UK

K. Sahami Program in Atmospheric and Oceanic Science, University of Colorado, Boulder Campus, Box 311, Boulder, CO 80304, USA

P. Santurette Météo-France, Centre National de Recherches Météorologiques, CNRM/GMME/RECYF, 42 av. G Coriolis, F-31057 Toulouse Cedex 1, France

C. Schär Atmospheric and Climate Science ETH, Winterthurerstr. 190, 8057 Zürich, Switzerland

K. P. Shine Department of Meteorology, University of Reading, Earley Gate, PO Box 243, Reading, RG6 6BB, UK

A. Slingo Hadley Centre for Climate Prediction and Research, UK Meteorological Office, London Road, Bracknell, RG12 2SZ, UK

S. A. Spall Hadley Centre for Climate Prediction and Research, UK Meteorological Office, Bracknell, RG12 2SZ, UK

G. R. Stuhne Scripps Institution of Oceanography (PORD Division), University of California, San Diego, 9500 Gilman Drive, La Jolla, CA 92093-0230, USA

A. J. Thorpe Hadley Centre for Climate Prediction and Research, UK Meteorological Office, London Road, Bracknell, RG12 2SZ, UK

I. J. Totterdell Southampton Oceanography Centre, Southampton SO14 3ZH, UK

P. J. Valdes Department of Meteorology, University of Reading, Earley Gate, PO Box 243, Reading RG6 6BB, UK

P. J. Webster Program in Atmospheric and Oceanic Science, University of Colorado, Boulder Campus, Box 311, Boulder, CO 80304, USA

D. L. Williamson National Center for Atmospheric Research, Boulder, CO 80307-3000, USA

J. T. H. Woods Huxley School of Environment, Earth Science and Engineering, Imperial College, London SW7 2BP, UK

C. Wunsch Program in Atmospheres, Oceans and Climate, Department of Earth, Atmospheric and Planetary Sciences, Massachusetts Institute of Technology, 541524, 77 Massachusetts Avenue, Cambridge, MA 02139, USA

I. Zawadzki Department of Atmospheric and Oceanic Sciences, McGill University, 805 Sherbrooke St. West, Montreal, QCH3A 2K6, Canada

Part 1. Weather Prediction at the Millennium

Predicting Uncertainty in Numerical Weather Forecasts
T. N. Palmer

Extratropical Cyclones: an Historical Perspective
A. J. Thorpe

Numerical Methods for Atmospheric General Circulation Models: Refinements or Fundamental Advances?
D. L. Williamson

Mesoscale Mountains and the Larger-scale Atmospheric Dynamics: a Review
C. Schär

The Upscale Turbulent Cascade: Shear Layers, Cyclones and Gas Giant Bands
W. R. Peltier and G. R. Stuhne

The Role of Water in Atmospheric Dynamics and Climate
K. A. Emanuel

New Observational Technologies: Scientific and Societal Impacts
F. Fabry and I. Zawadzki

Turning Dynamical Ideas into Forecast Practice: a Proposal for a Renewed Graphical Summary of the Synoptic Scale Situation
A. Joly and P. Santurette

Weather Forecasting: From Woolly Art to Solid Science
P. Lynch

Predicting Uncertainty in Numerical Weather Forecasts

T. N. Palmer

ECMWF, Reading, UK

The predictability of weather and climate forecasts is determined by the projection of uncertainties in both initial conditions and model formulation, onto flow-dependent instabilities of the chaotic climate attractor. Since it is essential to be able to estimate the impact of such uncertainties on forecast accuracy, no weather or climate prediction can be considered complete without a forecast of the associated flow-dependent predictability. The problem of predicting uncertainty can be posed in terms of the Liouville equation for the growth of initial uncertainty. However, in practice, the problem is approached using ensembles of integrations of comprehensive weather and climate prediction models, with explicit perturbations to both initial conditions and model formulation; the resulting ensemble of forecasts can be interpreted as a probabilistic prediction.

Many of the difficulties in forecasting predictability arise from the large dimensionality of the climate system, and special techniques to generate ensemble perturbations have been developed. Methods to sample uncertainties in model formulation are also described. Practical ensemble prediction systems are described, and examples of resulting probabilistic weather forecast products shown. Methods to evaluate the skill of these probabilistic forecasts are outlined. By using ensemble forecasts as input to a simple decision-model analysis, it is shown that that probability forecasts of weather have greater potential economic value than corresponding single deterministic forecasts with uncertain accuracy.

> He believed in the primacy of doubt, not as a blemish on our ability to know, but as the essence of knowing.
>
> *Gleick (1992) on Richard Feynman*

A. INTRODUCTION

This chapter deals with probabilistic weather prediction from its theoretical basis, through an outline of practical methodologies, to an analysis of validation techniques.

Why bother to estimate forecast uncertainty? There are at least two reasons. First, an ability to estimate the likelihood of possible errors and uncertainties distinguishes science from pseudo-science, and it is a fundamental tenet of experimental science that errors and uncertainties be so described. However, it is not enough to give flow-independent time-averaged forecast error bars; since the atmosphere is a chaotic system, and hence nonlinear, the growth of initial error is itself dependent on initial conditions.

As ensemble prediction techniques mature, there is a second important reason for estimating uncertainty: it provides the forecast user with more complete information for assessing weather-related risk, be it in commercial or humanitarian terms. In these days where weather prediction is increasingly a commercial activity, the ability to quantify risk is all important. It is explicitly and quantitatively shown towards the end of this chapter that ensemble forecasts have greater potential economic value than single deterministic forecasts as a tool for risk assessment.

B. THE LIOUVILLE EQUATION

The evolution equations in a climate or weather prediction model are conventionally treated as deterministic. These (N-dimensional) equations, based on spatially truncated momentum, energy, mass and composition conservation equations will be written schematically as

$$\dot{X} = F[X] \quad (1)$$

where X describes an instantaneous state of the climate sytem in N-dimensional phase space. Equation (1) is fundamentally nonlinear and deterministic in the sense that, given an initial state X_a, the equation determines a unique forecast state X_f.

The meteorological and oceanic observing network is sparse over many parts of the world, and the observations themselves are obviously subject to measurement error. The resulting uncertainty in the initial state can be represented by the pdf $\rho(X, t_a)$; given a volume V of phase space, then $\int_V \rho(X, t_a)dV$ is the probability that the true initial state X_{true} at time t_a lies in V. If V is bounded by an isopleth of ρ (i.e. comoving in phase space), then, from the determinism of equation (1), the probability that X_{true} lies in V is time invariant. The evolution of ρ is given by the Liouville conservation equation (introduced in a meteorological context by Gleeson, 1966, and Epstein, 1969)

$$\frac{\partial \rho}{\partial t} + \frac{\partial}{\partial X}(\dot{X}\rho) = 0 \quad (2)$$

where $\dot{X}$ is given by equation (1). In the second term of this equation, there is an implied summation over all the components of X. (The Liouville equation is for probability density in phase space what the continuity equation is for mass density in physical space. The Liouville equation rests on the fact that the equations of motion are deterministic, so that two trajectories in phase space cannot split apart or coalesce. Similarly, mass continuity is based on the fact that two Lagrangian trajectories in physical space cannot split apart or coalesce.)

Figure 1 illustrates schematically the evolution of an isopleth of $\rho(X, t)$. For simplicity we assume the initial pdf is isotropic (e.g. by applying a suitable coordinate transformation). In the early part of the forecast, the isopleth evolves in a way consistent with linearized dynamics; the N-ball at initial time has evolved to an N-ellipsoid at forecast time t_1. For weather scales of $O(10^6)$ km, this linear phase lasts for about 1–2 days into the forecast. Beyond this time, the isopleth starts to deform nonlinearly. The third schematic shows the isopleth at a forecast range in which errors are growing nonlinearly. Predictability is finally lost when the forecast pdf $\rho(X,t)$ has evolved irreversibly to the invariant distribution ρ_{inv} of the attractor. This is shown schematically in Fig. 1 using the Lorenz (1963) attractor.

The growth of the pdf through the forecast range is a function of the initial state. This can be seen by considering a small perturbation δx to the initial state X_a. From equation (1), the evolution equation for δx is given by

$$\delta \dot{x} = J\delta x \quad (3)$$

where the Jacobian is defined as

$$J = dF/dX \quad (4)$$

Since $F[X]$ is at least quadratic in X, then J is at least linearly dependent on X. This dependency is illustrated in Fig. 2 showing the growth of an initial isopleth of an idealized pdf at three different positions on the Lorenz (1963) attractor. In the first position, there is little growth, and hence large local predictability. In the second position there is some growth as the pdf evolves towards the lower middle half of the attractor. In the third position, initial growth is large, and the resulting predictability is correspondingly small.

The Liouville equation can be formally solved to give the value of ρ at a given point X in phase space at

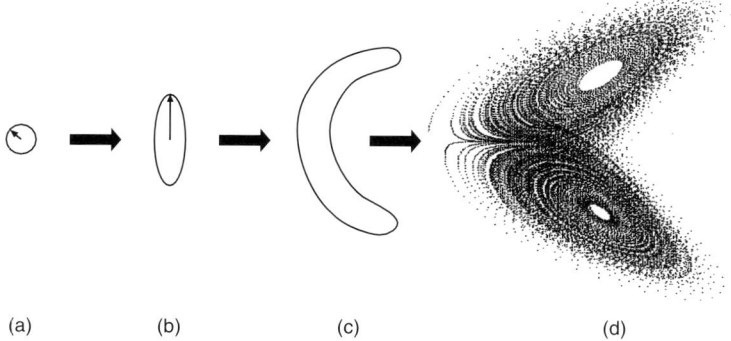

(a) (b) (c) (d)

FIGURE 1 Schematic evolution of an isopleth of the probability density function (pdf) of initial and forecast error in N-dimensional phase space. (a) At initial time; (b) during the linearized stage of evolution. A (singular) vector pointing along the major axis of the pdf ellipsoid is shown in (b), and its pre-image at initial time is shown in (a). (c) The evolution of the isopleth during the nonlinear phase; there is still predictability, though the pdf is no longer Gaussian. (d) Total loss of predictability, occurring when the forecast pdf is indistinguishable from the attractor's invariant pdf.

A simple example which illustrates this solution to the Liouville equation is given in Fig. 3, for a one-dimensional Riccati equation (Ehrendorfer, 1994a)

$$\dot{X} = aX^2 + bX + c \qquad (8)$$

where $b^2 > 4ac$, based on an initial Gaussian pdf. The pdf evolves away from the unstable equilibrium point at $X = -1$ and therefore reflects the dynamical properties of equation (8). Within the integration period, this pdf has evolved to the nonlinear phase.

The forward tangent propagator plays an important role in meteorological data assimilation systems. However, even though the forward tangent propagator may exist as a piece of computer code, this does not mean that the Liouville equation can be readily solved for the weather prediction problem. For example, the determinant of the forward tangent propagator is determined by the product of all its singular values. For a comprehensive weather prediction model, a determination of the full set of $O(10^7)$ singular values is currently impossible. More importantly, we do not have a very precise estimate of the initial pdf with which to initialize the Liouville equation. Finally, the Liouville equation takes no account of model error. This latter point is discussed in the next section.

C. THE PROBABILITY DENSITY FUNCTION OF MODEL ERROR

So far, we have assumed the "classical" chaotic paradigm, that loss of predictability occurs only because of inevitable uncertainty in initial conditions. However,

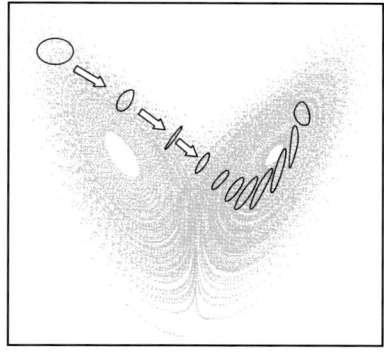
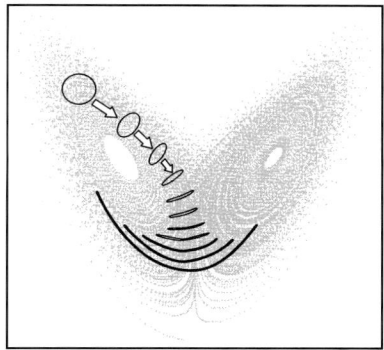
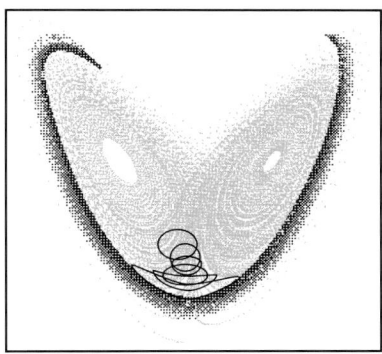

FIGURE 2 Phase-space evolution of an ensemble of initial points on the Lorenz (1963) attractor, for three different sets of initial conditions. Predictability is a function of initial state.

forecast time t (Ehrendorfer, 1994a,b; Palmer, 2000). Specifically

$$\rho(X, t) = \rho(X', t_a)/\det M(t, t_a) \qquad (5)$$

where

$$M(t, t_a) = \exp \int_{t_a}^{t} J(t')dt' \qquad (6)$$

is the so-called forward tangent propagator, mapping a perturbation $\delta x(t_a)$, along the nonlinear trajectory from X' to X', to

$$\delta x(t) = M(t, t_a)\delta(t_a) \qquad (7)$$

and X' in this equation corresponds to that initial point, which, under the action of equation (1) evolves to the given point X at time t.

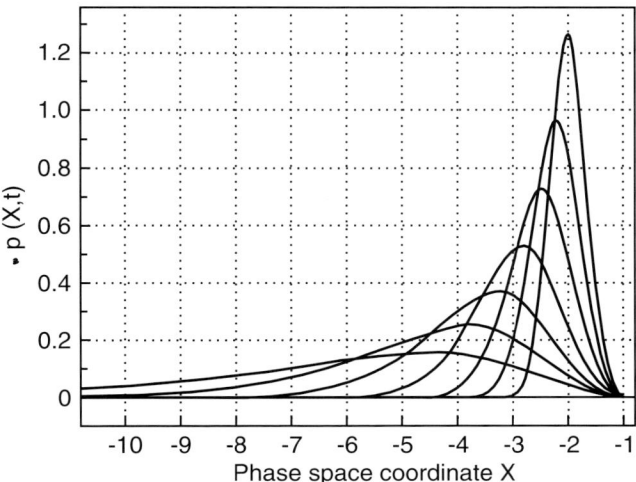

FIGURE 3 An analytical solution to the Liouville equation for an initial Gaussian pdf (shown peaked on the right-hand side of the figure) evolved using the Riccati equation (see text). (From Ehrendorfer, 1994a,b.)

there are also inevitable uncertainties in our ability to represent computationally the governing equations of climate. These uncertainties can contribute both to random and systematic error in model performance. In practice, as discussed below, it is not easy to separate the predictability problem into a component associated with initial error and a component associated with model error.

Weather and climate models have a resolution of $O(100)$ km in the horizontal. This immediately raises the problem of closure—how to represent the effect of partially resolved or unresolved processes onto the resolved state vector X. The effects of topography and clouds are examples. In weather and climate prediction models, equation (1) is generally expressed as

$$\dot{X} = G[X] + P[X; \alpha(X)] \qquad (9)$$

where $P[X; \alpha(X)]$ stands for some parametrized representation of unresolved processes, and $G[X]$ represents terms in the equations of motion associated directly with resolved scales. Conceptually, a parametrization is usually based on the notion of a statistical ensemble of subgrid scale processes within a grid box, in some secular equilibrium with the grid-box mean flow. This allows equation (9) to be written

$$\dot{X}_j = G_j[X] + P[X_j; \alpha(X_j)] \qquad (10)$$

where X_j and G_j represent the projection of X and G into the subspace associated with a single grid box x_j in physical space. Borrowing ideas from statistical mechanics, a familiar parametrization might involve the diffusive approximation, where α would be a diffusion coefficient which might depend on the Richardson number of the large-scale flow. However, whilst diffusive closures do in fact play a role for example in representing the effects of the turbulent planetary boundary in the lower kilometre of the atmosphere, they are certainly insufficient, and, in some circumstances, may be fundamentally flawed. To see this, it is enough to concentrate on one relevant process—atmospheric convection.

Often, such instability is released through overturning circulations whose horizontal scales are small compared with the smallest resolved scale of a global weather or climate model. However, at the large-scale end of the spectrum of convectively driven circulations is the organized mesoscale convective complex (Moncrieff, 1992), with horizontal scales of perhaps 100 km. They can be simulated explicitly in regional models with $O(1)$ km resolution, but such resolution is not practicable for global weather and climate models.

The form of parametrization given in equation (10) is appropriate for describing cumulus and simple cumulonimbus. For example, in a contemporary convective parametrization (e.g. Betts and Miller, 1986) if the resolved-scale vertical temperature gradient at x_j is convectively unstable, then over some prescribed timescale (given by α) P will operate to relax X_j back to stability. On the other hand, the existence of organized mesoscale convective complexes poses a problem for parametrizations of this form. In particular, the basic assumption of a quasi-equilibrium of subgrid-scale convectively forced motions (with the implication that the kinetic energy released by overturning circulations is dissipated on subgrid scales, rather than injected into the large scale) cannot be fully justified.

One means of addressing the erroneous assumption of deterministic locality in equation (10) would be to add to equation (10) a stochastic energy source term $S(X_j; \alpha)$. Recognizing this, Buizza et al. (1999a) have proposed the simple stochastic form

$$\dot{X}_j = G_j[X] + \beta P[X_j; \alpha(X_j)] \qquad (11)$$

where β is a stochastic variable representing a random variable drawn from a uniform pdf between 0.5 and 1.5. Stochastic representation of subgrid processes is a technique already utilized in turbulent flow simulations (Mason and Thompson, 1992)

An example of the impact of the random effect of the stochastic parametrization represented by equation (11) is given in Fig. 4, which shows sea-level pressure over part of Australia and the west Pacific from four 2-day integrations of the ECMWF model. The integrations have identical starting conditions, but different realizations of the pdf represented by β. The figure shows two tropical cyclones. The intensity of the cyclones can be seen to be very sensitive to the realization of the stochastic parametrization. In Fig. 4a the western cyclone is intense; in Fig. 4b the eastern cyclone is intense; in Fig. 4c they are both intense; in Fig. 4d neither are intense. This rather extreme example clearly shows the difficulty in predicting tropical cyclone development, and its sensitivity to model parametrization.

The purpose of the discussion above is to point out that although current parametrizations have been enormously successful in representing subgrid processes, there are inevitable uncertainties in the representation of such processes. In Section B we represented the evolution of the initial pdf given the deterministic equation (1) in terms of a Liouville equation. In the idealized case where model uncertainties are represented by an additive Gaussian white noise with zero mean and variance Γ, then equation (2) becomes a Fokker–Planck equation (e.g. Hasselmann, 1976; Moss and McClintock, 1989). However, in practice, a realistic state-dependent stochastic forcing (e.g. in equation (11)) would be too complex for this simple representation to be directly relevant.

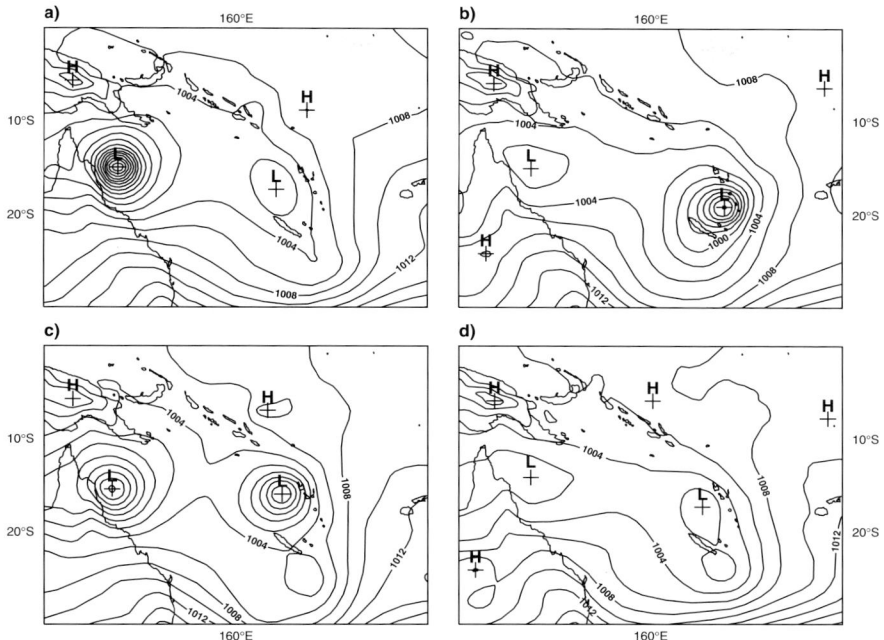

FIGURE 4 Four 2-day integrations of the ECMWF model from identical starting conditions but different realizations of the stochastic parametrization scheme represented by equation (11) with parameter settings as given in Buizza *et al.* (1999a). The field shown is sea-level pressure over parts of Australia and the west Pacific. The depressions in the pressure field represent potential tropical cyclones.

For ensemble forecasting (discussed in Section D), there are two other commonly used techniques for representing model uncertainty. The first technique is the multimodel ensemble (Harrison *et al.*, 1999; Palmer *et al.*, 2000). In the second technique, the values α of the parameters used in the deterministic equation (9) of one particular model, are perturbed in some stochastic manner (cf. Houtekamer *et al.*, 1996).

On the other hand, these latter techniques should be seen as conceptually distinct from the type of stochastic physics scheme described schematically in equation (11). In multimodel ensembles, and ensembles with stochastic α, the model perturbations account for the fact that the expected value of the pdf of subgrid processes is not itself well known. (Hence, for example, there are many different atmospheric convection parametrization schemes in use around the world; the Betts–Miller scheme described above is but one of these. The existence of this ensemble of convection schemes is an indication that the expectation value of the pdf of the effects of subgrid convection is not known with complete confidence!) By contrast, the stochastic physics scheme described in equation (11) is an attempt to account for the fact that in circumstances of convective organization, the pdf of subgrid processes is not especially sharp around the mean. This would argue for the combined use of multimodel ensembles and stochastic physics parametrization.

D. PROBABILITY FORECASTING BY ENSEMBLE PREDICTION

In the discussion above, the problem of forecasting uncertainty in weather and climate prediction has been formulated in terms of a Liouville, or, including model uncertainty, a form of Fokker–Planck equation. In practice these equations are solved using ensemble techniques, i.e. from multiple integrations of the governing equations from perturbed initial conditions, using either multiple models, or stochastic parametrizations, to represent model uncertainty. Insofar as the perturbed initial conditions constitute a random sampling of the initial pdf, this could be described as a Monte Carlo approach. However, the methodologies discussed below for estimating the initial perturbations do not necessarily constitute such a random sampling, for reasons to be discussed. As such, the more general term "ensemble prediction" is used.

On the timescale of 10 days, it is important to be able to estimate uncertainty in predictions of persistence or change in regime type. In the last few years, ensemble forecasting on this 10-day timescale has become an established part of operational global weather prediction (Palmer *et al.*, 1993; Toth and Kalnay, 1993; Houtekamer *et al.*, 1996; Molteni *et al.*, 1996). Different strategies have been proposed for determining the ensemble of starting conditions. Conceptually, the strategies can be delineated according to whether the initial perturbations merely sample observation error, or whether (given the dimension of phase space and the uncertainty in our knowledge of the initial pdf) the initial perturbations are constrained to lie on some dynamically unstable subspace. In the latter case, this dynamical subspace is defined in different ways. Within the meteorological community, there has been a very lively debate on the merits and deficiencies of the different strategies.

The strategies that use dynamically constrained perturbations, in some sense bypass the quantification of the many uncertainties that exist in specifying the pdf of initial error, and focus on perturbations that are likely to contribute to significant forecast error. Two types of dynamically constrained perturbation have been proposed: the first based on "bred vectors" (Toth and Kalnay, 1997) and the singular vectors (Buizza and Palmer, 1995) discussed above.

This strategy is relevant for user confidence in probability forecasts. As discussed in more detail in Section F, consider a user who could potentially suffer a catastrophic loss L if an event E occurs (in this case a circulation regime change bringing severe weather of some type). The user can take protective action at cost C, which might itself be substantial, but can be presumed to be less than L. Based on the ensemble forecast, one can estimate a probability p that E will occur. If p is sufficiently low, then the user should be able to assume that protective action is not necessary. On these occasions of very low p, the incurrence of a loss L would be seen as a failure of the forecast system, and user confidence in the system would be seriously compromised. Of course, an *a posteriori* constant offset could be added to the forecast probability, so that very small probabilities would never be issued. But in this case, the ensemble would be no realistic value to the user as a decision tool, since the issued probabilities would always be sufficiently high that protective action would always be taken.

The implication of this analysis is that it is important, arguably paramount, to be sure that the user can be confident in a forecast where the ensemble spread is small. For this reason, it is important in an operational environment to ensure that the initial perturbations are not conservative in the sense of not spanning phase-space directions where the pdf is underestimated (due to unquantified errors in the data analysis procedures), and where forecast error growth could be large because of dynamical instability. For these reasons, the ECMWF ensemble system is based in part on initial perturbations using rapidly growing singular vectors. Examples of the beneficial impact of singular vector perturbations over other types of perturbation are given in Mureau *et al.* (1993) and Gelaro *et al.* (1998).

At the time of writing, the ECMWF Ensemble Prediction System (EPS) comprises 51 forecasts of the ECMWF forecast model (see Buizza *et al.*, 1998; 1999b). As above, the control forecast is run from the operational ECMWF analysis. The 50 perturbed initial conditions are made by adding and subtracting linear combinations of the dominant 25 singular vector perturbations to the operational analysis. In addition, the 50 perturbed forecasts are also run with the stochastic parametrization defined in equation (11). The EPS is run every day and basic meteorological products are disseminated to all the national meteorological services of the ECMWF Member States. These products often take the form of probability forecasts for different events E, based on the fraction of ensemble members for which E is forecast. For example, Fig. 5 shows 6-day probability forecasts over Europe for the events $E_{>8}$, $E_{>4}$ defined as: difference of lower tropospheric temperature from a long-term climatology is $> 8C$, $> 4C$, respectively.

Based on the EPS, many of the European national meteorological services are providing their customers detailed probability forecasts of possible weather parameters for site-specific locations. For example, based on the EPS, the United Kingdom Meteorological Office provides forecast pdfs of surface temperature and rainfall for various specific locations in the UK. Joint probability distributions are also estimated (e.g. the probability of precipitation with surface temperature near or below freezing).

As discussed above (cf. Fig. 2), a feature of nonlinear dynamical systems is the dependence of error growth on initial state. Figure 6 illustrates this, based on two 10-day EPS integrations from starting dates exactly one year apart, the meteorological forecast variable being surface temperature over London. The unperturbed control forecast and verification are also shown. In the first example, the growth of the initial perturbations is relatively modest, and the associated forecast temperature pdf is relatively sharp. In the second example, the initial perturbations grow rapidly and the associated forecast temperature pdf is broad. Notice in the second example that the control integration is already very unskilful 3 days into the forecast. The large ensemble spread provides an *a priori* warning to the user that decisions made from single deterministic forecasts during this period could be extremely misleading.

Although forecasters have traditionally viewed weather prediction as deterministic, a culture change towards probabilistic forecasting is in progress. On the other hand, it is still necessary to demonstrate to users of weather forecasts that reliable probability forecasts

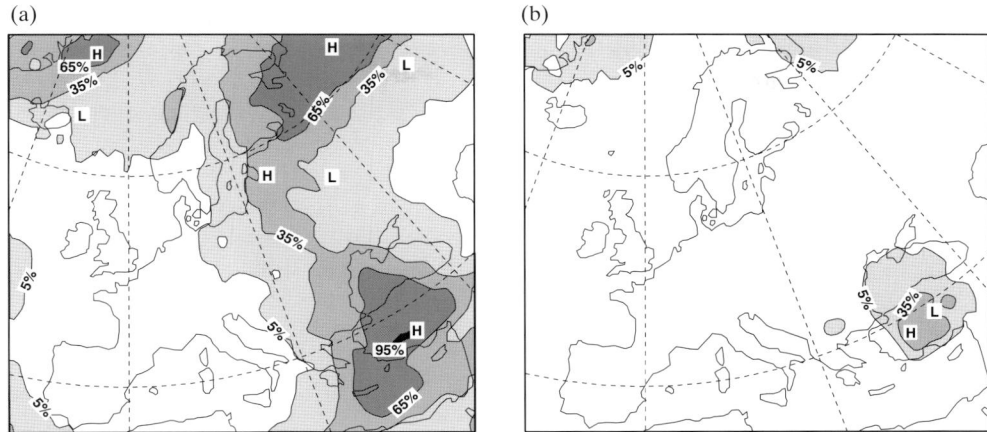

FIGURE 5 An example of an operational forecast product from the 50-member ECMWF ensemble prediction system. Probability that (a) 850 hPa temperature is at least 4°C above normal; (b) 8°C above normal, based on a 6-day forecast.

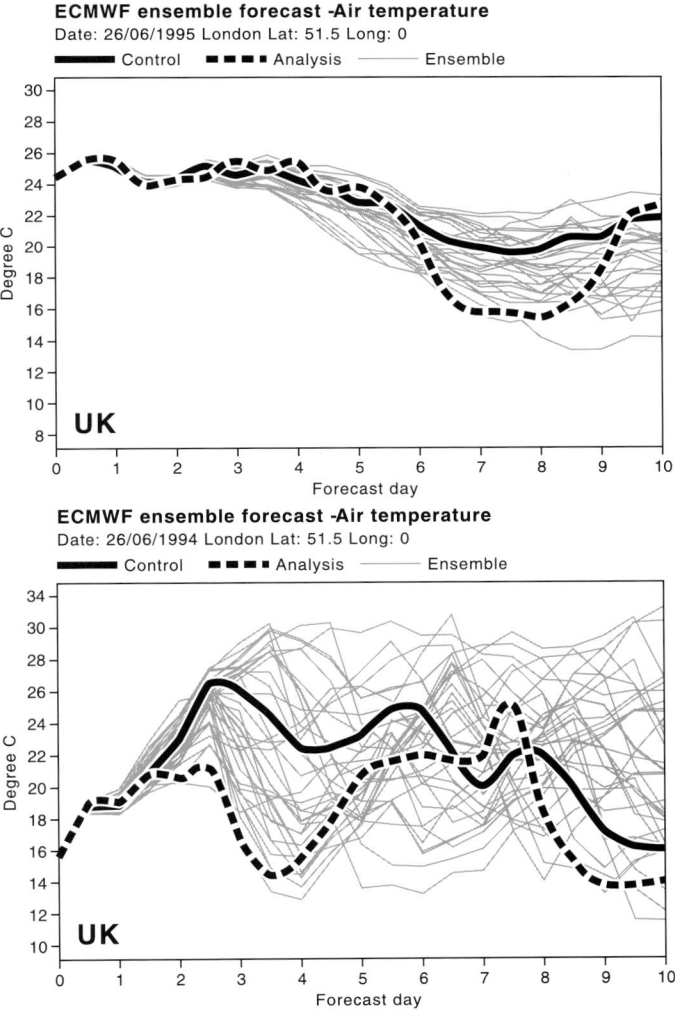

FIGURE 6 Time series of ensemble forecast integrations for temperature over London from starting conditions exactly one year apart. The unperturbed control forecast (heavy solid) and the verifying analysis (heavy dashed) are also shown. (Top) a relatively predictable period. (Bottom) an unpredictable period. (In these examples, the ensembles comprised 32 perturbed forecasts and the model did not include the stochastic representation as described in equation (11).)

provide greater value than imperfect deterministic forecasts. Such a demonstration will be given in Section F.

E. VERIFYING PROBABILITY FORECASTS

Consider a binary event E (e.g. as discussed in section D) which, for a particular ensemble forecast, occurs a fraction p of times within the ensemble. If E actually occurred then let $v = 1$, otherwise $v = 0$. Repeat this over a sample of N different ensemble forecasts, so that p_i is the probability of E in the ith ensemble forecast and $v_i = 1$ or $v_i = 0$, depending on whether E occurred or not in the ith verification ($i = 1, 2, \ldots, N$).

The Brier score (Wilks, 1995) is defined by

$$b = \frac{1}{N}\sum_{i=1}^{N}(p_i - v_i)^2, 0 \leq p_i \leq 1, v_i \in \{0, 1\} \quad (12)$$

From its definition $0 \leq b \leq 1$, equalling zero only in the ideal limit of a perfect deterministic forecast. For a large enough sample, the Brier score can be written as

$$b = \int_0^1 [p - 1]^2 o(p)\rho_{ens}(p)\,dp + \int_0^1 p^2[1 - o(p)]\rho_{ens}(p)dp \quad (13)$$

where $\rho_{ens}(p)dp$ is the relative frequency that E was forecast with probability between p and $p + dp$, and $o(p)$ gives the proportion of such cases when E actually occurred.

A reliability diagram (Wilks, 1995) is one in which $o(p)$ is plotted against p for some finite binning of width δp. In a perfectly reliable system $o(p) = p$ and the graph is a straight line oriented at 45° to the axes. Reliability measures the mean square distance of the graph of $o(p)$ to the diagonal line.

When assessing the skill of a forecast system, it is often desirable to compare it with the skill of a forecast where the climatological probability $\bar{o}$ is always predicted (so $\rho_{ens}(p) = \delta(p - \bar{o})$). In terms of this, the Brier skill score, B, of a given forecast system is defined by

$$B = 1 - \frac{b}{b_{cli}} \quad (14)$$

$B \leq 0$ for a forecast no better than climatology, and $B = 1$ for a perfect deterministic forecast.

Figure 7 shows two examples of reliability diagrams for the ECMWF EPS taken over all 6-day forecasts from December 1998 to February 1999 over Europe (cf. Fig. 5). The events are $E_{>4}$, $E_{>8}$, lower tropospheric temperature being at least 4K, 8K greater than normal. The Brier score, Brier skill score, and Murphy (1973) decomposition are shown in the figure.

The reliability skill score is extremely high for both events. However, the reliability diagrams indicate some overconfidence in the forecasts. For example, on those occasions where $E_{>4}$ was forecast with a probability

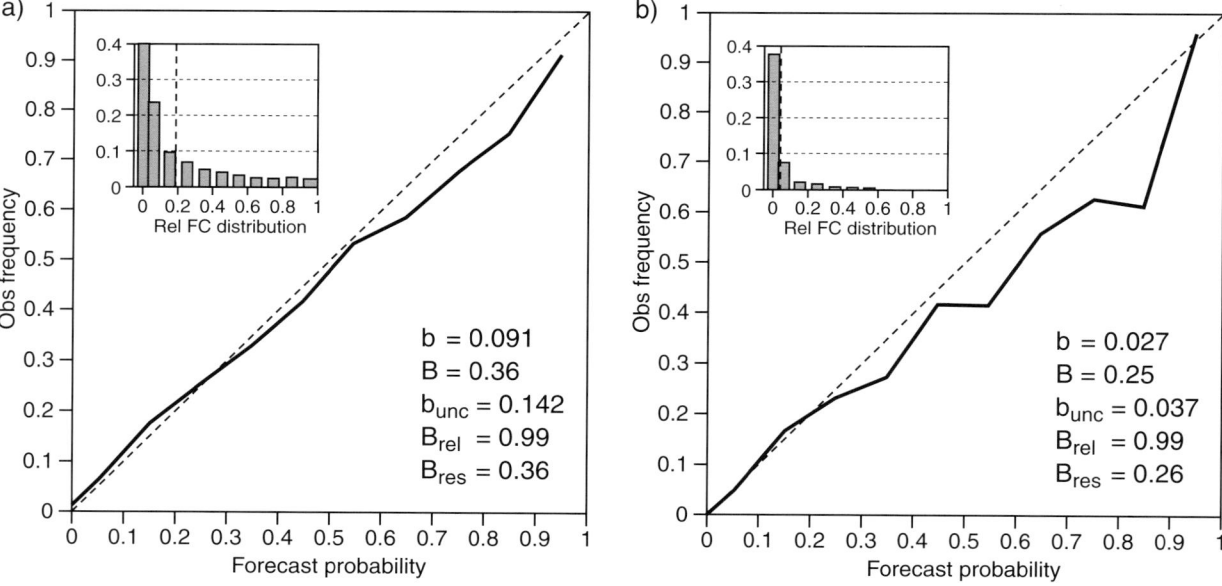

FIGURE 7 Reliability diagram and related Brier score, skill score and Murphy decomposition for the events: (a) 850 hPa temperature is at least 4 K above normal and (b) at least 8 K above normal, based on 6-day forecasts over Europe from the 50-member ECMWF ensemble prediction system from December 1998 to February 1999. Also shown is the pdf $\rho_{ens}(p)$ for the event in question.

between 80% and 90% of occasions, the event only verified about 72% of the time.

Reliability can be improved with increasing ensemble size. However, this raises a fundamental dilemma in ensemble forecasting given current computer resources. It would be meaningless to increase ensemble size by degrading the model (e.g. in terms of "physical" resolution) making it cheaper to run, if by doing so it could no longer simulate extreme weather events. Optimizing computer resources so that on the one hand, ensemble sizes are sufficiently large to give reliable probability forecasts of extreme but rare events, and on the other hand that the basic model has sufficient complexity to be able to simulate such events, is a very difficult balance to define.

F. THE ECONOMIC VALUE OF PROBABILITY FORECASTS

Although B provides an objective measure of skill for ensemble forecasts, it does not determine measures of usefulness for seasonal forecasts. In an attempt to define this notion objectively, we consider here a simple decision model (Murphy, 1977; Katz and Murphy, 1997) whose inputs are probabilistic forecast information and whose output is potential economic value.

Consider a potential forecast user who can take some specific precautionary action depending on the likelihood that E will occur. Let us take some simple examples relevant to seasonal forecasting. If the coming winter is mild (E = seasonal-mean temperature above normal), then overwintering crops may be destroyed by aphid growth. A farmer can take precautionary action by spraying. If the growing season is particularly dry (E = seasonal-mean rainfall at least one standard deviation below normal), then crops may be destroyed by drought. A farmer can take precautionary action by planting drought-resistant crops. In both cases taking precautionary action incurs a cost C irrespective of whether or not E occurs (cost of spraying, or cost associated with reduced yield and possibly with more expensive seed). However, if E occurs and no precautionary action has been taken, then a loss L is incurred (crop failure).

This simple "cost–loss" analysis is also applicable to much shorter range forecast problems (Richardson, 2000). For example, if the weather event was the occurrence of freezing conditions leading to ice on roads, and the precautionary action was to salt the roads, then C would correspond to the cost of salting, and L would be associated with the increase in road accidents, traffic delays, etc.

Figure 8 shows examples of optimal value as a func-

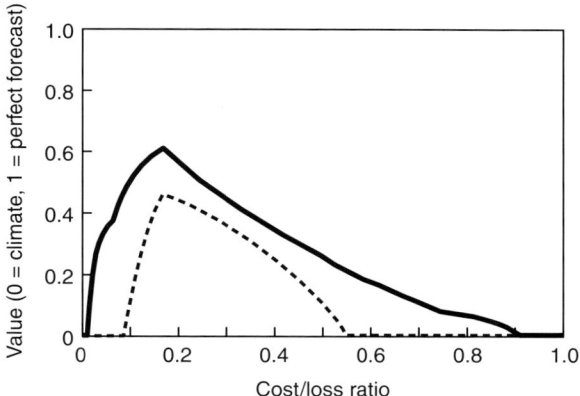

FIGURE 8 Potential economic value of the ECMWF ensemble prediction system as a function of user cost/loss ratio of 6-day weather forecasts over Europe (for the period December 1998 to February 1999) for the event $E_{<-4}$ = 850 hPa temperature at least 4°C below normal. (Solid) value of the ECMWF ensemble prediction system. (Dashed) value of a single deterministic forecast (the unperturbed "control" forecast of the EPS system). (From Richardson 2000.)

tion of user cost/loss ratio for the ECMWF 6-day ensemble weather prediction system and the event $E_{>4}$ (as in Fig. 7). The solid curve is the optimal value for the ensemble system, the dashed curve shows value for a single deterministic forecast (the unperturbed "control" integration in the ensemble). The figure illustrates the fact that the ensemble forecast has greater "value" than a single deterministic forecast. For some cost/loss ratios (e.g. $C/L > 0.6$), the deterministic forecast has no value, whereas the ensemble forecast does have value. The reason for this can be understood in terms of the fact that for a probabilistic forecast, different users (with different C/L) would take precautionary action for different forecast probability thresholds. A user who would suffer a catastrophic loss ($C/L \ll 1$) if E occurred, would take precautionary action even when a small probability of E was forecast. A user for whom precautionary action was expensive in comparison with any loss ($C/L \approx 1$) would take precautionary action only when a relatively large probability of E was forecast. The result demonstrates the value of a reliable probability forecast.

More generally, this analysis suggests that the economic value of ensemble prediction will be realized by coupling output of the individual members of the ensemble to quantitative user application models to predict pdfs of user-relevant variables. Successful examples of such a strategy have been realized in ship routing (Hoffschildt et al., 1999) and electricity demand (J. Taylor and R. Buizza, personal communication).

G. CONCLUDING REMARKS

Our climate is a complex nonlinear dynamical system, with spatial variability on scales ranging from individual clouds to global circulations in the atmosphere and oceans, and temporal variability ranging from hours to millennia. Weather and climate scientists interact with society through the latter's demands for accurate and detailed environmental forecasts: of weather, of El Niño and its impact on global rainfall patterns, and of man's effect on climate. The complexity of our climate system implies that quantitative predictions can only be made with comprehensive numerical models which encode the relevant laws of dynamics, thermodynamics and chemistry for a multi-constituent multiphase fluid. Typically such models comprise some millions of scalar equations, describing the interaction of circulations on scales ranging from tens of kilometres to tens of thousands of kilometres, from the ocean depth to the upper stratosphere. These equations can only be solved on the world's largest supercomputers.

However, a fundamental question that needs to be addressed, both by producers and users of such forecasts, is the extent to which weather and climate are predictable; after all, much of chaos theory developed from an attempt to demonstrate the limited predictability of atmospheric variations. In the past, the topic of predictability has been a somewhat theoretical one, somewhat removed from the practicalities of prediction. A famous climatologist remarked some years ago: "Predictability is to prediction as romance is to sex!" However, the remark is perhaps not so apposite today; the science of predictability of weather and climate is now an integral part of the practical prediction problem—the two cannot be separated. The predictability problem can be formulated through, e.g., a Liouville equation; however, in practice, estimates of predictability are made from multiple (ensemble) forecasts of comprehensive weather and climate prediction models. The individual members of the ensemble differ by small perturbations to quantities that are not well known. The predictability of weather is largely determined by uncertainty in a forecast's starting conditions, though the effects of uncertainty in representing computationally the equations that govern climate (e.g. how to represent the effects of convective instabilities in a model that cannot resolve individual clouds) are not negligible.

Chaos theory implies that all such environmental forecasts must be expressed probabilistically; the laws of physics dictate that society cannot expect arbitrarily accurate weather and climate forecasts. These probability forecasts quantify uncertainty in weather and climate prediction. The duty of the meteorologist is to strive to estimate reliable probabilities; not to disseminate predictions to society with a precision that cannot be justified scientifically. Examples showed that, in practice, the economic value of a reliable probability forecast (produced from an ensemble prediction system) exceeds the value of a single deterministic forecast with uncertain accuracy.

However, ensemble forecasting poses a fundamental dilemma given current computing resources. To be able to simulate extreme events requires models with considerable complexity and resolution. On the other hand, estimating reliably changes to the probability distributions of extreme and hence relatively rare events, requires large ensembles. One thing is certain; the more the need to provide reliable forecasts of uncertainty in our predictions of weather and climate, the more the demand for computer power exceeds availability, notwithstanding the unrelenting advance in computer technology. Indeed the need for quantitative predictions of uncertainty in weather and climate science is a relevant consideration in the design of future generations of supercomputers; ensemble prediction is a perfect application for parallel computing! There can be little doubt that the benefits to society of reliable regional probability forecasts of extreme weather events, seasonal variability and anthropogenic climate change justify such technological developments.

ACKNOWLEDGEMENTS

I would like to thank colleagues at ECMWF, Drs J. Barkmeijer, R. Buizza, F. Lalaurette, T. Petroliagis, K. Puri and D. Richardson for innumerable discussions and for help in producing many of the figures in this chapter.

References

Betts, A. K. and M. J. Miller, 1986: A new convective adjustment scheme. Part II: Single column tests using GATE wave, BOMEX, ATEX and arctic air-mass data sets. *Q. J. Meteorol. Soc.*, **112**, 693–709.

Buizza, R. and T. N. Palmer, 1995: The singular vector structure of the atmospheric global circulation. *J. Atmos. Sci.*, **52**, 1434–1456.

Buizza, R., T. Petroliagis, T. N. Palmer, J. Barkmeijer, M. Hamrud, A. Hollingsworth, A. Simmons and N. Wedi, 1998: Impact of model resolution and ensemble size on the performance of an ensemble prediction system. *Q. J. R. Meteorol. Soc.*, **124**, 1935–1960.

Buizza, R., M. J. Miller and T. N. Palmer, 1999a: Stochastic simulation of model uncertainties in the ECMWF ensemble prediction system. *Q. J. Meteorol. Soc.*, **125**, 2887–2908.

Buizza, R., A. Hollingsworth, A. Lalaurette and A. Ghelli, 1999b: Probabilistic predictions of precipitation using the ECMWF ensemble prediction system. *Weath. Forecast.*, **14**, 168–189.

Ehrendorfer, M., 1994a: The Liouville equation and its potential usefulness for the prediction of forecast skill. Part I: Theory. *Mon. Weath. Rev.*, **122**, 703–713.

Ehrendorfer, M., 1994b: The Liouville equation and its potential usefulness for the prediction of forecast skill. Part II: Applications. *Mon. Weath. Rev.*, **122**, 714–728.

Epstein, E. S., 1969: Stochastic dynamic prediction. *Tellus*, **21**, 739–759.

Gelaro, R., R. Buizza, T. N. Palmer and E. Klinker, 1998: Sensitivity analysis of forecast errors and the construction of optimal perturbations using singular vectors. *J. Atmos. Sci.*, **55**, 1012–1037.

Gleeson, T. A., 1966: A causal relation for probabilities in synoptic meteorology. *J. App. Meteorol.*, **5**, 365–368.

Gleick, J., 1992: *Genius: Richard Feynman and Modern Physics.* Little, Brown and Co.

Harrison, M., T. N. Palmer, D. S. Richardson and R. Buizza, 1999: Analysis and model dependencies in medium-range ensembles: two transplant cast studies. *Q. J. R. Meteorol. Soc.*, **125**, 2487–2515.

Hasselmann, K., 1976: Stochastic climate models. Part I. Theory. *Tellus*, **28**, 473–485.

Hoffschildt, M., J.-R. Bidlot, B. Hansen and P. A. E. M. Janssen, 1999: Potential benefit of ensemble forecasts for ship routing. ECMWF Technical Memorandum, 287.

Houtekamer, P. L., L. Lefaivre, J. Derome, H. Richie, and H. L. Mitchell, 1996. A system simulation approach to ensemble prediction. *Mon. Weath. Rev.*, **124**, 1225–1242.

Katz, R. W., and A. H. Murphy, 1997. Forecast value: prototype decision-making models. In *Economic value of Weather and Climate Forecasts* (R. W. Katz and A. H. Murphy, Eds) Cambridge University Press.

Lorenz, E. N., 1963: Deterministic nonperiodic flow. *J. Atmos. Sci.*, **42**, 433–471.

Mason, P. J. and D. J. Thompson, 1992: Stochastic backscatter in large-eddy simulations of boundary layers. *J. Fluid. Mech.*, **242**, 51–78.

Molteni, F., R. Buizza, T. N. Palmer and T. Petroliagis, 1996: The ECMWF ensemble prediction system: methodology and validation. *Q. J. R. Meteorol. Soc.*, **122**, 73–119.

Moncrieff, M. W., 1992: Organised convective systems: archetypal dynamical models, mass and momentum flux theory and parametrization. *Q. J. R. Meteorol. Soc.*, **118**, 819–850.

Moss, F. and P. V. E. McClintock, Eds., 1989: *Noise in Nonlinear Dynamical Systems.* Vol. 1, *Theory of Continuous Fokker–Planck Systems.* Cambridge University Press.

Mureau, R., F. Molteni and T. N. Palmer, 1993: Ensemble prediction using dynamically-conditioned perturbations. *Q. J. R. Meteorol. Soc.*, **119**, 299–323.

Murphy, A. H., 1973: A new vector partition of the probability score. *J. Appl. Meteorol.*, **12**, 595–600.

Murphy, A. H., 1977: The value of climatological, categorical and probabilistic forecasts in the cost-loss ratio situation. *Mon. Weath. Rev.*, **105**, 803–816.

Palmer, T. N., F. Molteni, R. Mureau, R. Buizza, P. Chapelet and J. Tribbia, 1993: Ensemble prediction. In ECMWF Seminar: *Validation of Models over Europe,* European Centre for Medium Range Weather Forecasts, Shinfield Park, Reading, UK, 1992.

Palmer, T. N., C. Brankovic and D. S. Richardson, 2000: A probability and decision model analysis of PROVOST seasonal multi-model ensemble integrations. *Q. J. R. Meteorol. Soc.*, **126**, 2013–2034.

Palmer, T. N., 2000: Predicting uncertainty in forecasts of weather and climate. *Rep. Prog. Phys.*, **63**, 71–116.

Richardson, D. S., 2000: Skill and economic value of the ECMWF ensemble prediction system. *Q. J. R. Meteorol. Soc.*, **126**, 649–668.

Toth, Z. and E. Kalnay, 1993: Ensemble forecasting at NMC: the generation of perturbations. *Bull. Am. Meteorol. Soc.*, **74**, 2317–2330.

Toth, Z. and E. Kalnay, 1997: Ensemble forecasting at NCEP and the breeding method. *Mon. Weath. Rev.*, **12**, 3297–3319.

Wilks, D. S., 1995: *Statistical Methods in the Atmospheric Sciences.* Academic Press.

Extratropical Cyclones
An Historical Perspective
Alan J. Thorpe
Met Office, Bracknell, UK

The history of ideas concerning extratropical cyclones has been intimately intertwined with that of meteorology itself. In this chapter, aspects of this history will be brought out with particular reference to the points in that history where the thinking has been influenced by, and has influenced, fundamental mathematics and physics. Many giants of physics and meteorology have contributed to the story—Fitzroy, Helmholtz, Shaw, Bjerknes, Charney, Eady, Sawyer, Eliassen and Kleinschmidt (to name only some prior to the more modern era). Specific examples of the free association between meteorological and basic physical ideas, relevant to the cyclone story, are: fluid dynamical analogies to electromagnetic phenomena, leading to the Bjerknes' circulation theorem; electrostatic analogies to guide potential vorticity thinking; geometrical concepts, such as transformations and rearrangement theory, illuminating quasi- and semi-geostrophic models of cyclones; and chaotic dynamics and the limits of cyclone predictability. Probably the most famous school of meteorology—the Bergen school—devoted itself to understanding and forecasting these phenomena. Their ideas have been pervasive in providing a useful conceptual framework, or an inflexible straightjacket depending on one's viewpoint.

Where has this rich history taken us to today? Whilst we have become accustomed to the routine successes of numerical weather prediction in forecasting these storms we remain struck by notable failures. Some say that this success has not obviously been driven by these profound theoretical ideas on cyclones although one would argue that it has influenced the human forecasters' contribution. On the other hand, the quest throughout this history for the all-encompassing "conceptual model" of the generic extratropical cyclone remains as elusive as ever, particularly as we probe into the mesoscale and small-scale structure of the cyclone. The future holds open the prospect, for example, of improved predictive skill by targeted observations being made in predicted "sensitive" zones by controllable robotic aircraft.

A. INTRODUCTION

Thinking about extratropical cyclones can be described as being in several epochs or phases that follow on from one another. One can relate these epochs on a time line (nineteenth century, twentieth century and modern era) or connect them with key individuals, or schools, whose thinking dominated changes from one view to another. In this short chapter I will develop the ideas roughly along a chronological line but with a special focus on the application of basic physics within meteorology addressing the cyclone problem. This will lead us to discuss: the thermal theory, the Bergen school, baroclinic instability theory, potential vorticity ideas and finally predictability, forecasting and observations. It is probably fair to say that extratropical cyclones are amongst the most studied of meteorological phenomena. So one can ask whether there are still unanswered questions and where this subject will go in the new century; an analysis of this concludes the chapter.

B. THE THERMAL THEORY

The thermal, or convective, theory of cyclones was developed during the 1860s and arguably became the first such theory of these ubiquitous extratropical phenomena properly based on at least some of the laws of physics. Essentially the idea was that cyclones derived their energy thermodynamically from the ascent of warm air and condensation of water vapour. This heating produced the circulation and development of low-pressure systems. The nineteenth century was a period of many advances in the theory of meteorology and of the improvement in observations of the atmosphere. It is no coincidence that the birth of professional societies, such as the Royal Meteorological Society, took place at the same time. In the early part of the nineteenth century, ideas about the atmosphere tended to be purely kinematical in nature. Then the application of the conservation of energy and the interchange between potential and kinetic energies was made to cyclones. In the 1830s, James Pollard Espy proposed the so-called convective theory of storms wherein latent heat release was hypothesized as the driving mechanism for storm development (Espy, 1841). The realization of the first law of thermodynamics, adiabatic changes and the definition of potential temperature allowed these ideas to be made more physically based. Examination of synoptic charts and observations led to a quantification that supported the idea that the huge source of energy from latent heating was the primary mechanism. (Many of the theories in 1860s and 1870s proposed what we might today call a "conditional instability of the second kind" theory of cyclone development.) Towards the end of the century, when upper air observations were becoming more available, certain contrary observations suggested that all was not well with the thermal theory. For example, it was found that there is frequently cold air above the near-surface warm air.

There was, however, an alternative view that had been developing throughout the nineteenth century. It was realized that there were often somewhat separate air currents (or masses) from polar and tropical regions connected with the large pole-to-equator temperature gradients. These gradients represent a reservoir of potential energy. Fitzroy (1863) suggested, mid-century, that the currents, as they moved past each other, might be responsible for cyclonic formation. This process was likened to the generation of eddies in a river "when one body of water passes another".

Both Helmholtz (1888) and Margules (1905) examined the steady-state configuration of two homogeneous rotating air masses of different density mimicking the polar and tropical streams. Margules deduced the necessity for this interface to slope with height and speculated about it being unstable in the sense that a perturbation of the interface might grow by using the available potential energy.

The beginning of the twentieth century was a period when the previously somewhat separate analyses of thermodynamics and dynamics started to come together. It was around this time that Ferrel derived the thermal wind relationship that makes explicit the relationship between the wind field and the mass field. A significant trigger for dynamical thinking at that time was the work of Vilhelm Bjerknes.

C. THE BERGEN SCHOOL

Observations at the turn of the twentieth century made it clear that cyclones have an open wave structure in the temperature field despite being isolated closed structures in the surface pressure field. This suggested that a useful approach to understanding their development might be gained by looking at the dynamic stability of (sloping) boundaries between warm and cold air masses. Various facets of this approach were to dominate twentieth-century thinking on cyclone formation. The transition between the nineteenth-century thinking and the twentieth-century "obsession" with dynamical instability was further developed within what is now perhaps the most famous single group of meteorologists in meteorological history, the Bergen school. Their thinking has been incredibly influential, particularly in synoptic practice and in forecasting, but the very essence of physical theory that their leader, Vilhelm Bjerknes, was so obsessed about ironically alluded them in their studies of the causes of cyclone growth.

Vilhelm Bjerknes was the son of a noted applied mathematician, Carl Anton Bjerknes. In his early career, Vilhelm assisted his father in his studies on electric and magnetic phenomena. Carl Anton wanted to show that the action-at-a-distance electric and magnetic forces were produced and propagated in a fluid-medium-filled space—the ether. He created various laboratory demonstrations of pulsating and rotating spheres in water, which seemed to be good analogies to electromagnetic phenomena. Vilhelm studied electromagnetic waves for his Ph.D., which was awarded in 1892. However, his father, who was prone to writer's block, wanted Vilhelm to help him write a manuscript on the hydrodynamic analogies and action-at-a-distance.

The Bjerknes' work was certainly not unique; Heinrich Hertz also aimed to find a mechanical basis for the action of electromagnetic forces via a theory of mechanics that would explain the physics of the ether. Of course, in hindsight this approach turned out to be flawed, as the existence of the ether was later to be disproved. However, for Vilhelm Bjerknes this work led him to what was his first major contribution to meteorology. Rather than use physical bodies rotating and pulsating within a fluid, he considered the problem of fluid bodies with densities different from their surroundings. This, he hoped, would lead him to some general conclusions about electromagnetic forces that did not depend on the geometry of the body. Helmholtz and Kelvin had also worked on the dynamical analogues between hydrodynamic and electromagnetic forces, but Vilhelm Bjerknes became increasingly convinced that Kelvin's circulation theorem was flawed. He realized that Kelvin's theorem only applied to fluids for which the density was a function only of pressure. In principle, and in practice for the atmosphere, the density can also be a function of other quantities such as temperature. This led Vilhelm to his circulation theorem, which shows that the circulation is not conserved in such circumstances. The theorem was interpreted geometrically by considering intersecting surfaces of pressure and density (Bjerknes, 1898). These intersections define tubes, which he called, for obvious reasons given the nature of his work at that time, solenoids. The circulation, necessarily generated from Bjerknes' theorem, rotate these surfaces towards one another. (Other terminology borrowed from electromagnetic theory was also introduced into meteorology, such as the idea of a flux.)

This was a fundamental result and Bjerknes felt it was a major contribution in his, and his father's, attempts to draw the analogy between hydrodynamic and electromagnetic phenomena. His published articles on the new theorem in 1898 and 1899 included, surprisingly, a section on applications to atmospheric and oceanic phenomena. These applications to the geophysical field, that he had not previously considered, were included due to his contact with Ekholm, Arrhenius and Pettersson in the Stockholm Physics Society at that time (Friedman, 1989).

Arrhenius, by training a chemist, was increasingly attracted to producing an exact science of the atmosphere and oceans. Ekholm was interested in cyclones and Pettersson in oceanography. In particular Ekholm, who was a balloonist, was developing ideas that winds in extratropical cyclones arise from unequal distributions of density and pressure, a fact that he had observed in his balloon flights. Ekholm told Bjerknes about meteorology and in particular about the controversy between the thermal theory of cyclones and the newer ideas about the "mixing" of different air masses. Ekholm felt that this surprising distribution of density and pressure was connected to the process by which cyclones formed. Bjerknes realized that his circulation theorem could explain the relationship between these solenoids and the development of circulation and thereby cyclones themselves. The American meteorologist Abbe provided some data, taken by kites, to Bjerknes, and his assistant Sandstrom analysed the data to show that indeed the size of the solenoids was large enough to account for the generation of strong winds within cyclones. This unexpected success of his circulation theorem to the entirely new field of meteorology, combined with the movement of basic theoretical physics away from the concept of the ether were instrumental in leading Bjerknes into his illustrious career in atmospheric science. Also at the turn of the century, oceanographers became interested in the new circulation theorem to help them understand ocean currents for fishery and for exploration. Ekman carried out these studies and the way forward was established for twentieth-century dynamical oceanography.

In the period from 1900 to 1906, Bjerknes came to regard the main direction of his research to lie in the creation of an exact mechanics or physics of the atmosphere so that "rational" predictions of its behaviour could be made. Bjerknes realized that the new advances in aeronautics not only created the possibility of obtaining the ever-important upper-air observations but also provided a market for weather forecasts. Bjerknes was intent also to make meteorology into a respected branch of physics. When Norway became independent from Sweden in 1905, a chair was created for Bjerknes in Christiania (now called Oslo) and when Vilhelm moved there in 1907, he tried to establish a group of assistants around him. However, working conditions were poor and Bjerknes left in 1913 to set up a geophysics institute in Leipzig. This was attractive, not least as a result of advances being made because of an increase in aerological activities in Germany. His group of assistants at that time included Sverdrup. The outbreak of World War I not only required Bjerknes to return to Norway but also brought even more significance to the weather forecasting enterprise that he was establishing. Military operations were always dependent on meteorological conditions.

It was these events that led Vilhelm Bjerknes to Bergen in 1917. In a way that presages more recent developments in meteorology, the science was driven forward by the realization of the importance of weather forecasts in many aspects of society including in the economic domain; commercial aviation, agriculture, fisheries and military action. (Indeed in 1921, Bjerknes referred to the "militarizing of meteorology" an idea that was abhorrent, for example, to L. F. Richardson.) Bjerknes began to set up a new weather service in Norway. It must have been rarer then than now that academic studies were so easily combined with an operational application. The weather forecasting problems for west Norway were clearly to do with cyclonic storms, and so it was that the Bergen school turned its attention to these phenomena with which their names are now so inextricably linked. Vilhelm's son Jacob became one of the driving forces in the activities in Bergen. Jacob wrote a paper "On the Structure of Moving Cyclones" in which the term "front" was coined (Bjerknes, 1919). Figure 1 shows the now classic, Bergen cyclone model. (A book, called "On the Origin of Rain" to mimic the title of Darwin's epic, was never completed. A paper "On the Structure of the Atmosphere when Rain is falling" was published in the *Quarterly Journal* (Bjerknes, 1920).)

The Bergen school constructed many of its ideas from observations of the lower atmosphere. By careful observation of the weather progression as cyclones came over, their attention was increasingly focused on lines of convergence, rain-stripes and discontinuities. This became known as an "indirect aerology" because

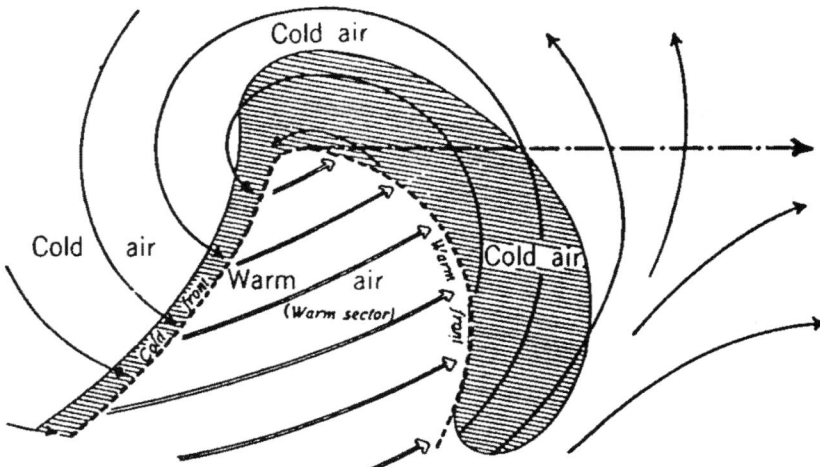

FIGURE 1 A version of the Bergen cyclone model—one of the most famous diagrams to emanate from the Bergen school. (From Bjerknes and Solberg, 1922).

of the lack of detailed *in situ* upper air observations. During 1919, the Bergen school established the idea of the "polar front" as a three-dimensional sloping boundary as well as a surface discontinuity. The analogy of the battling air masses on either side of the front led them to this terminology. The polar front was imagined as a single line stretching right around the Northern Hemisphere. These new ideas were not generally accepted immediately. The individual cyclones formed along the polar front as "pearls along a string". The cyclones developed warm, cold and occluded fronts conceived of as local perturbations to the polar front (Bjerknes and Solberg 1922). Solberg began to consider the dynamical problem of the development of these cyclones as a problem in the theory of dynamical instability of a sloping discontinuity (Solberg, 1928).

It is of some historical note that Vilhelm Bjerknes organized two conferences (separate German and non-German speaking) in Bergen in 1920 which brought together many leading meteorologists. They were held to persuade the international community of the efficacy of the Bergen school ideas. Davies (1997) notes that there were major contributions from the Bergen, British and Viennese Schools by their respective leaders Bjerknes, Shaw and Exner. The idea of "polar front meteorology" was proposed by the Bergen school and seemed to strike a chord with the other scientists. L. F. Richardson asked, with a large degree of insight into future developments in the subject, about the possibility of instability of the polar front. Although there was early resistance, by about 1935 the Bergen synoptic analysis methods were in wide usage. Some of the resistance was due to a perceived over concentration on surface features and a consequent downplaying of associated upper-level, and perhaps pre-existing, disturbances. Others saw it as breaking down the old order of Austrian/German hegemony in meteorology.

In 1926, Vilhelm Bjerknes moved from Bergen back to the University of Oslo and sought to develop a more academic flavour to the research. The polar front meteorology had been and continued to be promoted, with at times evangelical zeal, in the international arena. It has undoubtedly had an important influence on the subject. However, in hindsight, its focus on the lower atmosphere can be thought of as a temporary redirection of attention away from the twentieth century's preoccupation with the upper troposphere as the source of cyclonic development.

D. BAROCLINIC INSTABILITY MODEL

New observations and theorizing about the upper atmosphere (troposphere) were to enable a new thrust in cyclone research to develop post World War II. The time was evidently right for a consistent mathematical model to be produced of what became known as baroclinic instability. As so often happens in science two rather similar treatments were published at about the same time, by Eady (1949) and Charney (1947). The development of the new model was possible partly because the authors realized that it was crucial to start from equations that removed the short-timescale motions such as gravity waves. (This may indeed be the reason why Bergen school theorists such as Solberg were unable to produce such a consistent theory.) The wave perturbations sought were assumed to be in geostrophic balance and the framework within which the theories were embedded became known as the "quasi-geostrophic" theory, developed by Sutcliffe, Charney and Eliassen.

In essence the new baroclinic instability theory was deceptively simple. Imagine a simple westerly flow, the strength of which increased linearly with height associated with a constant pole-to-equator temperature gradient. On this basic state, a wavy perturbation of infinitesimal amplitude was imposed and the equations for the growth of special disturbances that retain their spatial structure in time were derived. The result is nowadays called a normal mode and the theory delivers a curve, which indicates the growth rate as a function of the horizontal scale of the wavy perturbation. It also delivers the spatial structure of all the meteorological fields in the normal mode. Faced with this growth rate curve, Eady used an interesting analogy borrowed from the Darwinian theory of evolution. He imagined a population of animals (the normal modes) all with slightly different characteristics (their horizontal scale) that were competing with each other to survive. The ones with the maximum growth rate would clearly quickly dominate the others and so become the dominant species. This theory of natural selection suggests that the dominant perturbations should have a horizontal wavelength of some 4000 km and these are now regarded as the embryonic cyclones of the extratropics. They are embryonic because the theory is linear and so can only describe these "cyclones" when they have very small amplitude. An important question then arose as to whether nonlinearity would alter the picture.

An intriguing aspect of the new baroclinic instability theory was that, in its simplest form it neglected the latent heating in clouds which the nineteenth-century thermal theory thought to be all important. The attention was thus firmly moved onto the potential energy available within the horizontal temperature contrasts. So whilst baroclinic instability was the first complete dynamical and thermodynamical theory it certainly reinforced the late nineteenth-century ideas about the importance of the transfer of potential energy into kinetic energy as the atmosphere overturns undergoing the baroclinic motions. However, this similarity with the thermal theory does not represent the major triumph of the baroclinic instability theory. This was that it allowed a clear appreciation that the requirement for these synoptic scale motions to maintain thermal wind balance has a crucial role in shaping the flow of energy from potential to kinetic. Features of cyclones such as the westward phase tilt with height is a necessary consequence of this constraint.

Another intriguing aspect of the theory is its division of the atmosphere into a steady basic state flow and an infinitesimal perturbation. These two features lay the theory open to two somewhat contrary criticisms. The first is that the atmosphere in the extratropics never exhibits such a basic state being, as it is, in a perpetual complex state of "turbulence" with weather systems ever present. The second is that there are always going to be infinitesimal perturbations, so there is an inevitability about cyclone growth. However, its basic elements, such as the typical scales, growth rates and wave structure, are clearly seen in observations of cyclones.

Much later, in the 1980s, attention focused on the limitations of the assumption of constant shape (normal mode) disturbances. This led to development of companion models describing the nonmodal components. In these models, attention was drawn away from infinitesimal perturbations towards finite-amplitude structures that can grow by the same baroclinic mechanisms, although at a greater rate. The reasoning was that such precursor finite-amplitude perturbations are ever present and in accord with synoptic experience. Also the idea was that the basic state flow envisaged in the Eady model would often be stable to normal mode growth due the presence of processes such as friction, but notwithstanding this, it would allow the growth of finite-amplitude (special) nonmodal disturbances.

This reasoning about precursor disturbance at upper levels has a long history (Davies, 1997). None other than Sir Napier Shaw (1914) concluded that "the proximate cause of the variation of pressure ... must be looked for at a height of about 7 to 9 km". This was linked to work by Dines and others that upper-level divergence must offset, or compensate, for low-level convergence. The importance of upper-level jets and confluence/diffluence at the jet entrances and exits as favoured locations for surface cyclogenesis was discussed in the 1940s. Pettersen and Smebye (1971) analysed synoptic charts and divided cyclogenesis into type A (akin to normal mode instability) and type B, where there is strong forcing by a pre-existing upper-level disturbance.

In the 1990s the baroclinic instability theories were properly extended to include latent heating in cloudy air, and this brought the theory into much better agreement with the observations, both in terms of scale and growth rate.

E. POTENTIAL VORTICITY IDEAS

One of the famous sons of the Bergen school was Carl-Gustav Rossby and he introduced into meteorology in 1940, the powerful concept of the conservation of potential vorticity (Rossby, 1940). This allowed the Bjerknes circulation theorem to be put within the context of a conservation principle. Although Bjerknes showed that circulation could be generated and destroyed, a fact of immense importance in figuring out the development of cyclones, Rossby showed that whilst these changes in circulation were going on, the air was subject to the important constraint of the conservation of potential vorticity. Hans Ertel in Germany published a full mathematical treatment of this new theorem (Ertel, 1942). The use of potential vorticity then followed two directions. The first was in the mathematical quasi-geostrophic theory (and the baroclinic instability theories of Charney and Eady), via the quasi-geostrophic potential vorticity, and the second was in analysis of weather charts via the Ertel–Rossby potential vorticity such as in the work of Ernst Kleinschmidt. Attention was somewhat drawn away from potential vorticity when it was realized in the 1950s that it was possible to use the so-called primitive equations instead of the quasi-geostrophic equations, to construct numerical weather forecasts. However, it became the common currency of theoretical studies and this has persisted to the present day.

Special mention must be made here of the pioneering work of Ernst Kleinschmidt. He quickly realized, in the early 1950s, the significance of the new quantity and proceeded to investigate its relevance and structure within extratropical cyclones; (Kleinschmidt 1950). He also understood that the latent heat release, beloved within the thermal cyclone theory, could produce local anomalies of potential vorticity within cyclones. These so-called "producing masses" could be held responsible for, at the least, increasing dramatically the growth of cyclones (see Fig. 2). He also identified the fact that it was possible to deduce the wind and temperature fields by a process of "inverting" the distribution of potential vorticity, were it to be known (Thorpe, 1993). This idea was revisited and reviewed by Hoskins et al. (1985). In that paper, unification was established between ideas such as baroclinic instability and potential vorticity, which has become known as "potential vorticity thinking". This has drawn together many of the strands of dynamical thinking on cyclones.

Potential vorticity thinking also curiously completes the circle begun by Vilhelm Bjerknes. The analogy between potential vorticity anomalies and electrostatic charges can be used to illuminate some aspects of the structure related to potential vorticity. The rationale behind doing this is interesting. Potential vorticity is considered to be a somewhat hypothetical or theoretical concept, being, as it is, an unobservable quantity at least by current direct meteorological measurements. It seems useful in this context to make it more tangible by drawing on analogies to simpler concepts in basic physics. These concepts are presumably common currency for the modern meteorologist who has become, as hoped for by Bjerknes, a respected physicist by trade. Assuming that electrostatic theory is such a concept for the modern meteorologist we can reverse the analogy established by Bjerknes. He wanted to use hydrodynamic analogies, as they were well established and widely understood, to illuminate the newer more abstract electromagnetic concepts such as the ether. In this reverse analogy the potential vorticity is akin to electric charge in being fundamental but only existing in terms of related fields such, as geopotential. These fields are related to the "charge" by the solution of a Laplacian equation. The atmosphere in which these potential vorticity charges are embedded can be thought of as a (nonlinear) dielectric medium. (Readers interested in pursuing this analogy might like to look at Thorpe and Bishop, 1995; see also Fig. 3.)

It is also now clear that it is possible to relate the criteria for the occurrence of various instabilities, such as baroclinic instability and conditional symmetric

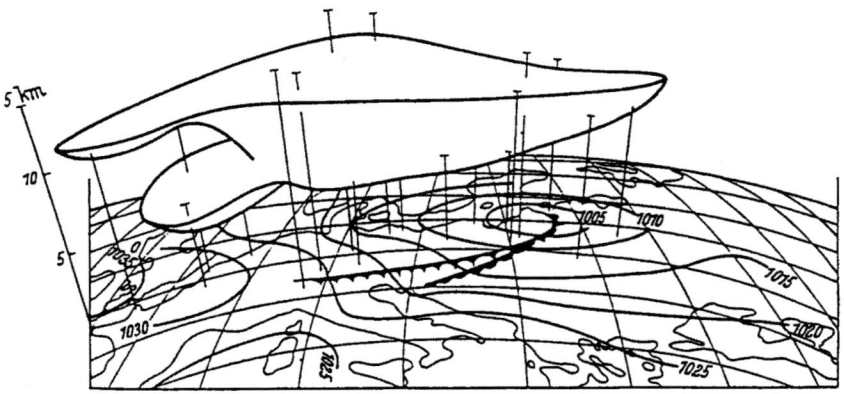

FIGURE 2 Ernst Kleinschmidt's conception of the upper-level potential vorticity "producing mass" superimposed on the surface synoptic chart analysed according to the Bergen school methodology. (From Kleinschmidt, 1950).

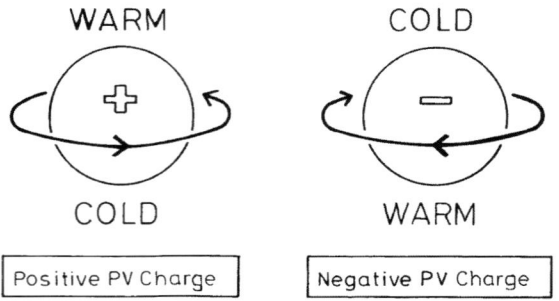

FIGURE 3 The "elementary" potential vorticity particles suggested by analogy with the theory of electrostatics. Each potential vorticity charge has circulation and a temperature dipole associated with it.

instability, to the potential vorticity distribution within the atmosphere. Hence we can see that the potential vorticity has a fundamental unifying significance within the theory of extratropical cyclones.

F. PREDICTABILITY, FORECASTING AND OBSERVATIONS

Considerable advances were made in the 1970s, and later, concerning modelling of the life cycles of idealized cyclones. These models could look into the change of amplitude and structure of cyclones as they grew to maturity. The computer "experiments" showed that in the nonlinear phase of the life cycle there were quasi-two-dimensional zones where horizontal temperature gradients increased substantially bringing the cold and warm air masses in close proximity. This is the process of frontogenesis. Fronts are therefore seen as a by-product of cyclone growth rather than as a cause. The nature of the frontogenetic process had in fact been outlined much earlier by Bergeron in the Bergen school. However, the dynamical/mathematical description of frontogenesis only occurred in the second half of the twentieth century following advances made by Sawyer, Eliassen and later, Hoskins. It was realized that the quasi-geostrophic model could be extended to the two-dimensional frontal zones undergoing intense frontogenesis using a new semi-geostrophic theory. This new theory had certain attractions in that it used Ertel–Rossby potential vorticity rather than the quasi-geostrophic counterpart and it predicted frontal collapse in a finite, rather than an infinite, time. Also semi-geostrophic theory is most simply described when using a coordinate transformation from Cartesian to geostrophic momentum coordinates. This transformation, it turns out, is of some considerable significance in branches of pure mathematics. The theory also has the advantage of exhibiting a set of equations that has a Hamiltonian form, unlike the primitive equations. This point of contact with basic mathematics and theoretical physics is proving a productive one in the attempt to establish the formal solvability of the equations that are commonly used for numerical weather prediction.

It has also been realized that the decay of cyclones presents at least as many problems of theoretical understanding as their growth mechanisms. Following the decay of one cyclone the storm-track is not immediately conducive to the growth of further cyclones due to the generation of a significant barotropic shear—the so-called barotropic governor mechanism.

In the 1980s, significant improvements took place in the techniques of data assimilation and in our understanding of the predictability of the atmosphere. A feature of the baroclinic instability theory was that it predicted that infinitesimal perturbations would grow rapidly. This is exactly what became crucial in predictability studies of the atmosphere carried out first by Lorenz in the development of what is now known as the theory of chaos. Any flow that is unstable becomes a real problem to predict because the growth of analysis errors will quickly pollute the forecast. In other words the very perturbations that lead to cyclone growth are, if misrepresented in the initial conditions for the forecast, responsible for the fundamental limitations in our ability to predict the weather. The cyclones will only be well forecast if we can carefully observe the atmosphere so as to reduce to an absolute minimum the amplitude of analysis errors, from observational error. It is clear now that the discovery that the atmosphere is baroclinically unstable on these scales had much wider implications than just for the understanding of extratropical cyclones. These crucial ideas of chaos are certainly one example where theory within meteorology has spawned new thinking within basic physics. The fields of chaos and complexity are now well established within physics degree courses, although there are still intriguing aspects such as the fact that at the microscopic level the Schrodinger equation is linear and so apparently does not exhibit chaos.

Our story has thus far dealt with the history of ideas concerning cyclones, but we now turn to the practitioner's art of forecasting these storms. Over the twentieth century the job of forecaster, of course, underwent a huge revolution due to the advent of numerical weather prediction and of satellite observations. It is sometimes said that the development of the ideas we have been describing here has had little effect on forecasting. This is because, on the one hand, the weather prediction models have not really been designed differently because of an appreciation of the theories, and on the other, forecasters seem to rely more on their experience than on an appreciation of the theory. Both these views are to some degree misguided but also contain an element of truth.

So what is the quality of cyclone forecasting at the start of the twenty-first century? One has to say that generally it is very good. However, there are still significant and persistent forecasting problems that occur when the numerical models and forecasters fail to predict particularly extreme and small-scale events. Clearly for the customers of these forecasts this is a substantial problem. How come we still have such problems after so much erudite theorizing over the last 150 and more years?

The history of cyclone research has tried to create a simple "conceptual" model of extratropical cyclones. Whilst one would have to say that many of the building blocks of this conceptual model are in place, there are still aspects of the problem that are unclear at the beginning of the twenty-first century. For example, there is the question of secondary cyclogenesis. Many of the forecast problems we have just been discussing are characterized by the growth of secondary cyclones that spawn from the frontal zones of the primary cyclone that had its genesis region close to the eastern coasts of the continents. These secondary cyclones may require a different conceptual model as they have certain characteristics that are very different from the primary ones. For example, it seems that the cold frontal zones of the primary cyclone sometimes break up into three-dimensional disturbances, one of which becomes a dominant secondary storm. On the other hand, on other occasions, such cold fronts can be remarkably undisturbed. These differences in behaviour can cause forecast models and forecasters some significant difficulties.

Recent ideas about secondary cyclones have focused on the nature of the environment in which such embryonic systems find themselves. The frontal environment is a very inhospitable one for a three-dimensional disturbance to grow in. It is often undergoing active frontogenesis; such as associated with a stretching deformation, or strain, flow. It may be that it is only when this active strain or frontogenesis ceases that the inherent dynamical instability of the frontal environment (associated in the lower troposphere with extrema of potential vorticity and potential temperature) can be realized. If so, then the ends of the storm tracks may be the preferred locations for the generation of such rapidly growing and small-scale secondary cyclones.

Such analyses of forecast problems have also promoted two further aspects of research into extratropical cyclones. The first is to gain a better appreciation of the smaller-scale features (mesoscale) embedded within the synoptic-scale cyclone. These are often cloud features, such as the cloud heads and they are usually the agents for the societal impact from the weather associated with extratropical cyclones. The second area of research is the realization that many examples of poor forecasts can be traced back to regions where, had there been additional observational data available, the forecast quality could have been significantly improved; for example, the so-called "Great Storm" of October 1987 over the UK (Burt and Mansfield, 1988). We will address these aspects in the next section when we look to the future in cyclone research.

G. THE FUTURE

The quest for a unifying "conceptual model" of extratropical cyclones has been a long and complex process. Various historical strands have been somewhat unified, but there are still notable scientific questions which remain to be answered. One concerns the relative influence of upper-level forcing and lower-tropospheric dynamical processes. Another concerns the relative roles of synoptic-scale dynamical forcing and the embedded mesoscale cloud processes. These strands have ebbed and flowed over at least the last 150 years. Even though we have self-consistent four-dimensional datasets from numerical models we still seek the simplified building blocks that can be encompassed into a conceptual model that qualifies as an "understanding" of the phenomena.

Scientists are probably not too good at predicting what the future in their subject holds, but I would like to suggest one possible exciting prospect for the future in cyclone research. Clearly, we know that errors in numerical forecasts of cyclone development and evolution must be attributable to either errors in the model formulation or to errors in the initial conditions. Our understanding of the physics of cyclones has led to, and will continue to lead to, improvements in the parametrizations of physical processes, such those as associated with clouds, in numerical forecast models. However, it is to the area of errors in initial conditions that I now turn. The cyclones that affect the western sides of the continents have a life cycle largely within what is generally called the oceanic storm tracks. Being over the oceans, cyclones develop in regions that certainly are not data void but are at least data sparse. Significant new data technologies have improved the situation tremendously. Meteorological satellites will continue to be important sources of data, as will the instrumented commercial aircraft and the "ships of opportunity". However, recent research associated with predictability concepts holds out the prospect that significant improvements in cyclone forecast skill may be available by carrying out additional targeted observations in so-called sensitive regions.

It is now possible to calculate, using the numerical forecast model itself, the perturbations that, if they were present in the initial conditions of the forecast, would grow the most rapidly and create a large forecast error associated with the cyclone in question. These are called targeted singular vectors, and a variant of these is also used to generate the various initial condition perturbations for the members of operational ensemble forecasts. The targeted singular vectors are located at analysis time in regions we can call sensitive in the sense that if there were to be an analysis error in that region it would create a large forecast error. The location of these sensitive regions is flow dependent and so will vary from cyclone to cyclone. Hence in future it will be possible to "forecast" ahead of time where the sensitive zone is for a given forecasted cyclone (Montani *et al.*, 1999). It would then be possible to utilize a mobile observing platform to take additional observations in the sensitive region. These extra observations will reduce the analysis errors in the sensitive region and thus lead to a more skilful forecast. An alternative method to locate the sensitive zone might be to compare the analysed potential vorticity distributions with the most recent and detailed satellite imagery for water vapour patterns to identify a potential analysis error. There are now mobile observing platforms, such as the aerosonde, which is a robotic small aircraft that can perform vertical sections to obtain soundings through the troposphere. These are a cost-effective additional source of data that can be flown to these critical sensitive regions. This may make the difference in that minority of cases that are currently poorly forecast and sometimes contain the most extreme weather.

References

Bjerknes, J., 1919: On the structure of moving cyclones. *Geofys. Publ.*, **1**(2), 8.

Bjerknes, J. and H. Solberg, 1922: Life cycles of cyclones and the polar front theory of atmospheric circulation. *Geofys. Publ.*, **3**(81), 18.

Bjerknes, V., 1898: Ueber einen hydrodynamischen Fundamentalsatz und seine Anwendung besonders auf die Mechanik der Atmosphäre und des Weltmeeres *Kongl. Svenska Vet.-Akad. Handl.*

Bjerknes, V., 1920: On the structure of the atmosphere when rain is falling. *Q. J. R. Meteorol. Soc.*, **46**, 119–140.

Burt, S. D. and D. A. Mansfield, 1988: The Great storm of 15–16 October 1987. *Weather*, **43**, 90–114.

Charney, J. G., 1947: The dynamics of long waves in a baroclinic westerly current. *J. Meteorol.* 4, 135–163.

Davies, H. C., 1997: Emergence of the mainstream cyclogenesis theories. *Meteorol. Z., N.F.*, **6**, 261–274.

Eady, E. T., 1949: Long waves and cyclone waves. *Tellus*, **1**, 33–52.

Ertel, H., 1942: Ein neuer hydrodynamischer Wirbelsatz. *Meteorol. Z.*, **59**, 277–281.

Espy J. P., 1841: *Philosophy of Storms*. Little and Brown.

Fitzroy, R., 1863: The Weather Book. *A Manual of Practical Meteorology* (2nd edn). London.

Friedman, R. M., 1989: *Appropriating the Weather: Vilhelm Bjerknes and the Construction of a Modern Meteorology*. Cornell University Press.

Helmholtz, H., 1888: Über atmosphärische Bewegungen. I. Mitteilung., *Sitz.-Ber. Preuss. Akad. Wiss.*, **5,** 647–663.

Hoskins, B. J., M. E. McIntyre and A. W. Robertson, 1985: On the use and interpretation of isentropic potential vorticity maps. *Q. J. R. Meteorol. Soc.*, **111**, 877–946.

Kleinschmidt, E. 1950: Über Aufbau und Enstehung von Zyklonen. *I. Meteorol. Rdsch.*, **3**, 1–6.

Kutzbach, G., 1979: *The Thermal Theory of Cyclones: A History of Meteorological Thought in the Nineteenth Century*. American Meteorological Society, Lancaster Press.

Margules, M., 1905: *Über die Energie der Stürme*. Jahrbuch der k.k., Central-Anstalt für Meteorologie und Erdmagnetismus.

Montani, A., A. J. Thorpe, R. B. Buizza and P. Unden, 1999: Forecast skill of the ECMWF model using targeted observations during FASTEX. *Q. J. R. Meteorol. Soc.*, **125**, 3219–3240.

Petterssen, S. and S. J. Smebye, 1971: On the development of extratropical cyclones. *Q. J. R. Meteorol. Soc.*, **97**, 457–482.

Rossby, C.-G., 1940: Planetary flow patterns in the atmosphere. *Q. J. R. Meteorol. Soc.*, **66** (suppl.), 68–87.

Shaw, W. N., 1914: Principi Atmospherica: A study of the circulation of the atmosphere. *Proc. R. Soc. Edinb., Ser. A.*, **34**, 77–112.

Solberg, H., 1928: Integration der atmosphärischen Strömungsgleichungen (I). *Geofys. Publ.*, **5**(9), 104–120.

Thorpe, A. J., 1993: An appreciation of the meteorological research of Ernst Kleinschmidt. *Meteorol. Z., N.F.*, **2**, 3–12.

Thorpe, A. J. and C. H. Bishop, 1995: Potential vorticity and the electrostatics analogy: Ertel-Rossby formulation. *Q. J. R. Meteorol. Soc.*, **121**, 1477–1495.

Numerical Methods for Atmospheric General Circulation Models
Refinements or Fundamental Advances?

David L. Williamson

*National Center for Atmospheric Research,[*]
Boulder, CO, USA*

This chapter focuses on developments in numerical methods for global atmospheric models over the last decade and speculates about future developments. Numerical Weather prediction (NWP) and climate modelling applications are contrasted. Ten years ago the common wisdom was that global NWP and climate models were becoming more and more similar and would ultimately merge into a single model. Taking the two communities as a whole, it is not clear that this is the case today, although on the surface it may appear to be so. By climate we mean long-term climate simulation rather than seasonal forecasting which falls somewhere between the two application extremes. In the following we will talk in sweeping generalities without necessarily pointing out the notable exceptions which do exist.

A. NUMERICAL WEATHER PREDICTION

Since 1987, NWP has seen an efficiency increase of a "factor of 72 by algorithmic changes in forecast model" (Hortal, 1999, p. 139; also Temperton, personal communication, 1999) allowing significantly higher resolution. For example, a model with T319 horizontal spectral truncation (smallest scale around 125 km) and 50 levels including the stratosphere can be run with the new methods for the same computer resources as one based on the older methods with T106 horizontal truncation (smallest scale around 375 km) and 19 levels concentrated in the troposphere. A factor of 72 is very remarkable and perhaps unprecedented in computational fluid dynamics. It far exceeds the gains expected with the acquisition of a next-generation computer, and is perhaps more like those provided by two generations.

Table 1 summarizes the sources of these gains in the column labelled NWP. Before the introduction of semi-Lagrangian approximations, most models were based on finite-difference or spectral-transform approximations to the atmospheric equations in Eulerian form, with centred or three-time-level temporal approximations. The time step of such approximations is restricted by a stability condition related to grid size. The smaller the grid size, the smaller the time step. The first major economical gain was the introduction of semi-Lagrangian approximations for the advection component in place of Eulerian which eliminated the time step restriction when coupled to semi-implicit approximations for the terms leading to gravity waves (Robert, 1982). This allowed an increase in the time step of a factor of 6. However, the methods require more computation per time step, yielding an overall gain of about 5. The time step for semi-Lagrangian models is limited

TABLE 1 Sources of Efficiency Gains from Numerical Algorithms

	NWP	Climate
Three-time-level Eulerian → semi-Lagrangian ($\Delta t \times 6$) but 20% overhead	5	1.2
Semi-Lagrangian Three-time-level → two-time-level Requires half the time steps	1.8	1.3
Full grid → reduced grid Reduction in number of grid points	1.4	1.4
Quadratic → linear grid $(3/2)^3$ for x, y, t	3.4	3.4
Troposphere → stratosphere and/or thinner PBL levels	1.7	1.7
	13.3	9

---

[*] The National Center for Atmospheric Research is sponsored by the National Science Foundation

by accuracy considerations with respect to the dynamical component and by the temporal approximations in the subgrid scale parameterizations rather than by the computational properties of the advection approximations. It is generally felt that 60 min is the maximum time step for subgrid scale parameterizations as currently formulated. This restriction is primarily an accuracy issue associated with linearizations made in the approximations, although occasionally stability can also be an issue. The first burst of activity in semi-Lagrangian developments is reviewed in Staniforth and Côté (1991).

The next efficiency gain was obtained by replacing the centred, or three-time-level semi-Lagrangian approximations with forward, or two-time-level versions (Côté and Staniforth, 1988; Temperton, 1997). These approximations eliminate every other time step of the centred scheme and offer a potential savings of a factor of two. However, in atmospheric models a few parameterized processes such as radiation, which can be very expensive, are not calculated every time step and in some forecast models not at every grid point. Instead they are calculated less frequently, such as every hour or every 3 h and the radiative heating rates are held fixed while the dynamics and other parameterizations advance through that period. Thus, the cost of the radiation calculation is not reduced by omitting every other time step, yielding a typical overall gain of 1.8 for NWP.

Additional savings were introduced by the reduced grid (Hortal and Simmons, 1991). These grids are applicable to spectral transform models and semi-Lagrangian models, but not in general to Eulerian grid point models based on spherical coordinates. A reduced grid is based on latitude and longitude circles, but the longitudinal angular grid increment increases at latitudes approaching the poles so that the longitudinal distance between grid points is reasonably constant. Depending on their actual definition, such grids may have up to 30% fewer points than a uniform latitude–longitude grid but produce a solution comparable to one produced on a uniform grid (Courtier and Naughton, 1994; Williamson and Rosinski, 2000).

Eulerian spectral transform models generally adopt quadratic unaliased computational Gaussian grids, even though higher-order nonlinearities arise in the baroclinic dynamical equations and in the suite of physical parameterizations. The models become unstable with coarser grids. The *quadratic grid* is the minimum latitude–longitude Gaussian grid for which the transform of the product of two fields, each of which is representable in spectral space, has no aliasing onto the waves of the underlying spectral representation (see Machenhauer, 1979, for a review of the spectral transform method). A linear grid is suitable for semi-Lagrangian spectral transform models (Côté and Staniforth, 1988; Williamson, 1997). The *linear grid* is defined to be the minimum grid required for transformations of a field from spectral space to grid point space and back again to spectral without loss of information. With such a grid, only linear terms are unaliased. The linear grid offers an increase in spectral resolution over the quadratic grid of 50% in each horizontal dimension, and an accompanying 50% increase in time step for schemes constrained by a stability condition.

When models are extended to include the stratosphere they experience stronger polar night jets which also impose an additional restriction on the time step of Eulerian models. Similarly, when the vertical resolution is increased in the planetary boundary layer, the vertical advection begins to impose a restriction on the time step of Eulerian models. Semi-Lagrangian models are not affected by these additional restrictions and offer proportional savings.

B. CLIMATE MODELS

With the modest resolutions commonly employed in climate models today, not all of the efficiency gain available to NWP models can be realized for climate simulations. Estimates of efficiency gains for climate models are given in the right column of Table 1. As mentioned above, the subgrid scale parameterizations restrict the time step to 60 min. With T42 spectral truncation, the time step is often taken as 20 min, which with centred time differencing for the dynamics yields an interval of 40 min ($2\Delta t$) over which the parameterizations, implemented as forward or backward approximations, are applied. Thus the basic time step gain is substantially less than 5.

Moving from three-time-level to two-time-level temporal approximations also yields less gain because the balance of resources devoted to dynamics and parameterizations is very different in the two applications. In climate models, a larger fraction of time is devoted to radiation, as large as 50%, which is unaffected by a dynamical time step increase since, as mentioned earlier, radiation is generally calculated on a longer time step than the dynamics. Thus only the remaining 50% of the model is sped up by a factor of 2.

Nevertheless, Table 1 shows that climate models could realize an efficiency gain of a factor of 13, so that, for example, a T75 model with the new approximations could be run for the cost of a T42 one with the old approximations. This gain is realizable by going to a resolution higher than T42, but most of it is not realizable starting from a resolution lower than T42. A T42-quality model is not possible for the cost of a T27 one based on the older approximations. At T27 the underlying grid is too coarse to properly define the surface characteristics and yields a simulation with the new approximations that is not as good as T42 with the older approximations (Williamson, 1997).

In spite of the increased resolution afforded by semi-Lagrangian approximations, most climate modelling applications continue to be made at T42 or coarser resolutions. For example, in the ongoing Atmospheric Model Intercomparison Project (AMIP II), 13 models are in the T42–T47 range and eight are T63.* The T63 set consists of models associated with numerical prediction centres rather than groups concentrating exclusively on climate modelling. The grid point models in AMIP II are all based on 4 × 5 degree grids except one finer at 2.5 × 3.75, again associated with an NWP centre.

Today, 100-year climate simulations are not uncommon. When climate-modelling groups obtain additional resources, such as from computer upgrades, they tend to devote them to longer simulations (e.g. 1000 years) or to ensembles of simulations (e.g. 10 members), rather than to increased resolution. The former choice is to determine if signals found in the 100-year runs are trends or simply a manifestation of low-frequency variability. The latter choice is to allow separation of signal from natural variability in nonequilibrium simulations associated with climate change scenarios.

C. NWP VERSUS CLIMATE MODELLING

On the surface it appears that climate and NWP groups use the same model components. However, they configure them very differently for the two applications. The goal in both is to minimize the "net error". However, the primary error measures of the two applications emphasize very different phenomena. Historically, NWP has concentrated on skill scores which measure the error in the detailed deterministic evolution of the predicted flow. Climate simulation, on the other hand, concentrates on statistics of the flow long after the initial conditions are forgotten. These statistics emphasize means and variances, but not the shape of the distribution. Much of the evolutionary and extreme aspects of the simulation are averaged out.

There are other differences between the two approaches. In NWP, forecasts are verified every day with an independent data set—the evolution of today's weather which is never identical to a past evolution. There is a definite measure of success or failure. The forecasts are available for everyone to see. In addition, the numerical forecasts are squeezed into the operational time slot available for running the model. On the other hand, in climate modelling, simulations are verified with the dependent data set used to develop the model. There is no equivalent independent detailed data set available for verification and the existing data set is too short to divide into two parts, one for development and one for validation. Climate forecasts are not verifiable in the lifetime of the model. Climate modellers are not restricted to an operational time slot; they can always use a little more computer time and get their results back a little later. In climate modelling, there is a subject judgement of which errors to minimize. The measures of success and failure are much more vague and personal. Finally, results of climate simulations are less openly available. This is improving with the various intercomparisons such as AMIP, but still many results are presented from model versions which are not well documented nor open to public scrutiny.

The different emphases of the two applications lead to different choices in model design. We provide here one example of how the different considerations led to very different choices made by the two approaches. The spectral transform method leads to unphysical characteristics in the solution of transported quantities, such as negative values for water vapour or chemical constituents (Rasch and Williamson, 1990). Shape-preserving semi-Lagrangian methods were developed to avoid this problem (Williamson and Rasch, 1989). In the early 1990s these methods were adopted in spectral-transform climate models to replace Eulerian spectral-transform moisture and chemistry transport while retaining the Eulerian spectral-transform dynamics (Hack et al., 1993; Feichter et al., 1996) even though they incurred a 20% overhead. They were adopted in the mid-1990s in some NWP models, when they came essentially free with semi-Lagrangian dynamics (Ritchie et al., 1995). Even so, some centers have still not adopted them.

Similarly, climate-modelling centres have not taken advantage of numerical efficiencies offered by advances in numerical approximations. Why? The advantage of resolution above T42 is not obvious in traditional climate statistics. In many atmospheric general circulation models, traditional large-scale statistics of the mean climate seem to converge in the range of T42 to T63 truncation (Boyle, 1993; Williamson et al., 1995) and the errors are not reduced by going to still higher resolution. Regional aspects of the climate, on the other hand, do not converge in many models (Déqué et al., 1994; Déqué and Piedelievre, 1995; Sperber et al., 1994; Stephenson et al., 1998). Although much of this nonconvergence with increasing horizontal resolution is associated with the addition of finer scales which are

---

* Details can be found at http://www-pcmdi.llnl.gov/amip/amiphome.html.

not included in lower resolution models, for some variables the larger scale components do not converge and are influenced by the finer scales (Williamson, 1999). In addition, the simulation is not universally improved with increased resolution.

Modellers reduce the net error by cancelling errors of opposite signs either deliberately, or more often inadvertently while tuning the model parameters to produce a simulation which matches observations and analyses as closely as possible. The errors of individual model components are very difficult to identify. Models are generally developed at the resolution at which they will be applied in production work. The parameterizations may be providing a nonlinear control at that resolution. Reducing the error in a single component may result in larger net error by disturbing a balance of errors. Williamson and Olson (1998) discuss an instance where one aspect of a Eulerian simulation was "better" than a matching semi-Lagrangian simulation, i.e. closer to the NCEP re-analysis, but argued that the "better" simulation was the result of a larger error in the dynamical component. The semi-Lagrangian approximations provide a more accurate solution to the vertical advection but that led to a larger total error. If increasing resolution reduces only one error, then the net error increases. There are indications that many simulation deficiencies are related to the parameterizations. In fact Déqué (1999, p. 23) concludes that "most of the problems of a GCM are present in the T42 resolution, and ... these problems will be solved by taking into account more realistic feedbacks and more complex equations, rather than by a more accurate numerical discretization in the horizontal". However, there are indications that parameterizations work differently at different resolutions and should be designed for a particular resolution or be resolution dependent (Chen and Bougeault, 1993; Williamson, 1999).

A more logical approach to model development might be to develop models first at high resolution, then develop a lower resolution "application version" which simulates the high-resolution version. The high-resolution version permits more complete sampling than possible in the atmosphere to guide the development of the application model. This approach is not infeasible, since much development work can be done with runs of length 6 months to 1 year, with 10-year simulations being the extreme needed. As mentioned above, the applications are likely to be 1000-year simulations or ensembles of 100-year simulations.

D. FUTURE EVOLUTION OF NUMERICS

As we have seen above, there has been a tremendous payoff in NWP from numerical algorithm development. The algorithms are much more efficient for the same quality solution, allowing increased resolution. However, the fundamental nature of the errors in the solution even at higher resolution is the same, hence the implied question in the subtitle: refinements or fundamental advances?

What do modellers want? Presumably NWP wants even higher-resolution models. The nature of current errors (e.g. degree of conservation, spectral ringing, both discussed below) seems acceptable and higher resolution is expected to provide more accurate forecasts of the detailed evolution of even smaller scale features which are critically important in weather forecasting, in particular of extreme events. Climate modellers, on the other hand, want approximations with fundamentally different characteristics when applied to large-scale flows.

Climate modellers are concerned with the maintenance of physical properties, e.g. positivity, conservation of mass, conservation of tracer correlations to a high degree. A small lack of conservation has little effect on a 10-day forecast but can accumulate to serious proportions in a 1000-year simulation. Conservation of tracer correlations is important to properly model the interactions of atmospheric constituents.

The forcing associated with the surface boundary conditions is more important in climate modelling, compared to initial boundary conditions in forecasting. Spectral models suffer from the well-known problem of spurious oscillations introduced by the underlying spectral basis functions. Sharp gradients in the fields being modelled lead to ripples spreading from the gradient. As a consequence, a field such as the surface orography, which should be flat over the oceans, may not be, and the ripples west of the Andes for example can be unreasonably large. The atmospheric flow responds to these ripples to produce upward and downward vertical motion which increases or decreases the clouds respectively, and which in turn modulate the surface fluxes. Examples of the noise in net surface fluxes can be seen in Holzer (1996, Fig. 9) and Kiehl (1998, Fig. 2). Increasing the horizontal resolution and simultaneously increasing the resolution of the surface orography as is usually done in climate modelling exacerbates the problem. There is some relief by filtering the mountains (Holzer, 1996) but a complete solution remains elusive.

The common pointwise semi-Lagrangian schemes do not conserve transported quantities *a priori*. Therefore arbitrary atmospheric and constituent mass fixers are often applied *a posteriori* to reclaim global conservation (Williamson and Olson, 1994). Although their local effect is relatively small for atmospheric mass and water vapour in tropospheric models, these computational processes have occasionally been pathological and troublesome in tropospheric–stratospheric models which include chemistry. Conservative interpolation methods have been developed (Leslie and Purser, 1995; Rančić, 1995) but have not gained wide usage.

Having locally consistent vertical velocities may become more desirable in the future. The different vertical velocities available in models, e.g. $\dot{\theta}$, $\dot{\sigma}$, ω, are all physically related, but can have very different time integrated properties as indicated by Lagrangian trajectory calculations. (Eluszkiewicz *et al.*, 2000).

Because of the emphasis on conservation and physical reality, climate modellers want flux-based schemes which provide local conservation of fluxes passing between grid cells, even though they might provide a false sense of security by not explicitly exposing errors. A variety of numerical methods, often referred to as modern advection algorithms, have been developed in other areas of computational fluid dynamics. These incorporate additional physical constraints besides simply approximating the derivatives in the governing equations. They include flux-corrected transport (Zalesak, 1979), total variation diminishing schemes (Harten, 1983) and piecewise parabolic method (Colella and Woodward, 1984). The challenge to atmospheric modelling is to apply these schemes to spherical geometry without sacrificing their attractive properties or accuracy and efficiency. Many of these new methods are based on directional splitting to break the multi-dimensional problem into a sequence of one-dimensional ones. However, the splitting error introduced can be significant. Rančić (1992) avoided splitting by developing a fully two-dimensional version of piecewise parabolic method, but the overall cost has discouraged others from following his approach. Lin and Rood (1996) and Leonard *et al.* (1996) developed multidimensional schemes from a directional split perspective but with the first-order error term eliminated. Extending their method to a full global model, Lin and Rood (1997) developed a flux-form finite-volume model without an advective time step restriction to which Lin (1997) added a Lagrangian vertical coordinate.

These schemes possess many properties that are attracting the attention of climate modellers. There is an overlooked problem, however. Schemes such as these which are nonlinearly corrected to maintain certain physical properties such as monotonicity will not conserve energy if energy is not a predicted quantity. The dynamical component of climate system models must conserve energy to tenths of a watt per square metre or better (Boville, 2000); a greater imbalance could produce drift of the deep ocean in a coupled system which is large enough to imply a nonequilibrium solution. In addition, the signal introduced by changing greenhouse gases is only a few watts per square metre, and the top-of-atmosphere imbalance in the nonequilibrium climate associated with the changing greenhouse gases during the later part of the twentieth century might be only a $0.5\,\mathrm{Wm^{-2}}$ (Boville, personal communication). Thus yet another *a posteriori* restoration will be required to achieve the level of conservation required for coupled climate system models.

It is still too early to know if these newer schemes look as attractive in global atmospheric applications as they do in theory and other applications. Experiments are currently underway in a controlled environment examining some of these newer approximations. However, preliminary experiments indicate that they are likely to be successful.

References

Boville, B. A., 2000: Toward a complete model of the climate system. In *Numerical Modeling of the Global Atmosphere in the Climate System*, (P. Mote and A. O'Neill, Eds), Kluwer Academic Publishers, pp. 419–442.

Boyle, J. S., 1993: Sensitivity of dynamical quantities to horizontal resolution for a climate simulation using the ECMWF (Cycle 33) model. *J. Climate*, **6**, 796–815.

Chen, D. and P. Bougeault, 1993: A simple prognostic closure assumption to deep convective parameterization: Part I. *Acta Meteorol. Sin.*, **7**, 1–18.

Colella, P. and P. R. Woodward, 1984: The piecewise parabolic method (PPM) for gas-dynamical simulations. *J. Comput. Phys.*, **54**, 174–201.

Côté, J. and A. Staniforth, 1988: A two-time-level semi-Lagrangian semi-implicit scheme for spectral models. *Mon. Weath. Rev.*, **116**, 2003–2012.

Courtier, P. and M. Naughton, 1994: A pole problem in the reduced Gaussian grid. *Q. J. R. Meteorol. Soc.*, **120**, 1389–1407.

Déqué, M., 1999: Final report of the project High Resolution Ten-Year Climate Simulations (HIRETYCS), EC Environment and Climate Program.

Déqué, M. and J.Ph. Piedelievre, 1995: High resolution climate simulation over Europe. *Climate. Dynam.*, **11**, 321–339.

Déqué, M., C. Dreveton, A. Braun and D. Cariolle, 1994: The ARPEGE/IFS atmosphere model: a contribution to the French community climate modelling. *Climate Dynam.*, **10**, 249–266.

Eluszkiewicz, J., R. S. Hemler, J. D. Mahlman, L. Bruhwiler, and L. Takacs, 2000: Sensitivity of age-of-air calculations to the choice of advection scheme. *J. Atmos. Sci.*, **57**, 3185–3201.

Feichter, J., E. Kjellström, H. Rodhe, F. Dentener, J. Lelieveld, and G.-J. Roelofs, 1996: Simulation of the tropospheric sulfur cycle in a global climate model. *Atmos. Envir.*, **30**, 1693–1707.

Hack, J. J., B. A. Boville, B. P. Briegleb, J. T. Kiehl, P. J. Rasch, and D. L. Williamson, 1993: Description of the NCAR Community Climate Model (CCM2). *NCAR Technical Note* NCAR/TN-382+STR.

Harten, A., 1983: High resolution schemes for conservation laws. *J. Comput. Phys.*, **49**, 357–393.

Holzer, M., 1996: Optimal spectral topography and its effect on model climate. *J. Climate*, **9**, 2443–2463.

Hortal, M., 1999: Aspects of the numerics of the ECMWF model. In *Recent Developments in Numerical Methods for Atmospheric*

Modelling, Proceedings of a seminar held at ECMWF, 7–11 Sept 1998, 127–143.

Hortal, M. and A. J. Simmons, 1991: Use of reduced Gaussian grids in spectral models. *Mon. Weath. Rev.*, **119**, 1057–1074.

Kiehl, J. T., 1998: Simulation of the Tropical Pacific Warm Pool with the NCAR Climate System Model. *J. Climate*, **11**, 1342–1355.

Leonard, B. P., A. P. Lock and M. K. MacVean, 1996: Conservative explicit unrestricted-time-step multidimensional constancy-preserving advection schemes. *Mon. Weath. Rev.*, **124**, 2588–2606.

Leslie, L. M. and R. J. Purser, 1995: Three-dimensional mass-conserving semi-Lagrangian scheme employing forward trajectories. *Mon. Weath. Rev.*, **123**, 2551–2655.

Lin, S.-J., 1997: A finite-volume integration method for computing pressure gradient force in general vertical coordinates. *Q. J. R. Meteorol. Soc.*, **123**, 1749–1762.

Lin, S.-J. and R. B. Rood, 1996: Multidimensional flux-form semi-Lagrangian transport schemes, *Mon. Weath. Rev.*, **124**, 2046–2070.

Lin, S.-J. and R. B. Rood, 1997: An explicit flux-form semi-Lagrangian shallow-water model on the sphere. *Q. J. R. Meteorol. Soc.*, **123**, 2477–2498.

Machenhauer, B., 1979: The spectral method. In *Numerical Methods Used in Atmospheric Models*, Vol. II, *Global Atmospheric Research Programme*, WMO-ICSU Joint Organizing Committee.

Rančić, M., 1992: Semi-lagrangian piecewise biparabolic scheme for two-dimensional horizontal advection of passive scalar. *Mon. Weath. Rev.*, **120**, 1394–1406.

Rančić, M., 1995: An efficient conservative, monotonic remapping as a semi-Lagrangian transport algorithm. *Mon. Weath. Rev.*, **123**, 1213–1217.

Rasch, P. J. and D. L. Williamson, 1990: Computational aspects of moisture transport in global models of the atmosphere. *Q. J. R. Meteorol. Soc.*, **116**, 1071–1090.

Ritchie, H., C. Temperton, A. Simmons, M. Hortal, T. Davies, D. Dent, and M. Hamrud, 1995: Implementation of the semi-Lagrangian method in a high-resolution version of the ECMWF forecast model. *Mon. Weath. Rev.*, **123**, 489-514.

Robert, A. J., 1982: A semi-Lagrangian and semi-implicit numerical integration scheme for the primitive meteorological equations. *J. Meteorol. Soc. Japan*, **60**, 319–325.

Sperber, K. R., S. Hameed, G. L. Potter and J. S. Boyle, 1994: Simulation of the northern summer monsoon in the ECMWF model: Sensitivity to horizontal resolution. *Mon. Weath. Rev.*, **122**, 2461–2481.

Staniforth, A. and J. Côté, 1991: Semi-Lagrangian integration schemes for atmospheric models—A Review. *Mon. Weath. Rev.*, **119**, 2206–2223.

Stephenson, D. B., F. Chauvin and J.-F. Royer, 1998: Simulation of the Asian summer monsoon and its dependence on model horizontal resolution. *J. Meteorol. Soc. Japan*, **76**, 237–265.

Temperton, C., 1997: Treatment of the Coriolis terms in semi-Lagrangian spectral models. In *Numerical Methods in Atmospheric and Ocean Modelling. The André J. Robert Memorial Volume* (C. Lin, R. Laprise and H. Ritchie, Eds), Canadian Meteorological and Oceanographic Society, Ottawa, Canada, pp. 293–302.

Williamson, D. L., 1997: Climate simulations with a spectral, semi-Lagrangian model with linear grids. In *Numerical Methods in Atmospheric and Ocean Modelling. The André J. Robert Memorial Volume* (C. Lin, R. Laprise and H. Ritchie, Eds), Canadian Meteorological and Oceanographic Society, Ottawa, Canada, pp. 279–292.

Williamson, D. L., 1999: Convergence of atmospheric simulations with increasing horizontal resolution and fixed forcing scales. *Tellus*, **51**, 663–673.

Williamson, D. L., and J. G. Olson, 1994: Climate simulations with a semi-Lagrangian version of the NCAR Community Climate Model. *Mon. Weath. Rev.*, **122**, 1594–1610.

Williamson, D. L. and J. G. Olson, 1998: A comparison of semi-Lagrangian and Eulerian polar climate simulations. *Mon. Weath. Rev.*, **126**, 991–1000.

Williamson, D. L. and P. J. Rasch, 1989: Two-dimensional semi-Lagrangian transport with shape-preserving interpolation. *Mon. Weath. Rev.*, **117**, 102–129.

Williamson, D. L. and J. M. Rosinski, 2000: Accuracy of reduced grid calculations. *Q. J. R. Meteorol. Soc.*, **126**, 1619–1640.

Williamson, D. L., J. J. Hack and J. T. Kiehl, 1995: Climate sensitivity of the NCAR Community Climate Model (CCM2) to horizontal resolution. *Climate Dynam.*, **11**, 377–397.

Zalesak, S. T., 1979: Fully multidimensional flux-corrected transport algorithms for fluids. *J. Comput. Phys.*, **31**, 335–362.

Mesoscale Mountains and the Larger-Scale Atmospheric Dynamics:
A Review
Christoph Schär
Atmospheric and Climate Science ETH, Zürich, Switzerland

Mountains play an important role in shaping the general circulation of the atmosphere and in affecting the amplitude and location of the northern hemisphere midlatitude storm tracks. Here a review is provided of mesoscale orographic processes that can affect the larger-scale circulation. An analysis of mountain-airflow regimes demonstrates that much of the earth's topography is too steep to allow for direct application of the balanced dynamics. Thus the airflow–mountain interaction is mostly on the mesoscale. Large-scale effects may, however, result if balanced-flow anomalies—such as surface thermal anomaly and/or internal potential vorticity (PV) anomalies—are generated by mesoscale processes. These ingredients represent a conceptual intermediary between the (unbalanced) mesoscale dynamics, and their subsequent (balanced) interaction with the synoptic-scale environment. The formation of balanced flow anomalies by mesoscale dynamical processes is reviewed. Particular attention is devoted to the formation of surface potential temperature anomalies by orographic retardation of cold-air advection, and the generation of PV anomalies by surface friction, gravity-wave breaking, and flow separation. For illustration, consideration will be given to observations and numerical simulations of the Alpine wake.

A. INTRODUCTION

Mountains play a key role not only in shaping the local weather and climate in response to the larger-scale forcing, but also by affecting the general circulation of our atmosphere on the synoptic and planetary scale. This aspect is highly relevant for weather prediction, seasonal forecasting and climate research. Despite the early recognition of the role of mountains on cyclonic activity and the structure of planetary-scale waves (e.g. Charney and Eliassen, 1949), the extent to which topography controls weather and climate has only become evident from general circulation numerical experiments. The typical procedure in such experiments invokes the comparison of a pair of climate integrations. In both, the distribution of land and sea is prescribed according to the earth's geography, while the topography is removed in the sensitivity experiment. A spectacular example of such an investigation is provided by Broccoli and Manabe (1992). Their results demonstrate that mountains imply remarkable remote effects through to the topographic control of the phase and amplitude of planetary-scale waves and the location and amplitude of the midlatitude storm tracks. For example, in the simulation of Broccoli and Manabe, large parts of Siberia are classified as dry in the topography run (in agreement with the observed climatology), but these would experience substantially larger rainfall amounts in the absence of topography.

In state-of-the-art atmospheric general circulation and climate models, the dynamical effects of mountains are represented by a wide range of processes and parameterizations. First, the dynamical and physical effects of large-scale topography are explicitly resolved, to an extent which depends primarily upon the underlying computational resolution. Second, boundary-layer effects over rough terrain are approximated by the use of special roughness length formulations (e.g. Georgelin et al., 1994). Third, the effects of deep gravity-wave propagation on the momentum budget of the upper troposphere and stratosphere is included through gravity-wave drag parameterization schemes (e.g. Palmer et al., 1986; Warner and McIntyre, 1999). Fourth, drag effects associated with low-level wakes have also been included in several large-scale models (e.g. Lott and Miller, 1997).

The purpose of this chapter is to review and assess dynamical processes contributing to synoptic-scale effects of mountains. The issue is pursued by asking how the (unbalanced) mesoscale mountain airflow is able to affect the (balanced) larger-scale dynamics. In fact, overcoming the apparent scale incompatibility between the direct mesoscale effects and the indirect synoptic and planetary-scale implications is a key to understand the interaction, since much of our understanding on baroclinic waves and the life cycle of baroclinic systems has been derived from the balanced dynamics (e.g. Hoskins et al., 1985).

B. REGIME DIAGRAM FOR FLOW PAST TOPOGRAPHY

We begin by reviewing the regime diagram of flow past isolated circular topography with height H and horizontal half-width L. The emphasis will be put on major topographic obstacles, such that the interaction with the boundary layer may be neglected. Specifically, consideration is given to the dry flow of a stratified atmosphere of uniform upstream velocity U and constant Brunt–Väisälla frequency N on an f-plane, using a free-slip lower boundary condition. Under these conditions, the flow response is governed by three dimensionless parameters. These are a Rossby number based on the mountain width, i.e.

$$R_0 = U/Lf \qquad (1)$$

with f denoting the Coriolis parameter, a dimensionless mountain height

$$\epsilon = NH/U \qquad (2)$$

which is sometimes referred to as inverse Froude number, and a nonhydrostatic scaling parameter

$$S = U/LN \qquad (3)$$

These parameters are sufficient for a rough layout of the regime diagram.

The regime diagram to be considered is presented as Fig. 1 and is spun up by R_0 and ϵ. Unless mentioned otherwise, an idealized circular topographic obstacle of the form

$$h(x, y) = H\left[(r/L)^2 + 1\right]^{-3/2} \qquad (4)$$

with $r^2 = x^2 + y^2$ will be considered. Whenever using the parameter S, we will assume a midlatitude ratio of $f/N \approx 1/100$ such that $S = R_0/100$. Below a few specific aspects of the regime diagram are discussed in some more detail. Further characteristics of the regime diagram relating to the aspect ratio of the obstacle are discussed in Smith (1989a) and Ólafsson and Bougeault (1997).

1. Balanced Solutions

For the flow system described above, quasi-geostrophic solutions for an unbounded atmosphere exist provided that the dimensionless slope of the obstacle satisfies

$$R_0\epsilon = \frac{NH}{Lf} < \frac{1}{2} \qquad (5)$$

Examples of respective analytical solutions are given in Fig. 2. The main features of the flow response are an isentropic lower boundary (i.e. flow *over* the mountain) and the presence of an anticyclone sitting over the mountain top. The latter is generated by vortex tube compression as the flow is directed over the obstacle. In the Northern Hemisphere, the presence of the mountain anticyclone implies an accelerated (decelerated) flow on the left-hand (right-hand) flank of the mountain, when looking into the downstream direction. For our mountain profile (4), flows with

$$\epsilon > \epsilon_{cap} = \frac{3\sqrt{3}}{2} \qquad (6)$$

are in addition characterized by a stagnation point and a concomitant Taylor cap, i.e. a vertically confined region with closed streamlines. These Taylor caps may be viewed as stratified analogue of Taylor columns in homogeneous flows (Huppert, 1975).

When using a linearized lower boundary condition (Smith, 1979d; Eliassen, 1980; Bannon, 1986), the breakdown of the quasi-geostrophic solutions is signalled by the generation of vorticity exceeding f. When using finite-amplitude topography in isentropic coordinates (Schär and Davies, 1988), there is in addition infinite stability $\partial\theta/\partial z \to \infty$ at the mountain top. Numerical simulations using primitive equation models in three dimensions indicate that the criterion (5) appears realistic, despite the fact that it is obtained from purely balanced considerations. In the limit of these nonlinear integrations, the breakdown of the balanced dynamics occurs as inertia gravity waves strengthen and disrupt the validity of the balanced dynamics (Trüb and Davies, 1995).

It is interesting to note that the consideration of higher-order balance beyond the quasi-geostrophic system does not notably extend the validity of the respective dynamics (e.g. Davies and Horn, 1988; Pierrehumbert, 1985). Thus, flows with $R_0\epsilon \gtrsim 1/2$ imply the breakdown of balanced solutions, either due to pronounced gravity wave generation, and/or due to the splitting of the incident low-level flow upstream.

2. Wake Formation and Transition into the Dissipative Regime

For hydrostatic flows with a free-slip lower boundary condition and in absence of background rotation ($R_0 = \infty$), the dimensionless mountain height is the

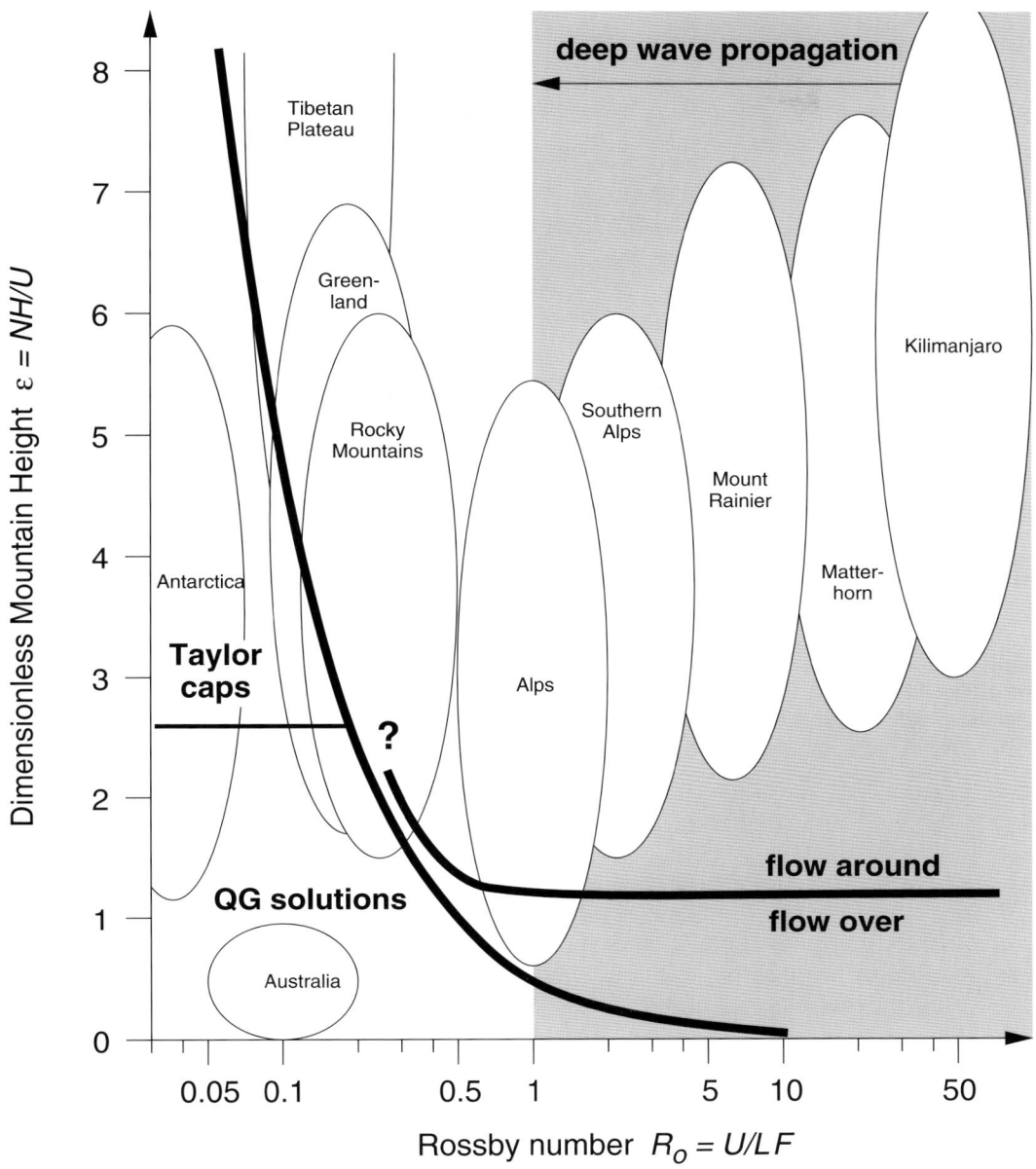

FIGURE 1 Regime diagram for idealized stratified flows past an isolated obstacle of height H, halfwidth L and shape $H(1 + r^2/L^2)^{-3/2}$, for an upstream profile characterized by uniform values of U and N. Regime boundaries are indicated by bold lines (see text for details) and the typical parameter range of several major mountain ranges is also included.

single control parameter that determines the steady-state flow response in the limit of the inviscid and adiabatic dynamics with a free-slip lower boundary condition. Thus the flow response does not depend upon the slope of the obstacle, but on its height alone. Idealized numerical simulations have served to identify two major flow regimes (Smolarkiewicz and Rotunno, 1989; Smith and Grønås, 1993). For configurations with $\epsilon < \epsilon_{crit}$, where

$$\epsilon_{crit} \approx 1.2 \tag{7}$$

is the critical mountain height, the flow is essentially inviscid and adiabatic, and predominantly directed *over* (rather than around) the obstacle (see Fig. 3, left-hand panels). The deviation of incident trajectories takes place approximately in vertical planes oriented along the flow, and there are pronounced foehn effects to the lee. The lateral deflection of the incoming air parcels is

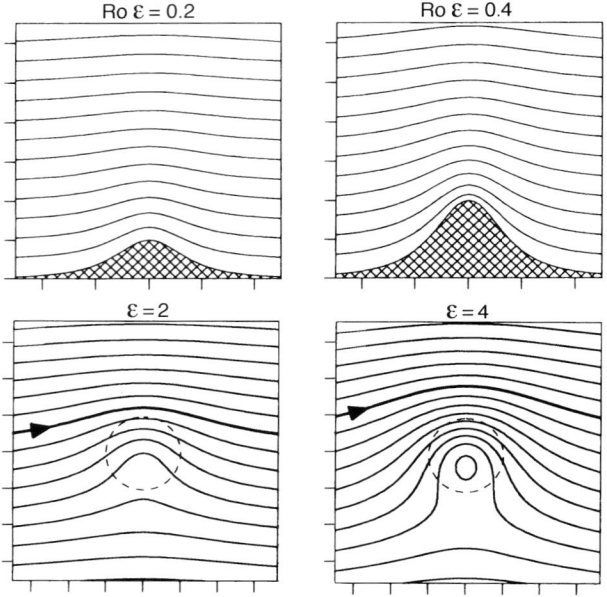

FIGURE 2 Quasi-geostrophic flow past circular finite-amplitude topography for Rossby number $R_0 = 0.1$ and dimensionless mountain heights of $\epsilon = 2$ (left-hand panels) and $\epsilon = 4$ (right-hand panels). The incident flow is from left to right. Upper and lower panels show the vertical distribution of isentropes and the stream function on the surface level, respectively. The dashed curves in the lower panels shows a topographic contour at radius $r = L$. The scaling on the horizontal and vertical axes is in terms of L and NL/f, respectively.

small. In contrast, for $\epsilon > \epsilon_{crit}$, the low-level flow largely circumscribes the obstacle quasi-horizontally, and the incident air mass experiences flow splitting (see Fig. 3, right-hand panels). This may lead to the formation of stationary or shedding lee vortices downstream. For elongated obstacles (Smith, 1989a), or when the upstream profile permits large-amplitude gravity wave propagation at lower levels (e.g. Grubisic and Smolarkiewicz, 1997) the flow may be approximately inviscid and adiabatic at low levels and primarily over the mountains, yet gravity-wave breaking and dissipation can occur at some upper levels. In such situations, an upper-level wake is generated.

The transition of the flow response when the mountain height ϵ is increased past its critical value is associated with a bifurcation (Smith and Grønås, 1993), i.e. there is a dramatic change in flow response as the mountain height is slightly increased. For circular topography and uniform upstream flow, the transition implies a wide range of effects, including the formation of stagnation points at the surface and/or at elevated levels, the onset of gravity-wave breaking and flow splitting. The associated changes in the underlying dynamics imply a transition into the dissipative regime

(Schär and Durran, 1997). Numerical simulations with an impulsive initialization from rest demonstrate that the transient development of quasi-steady or shedding flows occurs in two phases. During the first phase, which occurs over a dimensionless time of $tU/L = O(1)$, the flow is essentially inviscid and adiabatic, and potential vorticity (PV) is conserved. The transient evolution of the flow during the second phase, which occurs over a dimensionless time of $O(10)$ to $O(100)$, is controlled by dissipation and accompanied by the generation of PV anomalies. Dissipation occurs either in a gravity-wave breaking region aloft, in a recirculating wake to the lee, or in a frictional boundary layer. The respective mechanisms will further be discussed below.

The description given above is expected to be qualitatively applicable provided the horizontal scale of the obstacle permits deep gravity-wave propagation. Linear theory demonstrates that for a two-dimensional and uniformly stratified Boussinesq fluid of velocity U, this is the case for the wave number window defined by

$$R_0 = \frac{Uk}{f} > 1 \quad \text{and} \quad S = \frac{Uk}{N} < 1 \qquad (8)$$

where k denotes the horizontal wavenumber (see the reviews by Smith, 1979b; Gill, 1982, Durran, 1990). For $R_0 < 1$ and $S > 1$, gravity waves decay vertically with height as a result of rotational and nonhydrostatic effects, respectively.

In the presence of background rotation ($R_0 < \infty$), the critical mountain height becomes a function of the Rossby number, i.e. $\epsilon_{crit} = \epsilon_{crit}(R_0)$. Several studies have been conducted for this range of parameters (Thorsteinsson and Sigurdsson, 1996; Ólafsson and Bougeault, 1997; Trüb, 1993), and these demonstrate that the critical mountain height is increased by the presence of background rotation. The flow is thus more easily able to climb over the obstacle. Available data from these simulations has been included in the respective curve in Fig. 1. Rotational effects become particularly important for $R_0 < 1$. It appears, however, that no studies have addressed the range $R_0 < 1/4$ within the limit of the primitive equations, and it is unclear how the critical mountain height relates to the Taylor cone condition (6) discussed above.

Some studies have also addressed the nonhydrostatic effects near $S \approx 1$. So far these have found a comparatively small influence of nonhydrostatic effects upon the regime boundary separating flow over and around isolated mountains, while there is a notable reduction of deep gravity-wave activity (Miranda and James, 1992; Héreil and Stein, 1999).

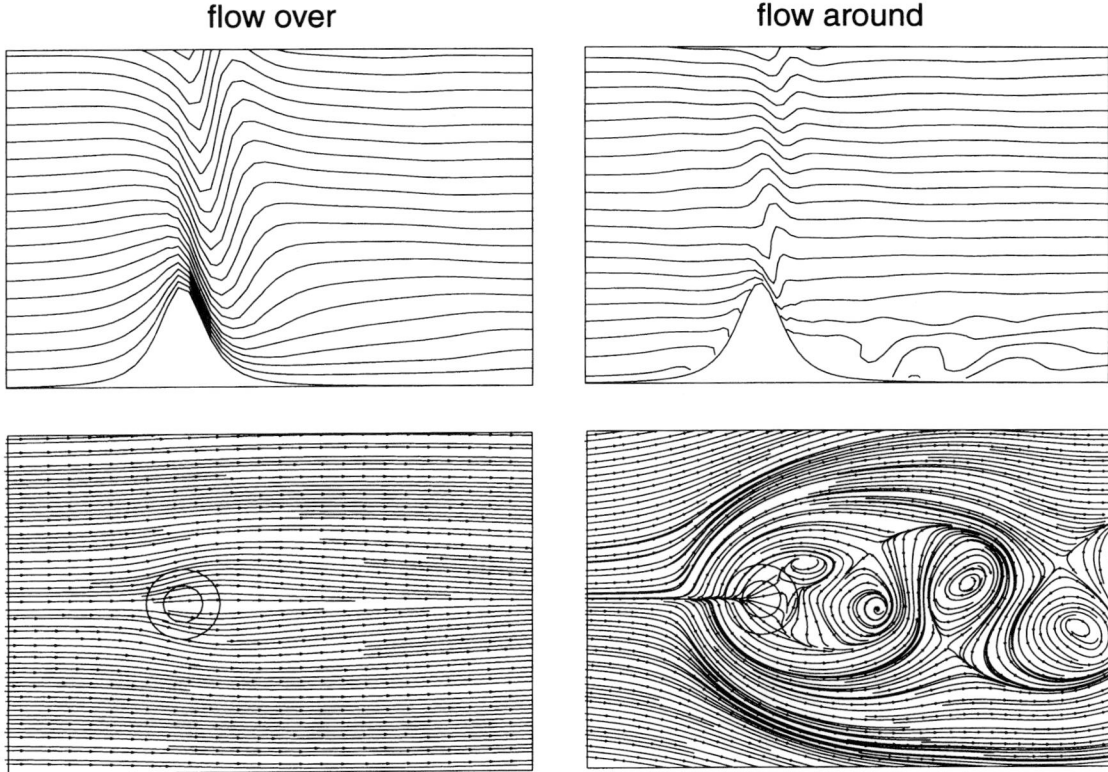

FIGURE 3 Hydrostatic nonrotating flow past an idealized mountain, depicting the "flow over" and "flow around" regimes for dimensionless mountain heights $\epsilon = 1$ (left-hand panels) and $\epsilon = 3$ (right-hand panels). The top panels show contours of potential temperature in a vertical section oriented along the flow across the centre of the mountain. The bottom panels display instantaneous surface streamlines. The incident flow is from left to right. (From Schär and Durran, 1997.)

3. Flow Regimes for Major Topographic Obstacles

To conclude the discussion of the previous subsections, the typical range of the flow parameters (R_0, ϵ) has been included in Fig. 1 for several major mountains. In doing so, it has been assumed that $N \approx 10^{-2}$ s^{-1} and $U \approx 10$ m s^{-1}. The horizontal scale of the mountains considered has been taken from topographic maps. In case of elongated obstacles, the assigned horizontal scale L is based upon the width (rather than the length) of the mountain, and it measures the distance from the mountain base to the mountain peak. Compared to the idealized obstacle (4), this yields a conservative estimate, and the real Rossby number might be somewhat larger than indicated in Fig. 1. For the mountain height, the maximum height has been selected in the case of isolated mountain peaks (e.g. for the Matterhorn), and the mean ridge height in the case of complex mountain ranges (e.g. for the Rocky Mountains and the Alps), respectively.

For each of the mountains considered, a typical parameter range in (R_0, ϵ)-space has been indicated to match the atmospheric variability. This range in parameter space is based upon an arbitrarily chosen factor 2. In reality, atmospheric conditions in the vicinity of many of the mountains are variable enough to span an even wider range in parameter space than is indicated in the figure.

There are several interesting inferences to be drawn from Fig. 1. First, most mesoscale mountain ranges (such as the European or southern Alps) do not allow for a confident application of the quasi-geostrophic or balanced dynamics. According to (5) and for typical midlatitude values of $f/N \approx 0.01$, the breakdown of the quasi-geostrophic dynamics occurs for obstacle slopes of $H/L \approx 0.005$. Mesoscale mountain ranges thus have horizontal scales which are about half an order of magnitude too small for confident application of quasi-geostrophic theory. In the case of strong flow or weak stability, quasi-geostrophic theory may (for episodes with small dimensionless mountain height ϵ) be applicable to the flow past the Rocky mountains or Greenland. Balanced solutions may here provide useful guidance but certainly not a realistic description of the flow field (Buzzi *et al.*, 1987; Bannon, 1992). If Antarctica were located in the midlatitude, the flow might well show the characteristics of the balanced flow solutions in Fig. 2, and sometimes even include the formation of a Taylor cap. However, due to the specific location of this obstacle at the pole, the quasi-geostrophic solutions mentioned above do not apply. The only "topographic obstacle" listed, which well satisfies the conditions for balanced flows, is Australia—if viewed as an isolated obstacle with a height of ~500 m and a radius of ~1500 km. However, here the height of the obstacle is smaller than the mean depth of the boundary layer, thus making the value of quasi-geostrophic mountain flow solutions questionable.

Second, most major mountain ranges are characterized by a height which often or always exceeds the critical dimensionless mountain height for flow splitting. For many mesoscale mountain ranges both flow regimes may occur, depending on the ambient atmospheric conditions. In the case of the Alps for instance, both flow regimes are frequently observed (Chen and Smith, 1987, Binder *et al.*, 1989). Flow over the Alps may in particular occur in strong southerly flow of weak stability. Here the reduction of the effective stability due to moist processes may play an important role, and this may also have some implications for heavy Alpine south-side precipitation events (Buzzi *et al.*, 1998; Schneidereit and Schär, 2000). However, the more common range of atmospheric conditions puts the Alps into the "flow-around" regime.

The inferences provided above are well supported by detailed analysis of primitive equation numerical simulations. For instance, a recent study by Kristjansson and McInnes (1999) has given consideration to the flow past Greenland. By conducting integrations both with and without the obstacle, a difference field "topography minus no topography" was isolated that qualitatively agrees with the onset of flow splitting and the formation of vertically oriented lee vortices to the east of Greenland. In agreement with this, observations (e.g. Scorer, 1988) and numerical simulations (Doyle and Shapiro, 1999) provide evidence for the formation of a tip jet at the southern end of Greenland, which supports the idea of rigorous flow splitting on a scale as large as Greenland.

C. INTERACTIONS WITH THE BALANCED LARGER-SCALE DYNAMICS

The discussion in the preceding section highlights the serious difficulty in identifying the mechanisms responsible for the influence of mountains upon the larger-scale dynamics. On the one hand, much of our understanding of the synoptic-scale dynamics has been driven by the framework of the balanced dynamics (Hoskins *et al.*, 1985). On the other hand, the balanced dynamics is unable to describe the dynamical processes around mesoscale topography.

An attractive interpretation may nevertheless be provided with the help of the invertibility principle, by considering the unbalanced generation of orographic flow perturbations, but assuming that their interaction with the synoptic-scale dynamics is approximately balanced. It then follows from the invertibility principle that relevant orographic perturbations must be associated with (i) a surface thermal anomaly and/or (ii) an internal PV anomaly. These ingredients thereby represent a conceptual intermediary between the (unbalanced) mesoscale formation of orographic flow anomalies, and their subsequent (essentially balanced) interaction with the synoptic-scale environment.

1. Surface Potential Temperature Anomalies

A well-documented mechanism for the generation of surface-θ anomalies is the retardation of an approaching cold front. In effect it results in the generation of a wake characterized by a warm surface anomaly. This process is only effective when flow splitting occurs, as often is the case when strongly stratified cold air behind a cold front is involved.

In the Alps, orographic retardation of cold fronts has been investigated thoroughly within the Alpine Experiment ALPEX in 1982 and the German fronts experiment in 1987. A succinct review may be found in Egger and Hoinka (1992). The dynamics of frontogenesis over, and frontal retardation by, mountains has been investigated in a wide range of frameworks. Shallow-water cold fronts were investigated in the semi-geostrophic (Davies, 1984) and primitive equation framework (Schumann, 1987), two-dimensional continuous fronts in the quasi-geostrophic (Bannon, 1983), semi-geostrophic (Bannon, 1984) and primitive equation limit (Garner, 1999). These studies, however, were restricted to two-dimensional geometry and thus to the absence of flow splitting. Several idealized and real-case studies with three-dimensional geometry were conducted using numerical models (e.g. Blumen and Gross, 1987; Heimann, 1992; Williams *et al.*, 1992; Colle *et al.*, 1999).

It is important to realize that the scale of the resulting surface-θ anomaly may differ from that of the underlying topography. While the spanwise scale of the wake is approximately determined by the scale of the obstacle, its streamwise extent is controlled by

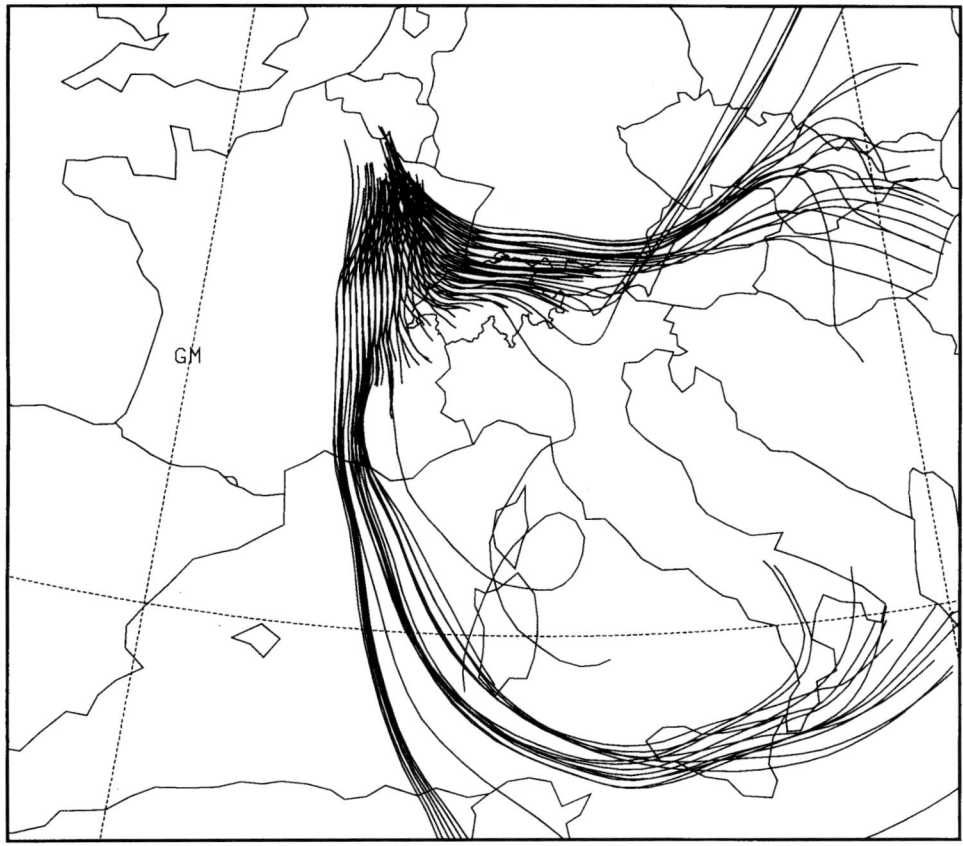

FIGURE 4 Splitting of an incident low-level airstream (initially located at the 850 hPa level) by the Alps. The figure derives from three-dimensional 54-h trajectory computations driven by a 14-km resolution numerical simulation of a cold-frontal passage on 30 April 1982. (From Kljun *et al.*, 2001.)

the advective timescale and may exceed the width of the mountain by a large factor. This aspect is well illustrated in the Alpine trajectory analysis shown in Fig. 4, which is based upon a real-case numerical simulation using 14 km horizontal resolution. The starting points for the relevant trajectories were selected immediately behind an impinging cold front. Pronounced flow splitting occurs and the track of the trajectories encompasses an extended wake region which is about 5 K warmer than the surrounding air. The scale of the wake anomaly has an extension of almost 1000 km in the downstream direction, exceeds the Alpine width by a large factor, and is of a size that can easily interact with the balanced part of the flow evolution.

The orographic retardation also implies geostrophic (or higher-order balanced) adjustment of the heavily distorted baroclinic configuration. This may induce a pressure drop through upper-level mass-flux divergence (Mattocks and Bleck, 1986; Tafferner, 1990) and vertical motion through Q-vector forcing (McGinley, 1982). However, from the viewpoint of the balanced dynamics it is the warm low-level air by itself which constitutes the driving agent of this development.

2. Potential Vorticity Anomalies

Early studies on flow splitting had isolated baroclinic generation of vorticity as an important ingredient of such flows (Smolarkiewicz and Rotunno, 1989). This process implies a distribution of vorticity vectors that are aligned along isentropic surfaces, which implies the absence of PV anomalies. However, PV anomalies may be generated by surface friction (Hunt and Snyder, 1980; Snyder *et al.*, 1985; Thorpe *et al.*, 1993), dissipation (Smith, 1989b; Schär and Smith, 1993a; Schär and Durran, 1997) and diabatic heating associated with orographic precipitation (Buzzi *et al.*, 1998).

Here further consideration is given to the generation of PV anomalies by frictional and dynamical processes. The inherent linkage between dissipation and PV generation can be demonstrated using the generalized Bernoulli theorem (Schär, 1993). This theorem provides a link between variations of the Bernoulli function on isentropic surfaces, on the one hand, and the total flux of potential vorticity, on the other. The theorem is valid for in the presence of arbitrary frictional and diabatic effects. For brevity, attention is restricted to the isentropic hydrostatic and steady-state version given by

$$\mathbf{J} = \mathbf{k} \times \nabla B \qquad (9)$$

Here $\mathbf{k}$ is the vertical unit vector,

$$B = c_p T + \frac{1}{2}\mathbf{v}^2 + gz \quad \text{with} \quad \mathbf{v} = (u, v) \qquad (10)$$

is the hydrostatic form of the Bernoulli function,

$$\mathbf{J} = \mathbf{v}\xi + \mathbf{J}_{NA} \qquad (11)$$

is the total flux of potential vorticity with its advective and nonadvective contributions (see Haynes and McIntyre, 1987, 1990), and

$$\xi = f + \left(\frac{\partial v}{\partial x}\right)_\theta - \left(\frac{\partial u}{\partial y}\right)_\theta \qquad (12)$$

is the absolute isentropic vorticity, which is the component of vorticity oriented perpendicular to θ-surfaces. As can be inferred from (9), variations of B on θ-surfaces imply the presence of nonzero fluxes of potential vorticity.

For further discussion, we assume a nonrotating framework and consider flow past an isolated circular obstacle. Three specific cases are illustrated in Fig. 5. Upstream, we assume zero PV and $\xi = 0$, thus B and θ-surfaces must coincide. Over the obstacle, however, any process that is able to generate variations of B on an isentropic surface implies nonzero fluxes of potential vorticity. For further analysis of Fig. 5, selected streamlines are considered in each panel which reside on the same isentropic surface downstream of the topography. As we may assume approximately inviscid and adiabatic flow conditions far downstream, (9) and (11) may be expressed as

$$\mathbf{v}\xi \approx \mathbf{k} \times \nabla B \qquad (13)$$

Thus, variations of B on θ-surfaces imply the advection of nonzero PV, despite the presence of zero PV upstream.

A range of processes can contribute towards generating variations of the Bernoulli function. Three of them are discussed in the panels of Fig. 5:

- In (a), it is assumed that the central streamlines are affected by boundary-layer processes and surface friction. The respective airparcels experience a frictional reduction of the Bernoulli function. The result is a wake that is characterized by a negative Bernoulli anomaly, and the implied sense of isentropic vorticity is as indicated.
- In (b), we assume that the central streamlines traverse a gravity-wave breaking region, whereas the other streamlines remain in the adiabatic and inviscid part of the mountain wave. Again, a Bernoulli deficit results, implying the generation of PV anomalies.
- In (c) the case of flow separation is considered, which is fundamentally different. Downstream of the mountain, neighbouring air parcels may have a completely different origin. A contrast in Bernoulli function between the different trajectories here

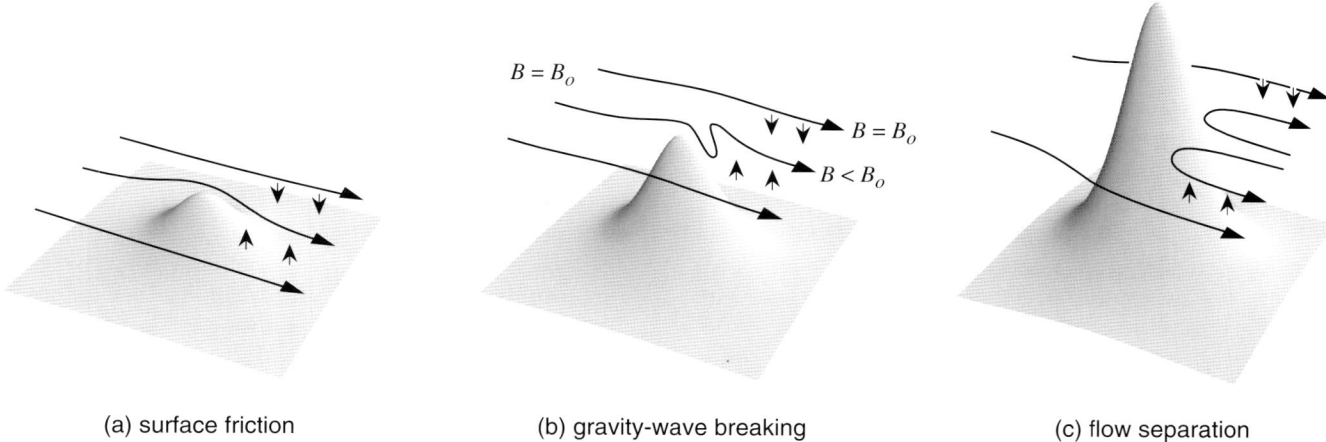

(a) surface friction (b) gravity-wave breaking (c) flow separation

FIGURE 5 Schematic display showing PV generation as a result of: (a) surface friction; (b) gravity-wave breaking; and (c) flow separation. In each case, the trajectories are selected such that they arrive on the same isentropic surface downstream of the obstacle. The wake region is characterized by a Bernoulli deficit, implying the presence of potential vorticity anomalies (see text for further details).

results either from weak dissipation or boundary-layer processes within the recirculating wake region. Dissipation in small regions of gravity-wave breaking over the flanks of the obstacle may contribute, provided that the respectively processed air does enter and remains for an extended period in the wake region.

The wakes in (b) and (c) do not require the presence of a boundary layer, but may evolve even when assuming a free-slip lower boundary condition. The reduction of B in the recirculating wake (c) follows from an argument of Batchelor (1957). He noted that steady wakes must always be influenced by dissipation. As the timescale over which an air parcel circulates around a closed contour approaches infinity, dissipative effects become significant no matter how small the dissipation rate. In addition, boundary-layer processes may often be relevant, but they do not necessarily play the same role as in flow separation of homogeneous fluid dynamics (i.e. vorticity generation in the boundary layer and subsequent flow separation), but may rather act by reducing the Bernoulli function in the wake region.

Observational studies of wakes of type (a) were conducted for several small hills, among them the Black Combe (height 600 m) in Cumbria (Vosper and Mobbs, 1997). Detailed case studies of observed atmospheric wakes of type (b) and (c) were carried out for comparatively simple obstacles such as mountainous islands. This includes studies relating to Hawaii (Smith and Grubisic, 1993), the island of St. Vincent in the tropical Caribbean (Smith *et al.*, 1997), and the Aleutian Islands (Pan and Smith, 1999). Real-case numerical studies of such wakes were conducted over the Alps (Aebischer and Schär, 1998) and Greenland (Doyle and Shapiro, 1999). For several of the obstacles considered (e.g. the islands of St. Vincent and Hawaii), the wake is composed of an approximately symmetrical quasi-steady recirculating region encompassed by two shearlines of negative and positive PV, and there is absence of vortex shedding. Understanding the apparent steadiness of these flows requires the consideration of the wake stability by application of absolute stability theory (Schär and Smith, 1993b), similar as in homogeneous fluid dynamics (Hannemann and Oertel, 1989; Huerre and Monkewitz, 1985). In the atmospheric case the stability of quasi-steady wakes may be increased by surface friction downstream of the obstacle (Grubisic *et al.*, 1995; Smith *et al.*, 1997).

Numerical studies of wakes in rotating atmospheres suggest that the basic mechanisms are similar as in the nonrotating limit (Ólafsson and Bougeault, 1997; Sun and Chern, 1994). As a result of rotation, however, there is a pronounced left–right asymmetry, and for blocked (unblocked) flow regimes the drag is increased (decreased).

D. NUMERICAL SIMULATIONS OF ALPINE WAKES

The Alpine wake is of particular interest for several reasons. First, the flow past the Alps rarely justifies the assumption of a balanced flow and is well known for its gravity-wave and foehn effects. Second, the Alps frequently affect their synoptic environment, and this has been documented in the long history of Alpine lee-cyclogenesis research (see, e.g., the reviews by Pierrehumbert, 1986; Tibaldi *et al.*, 1989). The Alps thus represent an ideal environment to address feedbacks between the mesoscale and synoptic-scale dynamics. Third, the structure of the Alpine orography is of very high complexity, as the Alps are intersected by many deep valleys and composed of a multitude of mountain massifs of differing scales.

The last aspect is of relevance when reconsidering our regime diagram shown in Fig. 1. In fact, when viewing the Alps as the superposition of a smooth larger-scale component (representing its main features) and a fine-scale component (representing the Alpine massifs, valleys and peaks), both components may be assigned a similar height of ~1.5 km (half the average Alpine ridge height) corresponding to $\epsilon \approx 1.5$, but the two components are characterized by very different horizontal scales. The large-scale component may be dominated by a Rossby number of $R_0 \approx 0.2$ (suggesting the predominance of the flow-over regime), while the small-scale component is associated with $R_0 > 4$ (suggesting the predominance of the flow-around regime). Thus, the two components fall into different regimes according to our regime diagram, with rotationally inflicted flow over the larger scales, and flow around the smaller-scale topographic features.

Analysis of high-resolution operational weather-forecasting products at least qualitatively confirms this supposition. Often the flow past the Alps exhibits the characteristics of flow both over and around the Alps. An example is shown in the left-hand panels of Fig. 6. The situation considered is that of a cold frontal passage. The panels show the formation of a large number of orographic PV streamers or "PV banners". Individual pairs of banners with anomalously positive and negative values of PV, respectively, can be attributed to individual flow-splitting events, either on the scale of the whole of the Alps (primary banners), or on that of individual massifs and peaks of the model topography (secondary banners).

The formation of a multitude of banners has implications on their far-field effects, and their ability to

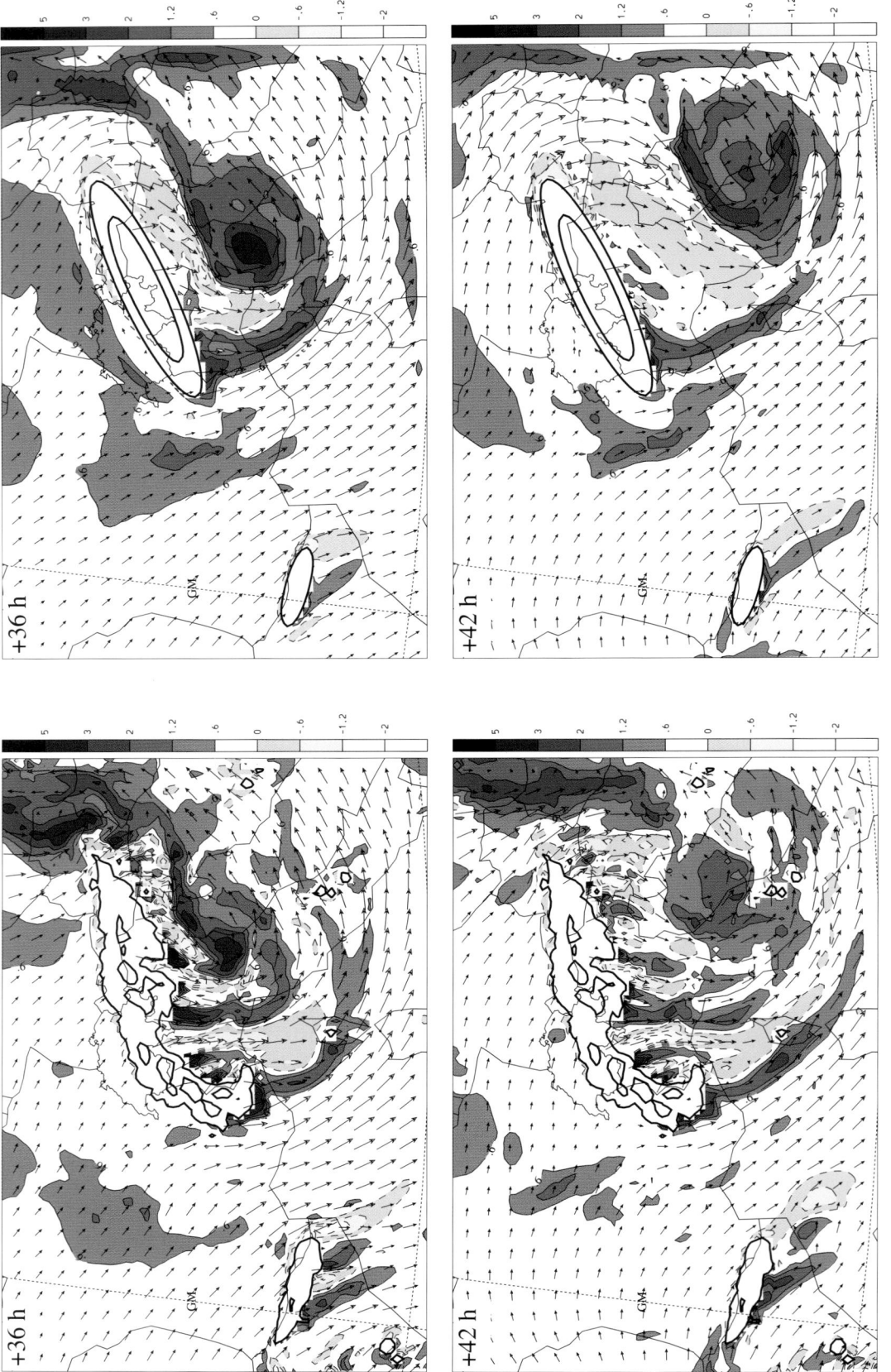

FIGURE 6 Potential vorticity in flow around the Alps for (left) a real-case simulation using the full topography and (right) the same flow evolution but using a simplified topography. (From Aebischer and Schär, 1998.)

affect the balanced dynamics. In essence, the far-field effects will be dominated by the primary banners (the outermost banners), while the effect of the secondary banners will decay quickly with height as governed by the associated Rossby depth of deformation, i.e. aN/f, with a denoting the distance between neighbouring banners. Only the primary banners may invoke deep interaction with other flow features, such as an approaching upper-level PV streamer. The right-hand panels of Fig. 6 suggest this kind of effect. The respective simulation is identical to that in the left-hand panels, except for the use of idealized ellipsoidal topography. In particular, the low-level PV of the developing lee cyclone is made up of a lee vortex that has rolled up from the primary banner which previously formed at the south-eastern tip of the Alps.

Simulations of the type shown in Fig. 6 are of substantial uncertainty. In particular, the banners have a width of a few grid increments, and this indicates that their width is determined by numerical rather than resolved physical processes. Furthermore, until recently there was no observational evidence for the quasi-two-dimensional structure of the banners and their apparent stability. This theme was addressed by the Mesoscale Alpine Programme (MAP). Specific scientific objectives formulated in the preparation of this major field campaign relate to the width and vertical structure of the banners, as well as their generation mechanism, stability and downstream development (Bougeault *et al.*, 1998, 2001). For many of these objectives, a two-stage strategy was developed, which is geared towards validating high-resolution numerical simulations against observational data, and subsequently using numerical models to address the dynamical issues under consideration. Here some preliminary results of this combined modelling and validation process are presented.

The model utilized is the Canadian Mesoscale Compressible Community Model (MC2) with a horizontal grid increment of 3 km. This model was run in real time during the MAP Special Observing Period (SOP) on a $343 \times 293 \times 50$ grid-point domain covering the whole of the Alpine area. The MC2 was nested within the Swiss Model (SM) with a horizontal grid increment of 14 km. Details on the model set-up are given in Benoit *et al.* (2001).

Here we analyse the 3 km resolution MC2 simulations for the North Foehn mission from 8 November 1999 (IOP 15). Figure 7 provides the 850 hPa potential vorticity and wind fields at 15 UTC. The synoptic weather situation during IOP 15 was characterized by northerly and north-easterly flow towards the Alps. A series of missions was flown to probe the Bora, the Mistral and the North Foehn region. On 8 November the MC2 and SM models predicted a pronounced Alpine-scale flow splitting, the generation of a Mistral to the west of the Alps, an associated (primary) shear-line, the onset of North Foehn across the Alpine passes, Bora to the east of the Alps and over the Dinaric region, as well as a complicated wake structure composed of dozens of secondary banners over the Po valley. The high computational resolution of the MC2 yields more PV banners than in the lower-resolution SM simulation, and these are of substantially finer scale.

Preliminary validation against *in situ* aircraft observations has generally shown satisfactory agreement with the simulations. In particular, the observations have confirmed the presence of the banners, their approximately two-dimensional structure and stability. In some cases there was agreement concerning the two-dimensional structure of the banners.

E. OUTLOOK

Surface friction, wake formation, and upper-level gravity-wave breaking show pronounced similarities in terms of how dissipation acts to generate a vertically oriented dipole of potential vorticity. This similarity prevails in spite of the completely different underlying mesoscale processes, and the net effect upon the flow is that of a force that is directed towards the low-level flow.

For low-level wakes, high-resolution numerical simulations suggest that the wakes consist of numerous fine-scale PV banners. These simulations appear to agree—at least qualitatively—with observations from field campaigns. In contrast, however, there is still substantial uncertainty about the structure of upper-level wakes, which are expected to form when gravity waves are able to propagate deep into the troposphere and stratosphere and break far aloft. Addressing the dynamics of these wakes is highly relevant given the importance of gravity-wave breaking for the momentum balance of the upper atmosphere. Particular difficulties relate to the wave-generation and trapping mechanisms at low levels (that are linked to boundary-layer processes), the process of wave-breaking (that is likely to be associated with substantial transience), the transfer of momentum and energy in the wavenumber space (that may be caused by nonlinear steepening and dissipative processes), and the fine-scale structure of the resulting upper-level wake (which might show similar complexity to that of low-level wakes).

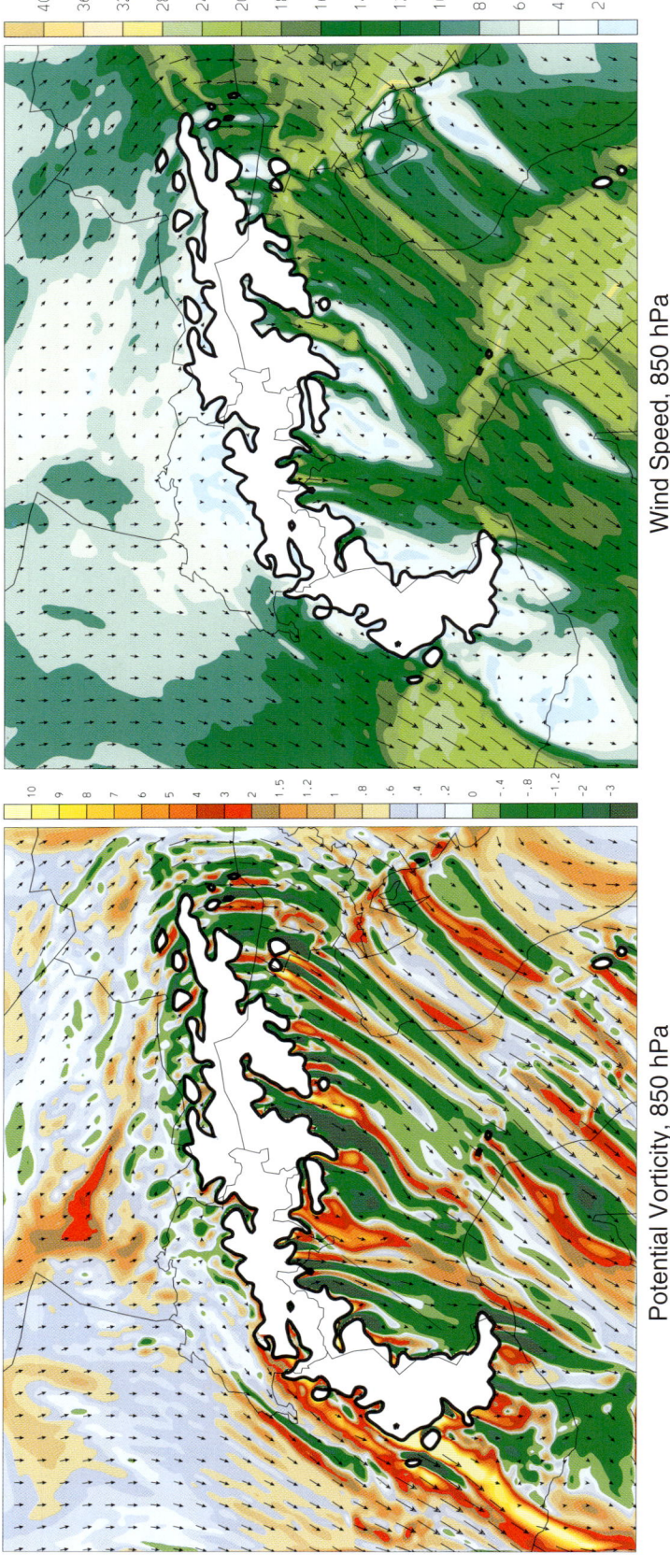

FIGURE 7 Potential vorticity (left) and horizontal wind velocity and vectors (right) at 850 hPa on 8 November 1999, 15 UTC, as simulated by a high-resolution numerical model. The bold solid line represents the intersection of the 850 hPa surface with the topography.

ACKNOWLEDGEMENTS

The author is indebted to Oliver Fuhrer, Daniel Lüthi and Michael Sprenger for their help with Figs 5 and 7; and to Robert Benoit and the whole MAP MC2 team for the fruitful collaboration during the MAP special observing period.

References

Aebischer, U. and C. Schär, 1998: Low-level potential vorticity and cyclogenesis to the lee of the Alps. *J. Atmos. Sci.*, **55**, 186–207.

Batchelor, G. K., 1957: On steady laminar flow with closed streamlines at large Reynolds numbers. *J. Fluid Mech.*, **1**, 177–190.

Bannon, P. R., 1983: Quasi-geostrophic frontogenesis over topography. *J. Atmos. Sci.*, **40**, 2266–2277.

Bannon, P. R., 1984: A semigeostrophic model of frontogenesis over topography. *Contrib. Atmos. Phys.*, **57**, 393–408.

Bannon, P. R., 1986: Deep and shallow quasi-geostrophic flow over mountains. *Tellus*, **38A**, 162–169.

Bannon, P. R., 1992: A model of rocky-mountain lee cyclogenesis. *J. Atmos. Sci.*, **49**, 1510–1522.

Benoit, R., C. Schär, P. Binder, S. Chamberland, H. C. Davies, M. Desgagné, C. Girard, C. Keil, N. Kouwen, D. Lüthi, D. Maric, E. Müller, P. Pellerin, J. Schmidli, F. Schubiger, C. Schwierz, M. Sprenger, A. Walser, S. Willemse, W. Yu and E. Zala, 2001: The realtime ultrafinescale forecast support during the Special Observing Period of the MAP. *Bull. Am. Meteorol. Soc.*, in press.

Binder, P., H. C. Davies and J. Horn, 1989: Free atmosphere kinematics above the northern Alpine foreland during ALPEX-SOP. *Contrib. Atmos. Phys.*, **62**, 30-45.

Blumen, W. and B. D. Gross, 1987: Advection of a passive scalar over a finite-amplitude ridge in a stratified rotating atmosphere. *J. Atmos. Sci.*, **44**, 1696-1705.

Bougeault, P., Binder, P. and Kuettner J. (Eds), 1998: *The Mesoscale Alpine Programme MAP: Science Plan.* http://www.map.ethz.ch/splan/spindex.html

Bougeault, P., P. Binder, A. Buzzi, R. Dirks, R. Houze, J. Kuettner, R. B. Smith, R. Steinacker and H. Volkert, 2001: The MAP special observing period. *Bull. Amer. Meteor. Soc.*, **82**, 433–462.

Broccoli, A. J. and S. Manabe, 1992: The effects of orography on midlatitude northern hemispheric dry climates. *J. Climate*, **5**, 1181–1201.

Buzzi, A., A. Speranza, S. Tibaldi and E. Tosi, 1987: A unified theory of orographic influences upon cyclogenesis. *Meteorol. Atmos. Phys.*, **36**, 91-107.

Buzzi, A., N. Tartaglione and P. Malguzzi, 1998: Numerical simulations of the 1994 Piedmont flood: role of orography and moist processes. *Mon. Weath. Rev.*, **126**, 2369–2383.

Charney, J. G. and A. Eliassen, 1949: A numerical method for predicting the perturbations of the middle latitude westerlies. *Tellus*, **1**(2), 38-54.

Chen, W.-D. and R. B. Smith, 1987: Blocking and deflection of airflow by the Alps. *Mon. Weath. Rev.*, **115**, 2578–2597.

Colle, B. A., C. F. Mass and B. F. Shull, 1999: An observational and numerical study of a cold front interacting with the Olympic Mountains during COAST IOP5. *Mon. Weath. Rev.*, **127**, 1310–1334.

Davies, H. C., 1984: On the orographic retardation of a cold front. *Contr. Atmos. Phys.*, **57**, 409–418.

Davies, H. C. and J. Horn, 1988: Semi-geostrophic flow of a stratified atmosphere over elongated isentropic valleys and ridges. *Tellus*, **40A**, 61–79.

Doyle, J. D. and M. A. Shapiro, 1999: Flow response to large-scale topography: the Greenland tip jet. *Tellus*, **51A**, 728–748.

Durran, D. R. 1990: Mountain waves and downslope winds. In *Atmospheric Processes over Complex Terrain* (B. Blumen, Ed.), Meteorological Monographs **23**, pp. 59–81. American Meteorological Society, Boston.

Egger, J. and K. P. Hoinka, 1992: Fronts and orography. *Meteorol. Atmos. Phys.*, **48**, 3–36.

Eliassen, A., 1980. Balanced motion of a stratified rotating fluid induced by bottom topography. *Tellus*, **32**, 537–547.

Garner, S. T., 1999: Blocking and frontogenesis by two-dimensional terrain in baroclinic flow. Part I: Numerical experiments. *J. Atmos. Sci.*, **56**, 1495–1508.

Georgelin, M., E. Richard, M. Petitdidier and A. Druilhet, 1994: Impact of subgrid-scale orography parameterization on the simulation of orographic flows. *Mon. Weath. Rev.*, **122**, 1509–1522.

Gill, A. E., 1982: *Atmosphere-Ocean Dynamics.* Academic Press.

Grubisic, V. and P. Smolarkiewicz, 1997: The effect of critical levels on 3D orographic flows: linear regime. *J. Atmos. Sci.*, **54**, 1943–1960.

Grubisic, V., R. B. Smith and C. Schär, 1995: The effect of bottom friction on shallow-water flow past an isolated obstacle. *J. Atmos. Sci.*, **52**, 1985–2005.

Hannemann, K. and H. Oertel, 1989: Numerical simulation of the absolutely and convectively unstable wake. *J. Fluid Mech.*, **199**, 55–88.

Haynes, P. H. and M. E. McIntyre, 1987: On the evolution of vorticity and potential vorticity in the presence of diabatic heating and frictional or other forces. *J. Atmos. Sci.*, **44**, 828–841.

Haynes, P. H. and M. E. McIntyre, 1990: On the conservation and impermeability theorems for potential vorticity. *J. Atmos. Sci.*, **47**, 2021–2031.

Heimann, D., 1992: 3-dimensional modeling of synthetid cold fronts interacting with northern Alpine foehn. *Meteorol. Atmos. Phys.*, **48**, 139–163.

Héreil, P. and J. Stein, 1999: Momentum budgets over idealized orography with a nonhydrostatic inelastic model. II: Three-dimensional flows. *Q. J. R. Meteorol. Soc.*, **125**, 2053–2073.

Hoskins, B. J., M. E. McIntyre and A. W. Robertson, 1985: On the use and significance of isentropic potential vorticity maps. *Q. J. R. Meteorol. Soc.*, **111**, 877–946.

Huerre, P. and A. M. Monkewitz, 1985: Absolute and convective instabilities in free shear layers. *J. Fluid Mech.*, **159**, 151–168.

Hunt, J. C. R. and W. H. Snyder, 1980: Experiments on stably stratified flow over a model three-dimensional hill. *J. Fluid Mech.*, **7**, 671–704.

Huppert, H. E., 1975: Some remarks on the initiation of inertial Taylor columns. *J. Fluid Mech.*, **67**, 397–412.

Kljun, N., M. Sprenger, and C. Schär, 2001. The modification of a front by the Alps: a case study using the ALPEX reanalysis data set. *Meteorol. Atmos. Phys.*, in press.

Kristjansson, J. E. and H. McInnes, 1999: The impact of Greenland on cyclone evolution in the North Atlantic. *Q. J. R. Meteorol. Soc.*, **125**, 2819–2834.

Lott, F. and M. Miller, 1997: A new subgrid scale orographic drag parameterization; its testing in the ECMWF model. *Q. J. R. Meteorol. Soc.*, **123**, 101–127.

Mattocks, C. and R. Bleck, 1986: Jet streak dynamics and geostrophic adjustment processes during the initial stage of lee cyclogenesis. *Mon. Weath. Rev.*, **114**, 2033–2056.

McGinley, J. A., 1982: A diagnosis of Alpine lee cyclogenesis. *Mon. Weath. Rev.*, **10**, 1271–1287.

Miranda, P. M. A. and I. N. James, 1992: Non-linear three-dimensional effects on gravity-wave drag: splitting flow and breaking waves. *Q. J. R. Meteorol. Soc.*, **118**, 1057–1081.

Ólafsson, H. and P. Bougeault, 1997: The effect of rotation and surface friction on orographic drag. *J. Atmos. Sci.*, **54**, 193–210.

Palmer, T. N., G. J. Shutts and R. Swinbank, 1986: Alleviation of a systematic westerly bias in general circulation and numerical weather prediction models through an orographic gravity wave drag parametrization. *Q. J. R. Meteorol. Soc.*, **112**, 1001–1039.

Pan, F. and R. Smith, 1999: Gap winds and wakes: SAR observations and numerical simulations. *J. Atmos. Sci.*, **56**, 905–923.

Pierrehumbert, R. T., 1985: Stratified ageostrophic flow over two-dimensional topography in an unbounded atmosphere. *J. Atmos. Sci.*, **42**, 523–526.

Pierrehumbert, R. T., 1986: Lee cyclogenesis. In *Mesoscale Meteorology and Forecasting* (P. S. Ray, Ed.), American Meteorological Society, pp. 493–515.

Schär, C., 1993: A generalization of Bernoulli's theorem. *J. Atmos. Sci.*, **50**, 1437–1443.

Schär, C. and H. C. Davies, 1988: Quasi-geostrophic stratified flow over finite amplitude topography. *Dynam. Atmos. Ocean*, **11**, 287–306.

Schär, C. and D. R. Durran, 1997: Vortex formation and vortex shedding in continously stratified flows past isolated topography. *J. Atmos. Sci.*, **54**, 534–554.

Schär, C. and R. B. Smith, 1993a: Shallow-water flow past isolated topography. Part I: Vorticity production and wake formation. *J. Atmos. Sci.*, **50**, 1373–1400.

Schär, C. and R. B. Smith, 1993b: Shallow-water flow past isolated topography. Part II: Transition to vortex shedding. *J. Atmos. Sci.*, **50**, 1401–1412.

Schneidereit, M. and C. Schär, 2000: Idealised numerical experiments of Alpine flow regimes and southside precipitation events. *Meteorol. Atmos. Phys.*, **72**, 233–250.

Schumann, U., 1987: Influence of mesoscale orography on idealized cold fronts. *J. Atmos. Sci.*, **44**, 3423–3441.

Scorer, R., 1988: Sunny Greenland. *Q. J. R. Meteorol. Soc.*, **114**, 3–29.

Smith, R. B., 1979a: Some aspects of the quasi-geostrophic flow over mountains. *J. Atmos. Sci.*, **36**, 2385–2393.

Smith, R. B., 1979b: The influence of mountains on the atmosphere. *Adv. Geophys.*, **21**, 87–230.

Smith, R. B., 1989a: Mountain-induced stagnation points in hydrostatic flow. *Tellus*, **41A**, 270–274.

Smith, R. B., 1989b: Hydrostatic flow over mountains. *Adv. Geophys.*, **31**, 1–41.

Smith, R. B. and S. Grønås, 1993: Stagnation points and bifurcation in 3-D mountain airflow. *Tellus*, **45A**, 28–43.

Smith, R. B. and V. Grubisic, 1993: Aerial observations of Hawaii's wake. *J. Atmos. Sci.*, **50**, 3728–3750.

Smith, R. B., A. C. Gleason, P. A. Gluhosky and V. Grubisic, 1997: The wake of St. Vincent. *J. Atmos. Sci.*, **54**, 606–623.

Smolarkiewicz, P. K. and R. Rotunno, 1989: Low froude number flow past three dimensional obstacles. Part I: Baroclinically generated lee vortices. *J. Atmos. Sci.*, **46**, 1154–1164.

Snyder, W. H., R. E. Thompson, R. E. Eskridge, I. P. Lawson, J. T. Castro, J. T. Lee, J. C. R. Hunt and Y. Ogawa, 1985: The structure of strongly stratified flow over hills: dividing streamline concept. *J. Fluid Mech.*, **152**, 249–288.

Sun, W.-Y. and J.-D. Chern, 1994: Numerical experiments of vortices in the wake of large idealized mountains. *J. Atmos. Sci.*, **51**, 191–209.

Tafferner, A., 1990: Lee cyclogenesis resulting from the combined outbreak of cold air and potential vorticity against the Alps. *Meteorol. Atmos. Phys.*, **43**, 31–47.

Thorpe, A. J., H. Volkert and D. Heimann, 1993: Potential vorticity of flow along the Alps. *J. Atmos. Sci.*, **50**, 1573–1590.

Thorsteinsson, S. and S. Sigurdsson, 1996: Orogenic blocking and deflection of stratified air flow on an f-plane. *Tellus*, **48A**, 572–583.

Tibaldi, S., A. Buzzi and A. Speranza, 1989: Orographic cyclogenesis. In *Proceedings Palmén Memorial Symposium on Extratropical Cyclones*, American Meterological Society, Helsinki 1988, pp. 107–128.

Trüb, J., 1993: Dynamical aspects of flow over Alpine-scale topography: A modeling perspective. Ph.D. Dissertation 10339, ETH Zürich.

Trüb, J. and H. C. Davies, 1995: Flow over a mesoscale ridge: pathways to regime transitions. *Tellus*, **47A**, 502–525.

Vosper, S. B. and S. D. Mobbs, 1997: Measurement of the pressure field on a mountain. *Q. J. R. Meteorol. Soc.*, **123**, 129–144.

Williams, R. T., M. S. Peng, D. A. Zankofski, 1992: Effects of topography on fronts. *J. Atmos. Sci.*, **49**, 287–305.

The Upscale Turbulent Cascade
Shear Layers, Cyclones and Gas Giant Bands
W. R. Peltier[1] and G. R. Stuhne[1,2]

[1]Department of Physics, University of Toronto, Toronto, Ontario, Canada
[2]Scripps Institution of Oceanography (PORD Division), University of California, San Diego, La Jolla, CA, USA

The upscale cascade of kinetic energy which characterizes the evolution of turbulent flows that are restricted to two spatial dimensions is mediated by the vortex-pairing interaction. In randomly initialized flows on the f-plane, this interaction coarsens the initial vorticity distribution by causing like-signed vortices to amalgamate and thereby leads to the creation of coherent vortical structures of ever-increasing spatial scale. In the sequence of analyses discussed herein, we analyse a series of problems of interest from an atmospheric dynamics perspective in which, in spite of their intrinsically three (or 2.5) dimensionality, the vortex-pairing interaction is nevertheless active and has important phenomenological consequences. These problems include the "classical" problems of parallel shear and baroclinic instability as well as the somewhat more exotic problem of vortex dynamics in shallow water flow on the surface of the sphere.

A. INTRODUCTION

Ever since the work of Fjortoft (1953; see also Salmon, 1998) it has been well understood that the inviscid flow of a classical fluid (whether liquid or gas), if confined to two space dimensions, is constrained to evolve such that kinetic energy "cascades" from smaller to larger spatial scales. This behaviour is a simple mathematical consequence of the joint constraints of kinetic energy and enstrophy (mean-square vorticity) conservation. Charney (1971) extended Fjortoft's result to the case of three-dimensional quasi-geostrophic flow. Although the validity of this result has been contradicted by both early numerical model results (e.g. Barros and Winn-Nielsen, 1974; Hua and Haidvogel, 1986) and by the analysis of atmospheric observations (e.g. see the analysis FGGE III data by Lambert, 1981), more results appear to confirm it.

In spite of the fact that vorticity is clearly not conserved in three-dimensional flows, in consequence of the action of the processes of vortex stretching and tilting through which vorticity may be generated or destroyed, there is nevertheless considerable interest in understanding the extent to which the fundamental interaction that supports the upscale cascade in two space dimensions may still manifest itself to some degree in three-dimensional reality. In two-dimensional flows on the f-plane (e.g. McWilliams, 1984; Maltrud and Vallis, 1991), this fundamental interaction is clearly seen to involve the "pairing" of vortices of like sign so as to produce coherent vortical structures of ever-increasing spatial scale. Since three-dimensional turbulent flows are clearly not constrained in general by this interaction and so may (and of course do) transfer kinetic energy downscale (Kolmogorov, 1941), it is an interesting question as to whether and in what circumstances there may nevertheless remain vestiges of the pairing interaction in three-dimensional flows. An equally interesting question, if such vestigal effects of the pairing interaction do remain in three-dimensional flows, concerns the processes that conspire to limit their spatiotemporal range of influence.

In this chapter our intention is to describe an exploration of these questions in which three different flows are analysed which have distinctly different geometries and physical properties but which share the characteristic that the pairing interaction exerts significant control on flow evolution, at least initially, even though the flows are intrinsically three dimensional (or at least 2.5 dimensional). In each of these three cases it will be argued that the vortex-merging interaction exerts intense though ephemeral control on flow evolution. Since our purpose herein is to compare the hydrodynamic phenomenology that occurs in these three distinctly different circumstances and since the analyses that have been performed of each flow geometry are each fairly complex in themselves, space will allow only the most cursory presentation of the methodology employed and results obtained in each case.

As implied in the title of this chapter, we will begin with a discussion of the simplest circumstance in which the pairing interaction occurs, namely that of a sheared parallel flow containing an "inflection point" in the velocity profile. The classical problem of the baroclinic instability of a zonal flow that is initially in thermal wind balance is considered subsequently, analysis of which suggests the existence of a route to deep development that does not begin on the Charney–Eady branch of unstable modes but rather with a subsynoptic-scale instability through which finite-amplitude vortices are engendered and which, through the pairing interaction, subsequently produces a deep energetic disturbance of wavelength that is still significantly less than that of the most unstable Charney–Eady wave. Finally, we will consider the spherical problem of vortex merging in which, through the Rossby-wave arrest of the upscale cascade (Rhines, 1975), a banded zonal velocity profile may develop that very well mimics the observed zonal velocity patterns visible on the discs of the Gas Giant planets Jupiter, Saturn, Neptune and Uranus. In the latter context the focus will be upon the mechanism whereby the upscale cascade is arrested and the way in which this determines the sign of the equatorial jet that emerges in the limit of long time. Conclusions based upon this intercomparison of the three model problems are offered in the final section.

B. PAIRING IN UNSTRATIFIED SHEAR FLOW

If confined so as to evolve in two space dimensions, an unstratified flow with vorticity $\omega = \nabla \times \mathbf{u}$, with $\mathbf{u}$ the velocity field, will advance in time so as to satisy the system

$$\frac{\partial \omega}{\partial t} = \frac{\partial \omega}{\partial x}\frac{\partial \Psi}{\partial z} - \frac{\partial \omega}{\partial z}\frac{\partial \Psi}{\partial x} + \frac{1}{R_e}\nabla^2 \omega \quad (1a)$$

$$\nabla^2 \Psi = -\omega \quad (1b)$$

in which Ψ is the stream function such that $u = -\partial \Psi/\partial z$, $w = \partial \Psi/\partial x$, and $Re = U_o h/\nu$ is the Reynolds number with U_o and h characteristic velocity and length scales respectively and ν the kinematic viscosity. Suppose we assume initial conditions such that $\mathbf{u} = U(z) = U_o \tanh(z/h)$ so that nondimensional $\omega(z) = 1/\cosh^2 z$ initially. If R_e is $\gtrsim 50$, then the initially parallel flow is Kelvin–Helmholtz (K-H) unstable and will evolve in the presence of weak noise so as to generate a train of like-signed vortices in which the fundamental wavelength λ_f is approximately seven times the initial depth of the shear layer. This length scale is of course precisely predictable on the basis of the solution of the associated two-point boundary-value problem for linear stability. Even in the presence of stable density stratification (Smyth and Peltier, 1990), the primary instability of the initial parallel flow is inevitably two dimensional unless the density stratification is extreme. Figure 1 illustrates the evolution of an unstratified flow according to equations (1) computed in a rectangular domain of length $2\lambda_f$ in which the spatial boundary conditions are periodic in x and impermeable at $z = \pm H$, with $H >> h$ so that the horizontal boundaries exert no control on flow development. Inspection of Fig. 1 demonstrates that once the initial K-H instability saturates the two like-signed vortices in the domain thereafter begin to orbit one-another and eventually fully merge, a process which ends in the creation of a single vortex of double the length scale of the initial train of K-H waves. This is the fundamental interaction that supports the upscale cascade of energy in two-dimensional turbulence. If we were to sufficiently increase both the horizontal and vertical scales of the domain, the merger process would continue unimpeded (Smyth and Peltier, 1993), leading to the creation of vortical structures of ever-increasing scale.

The pairing interaction itself may be very simply understood on the basis of a linear stability analysis of the finite-amplitude vorticity distribution that exists at the moment that the initial K-H instability saturates. Approximating this vorticity distribution by the analytical Stuart (1967) solution to the time-independent Euler equations, we have the solution for the stream function Ψ as:

$$\Psi = -\frac{(1-e)^2}{1-e^2}\log[\cosh(z-h/2) + e\cos x] \quad (2a)$$

which corresponds to the vorticity distribution

$$\omega = \frac{1-e}{\cosh(z-h/2) + e\cos x} \quad (2b)$$

in which e is the ellipticity parameter of the Stuart model. Figure 2 shows ω and Ψ for the Stuart solution for several values of the ellipticity parameter. Initial analyses of the stability of the Stuart solution were presented by Kelly (1967) for small amplitude and later by Pierrehumbert and Widnall (1982) for arbitrary amplitude, but at overly low spatial resolution, and by Klaassen and Peltier (1989) at spatial resolution sufficiently high to capture the stability characteristics accurately. Potylitsin and Peltier (1998, 1999) have more recently employed the same model to represent a columnar vortex in a rotating

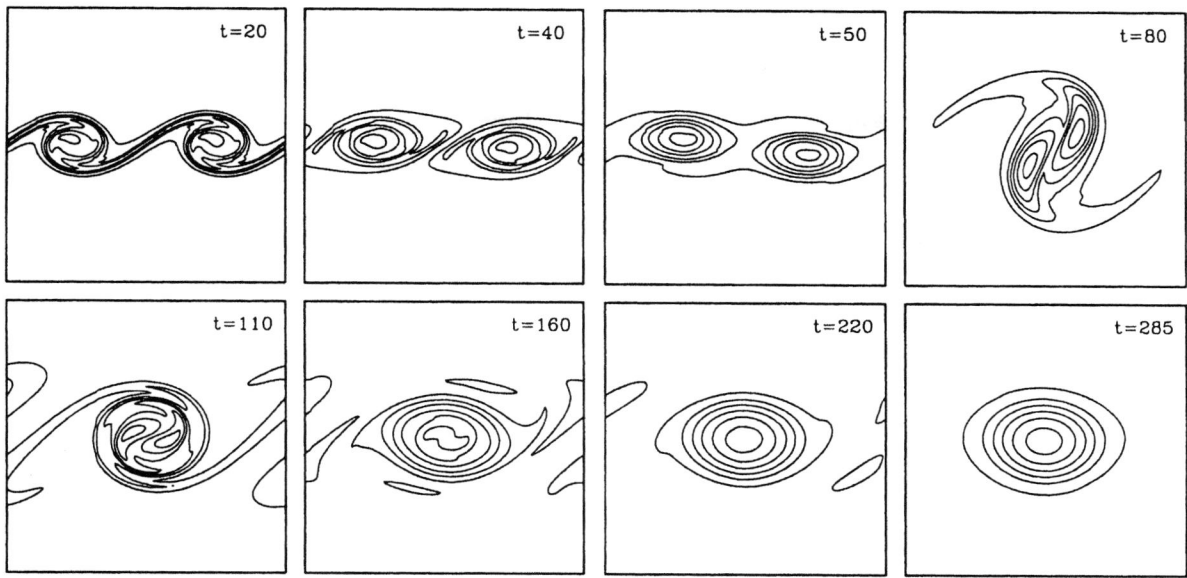

FIGURE 1 The evolution of the vorticity field in two-dimensional homogeneous fluid for a sequence of eight times, beginning with the time of saturation of the primary Kelvin–Helmholtz wave. The horizontal dimension of the domain in which the DNS integration is performed is twice the wavelength λ_f of the fastest growing linear instability. In the limit of long time the vorticity field engendered by the vortex pairing instability consists of a train of vortices with twice the wavelength of the initial train of K-H waves.

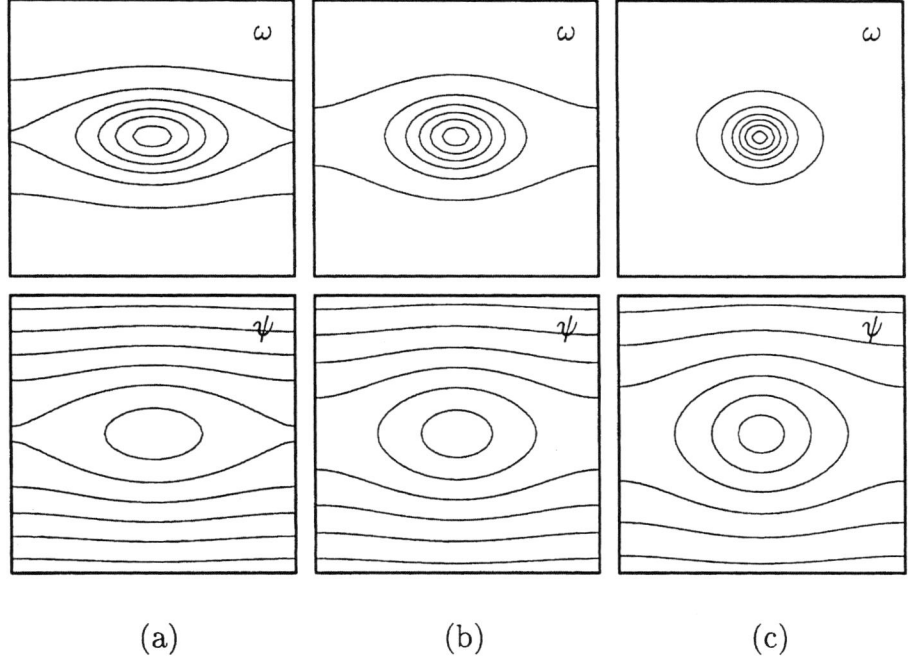

FIGURE 2 Three examples of the vorticity and stream-function fields for the steady Stuart (1967) solution of the Euler equations. In (a); (b) and (c) the values of the ellipticity parameter of the Stuart model are respectively $e = 0.33$ (highly elliptical vortex), $e = 0.5$ (slightly elliptical vortex) and $e = 0.75$ (almost circular vortex).

environment in their investigation of the destabilizing influence of rotation upon upright anticyclones.

The problem of determining the linear stability of a spatially periodic flow such as that represented by equations (2) is a problem in Floquet theory. If we represent the velocity field u_i by the superposition of the two-dimensional Stuart solution, denoted by $\tilde{u}_i(x, z)$ and a perturbation $u'_i(x, y, z, t)$ as:

$$u_i(x, y, z, t) = U_o\tilde{u}_i(x, z, t)(1 - \delta_{i2}) + U_o u'_i(x, y, z, t) \quad (3)$$

then Floquet theory dictates that the Navier–Stokes equation for the solution u'_i, with u'_i assumed small, should take the form:

$$u'_i(x, y, z, t) = \hat{u}_i(x, z, t) \exp[i(bx + dy)] \quad (4)$$

in which $\hat{u}_i$, in general, is a complex function that is periodic in x with fundamental wavelength λ_f fixed to that of the Stuart solution. Since the Stuart solution is precisely steady we may write

$$\hat{u}_i(x, z, t) = u_i^\dagger(x, z)e^{st} \quad (5)$$

with $s = \sigma + i\omega$ the complex eigenvalue of the resulting stability problem. Substituting (3), (4) and (5) into the linearized Boussinesq equations these become:

$$su_x^\dagger + \tilde{u}_x(\partial_x + ib)u_x^\dagger + \tilde{u}_z\partial_z u_x^\dagger + (\partial_x \tilde{u}_x)u_x^\dagger +$$
$$(\partial_z \tilde{u}_x) u_z^\dagger = -(\partial_x + ib)p^\dagger + \frac{1}{R_e} L_2 u_x^\dagger \quad (6a)$$

$$su_y^\dagger + \tilde{u}_x(\partial_x + ib)u_y^\dagger + \tilde{u}_z\partial_z u_y^\dagger = -idp^\dagger + \frac{1}{R_e} L_2 u_y^\dagger \quad (6b)$$

$$su_z^\dagger + \tilde{u}_x(\partial_x + ib)u_z^\dagger + \tilde{u}_z\partial_z u_z^\dagger + (\partial_x \tilde{u}_z)u_x^\dagger +$$
$$(\partial_z \tilde{u}_z) u_z^\dagger = -\partial_z p^\dagger + \frac{1}{R_e} L_2 u_z^\dagger \quad (6c)$$

$$(\partial_x + ib)u_x^\dagger + idu_y^\dagger + \partial_z u_z^\dagger = 0 \quad (6d)$$

with the operator $L_2 = (\partial_x + ib)^2 + \partial_z^2 - d^2$. Since the pairing interaction involves a two-dimensional instability of a two-dimensional flow we may take $d = u_y^\dagger \equiv 0$ so that equation (6b) is redundant and the stability of the Stuart solution will be determined by the simultaneous solution of (6a), (6c) and (6d) with unknowns $u_x^\dagger, u_z^\dagger$ and $p^\dagger$, a system with eigenvalue $s = \sigma + i\omega$. Since we may obtain a diagnostic equation for $p^\dagger$ such that

$$L_2 p^\dagger = -2[(\partial_x\tilde{u}_x)(\partial_x + ib)u_x^\dagger + (\partial_z\tilde{u}_z)\partial_z u_z^\dagger +$$
$$(\partial_x \tilde{u}_z)\partial_z u_x^\dagger + (\partial_z\tilde{u}_z)(\partial_x + ib)u_x^\dagger] \quad (7)$$

$p^\dagger$ is determined by $u_x^\dagger$ and $u_z^\dagger$, for which we may seek solutions in the form of Galerkin expansions as:

$$u_x^\dagger = \sum_{\lambda=-\infty}^{+\infty} \sum_{\nu=0}^{\infty} a_{\lambda\nu} e^{i\alpha(2\pi/\lambda)x} \cos \frac{\nu\pi z}{H} \quad (8a)$$

$$u_z^\dagger = \sum_{\alpha=-\infty}^{+\infty} \sum_{\nu=0}^{\infty} b_{\lambda\nu} e^{i\alpha(2\pi/\lambda)x} \sin \frac{\nu\pi z}{H} \quad (8b)$$

Upon substitution of (8) into (6) and computation of the inner product of each of equations (6) with the complex conjugates of each of the elements in the Galerkin basis employed in the expansions (8), we obtain the algebraic system:

$$s\, a_{\kappa\mu} = \left(I^{(1)}_{\kappa\mu\alpha\nu} + \frac{A\alpha\nu}{R_e}\delta_{\kappa\alpha}\delta_{\mu\nu}\right) a_{\alpha\nu} + I^{(2)}_{\kappa\mu\alpha\nu}b_{\alpha\nu} \quad (9a)$$

$$s\, b_{\kappa\mu} = I^{(3)}_{\kappa\mu\alpha\nu} + \left(I^{(4)}_{\kappa\mu\alpha\nu} + \frac{A\alpha\nu}{R_e}\delta_{\kappa\alpha}\delta_{\mu\nu}\right) b_{\alpha\nu} \quad (9b)$$

with $A_{\alpha\nu} = B_\alpha^2$, $B_\alpha = \alpha(2\pi/\lambda_f) + b$. By concatenating the coefficients $a_{\kappa\mu}$ and $b_{\kappa\mu}$ so as to form a single subscripted quantity, it will be clear that (9) constitutes a standard matrix eigenvalue problem. The four-subscripted arrays $I^{(1,2,3,4)}_{\kappa\mu\alpha\nu}$ consist of projections of the nonlinear two-dimensional fields onto the Galerkin basis. By appropriate truncation of the Galerkin expansions (8), the algebraic eigenvalue problem (9) is reduced to one of finite dimension and we may compute the eigenvalue s as a function of the Floquet exponent b which describes the wavenumber of the modulation in the x-direction (see (4)) and which is clearly restricted such that $b \leq 1$.

Direct calculation shows that the eigenvalues of (9) are such that $\omega \equiv 0$. Figure 3 then shows growth rate σ determined by solving (9) as a function of b for two different truncation schemes, respectively labelled $\beta = 0$ and $\beta = 1/2$. These are:

$$2|\alpha| + \nu \leq N \quad (\beta = 0) \quad (10a)$$

$$2|\alpha + \frac{1}{2}| + \nu \leq N \quad (\beta = 1\backslash2) \quad (10b)$$

Inspection of the growth rate spectrum for these two schemes shown in Fig. 3, for the sequence $N = 15, 19, 23$, demonstrates that there are two branches of the modal spectrum, denoted respectively by the solid and dashed lines. In both branches the maximum growth rate occurs at $b = 1/2$, signifying that the fastest-growing instabilities are occurring at the wavelength of the first spatial subharmonic of the initial K-H instability. The eigenfunctions of these two fastest-growing modes are illustrated in Fig. 4 in terms of a superposition of the stream function of the Stuart basic state together with the growing perturbation which is assumed to have an amplitude equal to 1/4 the maximum vorticity. Inspection of these

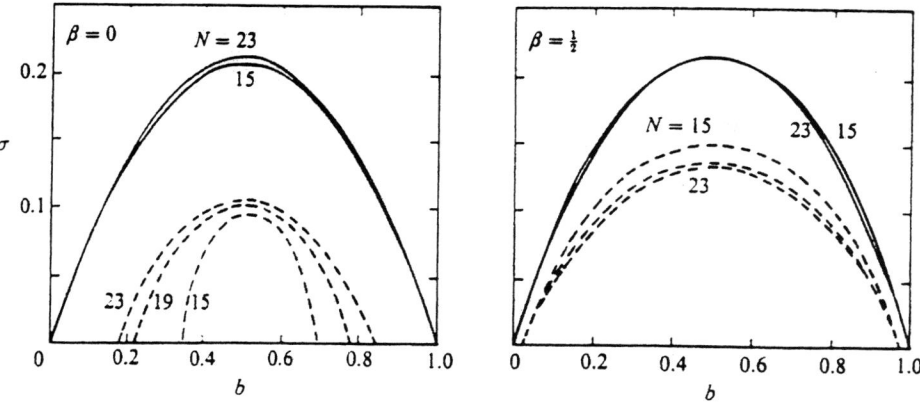

FIGURE 3 Growth rate spectra for the modes of instability of the Stuart solution with $e = 0.5$ as a function of the Floquet parameter b. Results are show for the two different truncation schemes labelled $\beta = 0$ and $\beta = \frac{1}{2}$ which are defined in the text. Results are also shown for three different truncation levels $N = 15$, $N = 19$, and $N = 23$.

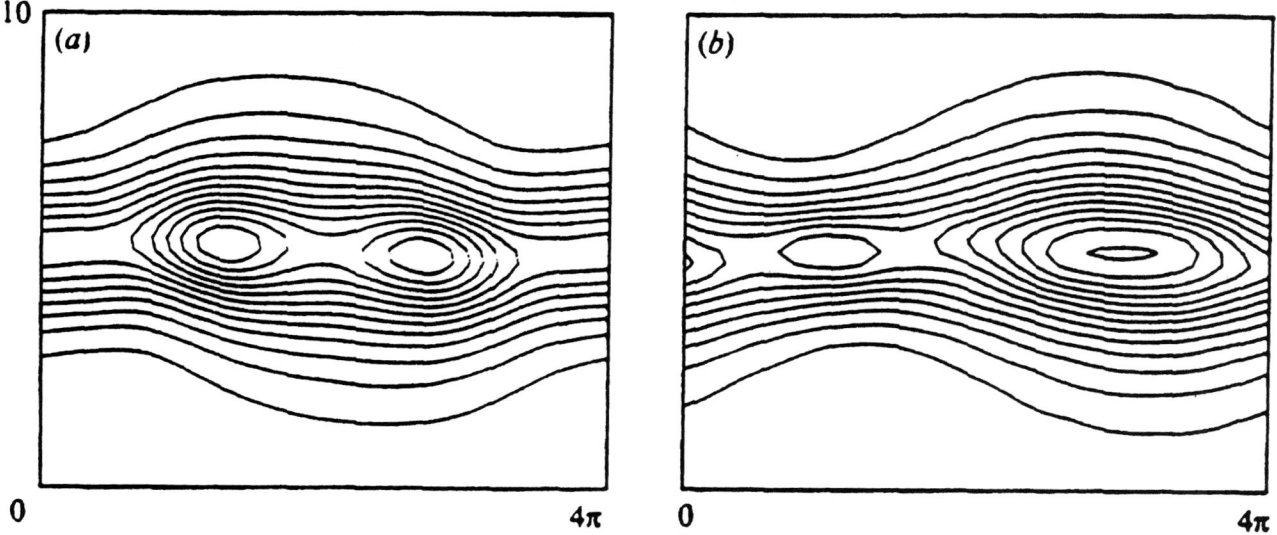

FIGURE 4 Structures of the fastest-growing (subharmonic) instabilities of the Stuart solution for both the orbital-merging mode (a) and for the "draining" mode (b) discussed in the text.

structures will show that the fastest-growing instability at $b = \frac{1}{2}$ shown in (a) corresponds to the first stage of the vortex amalgamation process illustrated previously in Fig. 1. This is the pairing interaction that involves orbital merger. The subdominant second mode at $b = \frac{1}{2}$, whose eigenfunction has been employed to produce the result shown in (b), corresponds conversely, to an interaction in which one vortex grows at the expense of the other by a non-orbital "draining" interaction.

Even in simple two-dimensional shear flows, therefore, the upscale cascade of energy that is mediated by vortex merger, may occur via at least two fundamentally different processes. Furthermore, the bandwidth (in the *Floquet parameter* b) of the merging instabilities is broad, so that the theory explains not only the primary two-vortex interaction but also the more exotic mergers involving $3 \to 1$ and $3 \to 2$ interactions that are also observed to occur in laboratory flows, albeit with significantly lower probability (see Klaassen and Peltier 1989 for more detailed discussion). However, when such parallel flows are allowed to access the third space coordinate, the subharmonic pairing interaction must eventually compete with a broad spectrum of fully three-dimensional instabilities for control of flow evolution. These more complex mechanisms may occur on shorter timescales than the pairing interaction and inhibit it completely. Detailed discussions of these mechanisms will be found in Klaassen and Peltier (1985, 1991), Smyth and Peltier (1991, 1994), Potylitsin and Peltier (1998, 1999), and Caulfield and Peltier (1994, 2000). In the next section we will focus upon the mechanics involved in the baroclinic instability process itself and find further evidence of the importance in fully three-dimensional stratified rotating flow of the action of pairing.

C. PAIRING IN ROTATING STRATIFIED FLOW: SUBSYNOPTIC BAROCLINIC INSTABILITY

Of all of the hydrodynamic interactions that have been studied in the context of our continuing attempts to understand the dynamics of the atmosphere, it is unlikely that any have attracted the same degree of attention as has the baroclinic instability process. Here we will describe a series of results recently obtained in the course of an investigation of the issue of the origins of subsynoptic scale instability (see Yamasaki and Peltier, 2001a,b, for a much more detailed analysis than space will allow herein). Figure 5d–f illustrates the properties of a zonal mean state determined on the basis of Reanalysis Project data from the entrance of the Pacific storm track in winter, while Fig. 5a–c shows the same characterizations of the mean state for an analytical model designed so as to adequately mimic the observations. A notable property of both the observations and the analytical model of them is that there exists a weak horizontal gradient of potential vorticity in the cross-front direction.

Figure 5 shows the results obtained by analysing the linear stability of this two-dimensional basic state to three-dimensional perturbations using exactly the same Galerkin methodology employed in the analysis of the two-dimensional pairing interaction described in the last section. In Fig. 6a,b the modes of instability are characterized respectively by their growth rate and phase speed as a function of the modal wavenumber in the zonal direction parallel to the mean flow. Focusing first on the growth rate spectrum in Fig. 6a, we note that there appears to exist two distinct modal branches. The first, with maximum growth rate at a wavelength near 4000 km, clearly corresponds to the classical Charney–Eady mode. The second modal branch appears to have most rapid growth at a wavelength that is considerably shorter, near 1000 km, although, as the truncation level in the Galerkin representation of the perturbation is increased, there is no evidence of a short-wavelength cutoff for the modes in this branch and the convergence as a function of truncation level is slow. Also shown in Fig. 6a as open circles, with or without an interior "×", are growth rates determined on the basis of DNS using a nonlinear, nonhydrostatic, anelastic model to describe the evolution of the unstable mean state shown in Fig. 5. The growth rates delivered by the nonlinear model, using a "no-noise" initialization procedure (see Yamasaki and Peltier, 2000a, for discussion), are very similar in magnitude but not identical to those delivered by the Galerkin-based analyses of linear stability. Inspection of the phase speed spectra shown in Fig. 6B demonstrates that convergence in this regard is excellent at all wavelengths.

Figure 7 presents the structure of the subsynoptic normal mode determined on the basis of the DNS solution of the initial-value problem in a domain of zonal length $\lambda_f = 1000$ km using the complete nonlinear anelastic model. It will be noted that this structure consists of a boundary-trapped disturbance rather than of a deep troposphere filling structure as is characteristic of the conventional synoptic scale Charney-Eady mode. Detailed diagnostic analysis of the energy conversions that are responsible for the growth of the subsynoptic scale structure, perturbation kinetic energy and growth rate time series for which are shown in Figs 8a and 8b respectively, demonstrates that the mode is supported by the baroclinic instability process. The level of perturbation kinetic energy at which this mode saturates, shown in Fig. 8a, however, is clearly rather low, sufficiently low that if no further process were available whereby it might continue to amplify, one could not reasonably expect it to be physically important. In Fig. 9, however, we show a sequence of planform views of the surface potential temperature illustrating the further evolution of this structure that occurs when the zonal length of the domain is simply doubled to $2\lambda_f$. Kinetic energy and growth rate time series from this simulation are shown in Figs 8c and 8d respectively. Evident by inspection of both Figs 8c and 9 is the fact that in the longer zonal domain, the weak subsynoptic scale cyclonic structures undergo a pairing instability after the initial shallow instability saturates, in close analogy with the temporal evolution of the K-H instability discussed in the last section.

After pairing is complete, apparently through a "draining" interaction rather than through orbital merger, the baroclinic disturbance has deepened so as to fill the troposphere and its energy has increased by almost two orders of magnitude, as evidenced by Fig. 8c. The deep structure that develops subsequent to pairing is illustrated in Fig. 10 a,b,c through the zonal average of the squared meridional velocity, the planform of vertically averaged meridional velocity at the surface and a zonal/height cross-section of the same field meridionally averaged. Although this structure is

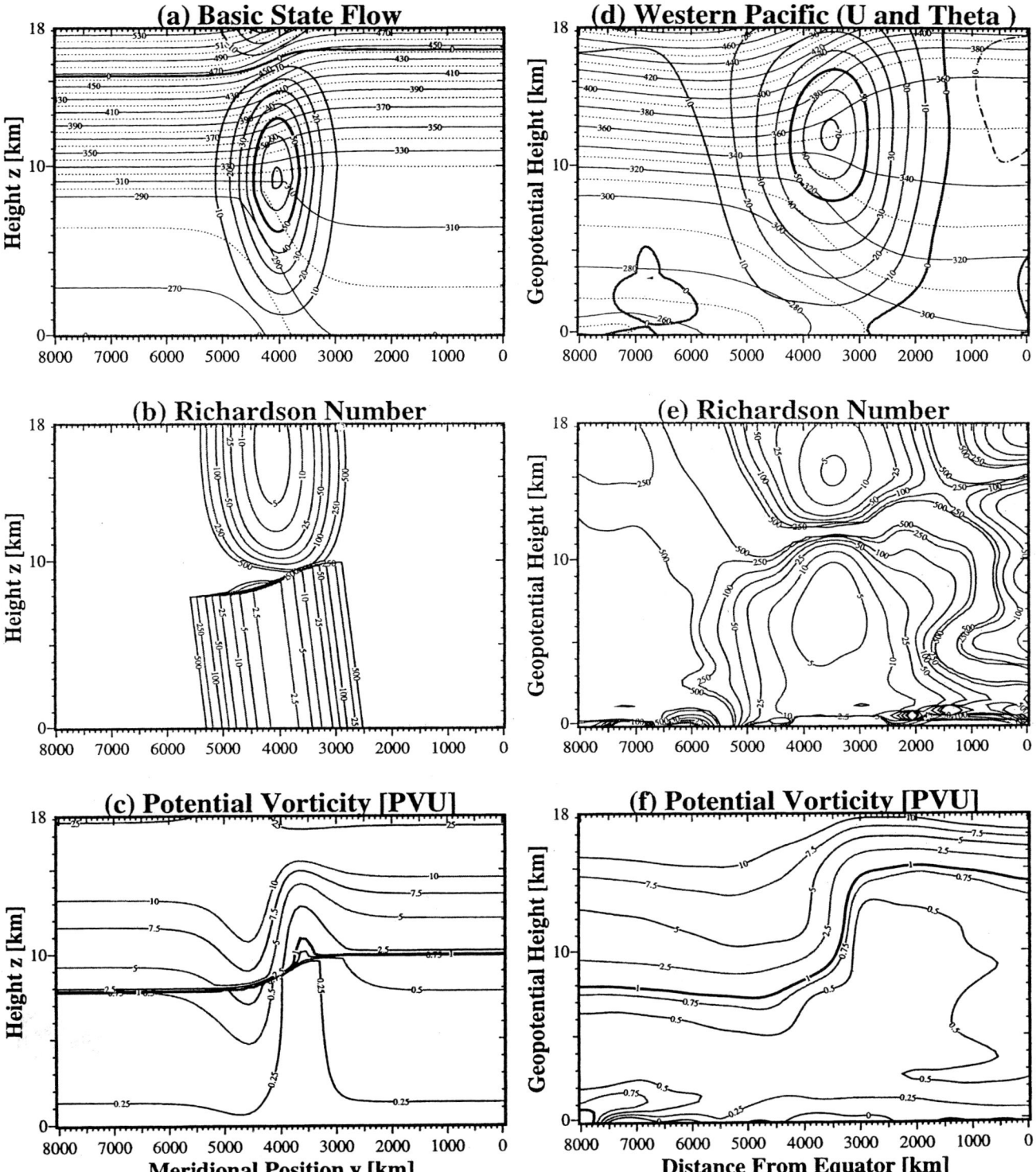

FIGURE 5 Meridional cross-sections of the zonal flow observed at the entrance to the Pacific storm track based upon reanalysis project data expressed in terms of: (d) zonal velocity (u) and potential temperature (θ); (e) Richardson number and (f) potential vorticity. The same fields are shown in (a)–(c) for the analytical model of Yamasaki and Peltier (2000a).

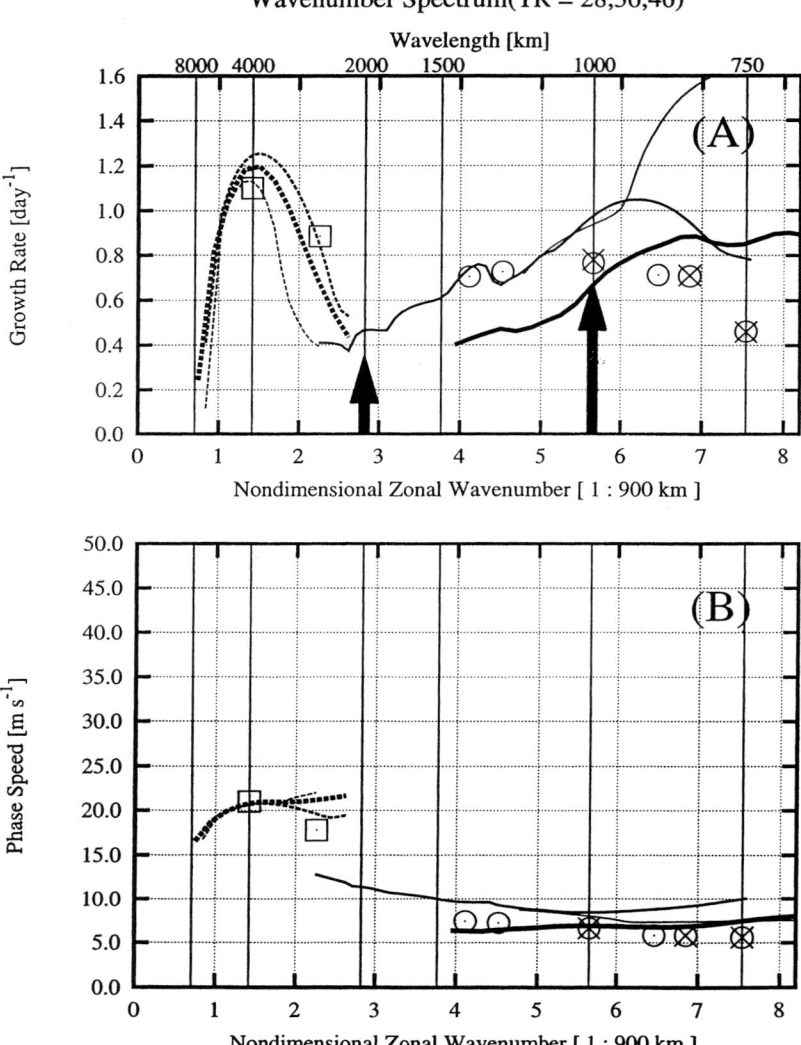

FIGURE 6 (a) Growth rate as a function of zonal wavenumber for the spectrum of instabilities supported by the analytical model meridional cross-section shown in Fig. 5a. (B) Phase speed spectrum for the same model. The open circles superimposed are growth rates and phase speeds determined from the early stages of the nonlinear DNS integrations for comparison. The parameter TR denotes the wave number truncation of the Galerkin expansions employed to represent the hydrodynamic percurbations, the light to heavy weighted dashed and solid lines in the long wavelength and short wavelength regimes corresponding to TR = 28, 36, 46 respectively.

now very similar in form to a conventional Charney-Eady mode, it has a wavelength of only 2000 km, half that of this conventional synoptic scale disturbance. Clearly, there appears to exist a route to "deep development" that does not begin with linear instability on the Charney–Eady branch of synoptic scale instability but rather with the growth of a subsynoptic scale structure which later amplifies through an upscale cascade of energy supported by the pairing interaction. Of course if the initial frontal structure were sufficiently elongate zonally so as to support the Charney-Eady wave then this disturbance would appear immediately. In nature, however, such elongate frontal structures do not always exist and not all cyclones occur on scales as large as that of the fundamental Charney-Eady mode. The upscale cascade route to deep development revealed by the analyses discussed herein may be rather important to the understanding of cyclogenesis in such circumstances.

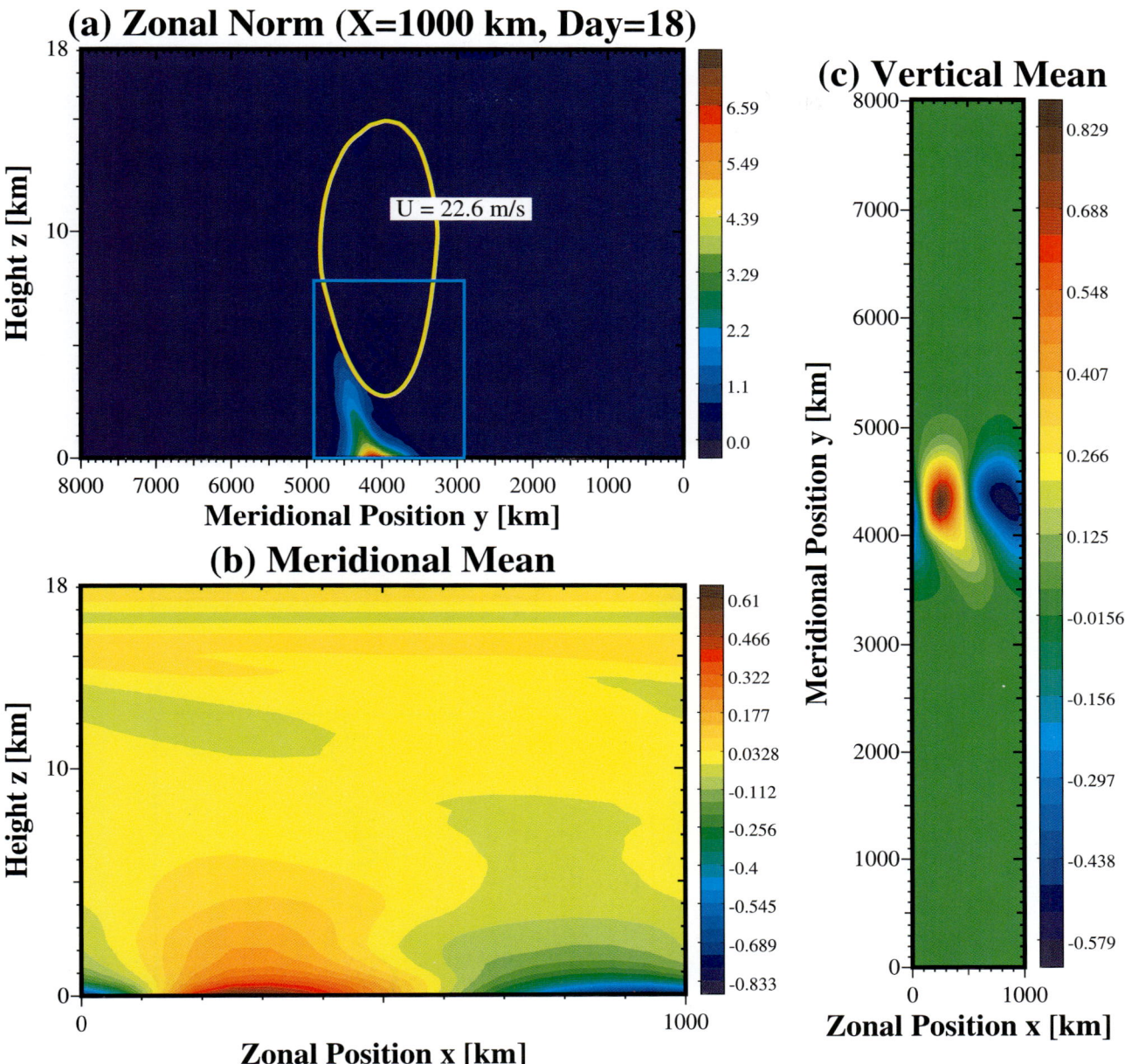

FIGURE 7 Structure of the fastest growing subsynoptic scale mode of baroclinic instability delivered by the initial value integration using the zonal mean state with structure shown in Fig. 5a. This subsynoptic scale structure is illustrated by (a) the zonal norm of the meridional velocity, (b) the meridional mean of the meridional velocity and (c) the vertical mean of the meridional velocity.

D. PAIRING IN VORTEX DYNAMICS ON THE SPHERE: ROSSBY-WAVE ARREST OF THE UPSCALE CASCADE AND THE "PARITY" OF GAS GIANT BANDS

One of the most compelling problems in planetary fluid dynamics, in which the upscale cascade of energy leads to the creation of coherent vortical structure that is observationally significant, is that concerning the origins of the latitudinally banded pattern of zonal velocity that characterizes the flow on the discs of the Gas Giant planets Jupiter, Saturn, Neptune and Uranus. Figure 11 illustrates this zonal wind structure for Jupiter based upon the analysis of the Voyager I and II observations by Beebe (1994), together with the fit to the data based

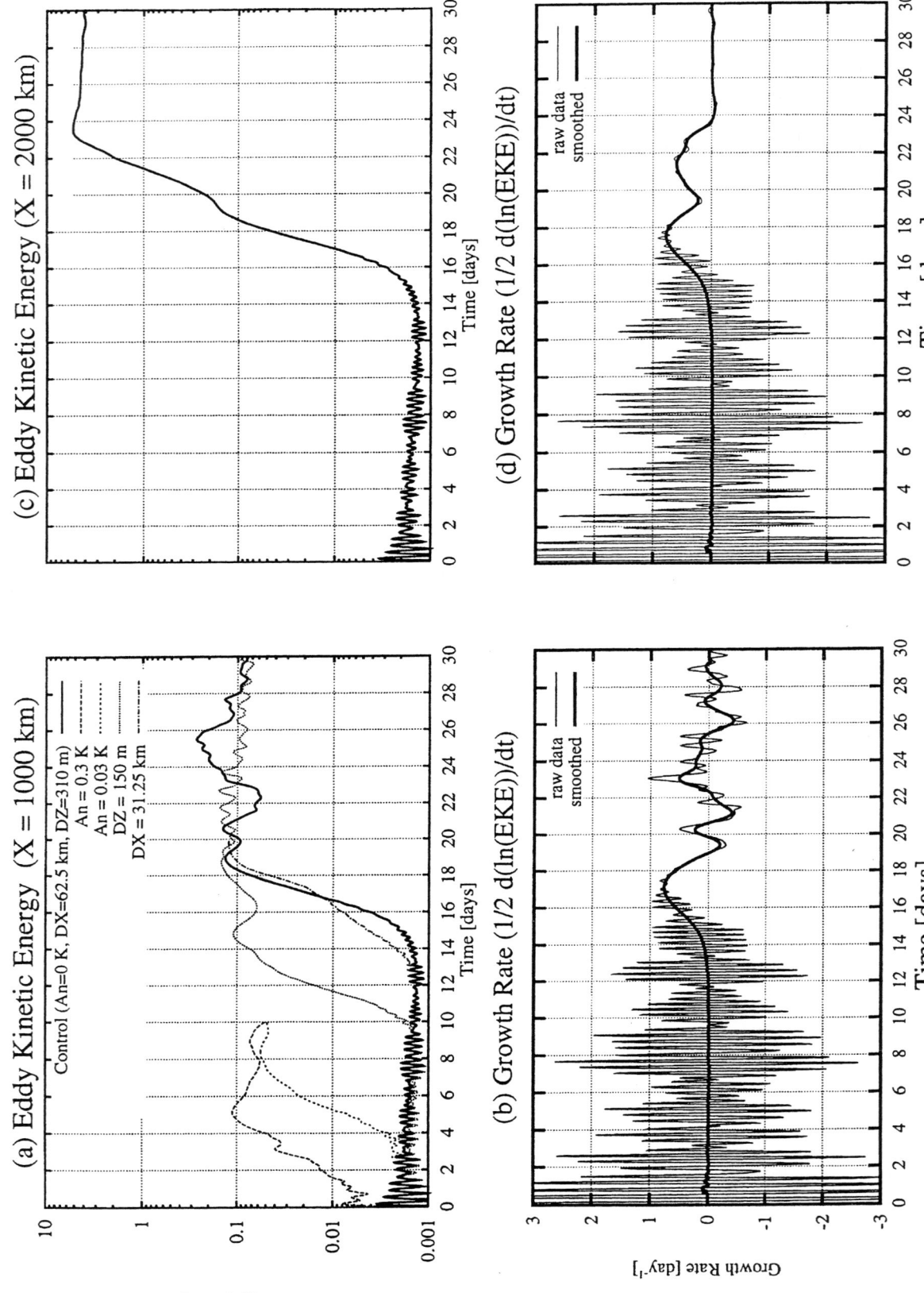

FIGURE 8 (a), (b) Kinetic energy and growth rate time series from the initial-value integrations of the nonlinear inelastic model when the zonal length of the domain is equal to 1000 km. (c), (d) same as (a), (b) but for the integration with domain length of 2000 km.

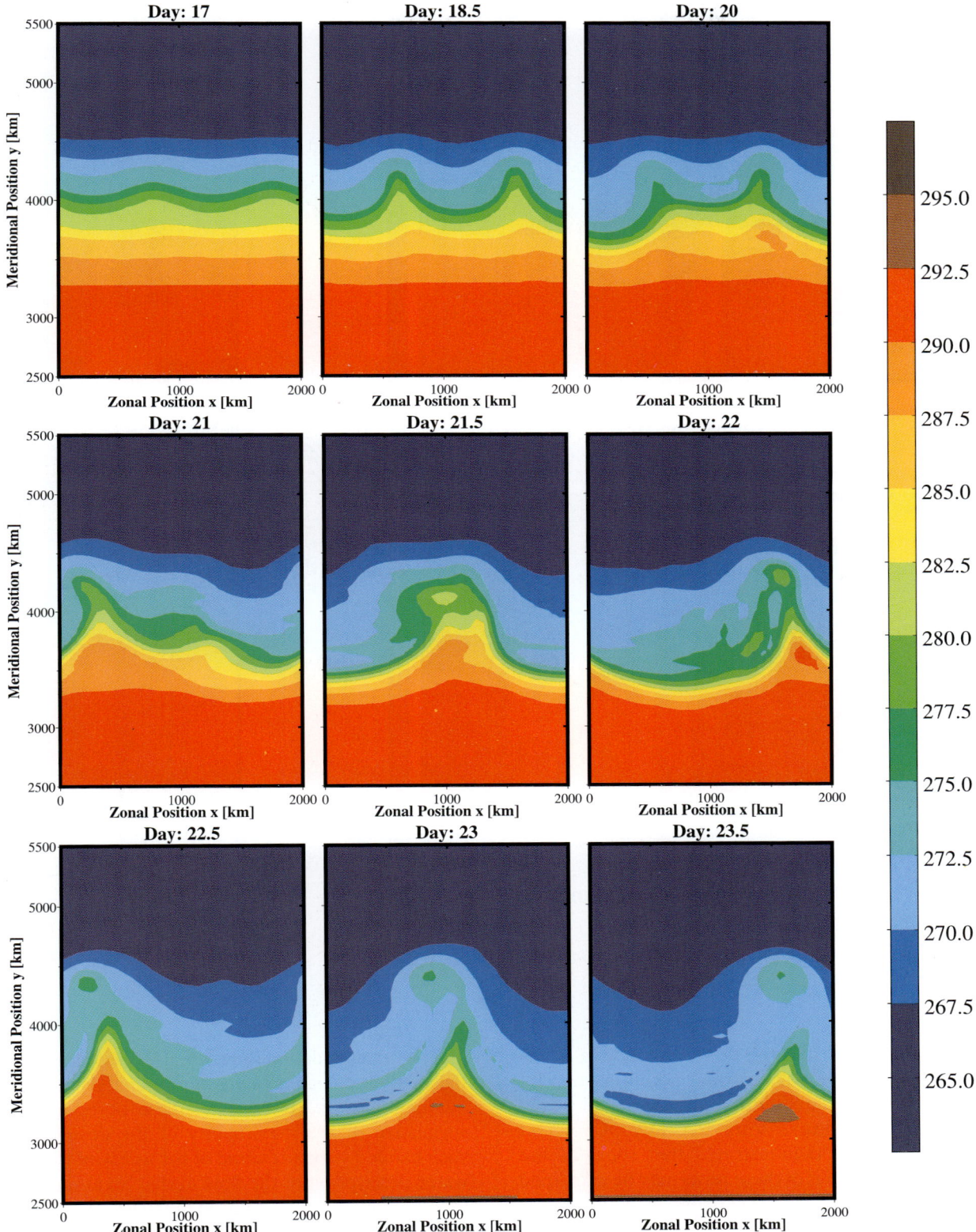

FIGURE 9 Surface potential temperature fields from the nonlinear time-dependent integration at a sequence of nine times in the evolution of the flow away from the parallel flow with meridional structure shown in Fig. 5a. Illustrates the pairing interaction.

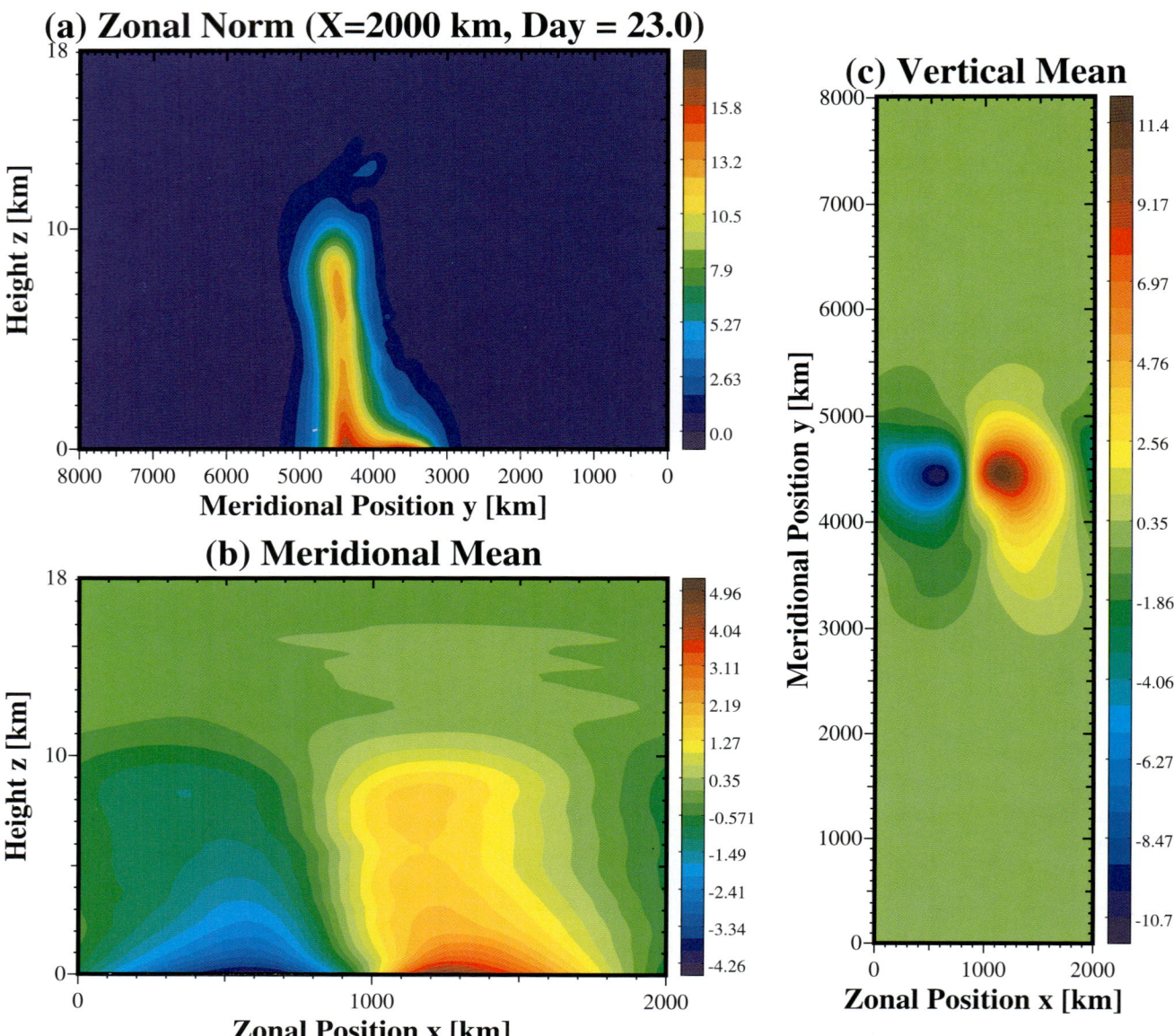

FIGURE 10 Same as Fig. 8 but for a time subsequent to completion of the pairing interaction.

upon the assumption that the flow corresponds to the pole-to-pole "staircase" in barotropic potential vorticity shown in Figure 11 (see Stuhne, 1999, for details of the construction procedure). Very recently, Cho and Polvani (1996) have demonstrated that some of the observed characteristics of these flows may be explained on the basis of simple unforced spherical shallow water integrations initialized with a "balanced" but otherwise random initial distribution of vorticity. The statistically stationary equilibrium states into which these flows evolve, when scaled to any one of the four Gas Giant planets, are banded in their zonal velocity patterns with the number of bands being approximately equal to the number observed on each of these objects. Although suggestive of the inevitability of the development of such banded structures, and of the utility of the shallow-water description of the flow in the "weather layer" on these planets, this explanation of the development of the bands that is embodied in such solutions of the unforced initial-value problem is significantly flawed.

In order to illustrate the nature of this flaw and to demonstrate that it is not due to any merely numerical

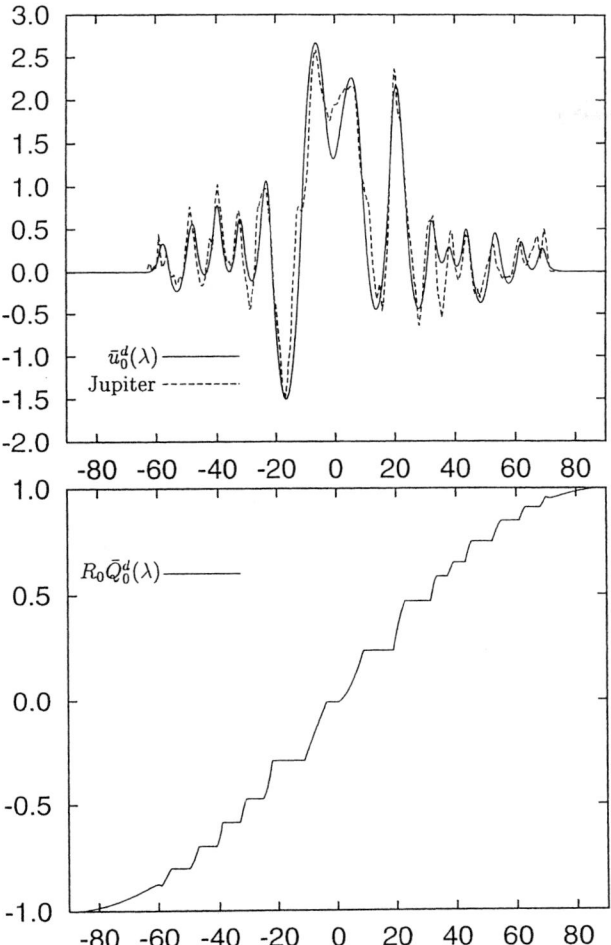

FIGURE 11 (Top) zonal velocity field as a function of latitude on the disc of Jupiter according to the analyses of Beebe (1994), compared to the zonal velocity field predicted by a model which assumes that barotropic potential vorticity varies over the disc according to the "monotonic staircase" model (bottom).

aspect of the shallow-water integrations reported by Cho and Polvani, it is useful to first reproduce their results using an entirely different numerical methodology. Whereas Cho and Polvani employed a shallow-water reduced form of the NCAR spectral model of the atmospheric general circulation as a basis for their numerical analyses, we have further tested the robustness of these results by employing the recently constructed shallow-water model of Stuhne and Peltier (1996, 1999), a model based upon regular icosahedral tiling of the sphere and the use of the multigrid method to evolve the shallow-water system of equations forward in time. With initial conditions corresponding to a random distribution of barotropic vorticity over the surface of the sphere, assumed to be in a state of Charney (1955) balance

rather than in the state of higher-order balance assumed by Cho and Polvani (1996), Fig. 12 shows the results obtained for each of the Gas Giant planets when the shallow-water model is appropriately scaled to them, in terms of views of the vorticity field on their discs seen from a point in the equatorial plane. Inspection of these results demonstrates very clearly that in the initial phase of the evolution of these flows the dominant interaction that is occurring involves the same vortex-merging instability initially explored by McWilliams (1984), a pairing-driven interaction that leads to the coarsening of the vorticity distribution as kinetic energy cascades upscale, leading to the development of coherent vortical structures of ever-increasing size. Eventually, however, this vorticity-coarsening phase of flow evolution is arrested, evidently at the scale first discussed by Rhines (1975), beyond which Rossby-wave effects become important. After the vorticity distribution begins to relax into a field of Rossby waves, the latitudinally banded structure in zonal velocity begins to develop. Figure 13 compares the mean zonal flows obtained from the Gas Giant integrations using the icosahedral model with those previously obtained by Cho and Polvani (1996). Evident upon inspection of this figure is that these two independent sets of analyses are entirely consistent with one another, demonstrating that the statistical equilibrium zonal flows are not significantly influenced by the details of the shallow-water numerical model employed to obtain them.

This is important because it demonstrates that the flaw in these results, as possible explanations of the zonal flows on the discs of the Gas Giants, is not due to numerical abberation. The flaw in question has to do with the fact that, for each of the four planetary scales and rotation rates analysed, the sign of the equatorial jet in the statistical equilibrium state is retrograde rather than prograde, as is observed to be the case for the two largest Gas Giants, Jupiter and Saturn. The observed zonal flow on the disc of Jupiter was shown previously in Fig. 11. Detailed diagnostic analysis of the results of the new shallow-water integrations suggests an explanation as to why the specific mechanism of the arrest of the upscale turbulent cascade of kinetic energy that occurs in the unforced shallow-water model fails to predict the correct sign for the equatorial jets that dominate the flow in the weather layers of the two largest Gas Giant planets. The vorticity equation of the shallow-water dynamical model may be written as follows, in the form of a forced barotropic vorticity equation in which the forcing consists of the "unbalanced" component of the flow as:

$$\frac{\partial Q}{\partial t} + \mathbf{u}_b \cdot \nabla Q = \mathcal{F}(\theta, \lambda, t) \qquad (11)$$

in which Q is the barotropic potential vorticity, $\mathbf{u}_b$ is the

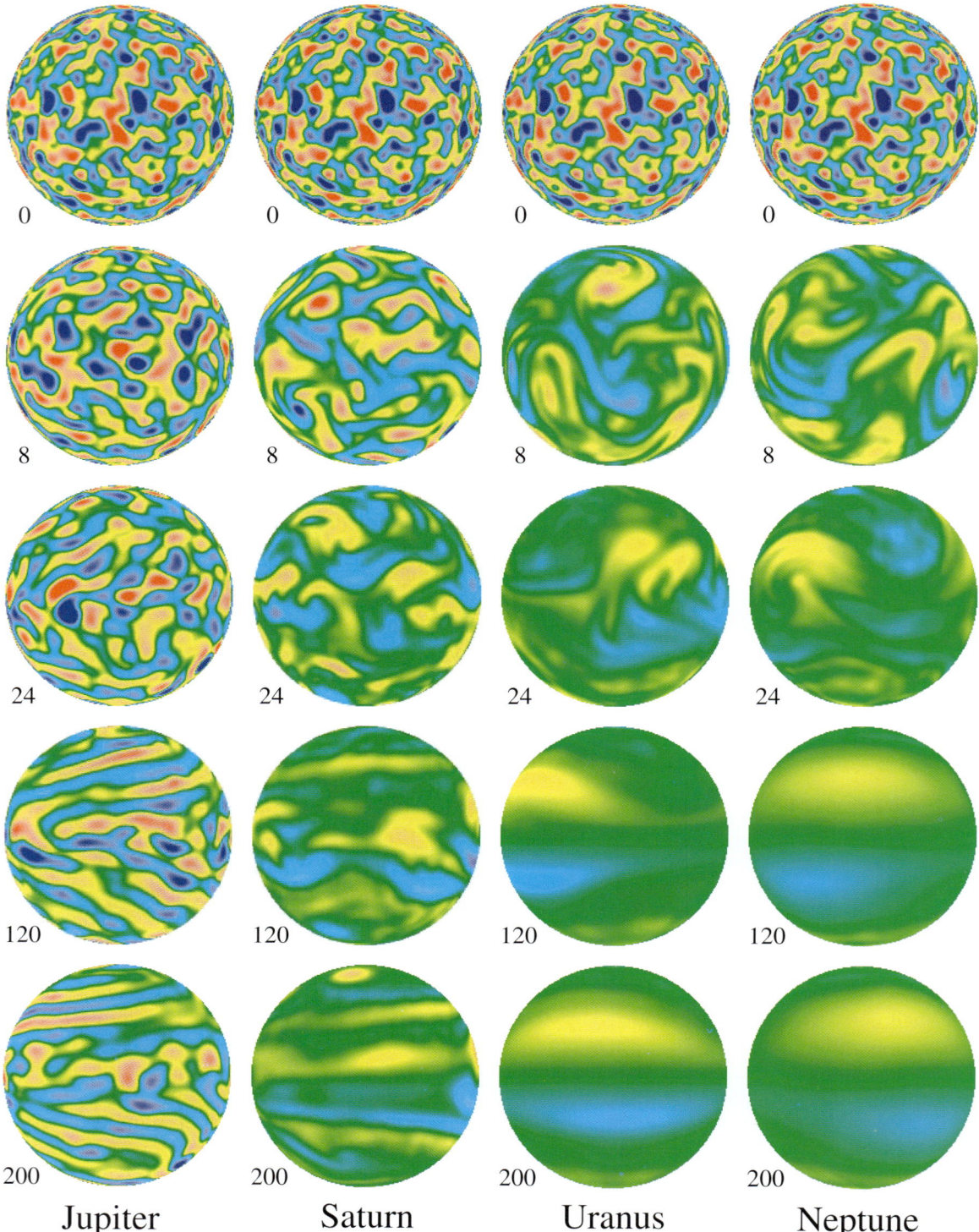

FIGURE 12 Evolution of the vorticity field delivered by freely evolving shallow-water integrations initialized with random but Charney (1959) balanced initial conditions. Results are shown for each of the Gas Giants with time measured in units of integral numbers of rotation periods shown adjacent to each image.

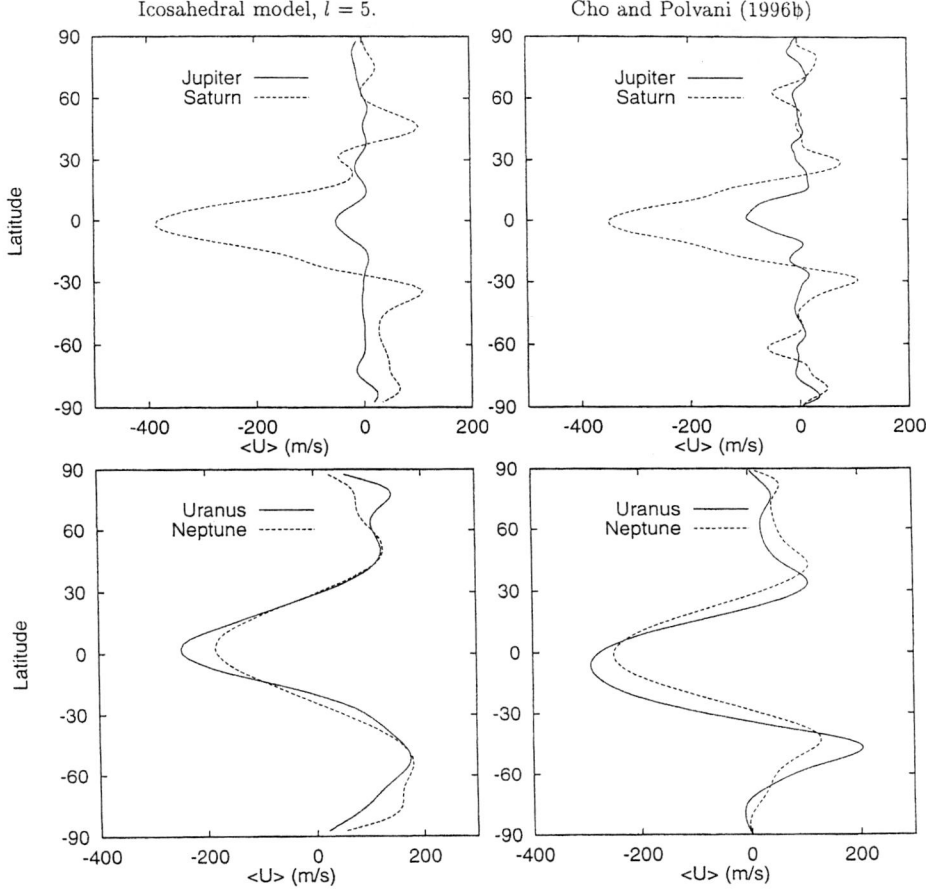

FIGURE 13 Statistical equilibrium zonal velocity fields from each of the four Gas Giant integrations performed using the new spherical icosahedral multigrid model compared with those previously obtained by Cho and Polvani (1996).

balanced (barotropic) component of the velocity field and $\mathcal{F}$ is the forcing of this balanced component by the evolving imbalance associated in part with surface gravity wave "activity". For the shallow-water model, the forcing associated with imbalance $\mathcal{F}$ has the specific form, with $\mathbf{u}_d$ the divergent component of the velocity field:

$$\mathcal{F}(\theta,\lambda,t) = -\nabla \cdot (\mathbf{u}_d\, Q) \qquad (12)$$

with Q the barotropic PV field defined as

$$Q = = \nabla^2 \Psi + R_o^{-1} \sin\theta \qquad (13)$$

where the nondimensional parameter $R_o = U/2\,\Omega_o r_s$, r_s and Ω_o being the observable radius of the planet in question and angular velocity respectively and $U = \sqrt{2E_k}$ being a characteristic velocity taken to be the RMS velocity over the surface of the sphere. In Fig. 14 is shown the specific spatial structure over the sphere of the imbalance $\mathcal{F}$ at one instant of time, specifically after 200 planetary rotations, taken from the Jupiter integ-

ration. Inspection of these results demonstrates that the imbalance is of a form that will tend to "mix" the barotropic vorticity distribution most strongly in the equatorial region. That intense homogenization of barotropic PV near the equator will inevitably produce a retrograde equatorial jet is demonstrated in Fig. 15 which illustrates the fact that a staircase distribution of barotropic PV, a structure which appears to best fit all of the observed Gas Giant zonal flows, in which a broad "step" of uniform PV exists on the equator itself, must correspond to a retrograde zonal jet. Conversely, a PV staircase in which one of the PV discontinuities on the staircase is coincident with the equator corresponds to a zonal flow in which the equatorial jet is prograde. The zonal flow in Jupiter's weather layer (Fig. 11a) is somewhat more complex than either of these extremes as it appears to have a weak retrograde element on the equator which introduces a slight diminution of strength of what is otherwise an intense prograde equatorial jet. That even this minor feature is well captured

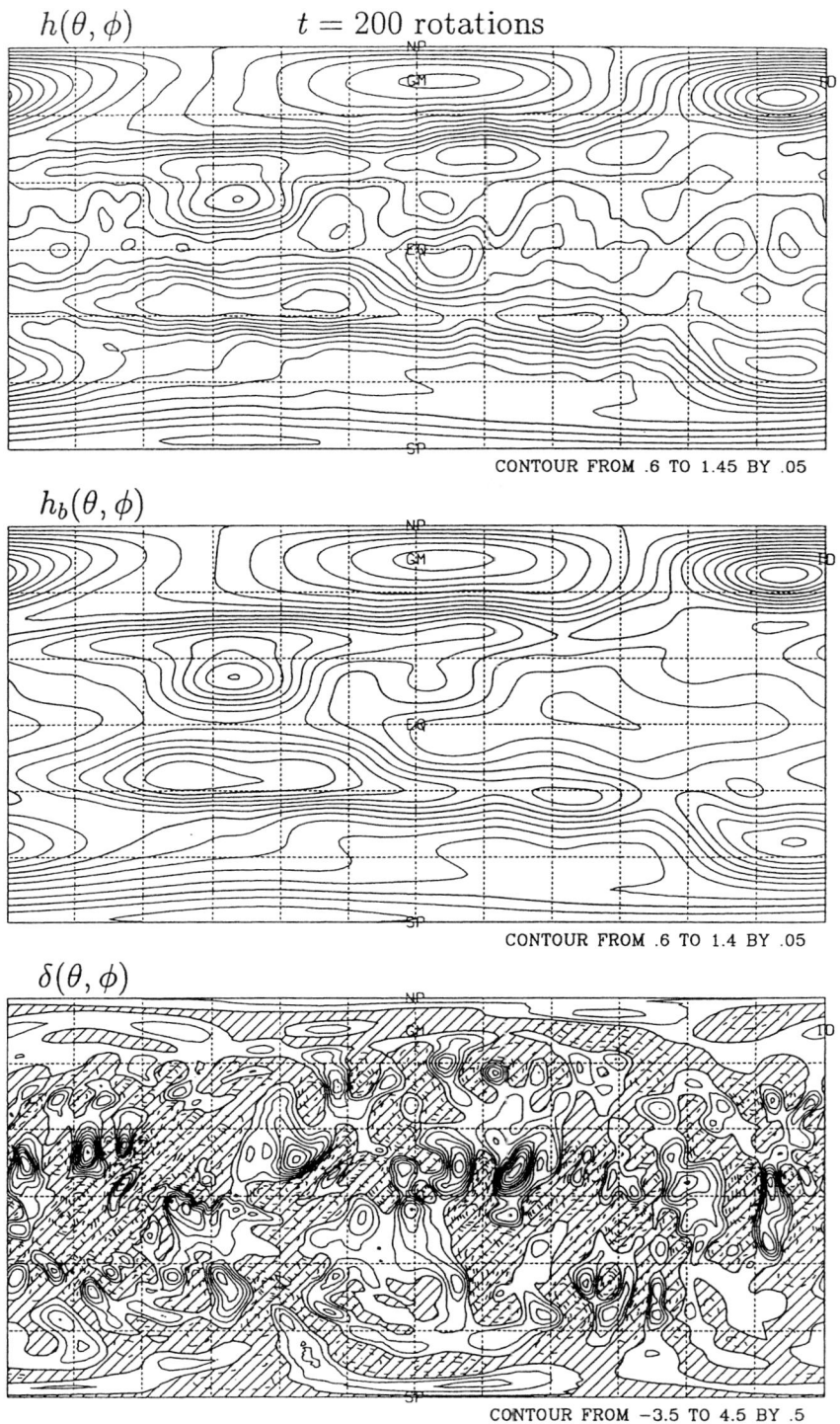

FIGURE 14 Three fields extracted from the shallow-water integration for a Jupiter scale object, respectively the height field h, the (Charney) balanced height field (h_b), and the divergence field δ (a measure of the imbalance in the model). These data correspond to a time 200 planetary rotations after initializing the flow with random (balanced) initial conditions in the vorticity field.

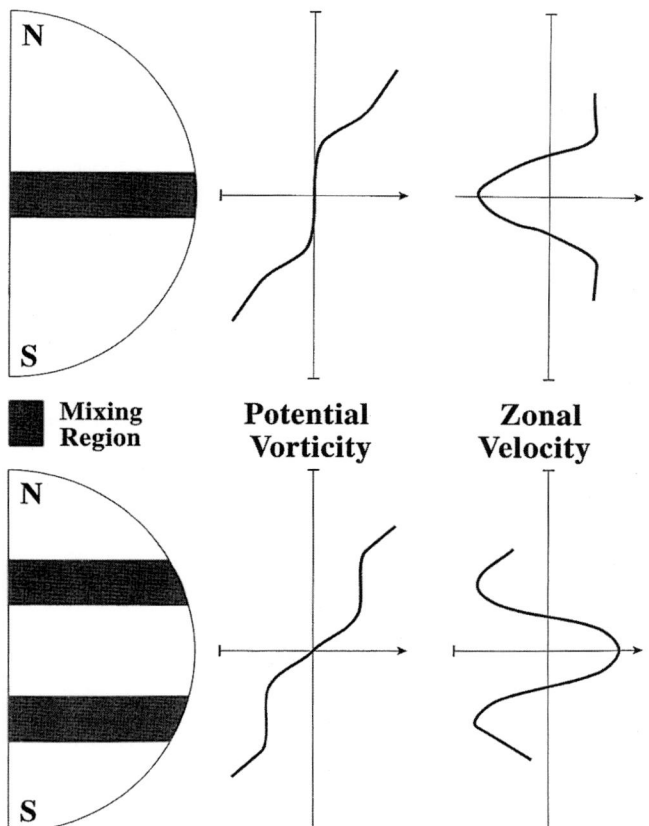

FIGURE 15 Sketch of the dependence of the sign of the equatorial jet in the balanced component of the flow as a function of the latitudinal regions over which the barotropic vorticity field is homogenized by mixing.

by our barotropic PV staircase construction of the latitudinal variation of zonal velocity is important. It is very well understood on theoretical grounds that such a monotonically increasing distribution of PV, from the south pole to the north pole across the disc, is finite-amplitude (Arnol'd, 1965, 1969) stable. This fact clearly explains, in principle, why the zonal flows observed by both Voyager I and II were essentially identical at large scale.

On the basis of the analysis described herein it should be clear that, although the unforced evolution of a shallow-water model from random vorticity initial conditions is successful in "explaining" the banded structure of the zonal flows observed on the discs of the Gas Giant planets, a model of this kind is unsuccessful in explaining what is arguably the most important characteristic of this zonal structure, namely the sign of atmospheric differential rotation in the equatorial region. In order to successfully predict this sign it is clear on the basis of the results described herein that the shallow-water model must be forced so as to homogenize barotropic potential vorticity in two primary zonal bands that are displaced from the equator itself, although some equatorial mixing is also required to explain the observed weakening of the prograde jet in this region. One simple mechanism through which strong forcing of the off-equatorial kind is quite likely to be realized is that which would naturally arise through the interaction of the thermal convection process in the convection zones of Jupiter and Saturn, which underlie the stably stratified weather layer in which the observed zonal flows develop, with the weather layer itself. If the convection process were to be organized into cells consisting of Taylor cylinders aligned with the rotation axis, as envisioned by Busse (1994), then the outcropping of these convection cells in the off-equatorial region would be expected to effect precisely the spatial homogenization of the barotropic vorticity field in the off-equatorial region that is required to explain the dominantly prograde nature of the observed equatorial jets. Since the smaller Gas Giants may not be as convectively active in their interiors as are Jupiter and Saturn, this would also explain why their equatorial jets are retrograde, the unforced shallow-water model therefore providing an appropriate description of their zonal structures.

E. CONCLUSIONS

We have described three different hydrodynamic problems, each of which is of fundamental interest from an atmospheric dynamics perspective, in which the same upscale cascade of energy occurs as that which exerts such profound influence on fluid turbulence in flows restricted to two space dimensions. In each of these problems the flow either had access to all three space dimensions (Parallel shear flows, baroclinic zonal flow) or to 2.5 dimensions (shallow water flow on the sphere). The upscale cascade of energy that may occur in each of these circumstances involves the vortex-pairing interaction that is most clearly describable in the case of parallel shear flow and for which a detailed mathematical theory of the interaction exists which was presented in Section B. There it was shown that vortex merger could occur in principle by either one or the other of two different mechanisms of pairing, namely orbital merger and draining respectively. In parallel flows the mechanism of orbital merger dominates. Whether this interaction actually occurs in nature is strongly dependent upon the density stratification which in general acts so as to suppress this mechanism. In many naturally occurring flows, as opposed to those realized in the laboratory, fast-growing three-dimensional instabilities may induce intense three-dimensional turbulent mixing of the shear layer before the pairing interaction can mature (e.g. see Caulfield and Peltier, 2000, for a detailed recent discussion of "the mixing transition" in stratified parallel flows).

In the discussion of the classical problem of baroclinic instability presented in Section C, we focused upon the fact that balanced baroclinic zonal flows containing weak cross-front gradients in potential vorticity may support subsynoptic scale modes of baroclinic instability with growth rates that are comparable to those of the classical Charney–Eady mode. So long as the flow in which a mode of this kind develops is constrained such that energy cannot cascade to longer zonal wavelength then disturbances of this kind should not (under dry atmospheric conditions) develop sufficiently so as to be important constituents of "weather". However, these waves are subject to a pairing interaction through which their energy level dramatically increases even when horizontal scale is restricted to be considerably less than the scale of maximum growth in the Charney-Eady branch. Yamasaki and Peltier (2001a,b) have recently suggested on this basis that there is a root to deep development that relies on precisely this upscale cascade scenario.

The final example of the upscale cascade discussed herein concerned the understanding of phenomenology arising in the integration of a shallow-water model in spherical geometry in the absence of forcing and from random balanced vorticity initial conditions. In such flows, so long as the initial vorticity distribution is characterized by vortices of dominant scale less than the Rhines scale, the flow initially evolves under the control of the same pairing-mediated upscale cascade of energy as that displayed so clearly in the two-dimensional Cartesian integrations of McWilliams (1984). This cascade is eventually arrested once the vorticity distribution begins to coarsen beyond the Rhines scale, the process of arrest being associated with the increasing influence of Rossby-wave activity. In unforced shallow-water integrations this mechanism of arrest delivers banded zonal flows of the kind observed on the discs of the Gas Giant planets. The equatorial jets in these banded zonal flows are, however, inevitably retrograde, whereas on the largest Gas Giants the equatorial jets are observed to be prograde. Analysis revealed that this flaw in the predicted zonal flows for the Gas Giants is due to the latitudinal distribution of mixing to which the barotropic vorticity field is subject in the unforced shallow-water integration due to the latitudinal distribution of imbalance. Based upon these results we were able to suggest a specific forcing mechanism which, when applied to the shallow-water model through an appropriate parameterization, should suffice to correct what is otherwise a fatal flaw in the shallow-water model of the discs of the largest of the Gas Giant planets. Results of such further analyses will be presented elsewhere.

References

Arnol'd, V. I., 1965: Conditions for nonlinear stability of stationary plane curvilinear flows of an ideal fluid. *Soviet Math.*, **6**, 773–777.

Arnol'd, V. I., 1969: On an a-priori estimate in the theory of hydrodynamic stability. *Am. Math. Soc. Trans.*, **79**, 267–275.

Baros, V. R. and A. Winn-Nielsen, 1974: On quasi-geostrophic turbulence: A numerical experiment. *J. Atmos. Sci.*, **31**, 609–621.

Beebe, R., 1994: Characteristic zonal winds and long lived vortices in the atmospheres of the outer planets, *Chaos*, **4**, 113–127.

Busse, F. H., 1994: A simple model of convection in the Jovian atmosphere, *Icarus*, **29**, 123–133.

Caulfield, C. and W. R. Peltier, 1994: Three dimensionalization of the stratified mixing layer. *Phys. Fluids*, **6**, 3803–3805.

Caulfield, C. and W. R. Peltier, 2000: The anatomy of the mixing transition in homogeneous and stratified parallel flows. *J. Fluid Mech.*, **43**, 1–47.

Charney, J. G., 1955: The use of the primivite equations in numerical prediction. *Tellus*, **7**, 22–35.

Charney, J. G., 1971: Geostrophic turbulence. *J. Atmos. Sci.*, **28**, 1087–1095.

Cho, Y.-K. and L. M. Polvani, 1996: The emergence of jets and vortices in freely evolving shallow water turbulence on the sphere. *Phys. Fluids*, **8**, 1531–1540.

Fjortoft, R., 1953: On the changes in the spectral distribution of kinetic energy for two dimensional non-divergent flow. *Tellus*, **5**, 225–230.

Hua, B. L. and D. B. Haidvogel, 1986: Numerical simulations of the vertical structure of quasi-geostrophic turbulence. *J. Atmos. Sci.*, **43**, 2923–2936.

Kelley, R. E., 1967: On the stability of an inviscid shear layer which is periodic in space and time. *J. Fluid Mech.*, **27**, 657–689.

Klaassen, G. P. and W. R. Peltier, 1985: The onset of turbulence in finite amplitude Kelvin–Helmholtz billows. *J. Fluid Mech.*, **155**, 1–35.

Klaassen, G. P. and W. R. Peltier, 1989: The role of transverse instabilities in the evolution of free shear layers, *J. Fluid Mech.*, **202**, 367–402.

Klaassen, G. P. and W. R. Peltier, 1991: The influence of stratification on secondary instability in free shear layers, *J. Fluid Mech.*, **227**, 71–106.

Kolmogorov, A. N., 1941: The local structure of turbulence in incompressible viscous fluid over very large Reynolds numbers, and dissipation of energy in the locally isotropic turbulence. (English translation: *Proc. R. Soc. Lon. A434*, Turbulence and Stochastic Processes: Kolmogorov's Ideas, 50 Years On, pp. 9–17.)

Lambert, S. J., 1981: A diagnostic study of global energy and enstrophy fluxes and spectra, *Tellus*, **33**, 411–414.

Maltrud, M. E. and G. K. Vallis, 1991: Energy spectra and coherent structures in forced two-dimensional and beta-plane turbulence, *J. Fluid Mech.*, **228**, 321–342.

McWilliams, J. C., 1984: The emergence of isolated coherent vortices in turbulent flow, *J. Fluid Mech.*, **146**, 21–43.

Pierrehumbert, R. T. and S. E. Widnall, 1982: The two and three dimensional instabilities of a spatially periodic shear layer, *J. Fluid Mech.*, **114**, 59–82.

Potylitsin, P. G. and W. R. Peltier, 1998: Stratification effects on the stability of columnar vortices on the f-plane, *J. Fluid Mech.*, **355**, 45–79.

Potylitsin, P. G. and W. R. Peltier, 1999: Three dimensional destabilization of Stuart vortices: the influence of rotation and ellipticity, *J. Fluid Mech.*, **387**, 205–226.

Rhines, P. B., 1975: Waves and turbulence on a β-plane, *J. Fluid Mech.*, **69**, 417–443.

Salmon, R., 1998: *Lectures on Geophysical Fluid Dynamics*, Oxford University Press.

Smyth, W. D. and W. R. Peltier, 1990: Three dimensional primary instabilities of a stratified, dissipative, parallel flow. *Geophys. Astrophys. Fluid Dynam.*, **52**, 249–261.

Smyth, W. D. and W. R. Peltier, 1991: Instability and transition in finite amplitude Kelvin–Helmholtz and Homboe waves, *J. Fluid Mech.*, **228**, 387–415.

Smyth, W. D. and W. R. Peltier, 1993: Two dimensional turbulence in homogeneous and stratified shear layers, *Geophys. Astrophys. Fluid Dynam.*, **69**, 1–32.

Smyth, W. D. and W. R. Peltier, 1994: Three dimensionalization of barotropic vortices on the f-plane, *J. Fluid Mech.*, **265**, 25–64.

Stuart, J. T., 1967: On finite amplitude oscillations in laminar mixing layers, *J. Fluid Mech.*, **29**, 417–440.

Stuhne, G. R., 1999: Classical hydrodynamics on the sphere: Gas giant phenomenology and novel numerical methodology. Ph.D. Dissertation, Department of Physics, University of Toronto.

Stuhne, G. R. and W. R. Peltier, 1996: Vortex erosion and amalgamation in a new model of large scale flow on the sphere, *J. Comput. Phys.*, **121**, 58–81.

Stuhne, G. R. and W. R. Peltier, 1999: New icosahedral grid-point discretizations of the shallow water equations on the sphere. *J. Comput. Phys.*, **148**, 23–55.

Yamasaki, Y. H. and W. R. Peltier, 2001a: On the existence of subsynoptic scale baroclinic instability and the nonlinear evolution of shallow disturbances, *J. Atmos. Sci.*, **58**, 657–683.

Yamasaki, Y. H. and W. R. Peltier, 2001b: Baroclinic instability in an Euler equations based column model: the coexistence of a deep synoptic scale mode and a shallow subsynoptic scale mode, *J. Atmos, Sci.*, **58**, 780–792.

The Role of Water in Atmospheric Dynamics and Climate

Kerry Emanuel

Program in Atmospheres, Oceans, and Climate, Massachusetts Institute of Technology, Cambridge, MA, USA

From its beginnings nearly 2000 years ago until the end of the nineteenth century, the science of meteorology focused strongly on the role of water and its change of phase; indeed the very word "meteorology" betrays this focus. As synoptic observation networks began to reveal rotary weather systems and fronts, however, the conceptual picture of large-scale atmospheric dynamics changed dramatically to one in which water and its change of phase were considered embellishments on a world view of dry, adiabatic dynamics. This view persists today, even to some degree in the Tropics, where phase change of water is a dominant physical process. After reviewing this historical progression, I will highlight both the advances made possible by this dry, adiabatic world view, and the roadblocks to a more complete understanding that this has created. The chapter concludes with a discussion of the role of water in the climate system, and the need for scientists better trained in thermodynamics to solve many outstanding problems in weather and climate.

A. HISTORICAL OVERVIEW*

To the average person, the word "weather" entails the variation of wind, temperature, clouds and precipitation. Almost all textbooks on atmospheric science state that "weather" is confined to the troposphere because, although winds, pressure and temperature vary considerably above the tropopause, there are essentially no clouds or precipitation there. Water is clearly central to the notion of weather.

The earliest natural philosophers were preoccupied with explaining clouds and precipitation, which are of paramount significance in any agrarian economy. Thales, whom tradition regards as the first known philosopher of the natural world, lived in the seventh century BC, and held that water is the central element that lies at the basis of all creation, from the sky and the earth to all living things. It took more than 200 years for Greek philosophers to add earth, fire and air to the list of central elements. (It was Thales' follower, Anaximander, who first ascribed wind to the flow of air, a fact not accepted for centuries and still regarded with suspicion by many entering graduate students.) Anaxagoras (499–427 BC) sought to explain why hail often occurs in summer, in spite of high surface temperatures. From experience on mountains, he observed that temperature decreases with altitude and deduced that hail must form high up within clouds. Noting that air and debris tend to shoot upward above fires, Anaxagoras speculated that tall clouds result from vigorous heating of air near the ground, showing the value of logical inference from observation, even in the complete absence of any guiding theoretical principles. But theory would soon have its day in the sun.

In his famous treatise *Meteorologica*, written around 340 BC, Aristotle focused on the natural causes of clouds, rain, floods and thunder. Much like today's global-warming naysayers, Aristotle took much delight in attacking accepted wisdom. Using a somewhat convoluted argument, he categorically rejected the idea that wind is moving air and spent considerable effort debunking Anaxagoras's theory that hail forms high up within clouds. If hailstones formed so high within clouds, Aristotle argued, then surely their fall through so much air would render them perfect spheres by the time they reached the ground. Since this was clearly not the case, hail must therefore form near the ground. How to explain that hail is made from ice, which even then was generally recognized as being cold? Why, it was because water falling through the air must compress and thereby turn into ice. This was perhaps the first instance (but by no means the last) of what happens when the mechanistically inclined try to do thermodynamics. By denying that wind is moving air and

---

* In this overview, I have relied heavily on the historical work of Frisinger (1977) and Kutzbach (1979).

that hail forms above the freezing level, Aristotle, the theoretician, set back meteorology by more than 2000 years.

As global exploration and trade became increasingly important during the sixteenth and seventeenth centuries in western Europe, the global distribution of wind became a natural subject of attention among natural philosophers. It is perhaps not surprising that the well-known treatises on the origin of the trade winds by Edmund Halley and George Hadley, at the beginning of the eighteenth century, occurred near the prime of the Age of Sail. This, together with the publication of the Euler equations in 1755 and their extension to the case of motion on a rotating planet in 1860, marked the early advent of dynamical meteorology. But the problem of the origin of cyclones served to refocus attention on the importance of thermodynamics in general, and of phase change of water in particular.

The characterization of cyclones as rotary storms was well in place by the middle of the eighteenth century, owing mostly to the observational work of Benjamin Franklin and Elias Loomis. In an attempt to explain their physical origin, the American scientist James Pollard Espy began development of what was to be called the thermal theory of cyclones, beginning in the 1830s. Observing that cyclones are invariably associated with clouds and precipitation, Espy came to believe that such storms are powered by the liberation of the latent heat of vaporization, and undertook extensive laboratory experiments designed to determine what we now refer to as the moist adiabatic lapse rate. His work is particularly remarkable when it is recognized that the formulation of the science of thermodynamics took place during the 1840s and 1850s, and the first law of thermodynamics was not introduced to meteorology until the 1860s. But already in 1841, Espy stated that

> "When the air near the surface of the earth becomes more heated or more highly charged with aqueous vapor, its equilibrium is unstable, and up-moving columns or streams will be formed. As the columns rise, their upper parts will come under less pressure, and the air will therefore expand; as it expands, it will grow colder about one degree and a quarter for every one hundred yards of its ascent. The ascending columns will carry up with them the aqueous vapor which they contain, and, if they rise high enough, the cold produced by expansion from diminished pressure will condense some of its vapor into cloud. A soon as cloud begins to form, the caloric of elasticity of the vapor or steam is given out into the air in contact with the little particles of water formed by the condensation of the vapor. This will prevent the air, in its further progress upwards, from cooling so fast as it did up to that point; that is, about five-eighths of a degree for one hundred yards of ascent, when the dew point is about 70 degrees." (Espy, 1841)

Espy went on to deduce that if the environmental lapse rate is greater than the moist adiabatic lapse rate, convection may occur, and he believed that it was such convection that led to the formation of middle-latitude cyclones. He had an important ally in Elias Loomis, who in 1843 wrote that

> "The heat liberated in the formation of this cloud raises the thermometer, causing a more decided tendency of the air inward toward the region of condensation.... Relative elevation of temperature under the cloud gives increased velocity to the inward current of air. More cloud is thus formed, heat liberated.... Thus the storm gains violence by its own action." (Loomis, 1843)

This proposed explanation for the origin of middle-latitude cyclones, which depended entirely on latent heat release, bears an uncanny resemblance to the theory of Conditional Instability of the Second Kind (CISK), developed 120 years later by Charney and Eliassen (1964) to explain the development of tropical cyclones:

> "We should consider [the depression and the cumulus cell] as supporting one another—the cumulus cell by supplying the heat for driving the depression, and the depression by producing the low-level convergence of moisture into the cumulus cell."

Although Espy went too far in relying on latent heating as a power source for extratropical cyclones, his work on moist convection marked the first progress on this topic since that of Anaxagoras, more than 2200 years earlier and anticipated by more than a century a prominent theory of tropical cyclones.

By the late 1860s, it was generally accepted that the heat liberated when water vapor condenses is the major source of power for extratropical cyclones. For example, in 1868 the Scottish scientist Alexander Buchan wrote that

> "The chief disturbing influences at work in the atmosphere are the forces called into play by its aqueous vapour." (Buchan, 1868)

Clearly, water played a central role in conceptions of atmospheric physics through much of the ninteenth century. But beginning in the mid 1870s new observations began to cast the thermal theory into doubt. In 1874, Julius Hann, later the director of the Austrian weather service, pointed out that heavy tropical rains appeared to have no effect on the surface pressure, a fact also noted in the next century by Charney and Eliassen (1964). Loomis (1877), in stark contrast with his earlier beliefs, stated that

> "It seems safe to conclude that rainfall is not essential to the formation of areas of low barometer, and is not the principal cause of their formation or their progressive motion".

But at this time, scepticism about the thermal theory was not widespread, and attempts were made to reconcile the new observations with the theory. In 1874,

Hann pointed out that latent heat release must lead to inflow, but if that inflow were not given a vertical motion, the pressure would not fall. Here was perhaps the first hint that latent heating, as it is an *internal* process, need not be associated with the generation of kinetic energy, a fact that is difficult for some to accept even today.

The desiccation of atmospheric science began in earnest as a result of more detailed observations of extratropical cyclones, undertaken in Europe in the 1880s and 1890s. These observations revealed that extratropical cyclones are always associated with substantial horizontal temperature gradients. The fact that such gradients are associated with a reservoir of potential energy was established by the French scientist Margules at about this time. These developments paved the way to the 1916 work of Vilhelm Bjerknes, who is credited with the idea that extratropical cyclones result from the dynamical instability of the baroclinic currents found in middle latitudes. Bjerknes' work marked a transition away from a preoccupation with energetics and toward a focus on stability.

More than any other development, it was the application of linear stability theory to the problem of baroclinic instability that initiated a long period during which the effects of water in atmospheric dynamics were virtually ignored. In linear formulations, it is awkward, at best, to include the strong irreversibility of the process of precipitation, which removes water from the system and thereby causes descending air to be unsaturated. In his seminal paper on baroclinic instability, Charney (1947) argued that

> "As long as one is concerned with waves of small enough amplitude, the vertical motions will not be of sufficient magnitude to cause condensation, so that this factor may also be ignored."

Charney here appears to argue not that condensation is unimportant, but that it can be ignored during the early stages of baroclinic development. Nevertheless, he recognized that "condensation can produce appreciable errors" in calculations of baroclinic growth rates and structure. Eady (1949) noted that baroclinic growth rates calculated by ignoring condensation are too small. In a widely ignored portion of his otherwise well-known paper, Eady included the effects of a pre-existing, zonally oriented band of saturated air in his linear analysis of baroclinic growth, and noted the strong tendency toward frontogenesis at the boundaries of the cloudy zone as well as an increase in the rate of growth.

As is often true in science, it is the neglect of secondary effects that permits a focus on the essence of a phenomenon and leads to true progress in understanding it. This was clearly the case in the development of the theory of baroclinic instability, which led to the first coherent and physically consistent view of extratropical dynamics. By neglecting the secondary effects of condensation and friction, one can view the dynamics in terms of the evolution of the adiabatic invariants potential temperature and potential vorticity, as nicely summarized in the work of Hoskins *et al.* (1985).

A casual survey of the literature published during the 1960s and 1970s reveals far more concern with the effect of nonlinearity associated with the finite amplitude of real baroclinic systems than with irreversible processes such as phase change of water and friction. It was not until the 1980s that extensive attempts were made to incorporate these effects in idealized models and theory.

Water in its various phases plays a critical role in the climate system. As early as the late eighteenth century, Jean Baptiste-Joseph Fourier described the warmth of the earth's surface in terms of a "hothouse" effect, and by the middle of the nineteenth century, the Englishman John Tyndall recognized that much of the absorption of infrared radiation in the atmosphere is caused by the presence of water vapour. The central importance of water vapour in the greenhouse effect was established quantitatively by Svante Arrhenius (1896), using primitive measurements of absorption coefficients. In the twentieth century, it came to be realized that clouds are responsible for much of the reflection of solar radiation by the earth system, as well as being strong absorbers of infrared radiation.

B. WATER IN ATMOSPHERIC DYNAMICS

1. Zonally Symmetrical Equilibrium States

There is a long tradition in dynamical meteorology of considering the evolution of small perturbations to an equilibrium state. In their classical treatises on baroclinic instability, Charney (1947) and Eady (1949) took the equilibrium states to be zonally oriented flow with vertical shear, satisfying hydrostatic and geostrophic (and therefore thermal wind) balance.

It is interesting to note that while such states satisfy the equations of motion, they are not in thermodynamic equilibrium. In order to satisfy thermodynamic as well as dynamic balance, such states would have to be in radiative equilibrium. We know from detailed calculations of radiative equilibrium, holding water vapour and clouds at their climatological values, that most of the lower troposphere would have super-adiabatic lapse rates. Besides being unphysical, such states would render untenable the mathematical problem of linear baroclinic instability.

A more practical approach is to allow what we are calling the equilibrium state to contain small-scale convection and then consider the stability of such states to macromotion on scales much larger than those characterizing the convective eddies. Thus we may consider the equilibrium state to be one of *radiative–convective* equilibrium. The first detailed calculations of radiative–convective equilibrium that included moist convection were described by Manabe *et al.* (1965). Such states have nearly moist adiabatic lapse rates in the troposphere and are thus quite stable to dry adiabatic displacements. These states are not only baroclinically unstable, but unstable to slantwise moist convection everywhere that the horizontal temperature gradient is nonzero.

Slantwise moist convection has time and space scales that are intermediate between the scales of upright moist convection and of baroclinic instability (Emanuel, 1983). If we continue to be interested in the stability of equilibrium states to baroclinic motion, then we may also regard the slantwise convection in the same way that we handled upright moist convection: we assume that the convection adjusts the macroflow to nearly neutral equilibrium on timescales small compared to the timescales over which radiation, surface fluxes and synoptic-scale flows destabilize the atmosphere to slantwise convection. (This is not as good an assumption as in the case of vertical convection, and may break down in the ascent regions of strong extratropical cyclones.) Thus we may regard the equilibrium state as one of *radiative-slantwise convective* equilibrium. That the baroclinic atmosphere is close to such a state is shown in the observational analyses by Emanuel (1988). Note that in strongly baroclinic regions, such states are decidedly stable to purely vertical displacements.

Given the latitudinal distribution of moist entropy in the boundary layer together with the stipulation that the zonal winds vanish at the surface, the thermal and kinematic structure of the radiative-slantwise convective equilibrium state is completely determined. If the tropopause is determined by matching the troposphere to the radiative equilibrium of the stratosphere, then the tropopause height is also completely determined. Given the observed, zonally averaged distribution of moist entropy at the surface, the tropopause height, so determined, is very close to the observed height, as shown in Fig. 1. The sharp descent of the tropopause across the jet is captured, as well as the gradual decline in tropopause height toward the pole.

The dynamically inclined like to state that the height of the tropopause outside the tropics is determined by dynamics (e.g. see Lindzen, 1993). This sentiment is based in part on the false idea that moist convection is confined to the tropics. In fact, middle and high latitudes experience a great deal of deep convection, over the continents in summer and over the oceans in winter. The residence time of air over the oceans in winter or the continents in summer is more than sufficient for convection to establish moist adiabatic lapse rates (along angular momentum surfaces, as required by the theory of slantwise moist convection), while the transit of air over continents in winter or oceans in summer is not so long that radiative cooling can appreciably change the lapse rates, except near the surface. Thus the only places where one finds appreciably submoist adiabatic lapse rates (along angular momentum surfaces) are polar regions and boundary layers over continents in winter or oceans in summer.

2. Baroclinic Instability

In assessing the stability of the equilibrium zonal flow, to be consistent, one should consider flows that satisfy thermal wind balance *and* radiative–convective equilibrium. It is formally inconsistent to neglect phase change of water in the linear stability problem, when the flow whose stability is being assessed is itself convecting. Thus, the classical problem of baroclinic instability is, in an important sense, ill-posed. The static stability is taken to be the observed stability, which is established by moist convection, but phase changes of water are neglected in the perturbations.

The problem of the stability of convecting atmospheres to motions on scales much larger than those characterizing the spacing between cumulus clouds was

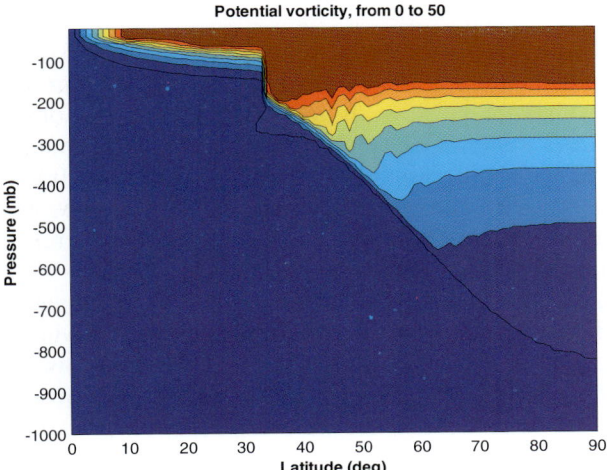

FIGURE 1 Potential vorticity in a radiative-slantwise convective equilibrium state, with an assumed isothermal stratosphere. The troposphere is assumed to be neutral to slantwise moist convection and to have vanishing zonal wind at the surface; the wind distribution above the surface is determined from the thermal wind equation.

addressed by Emanuel *et al.* (1994). They used as a starting point the observation that ascending regions are moister than descending regions, so that not all the moisture convergence into ascending regions is compensated by precipitation; some must be used to moisten the atmosphere. Thus not all of the adiabatic cooling associated with large-scale ascent can be compensated by latent heating and there is net cooling of ascending air. Emanuel et al. (1994) argued on this basis that to a first order of approximation, the dynamics of large-scale motions in convecting atmospheres are the same as those of dry atmospheres, but with a greatly reduced effective static stability. When applied to baroclinic instability, this would give horizontal scales less than, and growth rates greater than the classical values by about a factor of 5.

The problem with this approach is that observations show rather clearly that waves of even very small amplitude have sufficient downward motion in their descent phases to utterly wipe out convection there. (An interesting exception may be the case of a certain variety of "polar lows", which appear to be examples of baroclinic instability in strongly convecting atmospheres, usually in very cold air flowing out over relatively warm ocean or lake waters. In these cases, deep convection can often be found in all quadrants of the disturbances.) Thus, in this particular sense, the waves become nonlinear almost immediately. One way of dealing with this problem is to assume that the ascent regions of the growing disturbances are saturated, or convectively adjusted, while the descent regions are subsaturated. Formally, one has to assume in the "linear" models that this moist-up, dry-down condition is the *only* important nonlinearity in the system during its early growth. This approach was taken by Emanuel (1985) in an analytical model of frontal circulations, by Thorpe and Emanuel (1985) in nonlinear models of frontogenesis, and to linear and nonlinear baroclinic instability by Emanuel *et al.* (1987) and Fantini (1993). The first three of these works used semi-geostrophic models, for which the potential vorticity plays the same mathematical role as static stability in quasi-geostrophic models. Slantwise neutrality implies zero effective potential vorticity, so in these works the potential vorticity was taken to be very small in the ascent regions of the waves and fronts.

The main conclusion of these studies is that while condensation in the ascent regions of the waves increases frontogenesis and the growth rate of baroclinic waves (the latter by as much as a factor of 2), the main effect is on the structure of the waves, with the ascent regions generally confined to thin, sloping sheets, in contrast with broad regions of descent. The overall horizontal wavelength of baroclinic disturbances is reduced by roughly a factor of 2, and fronts form much more quickly at the lower than at the upper boundary when condensation is accounted for. Later work by Fantini (1993) showed that the sensitivity of growth rates to very small moist potential vorticity is somewhat exaggerated by semi-geostrophic models; when the primitive equations are used the increase in growth rates by condensation are even more modest. Nevertheless, the structural differences remain, and Fantini (1995) showed that condensation favours relatively stronger warm fronts.

Some aspects of the asymmetry introduced by condensation in baroclinic waves are illustrated in Fig. 2, which shows the distribution of vertical motion in developing baroclinic waves when the effective static stability is reduced by 90% in ascending regions of the disturbances. Note the concentration of upward motion into relatively thin bands and small patches surrounding the surface low pressure areas.

The complex and important role of water in baroclinic disturbances is nicely illustrated by the observational studies conducted by Browning (1986, 1990, 1994) and Parsons and Hobbs (1983), amongst many others. It is clear that, in addition to contracting the main regions of ascent associated with the fronts and the cyclones, the presence of moisture gives rise to a rich spectrum of mesoscale phenomena, including mesoscale precipitation bands and squall lines. These mesoscale phenomena constitute a large measure of what the average person considers to be "weather" and pose at least as great a challenge to forecasters as prediction of the surface pressure field did a generation ago.

3. The Tropics

Compared to the extratropical atmosphere, progress in understanding and predicting tropical weather systems has been slow. In contrast with extratropical weather systems, most tropical systems are dominated by diabatic processes such as condensation, radiation, and surface fluxes. The training of graduate students, which in many schools is dominated by instruction in adiabatic, quasi-geostrophic dynamics, leaves them largely unprepared to tackle the Tropics, for which a solid grounding in moist thermodynamics is essential. Progress in this century will be made by those who do not shrink from acquiring comprehensive backgrounds in both dynamics and thermodynamics.

Moist convection is a turbulent, chaotic process, with characteristic eddy turnover times on the order of an hour. For this reason, the predictability timescale of convection on the scale of the individual convective cell is no more than a few hours. As in the case of classical turbulence, there is some hope that larger-scale

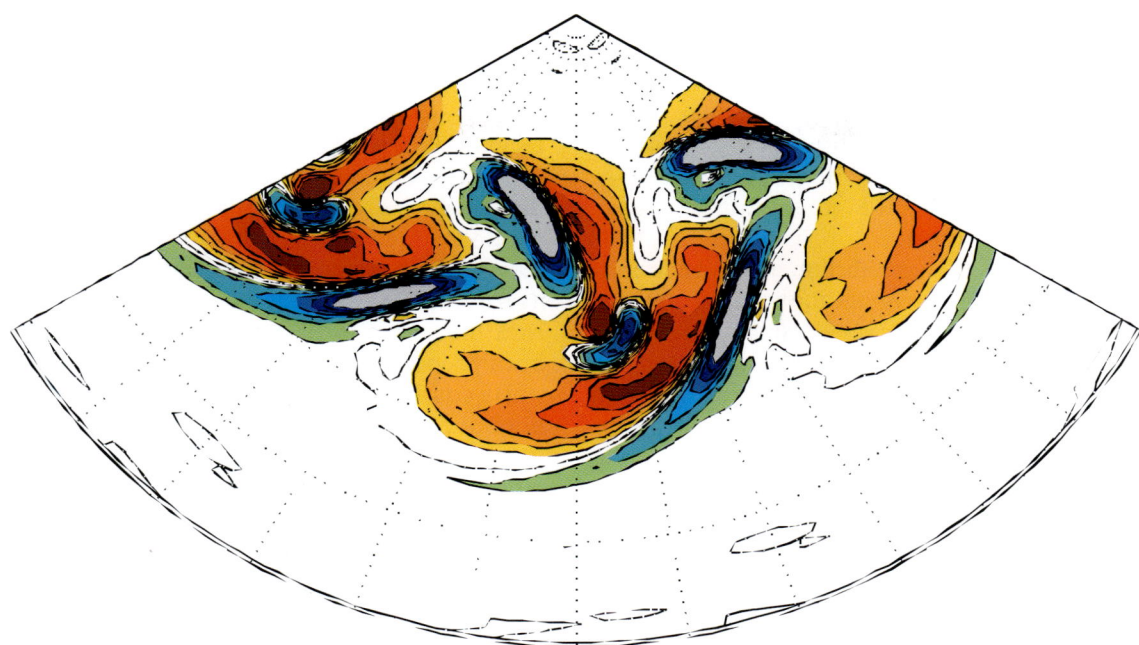

FIGURE 2 Distribution of vertical velocity over a third of the Northern Hemisphere in a primitive equation model simulation, in which condensation is represented by applying a reduced static stability to upward moving air. Cool colours represent upward motion; warm colours represent downward motion. (Model designed by and figure courtesy of Tieh Yong Koh.)

phenomena that interact with the turbulence can be predicted on longer timescales, if there is a well-defined relationship between the macroflow and the statistics of the turbulence. Part of the historical problem in treating the Tropics can be traced to a failure to distinguish between the behaviour of individual convective elements and that of large-scale systems that interact with whole ensembles of convective elements.

This problem of predictability is nicely illustrated by the case of radiative–convective equilibrium. In recent years, it has become possible to simulate such states explicitly, with three-dimensional, nonhydrostatic models that simulate the convection explicitly, albeit with marginally adequate horizontal resolution. In one such experiment, the radiative cooling of the troposphere is simply specified, and classical bulk aerodynamic formulae are used to calculate fluxes of heat, moisture and momentum from a sea surface of fixed temperature. The simulation is conducted in a doubly periodic domain of horizontal dimensions on the order of 100 km, so as to encompass many individual convective cells. Since the radiative cooling is specified, and since in equilibrium the domain average latent heating must equal the radiative cooling, the statistically averaged amount of convection is specified in this experiment. Thus, for example, the amount of precipitation integrated over the whole domain and averaged in time, is constant. If one considers space–time subdomains of the precipitation, on the other hand, there is considerable fluctuation of the precipitation. Figure 3 shows the ratio of the standard deviation of precipitation to the mean precipitation in such subdomains as a function of their size and length of averaging time. It would appear

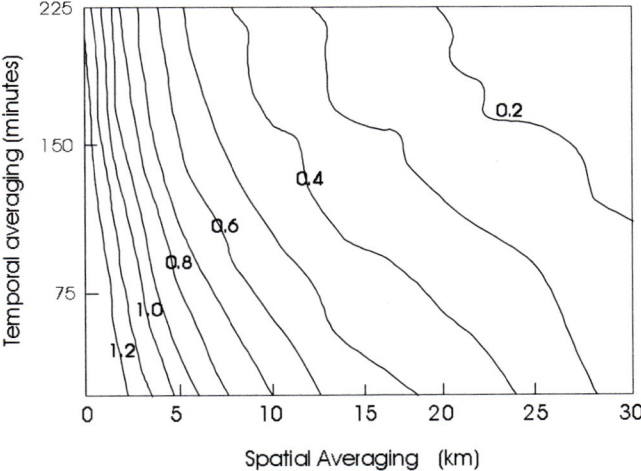

FIGURE 3 Ratio of the variance to the domain average of the precipitation in a three-dimensional numerical simulation of radiative–convective equilibrium over a water surface, as a function of space–time averaging. The ordinate is the length of time averaging; the abscissa is the length of averaging in space. This ratio asymptotes to $\sqrt{2}$ for short averaging intervals. (From Islam et al., 1993.)

from this experiment that to achieve a stable statistical relationship between the convection and the large-scale forcing (in this case, just the radiative cooling), one has to average over at least 30 km and several hours. Thus we may hope that for weather systems with space and timescales larger than these, there may be a stable statistical relationship between the convection and the weather system in question.

The radiative–convective equilibrium state is a natural basic state to use for thinking about the tropical atmosphere, and which to use as a basic state in linear perturbation analysis. The failure to recognize this and to regard the tropical atmosphere as basically dry with moist perturbations led to such historical anomalies as the theory of conditional instability of the second kind (CISK).

The problem of the behaviour of even linear perturbations to radiative–convective equilibrium is far from solved. Computer speeds and capacities are only just now becoming large enough to simulate radiative–convective equilibrium over large enough domains to examine the behaviour of synoptic-scale tropical disturbances while still explicitly simulating cumulus clouds. As mentioned in the preceding section, this problem was looked at from a theoretical standpoint by Emanuel *et al.* (1994), who argued that linear perturbations to radiative–convective equilibrium should behave very much like adiabatic linear perturbations, except that they would experience an effective static stability about an order-of-magnitude less than that of their adiabatic counterparts, and that the perturbations would be damped in proportion to their frequency, owing to the finite timescale over which convective clouds respond to changes in their environment. More recently, Wheeler and Kiladis (1999) have shown that observed anomalies of outgoing longwave radiation in the equatorial region indeed have dispersion characteristics very much like adiabatic modes of the equatorial waveguide, but with greatly reduced equivalent depths, as shown in Fig. 4. It remains to identify the mechanisms that serve to excite such modes, and many have been proposed in the literature.

By considering moist convection as a statistical equilibrium process, latent heating must be regarded as an internal process and large-scale circulations must be regarded as arising from dynamical processes and from thermodynamic energy input from the lower boundary or from interactions amongst clouds, water vapour and radiation. One must regard the conversion of thermodynamic to mechanical energy in terms of sources and sinks of the moist entropy, not the conversion between latent and sensible forms of enthalpy. Thus, tropical cyclones are driven by the wind-induced transfer of moist enthalpy from the sea surface, not by the release of latent heat in cumulus clouds, which is no more than a transfer process. (By analogy, automobiles are driven by engines, not by drive shafts, even though the latter are essential components for transferring motive power to the wheels.)

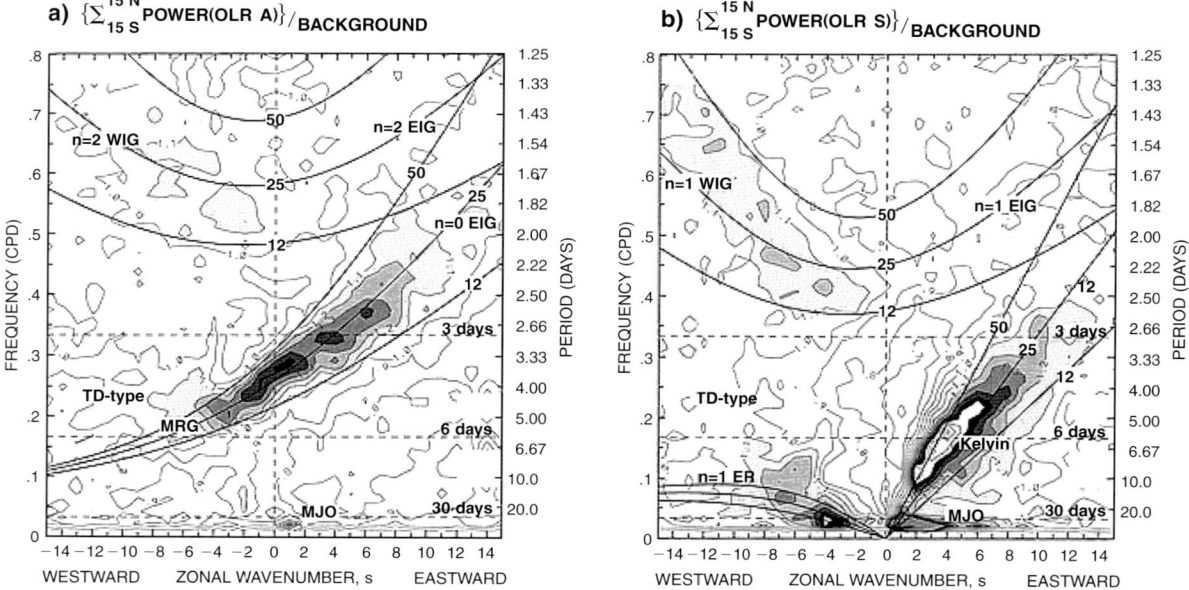

FIGURE 4 Spectra in the wavenumber–frequency domain of satellite-measure outgoing longwave radiation measured over a period of several years in a latitude belt extending from 5°S to 5°N. A background red-noise spectrum has been subtracted from these fields. (a) Asymmetrical modes. (b) Components that are symmetrical about the equator. The solid curves represent the dispersion curves of equatorially trapped waves, labelled by their equivalent depths. (From Wheeler and Kiladis, 1999.)

Clearly, water plays an essential role in tropical meteorology, and many tropical weather systems are gradually becoming better understood. Yet there remain fundamental problems. For example, there exists no theory for the Madden–Julian Oscillation that is general accepted and can account for all of its observed characteristics. Perhaps even more troubling, we lack a theory for moist convective spacing and vertical velocity scales, even in radiative–convective equilibrium states. (Theories have been proposed by Rennó and Ingersoll (1996) and Emanuel and Bister (1996), but these were based on the assumption that the main irreversible entropy source in radiative–convective equilibrium is mechanical dissipation. This was shown to be a bad assumption by Pauluis and Held (personal communication, 2000), thus leaving us without a viable theory for moist convective scaling.) We do not understand nor can we predict with any real skill the genesis of tropical cyclones. Are tropical cloud clusters examples of self-organized convection, as in the case of squall lines, or do they represent the ascent phase of tropical waves? Problems like these make the Tropics the last great frontier in meteorology.

C. WATER IN THE CLIMATE SYSTEM

Perhaps the most mysterious role of water in the earth system is its regulation of climate. Evaporation of water from the surface keeps the average surface temperature far cooler than it would otherwise be, but water vapour is also the most important greenhouse gas in the atmosphere and condensed water in the form of clouds is also an important absorber of longwave radiation. On the other hand, much of the earth's albedo is owing to clouds. The difficulty of correctly handling water in its various phases is probably the main source of uncertainty in model-based predictions of global warming.

Figure 5 illustrates the nature of the problem. This shows the distribution of clouds and of upper tropospheric water vapour as revealed by satellite-sensed radiation in the 6.7 μm band. Note the almost bimodal distribution of water in this picture: there are very dry areas interleaved with very wet areas, with sharp gradients separating the two. The dry areas denote broadly descending air; they are dry because precipitation has removed most (but not all) of the water from ascending air.

It is not, in fact, difficult to replicate this first-order characteristic of the water distribution: wet ascent and dry descent. Simple advective–diffusive models will suffice; it has even been shown that a simple model that does no more than remove supersaturation replicates

FIGURE 5 Upper tropospheric humidity patterns on 27 April, 1995, at 1745 UTC as measured by the GOES 8 satellite. The 6.7 μm data are related to water vapour, with a peak signal between about 550 and 350 mb. The data are influenced by high clouds. (Image courtesy of A. Gruber and R. Achutuni, NOAA National Environmental Satellite Data and Information Service.)

the moist–dry patterns seen in Fig. 5. But this seemingly simple solution masks the crux of the problem: How does water respond to changes in global temperature? Here we are looking for comparatively subtle, but climatically important, changes in the humidity of the dry and moist regions and in the relative areas occupied by each.

It is not difficult to see that the sensitivity of the water distribution to global climate change depends crucially on cloud microphysical processes. For example, if precipitation formation were microphysically impossible, then the atmosphere would be saturated everywhere and filled with cloud. The amount of water in the actual atmosphere depends on just how efficient the removal of water by precipitation is, as argued by Emanuel (1991) and demonstrated conclusively by Rennó et al. (1994). While the spatial distribution of water in the atmosphere is strongly affected by large-scale circulation, such circulation cannot

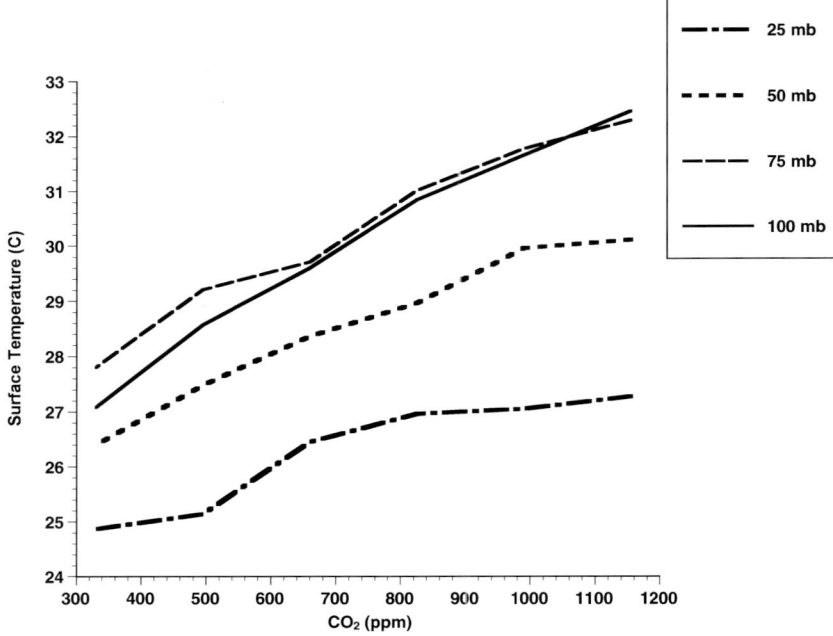

FIGURE 6 Dependence of surface temperature in radiative–convective equilibrium on atmospheric CO_2 content, for four different vertical resolutions, using the convection scheme described by Emanuel and Zivkovic-Rothman (1999) and the radiation scheme of Chou et al. (1991).

remove the sensitivity of the overall water content to cloud microphysics and turbulence. However, this sensitivity *is* effectively removed by the expedient of having insufficient vertical resolution in models, as demonstrated by Emanuel and Zivkovic-Rothman (1999), who showed that numerical resolution of water vapour requires at least 50 mb resolution through the troposphere, and that microphysical sensitivity declines with vertical resolution. Today's climate models do not achieve this resolution, and their water vapour feedback is thus similar among all the models (Cess et al., 1990) and insensitive to the way cumulus clouds are treated. Figure 6 shows the tropical climate sensitivity of radiative–convective equilibrium in a model using the microphysically based convective scheme of Emanuel and Zivkovic-Rothman, as a function of vertical resolution. The relatively low sensitivity seen in the high-resolution model is caused by the strong dependence of stochastic coalescence on adiabatic water content, which is in turn a strong function of temperature. Thus the warmer the climate, the greater the fraction of condensed water that is converted to precipitation. On this basis, one would expect the relative humidity of the upper troposphere in regions of large-scale ascent to decline with increasing temperature.

It is simply not credible that models that cannot replicate the observed water vapour distribution can none-the-less be expected to reproduce the correct distribution of, and climate sensitivity to, clouds. That climate models fail to replicate basic aspects of the water vapour and cloud distributions has been demonstrated by Sun and Oort (1995) and Lindzen et al. (2000).

Progress toward understanding climate today is impeded by a culture of climate modelling that has increasingly emphasized a single measure of success, the simulation of global mean temperature, over the sceptical attention to failures that has traditionally served as a critical component of scientific progress. One can only hope that the barrier that now exists between climate modellers and cloud physicists will gradually erode as the intricacies of the role of water in climate become more apparent.

D. SUMMARY

The critical role of water in weather and climate was a focus of natural science from the time of the ancient Greeks through the end of the nineteenth century. This role was temporarily de-emphasized by the success of adiabatic theories in explaining the basic behaviour of middle-latitude weather systems, beginning in the 1940s, but has enjoyed a resurgence of interest in recent decades owing to the obvious importance of clouds and precipitation in the characterization of weather, but also because of the important effect of latent heating on the dynamics and structure of weather systems. The marriage of research in cloud physics to numerical weather prediction is a happy one, producing increasingly fruitful results in the form of better weather forecasts. We are now just beginning to see the benefits of accounting for cloud physics in understanding and modelling climate.

References

Arrhenius, S., 1896: On the influence of carbonic acid in the air upon the temperature of the ground. *Phil. Mag.*, **41,** 237–276.

Browning, K. A., 1986: Conceptual models of precipitation systems. *Weath.. Forecast.* **1,** 23–41.

Browning, K. A., 1990: Organization and internal structure of synoptic and mesoscale precipitation systems in midlatitudes. In *Radar in Meteorology: Battan Memorial and 40th Anniversary Radar Meteorology Conference.* (D. Atlas, Ed.) Boston, MA, American Meteorological Society, pp. 433–460.

Browning, K. A., 1994: Structure of a frontal cyclone. *Q. J. R. Meteorol. Soc.*, **120,** 1535–1557.

Buchan, A., 1868: *A Handybook of Meteorology.* Blackwood Press, Edinburgh.

Cess, R. D., and 30 co-authors, 1990: Intercomparison and interpretation of climate feedback processes in 19 atmospheric general circulation models. *J. Geophys. Res.*, **95,** 16601–16615.

Chan, J. C.-L., and R. T. Williams, 1987: Analytical and numerical studies of the beta-effect in tropical cyclone motion. *J. Atmos. Sci.*, **44,** 1257–1264.

Charney, J. G., 1947: The dynamics of long waves in a westerly baroclinic current. *J. Meteor.*, **4,** 135–163.

Charney, J. G. and A. Eliassen, 1964: On the growth of the hurricane depression. *J. Atmos. Sci.*, **21,** 68–75.

Chou, M.-D., D. P. Krats and W. Ridgway. 1991: Infrared radiation parameterization in numerical climate models. *J. Climate*, **4,** 424–437.

Eady, E. T., 1949: Long waves and cyclone waves. *Tellus*, **1,** 33–52.

Emanuel, K. A., 1983: The Lagrangian parcel dynamics of moist symmetric instability. *J. Atmos. Sci.*, **40,** 2368–2376.

Emanuel, K. A., 1985: Frontal circulations in the presence of small moist symmetric stability. *J. Atmos. Sci.*, **42,** 1062–1071.

Emanuel, K. A., 1988: Observational evidence of slantwise convective adjustment. *Mon. Wealth. Rev.*, **116,** 1805–1816.

Emanuel, K. A., 1991: A scheme for representing cumulus convection in large-scale models. *J. Atmos. Sci.*, **48,** 2313–2335.

Emanuel, K. A. and M. Bister, 1996: Moist convective velocity and buoyancy scales. *J. Atmos. Sci.*, **53,** 3276–3285.

Emanuel, K. A. and M. Zivkovic-Rothman, 1999: Development and evaluation of a convection scheme for use in climate models. *J. Atmos. Sci.*, **56,** 1766–1782.

Emanuel, K. A., M. Fantini, and A. J. Thorpe, 1987: Baroclinic instability in an environment of small stability to slantwise moist convection. Part I: Two-dimensional models. *J. Atmos. Sci.,* **44,** 1559–1573.

Emanuel, K. A., J. D. Neelin, and C. S. Bretherton, 1994: On large-scale circulations in convecting atmospheres. *Q. J. R. Meteorol. Soc.,* **120,** 1111–1143.

Espy, J. P. 1841: *The Philosophy of Storms.* Little and Brown, Boston.

Fantini, M., 1993: A numerical study of two-dimensional moist baroclinic instability. *J. Atmos. Sci.*, **50,** 1199–1210.

Fantini, M., 1995: Moist Eady waves in a quasigeostrophic three-dimensional model. *J. Atmos. Sci.*, 52, 2473–2485.

Frisinger, H. H., 1977: *The history of meteorology: to 1800.* American Meteorological Society, 45 Beacon St., Boston.

Hoskins, B. J., M. E. McIntyre and A. W. Robertson, 1985: On the use and significance of isentropic potential vorticity maps. *Q. J. R. Meteorol. Soc.*, **111** , 877–946.

Islam, S., R. L. Bras, and K. A. Emanuel, 1993: Predictability of mesoscale rainfall in the tropics. *J. Appl. Meteorol.*, **32,** 297-310.

Kutzbach, G., 1979: *The thermal theory of cyclones.* American Meteorological Society, 45 Beacon St., Boston.

Lindzen, R. S., 1993: Baroclinic neutrality and the tropopause. *J. Atmos. Sci.*, **50,** 1148–1151.

Lindzen, R. S., M.-D. Chou and A. Y. Hou, 2000: Does the earth have an adaptive iris? *Bull. Am. Meteorol. Soc.*, **81,** in press.

Loomis, E., 1843: On two storms which were experienced throughout the United States in the month of February, 1842. *Trans. Am. Phil. Soc.*, **9,** 161–184.

Loomis, E., 1877: Contributions to meteorology. *Am. J. Sci.*, **14,** 16.

Manabe, S., J. Smagorinsky, and R. F. Strickler, 1965: Simulated climatology of a general circulation model with a hydrologic cycle. *Mon. Weath. Rev.*, **93,** 769–798.

Parsons, D. B. and P. V. Hobbs, 1983: Mesoscale and microscale structure and organization of clouds and precipitation in midlatitude cyclones. Pt 7: Formation, development, interaction, and dissipation of rainbands. *J. Atmos. Sci.*, **40,** 559–579.

Rennó, N. O. and A. P. Ingersoll, 1996: Natural convection as a heat engine: A theory for CAPE. *J. Atmos. Sci.*, **53,** 572–585.

Rennó, N. O., K. A. Emanuel, and P. H. Stone, 1994: Radiative-convective model with an explicit hydrological cycle. Part I: Formulation and sensitivity to model parameters. *J. Geophys. Res.,* **99,** 14429–14441.

Sun, D.-Z. and A. H. Oort, 1995: Humidity-temperature relationships in the tropical troposphere. *J. Climate,* **8,** 1974–1987.

Thorpe, A. J. and K. A. Emanuel, 1985: Frontogenesis in the presence of small stability to slantwise convection. *J. Atmos. Sci.*, **42,** 1809–1824.

Wheeler, M. and G. N. Kiladis, 1999: Convectively coupled equatorial waves: analysis of clouds and temperature in the wavenumber-frequency domian. *J. Atmos. Sci.*, **56,** 374–399.

New Observational Technologies
Scientific and Societal Impacts
Frédéric Fabry and Isztar Zawadzki
Department of Atmospheric and Oceanic Sciences, McGill University, Montreal, Canada

Atmospheric prediction is posed as an initial-value problem. Defining the initial atmospheric state is thus essential. The need to describe the atmospheric state at its vast range of scales encouraged the development of atmospheric remote sensing. The last 50 years of technological development produced an exponentially increasing diversity of instruments and of the number of parameters they can detect. The parallel development of numerical modelling and the possibility of combining observations and models through proper data assimilation bring us to the point where remote sensed observations and models can be merged into a single unit. These advances give unprecedented insight into the relationship between dynamics and microphysics. At the same time, the data from these instruments are being distributed more broadly to both the scientific community and the public in general. Some of these developments are changing the way economic activity relates to weather. For the general public, televised weather patterns bring some aspects of the atmospheric science into the realm of common knowledge.

A. INTRODUCTION

The place of observations in general, and of remote sensing in particular, within the atmospheric sciences is better understood in the more general context of atmospheric modelling. Thanks to the rapid development of information-processing technology, numerical prediction models of the atmosphere are continuously becoming more powerful and sophisticated, and can now simulate phenomena from the micro to the planetary scales. Since atmospheric prediction is an initial-value problem, data at all these scales are required to initialize the models. Partly driven by this need, there has been and continues to be a tremendous drive toward progress in our observational capabilities. New *in situ* sensors, in particular aircraft probes, are being built. Sensors are now being carried on new platforms like remotely piloted vehicles, or on improved ones like the new global positioning satellite (GPS) dropsondes and radiosondes. But these efforts alone cannot satisfy the needs for data of scientists and of models, as *in situ* data cannot always be available where it is needed and at the appropriate resolution. Hence, the most spectacular progress in observational technologies has occurred in remote sensing.

The same developments in computational technology that have made models possible have also had a tremendous impact on remote sensing. First, new types of sensors have been developed thanks to new ideas and technologies. But more importantly, our increased ability to store, process and communicate data has resulted in an improved use of existing sensors, and has made previously difficult-to-use equipment more accessible to a broader community. Consequently, the scientific as well as the broader impacts of these observational technologies have blossomed in recent years.

The nature of scientific and societal impacts is very different. Scientific impacts of new observational technologies are closely linked to the variables being measured or derived from remote-sensing measurements. Hence, they tend to depend strongly on the type of instrument being used, and they are most strongly felt by the community using this information. Some familiarity with the technology and what it can provide is required to fully understand these impacts. In contrast, societal impacts generally transcend the details of the technology, and reach, by their nature, broader communities. We therefore chose to first describe a few of these new observational technologies and their specific scientific impacts, and then conclude with a rapid overview of their societal impacts.

B. REMOTE SENSING OF THE ATMOSPHERE

Remote sensing refers to a variety of techniques used to obtain information about an object, or a portion of an object, without being in direct physical contact with the object of interest. In order to obtain information remotely, remote sensors measure properties of electromagnetic and acoustic waves that have been emitted, reflected, refracted, or delayed by the object. Knowledge about the nature and the degree of interaction between these waves and the object gives us clues that can be used to infer properties of the object that are of meteorological interest.

While *in situ* sensing is limited to data at one or a few points in time, remote sensing allows the collection of information over large distances or large volumes without disturbing the medium being observed. Though remote sensors are generally expensive instruments, their cost on a per-data-point basis is much lower than *in situ* sensors. This capability of repeatedly collecting large amounts of information over an extensive area explains the interest in remote-sensing by the geophysical sciences in general and by meteorology in particular. Unfortunately, remote-sensing measurements are often partial, or indirectly related to meteorological quantities of interest. For example, Doppler velocity gives us only the radial component of a three-dimensional wind, and brightness temperature at a given wavelength is a complex function of the temperature, concentration and radiative properties of solids, liquids and gases in the direction of the observation. For a long time, this limitation has prevented the use of remote-sensing data to its full potential. However, thanks to variational techniques for information retrieval and data assimilation methods, it is now possible to ingest partial information like that obtained by remote sensors into numerical models. Models can then use these observations to retrieve unobserved variables, considerably increasing our ability to study and predict meteorological phenomena. As a result, we are just entering an era where remotely sensed information, which was used qualitatively or semi-quantitatively until now, could become the main source of routine quantitative information.

It is now considerably easier to obtain more raw information from remote sensing than it has been in the past thanks to better instruments and signal processing. As a result, what was only possible in the most sophisticated of radars 10 years ago can now be achieved with relatively low-cost instruments. In parallel, the amount of data that remote sensors can collect is remarkable. For example, the U.S.–Taiwanese project COSMIC (Rocken *et al.*, 2000) alone will provide more than 3000 daily profiles of refractive index, a parameter closely linked to temperature and moisture, compared to the 1200 daily radiosonde launches that are currently used to define the state of the atmosphere above the surface.

C. REMOTE SENSORS AND THEIR SCIENTIFIC IMPACTS

Remote-sensing information comes from many types of instrumentation, and every one of them has undergone phenomenal progress in recent years. Rare are the atmospheric parameters for which one cannot get some type of measurement. While instruments like radars and geostationary imaging satellites have become routinely used, many other technologies are maturing rapidly or just being developed. Here is a quick overview of the parameters being measured by remote sensing.

1. Air Temperature and Moisture

Knowledge about air temperature and moisture is essential to understand and predict meteorological phenomena. These two parameters can be measured remotely using a variety of techniques. However, for most of these techniques, the distance over which the measurement can be made is limited to at most a few tens of kilometres. As a result, they are generally used to obtain only vertical profile information. Yet many of these sensors can achieve global coverage when carried on satellites. For example, infrared and microwave radiometers can measure coarse profiles of temperature or humidity, and geostationary satellites make such measurements routinely.

A variety of techniques and instruments exists to measure temperature and humidity (Weckwerth *et al.*, 1999). Ground-based and satellite-borne radiometers measure the emission of well-mixed gases like carbon dioxide or oxygen to obtain temperature information, and then the emission of water vapour to obtain moisture information. Despite the sophisticated measurement and processing techniques being used, the nature of this passive remote-sensing measurement limits the range resolution that can be achieved. Active techniques include the radio acoustic sounding system (RASS) and lidar-based measurements. RASS measurements rely on the combination of an acoustic source and of a wind profiler; the latter, generally used to measure wind speeds using the Doppler effect now measures the speed of sound waves, which is related to temperature (May *et al.*, 1990). Depending on the profiler, RASS measurements can be made for anywhere from several hundred metres to the whole troposphere at a resolution of a few hundred metres. Lidars, or light radars, can be used to measure the humidity of air up to cloud base using either differential attenuation (Bösenberg, 1998) or Raman techniques (Whiteman *et al.*, 1992). While lidar technology is more experimental than the others, its rapid development may make lidar the water vapour profiler of the future. In addition to remote sensing measurements, wind, temperature, and humidity data measured by sensors mounted on commercial aircraft (Fleming, 1996) are now routinely being used for data analysis and model initialization purposes.

A promising way to obtain information on temperature and particularly moisture is to measure the refractive index of air n. As electromagnetic waves travel through a medium other than a vacuum, their velocity decreases by a factor n. At radio frequencies, the refractive index of air is a function of density, hence temperature, as well as moisture. Thus, the measurement of refractive index can give precious information on these two parameters. Several techniques have been proposed to measure n. One measures the changes in the travel time of radar waves between a radar and fixed targets on the ground (Fabry et al, 1997). The others rely on the measurement of the propagation delay of signals from the global positioning satellites from ground stations or satellites to obtain refractive index measurements along a space–ground path or a space–limb–space path (Businger et al., 1996; Ware et al., 1996, 2000). While these techniques provide indirect information on temperature and moisture, the relative ease of these measurements will result in a considerable improvement in our ability to map temperature and moisture around the globe. On the other hand, the refractive index is a parameter that can directly be assimilated into numerical models.

Thanks to all these techniques, our knowledge of the spatial distribution of temperature and moisture is improving rapidly in both real-time and field-experiment environments. The interest in these measurements is particularly high among researchers working on convection initiation, precipitation prediction, and **boundary**-layer processes, where progress is being slowed by the limited availability of suitable measurements. For example, current forecasts of thunderstorm development are hampered by the lack of sufficient temperature and moisture information at mesoscale resolution that cannot be easily obtained by conventional *in situ* techniques (Crook, 1996). Yet, thanks to some of these new technologies, one can start to observe moisture at very high resolution (Fig. 1), and relate these observations with storm development (Fig. 2).

2. Clouds and Precipitation

Another domain where remote sensing has made remarkable progress is in the measurement of cloud and precipitation. A large variety of instruments are now sampling cloud and precipitation properties: radiometers and imagers on board satellites, ceilometers and other lidars, as well as radars are providing a far more detailed picture of the distribution and properties of hydrometeors than was available a decade or two ago (Serafin and Wilson, 2000).

Cloud measurements have probably registered the greatest progress, thanks to the conjunction between improved technological capability and the increased interest in clouds spurred by the uncertainty they cause in climate models. In addition to the cloud coverage data provided by ceilometers and satellite imagery, we can also obtain information on the microphysical and dynamical characteristics of clouds using lidars (Fig. 3) and radars (Fig. 4). Cloud information that can be obtained by specialized remote sensors includes their physical state (ice, liquid, or mixed cloud), clues about the size distribution of their hydrometeors, and the dominant crystal shape. Regular measurements made at several sites include cloud base and cloud top height, the detection of multilevel clouds, and path-integrated liquid water content. While we are far from obtaining a complete picture of clouds by remote sensing, there is not one aspect of clouds for which one cannot obtain at least some information by remote sensing.

None-the-less, characterization of precipitation remains one of the main topics of focus of atmospheric remote sensing. For this task, radar remains the tool of choice, as it is one of the rare instruments capable of providing meteorological information in three dimensions over large areas. Most industrialized countries now have radar networks for the nowcasting and very short-term predictions of flash floods and other severe weather events. Thanks to the networking of radars over continental scales, rainfall statistics and processes can be observed and studied in ways never possible before (Fig. 5). Precipitation radars are now just for the first time being carried on satellites (e.g. Tropical Rainfall Measuring Mission), providing a new look at the global distribution of precipitation especially over oceans and remote areas.

A metamorphosis of radar measurement of cloud and precipitation is now being realized by the use of dual-polarization radars. While conventional radars typically transmit and receive radio waves at one polarization, dual-polarization radars transmit at two polarizations, usually horizontal and vertical. This allows us to obtain information on the shape of the target such as its average horizontal-to-vertical axis ratio, and the variability of target shapes within the sampling volume. Although not without ambiguity, this information can then be used to identify the nature of the target being observed (rain, snow, hail, ground, etc.) much more reliably than was possible before. Equally important, polarization diversity separates well the meteorological and non meteorological targets. Studies of cloud and of precipitation physics and their applications like the

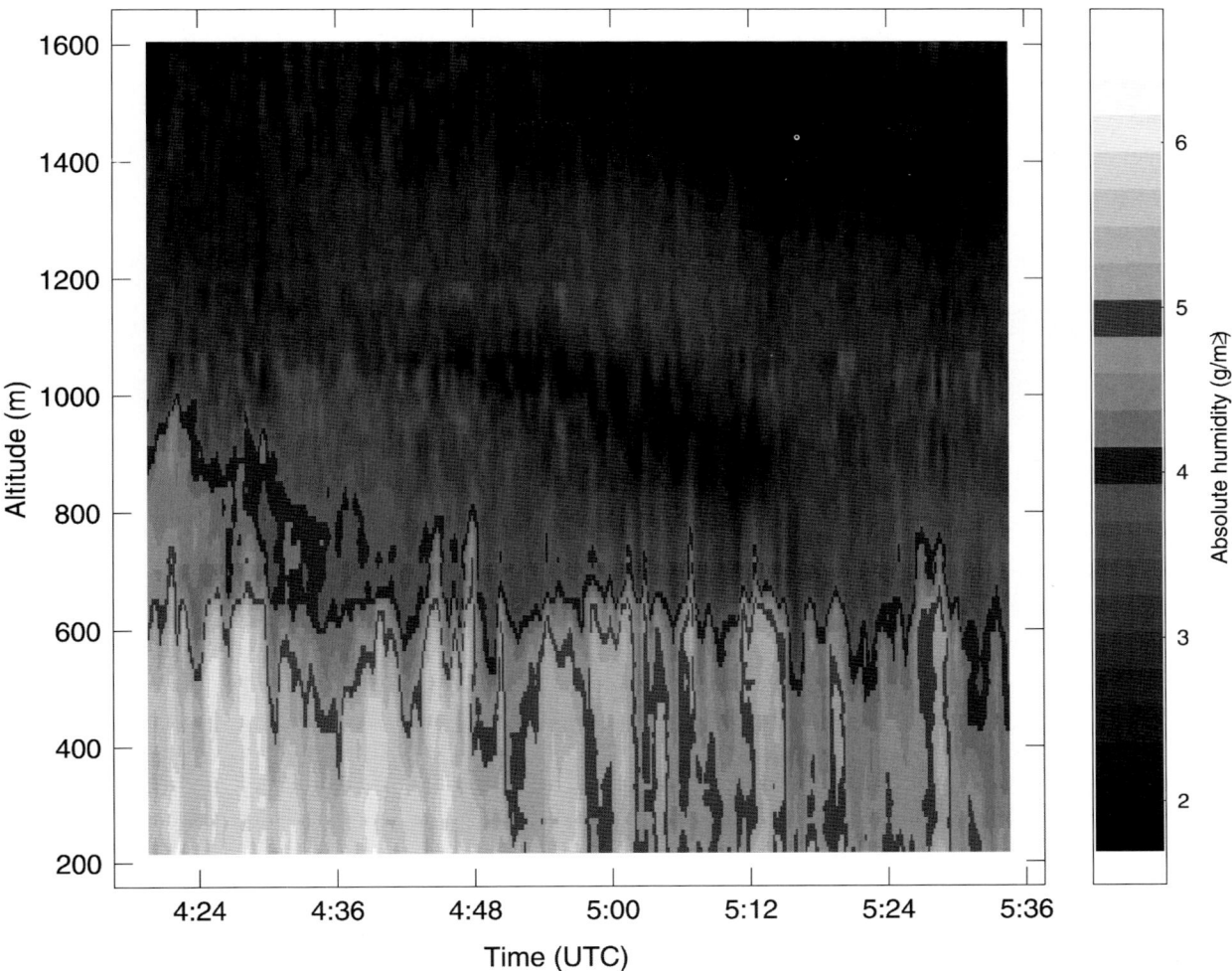

FIGURE 1 Time–height cross-section of the absolute humidity measured by a differential absorption lidar on the island of Gotland, Sweden, on 13 September 1996. Absolute humidity is obtained by measuring the difference in the lidar attenuation at two nearby frequencies, one on a water vapour absorption line, and one just outside. Strong variability in humidity can be observed below 600 m associated with convective activity, and around 700 m associated with dry air entrainment at the top of the boundary layer. (Adapted from Wulfmeyer, 1999.)

detection of hazardous conditions such as hail and freezing rain may have the most to gain from this technological improvement. These and other techniques are finally allowing us to automatically clean the artefact-filled radar images to the extent that they can be used directly by nonspecialists for a variety of applications like hydrological predictions and the initialization and verification of mesoscale meteorological models.

3. Wind

The measurement of air motion, or wind, over large volumes is an extremely difficult task for remote sensors. But remote sensors can easily obtain information on the effect of the wind on objects. Satellite estimates of wind are made by tracking clouds from one image to the next, or by monitoring the state of the surface of oceans using scatterometers, altimeters, and synthetic aperture radars. Lidars, radars, and wind profilers all obtain velocity information from the Doppler shift of precipitation, insects, dust, refractive index gradients, or ionization channels from meteorites that are all carried by the wind.

Detailed wind information is routinely obtained by either Doppler radars or wind profilers. Doppler radars measure over mesoscale areas the three-dimensional configuration of the radial velocity of targets. These measurements are both very detailed spatially yet very incomplete as only one component of the wind is measured at every point. Wind profilers are specialized

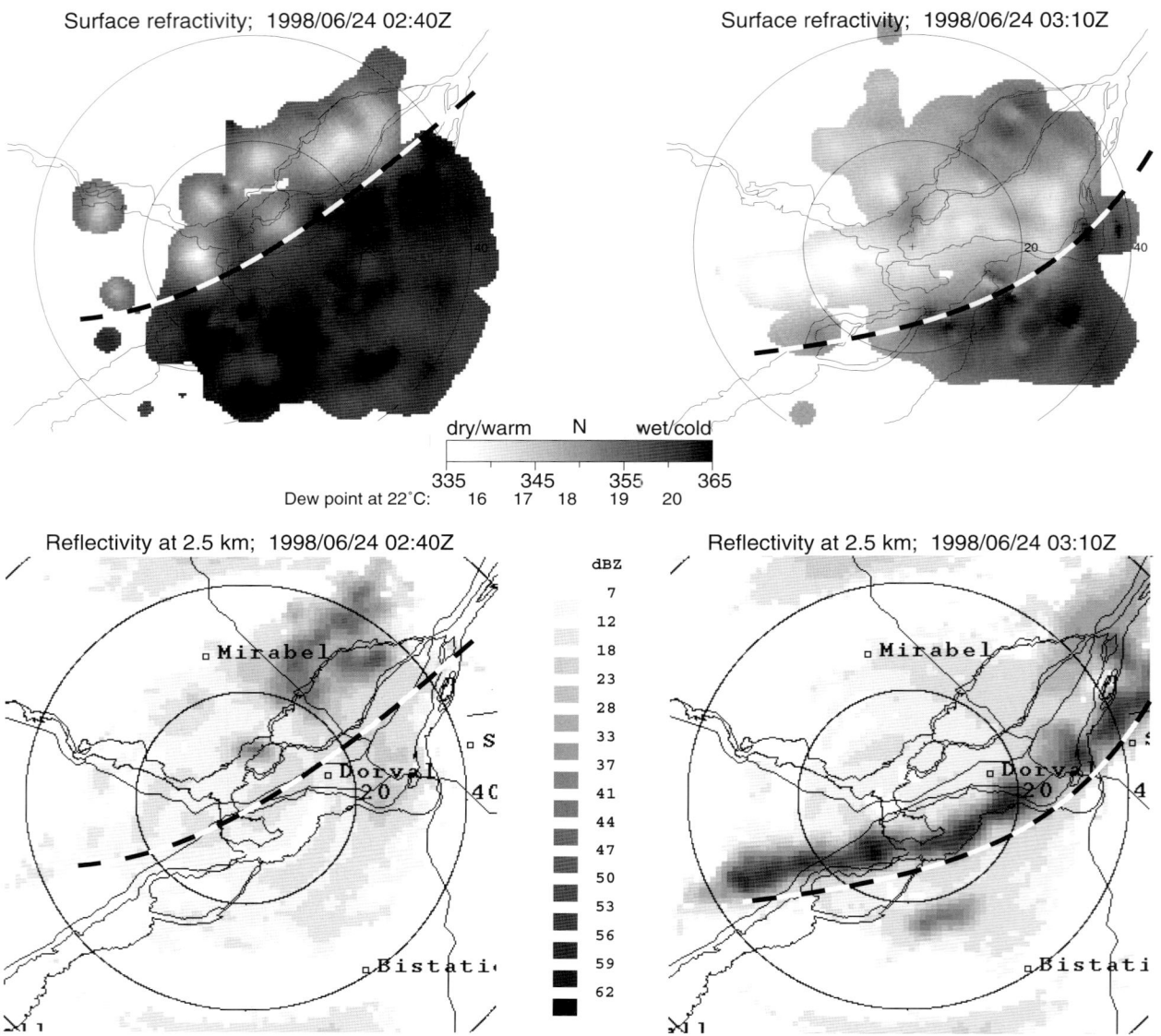

FIGURE 2 Maps of surface refractivity (top) and reflectivity of weather targets (bottom) taken before (left) and after (right) the formation of a squall line in Montreal, Canada, on 24 June 1998. Refractivity ($N = (n-1) \times 10^6$) was measured by monitoring the change in the travel time of radar waves between the radar and fixed ground targets like buildings and towers. Reflectivity is obtained from the strength of radar echoes caused by precipitation. Range rings from the radar are every 20 km. The squall line formed just behind the boundary between two air masses of **different** moisture (dashed line) whose presence could be observed on the refractivity map prior to the formation of the squall line (Fabry and Creese, 1999).

Doppler radars that are designed to measure the vertical profile of winds as a function of time. With these instruments, 3-D winds can be obtained, but only over a narrow column. They are used to detect dangerous wind conditions like tornadoes and downbursts, or to complement the information obtained by radiosondes.

Many more research system designs exist, each trying to fill a niche or overcome the limitations of single Doppler radar technology. For example, Doppler radars have been mounted on vehicles in order to get close to and obtain high-resolution data of the meteorological phenomenon of interest (Wurman et al., 1995). Often, during field experiments, two or more Doppler radars are set up to obtain additional components of the wind over selected areas, enabling the detailed measurement of 3-D winds. Another possible set up is to have one radar transmitter coupled with multiple receivers at several locations, each obtaining a different

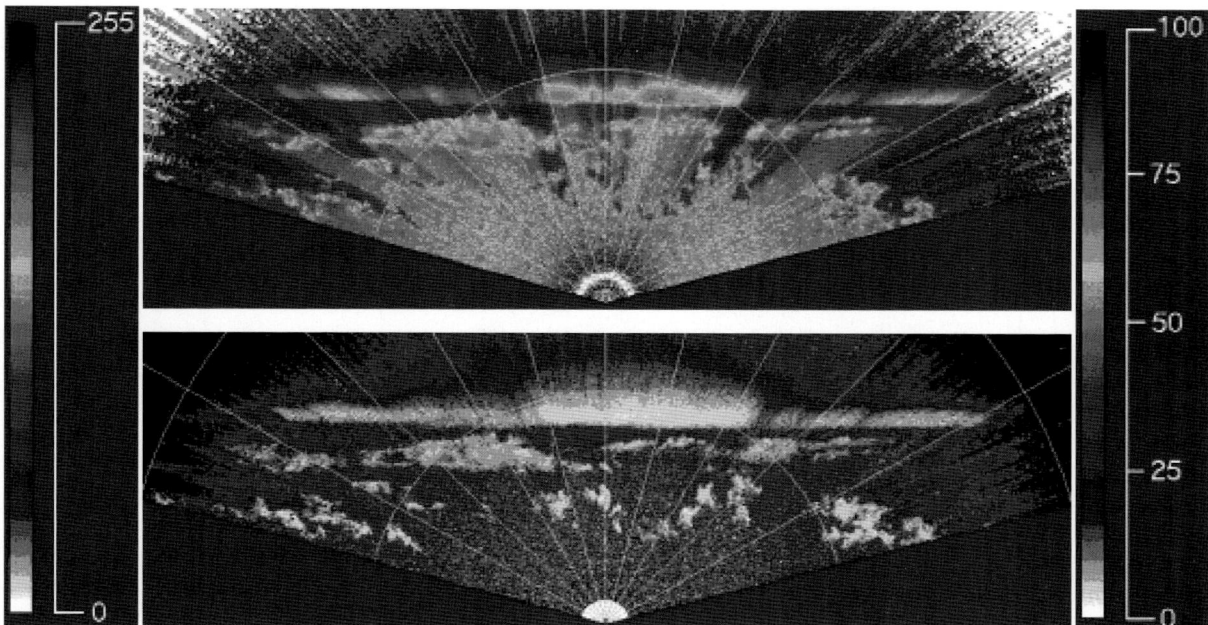

FIGURE 3 Range–height section of lidar echo strength (top) and depolarization ratio (bottom) taken in Mirabel, Canada, on 15 December 1999 at 11:47. Range rings are every 1 km. On the echo strength map, echoes can be observed nearly everywhere below 1 km altitude. Regions of enhanced echoes are caused by broken cloud layers near 900 m altitude as well as isolated patches around 500 m, while the weaker echoes are caused by precipitation. The signal beyond an altitude of 1 km has been attenuated by the cloud layers. In the depolarization ratio map, one can see that the echoes from precipitation is highly depolarized, implying that the targets have the irregular shapes of snowflakes, while the cloud echoes have low depolarization, implying that they are made of supercooled water droplets. (Image courtesy of L. Bissonnette, Canadian Defence Research Establishment.)

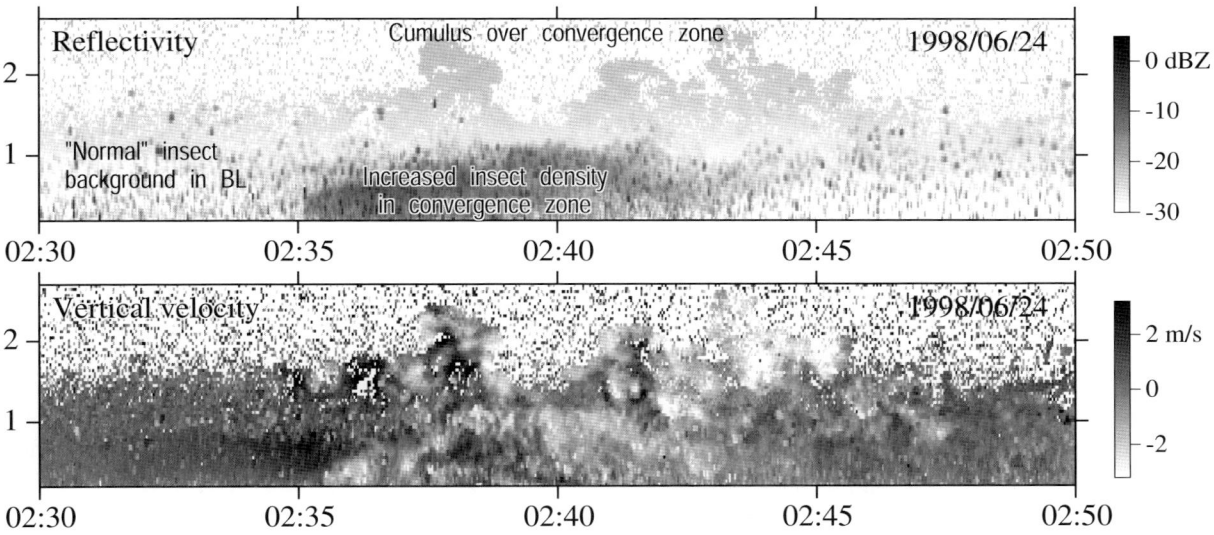

FIGURE 4 Time–height image of reflectivity and vertical velocity measured by a radar pointing vertically during the passage of the front illustrated in Fig. 2. Convergence at the frontal boundary (2:35) lifts air as well as insects. This lifting forces air parcels through the capping inversion of the boundary layer, and cumulus clouds form as a result. Five cumulus clouds can be observed at different stages of their evolution, the more mature clouds being the further away from the convergence line. The first two cumuli, barely visible on the velocity map at 2:35 and 2:36:30, are mere hot bubbles that are still within the low-level echoes caused by insects. The next one, centred around 2:38, is a vigorous cumulus cloud, with an updraft in its core and downdrafts at the periphery. The fourth one around 2:42 is at the early stages of dissipation with a few updrafts and many downdrafts, while the last one (2:44) is collapsing rapidly. Note that these data were collected by a simple, reliable system costing less than 50k euros, made possible by recent progresses in technology.

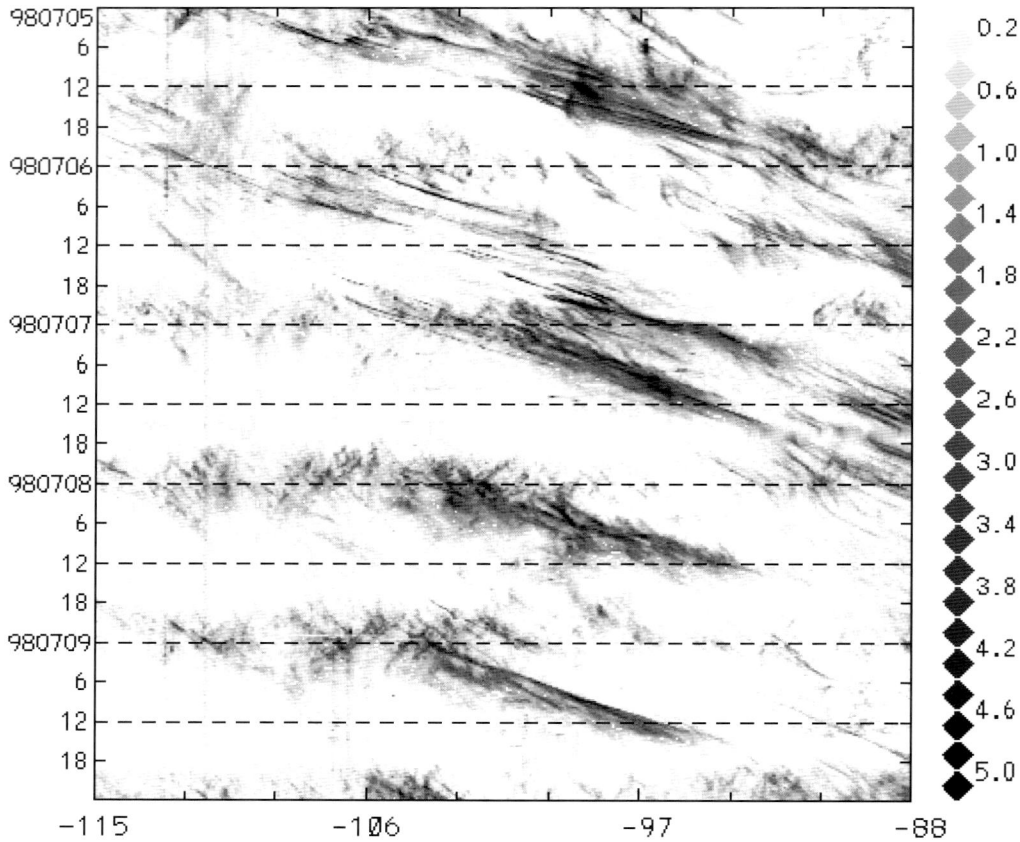

FIGURE 5 Hovmoller (time against longitude) diagram of area-averaged rainfall rate (mm/hr) over the continental United States for a 5-day period between 5 July 1998 and 9 July 1998. This image was obtained by averaging the rainfall rate measured by the U.S. radar network over 2-km strips for all latitudes between 29°N and 45°N every 5 min. Under this projection, horizontal bands of rain are associated with the non-propagating diurnal forcing of precipitation, best exhibited over the mountainous west. The sloping streaks are zonally propagating areas of rainfall having convective origin. Many of these streaks have their origin over the east slope of the Rockies (near 105°W) and a few originate over midcontinent (90–95°W). Note the nearly daily diurnal forcing of continental-scale coherence in rainfall by the elevated terrain of the west, influencing the probability of precipitation in the eastern 2/3 of continent. (Image courtesy of R.E. Carbone, National Center for Atmospheric Research.)

component of the wind (Wurman et al., 1993; Fig. 6). Finally, if the radar is airborne and moves rapidly, multiple views can be obtained from the same storm to complete the wind information, and the storm can be followed by the aircraft over a significant portion of its lifetime (Hildebrand et al., 1994). If one can monitor the time evolution of the wind, it is possible to retrieve the forces acting on that wind, from which temperature and pressure information can be obtained.

4. Others

Advancements in remote-sensing technology not only improved the measurement of standard meteorological variables but also that of other parameters like the radiative properties of the atmosphere and the characterization of trace components in the atmosphere. A considerable portion of this work was stimulated by the need to understand atmospheric chemistry in general and the ozone hole problem in particular, as well as by the need to quantify the forcing terms in the climate system.

Atmospheric chemistry in the lower troposphere and in the stratosphere is becoming a new frontier for remote sensors. A variety of specialized sensors based on radiometry and lidar technology is being developed to measure trace gas like ozone and nitrous oxides. These sensors complement the increasing number of *in situ* measurements being made on surface stations and with aircraft. A considerable proportion of these sensors is satellite-borne in order to obtain global **coverage**. Furthermore, lidars are also used to characterize aerosols, primarily those associated with the boundary

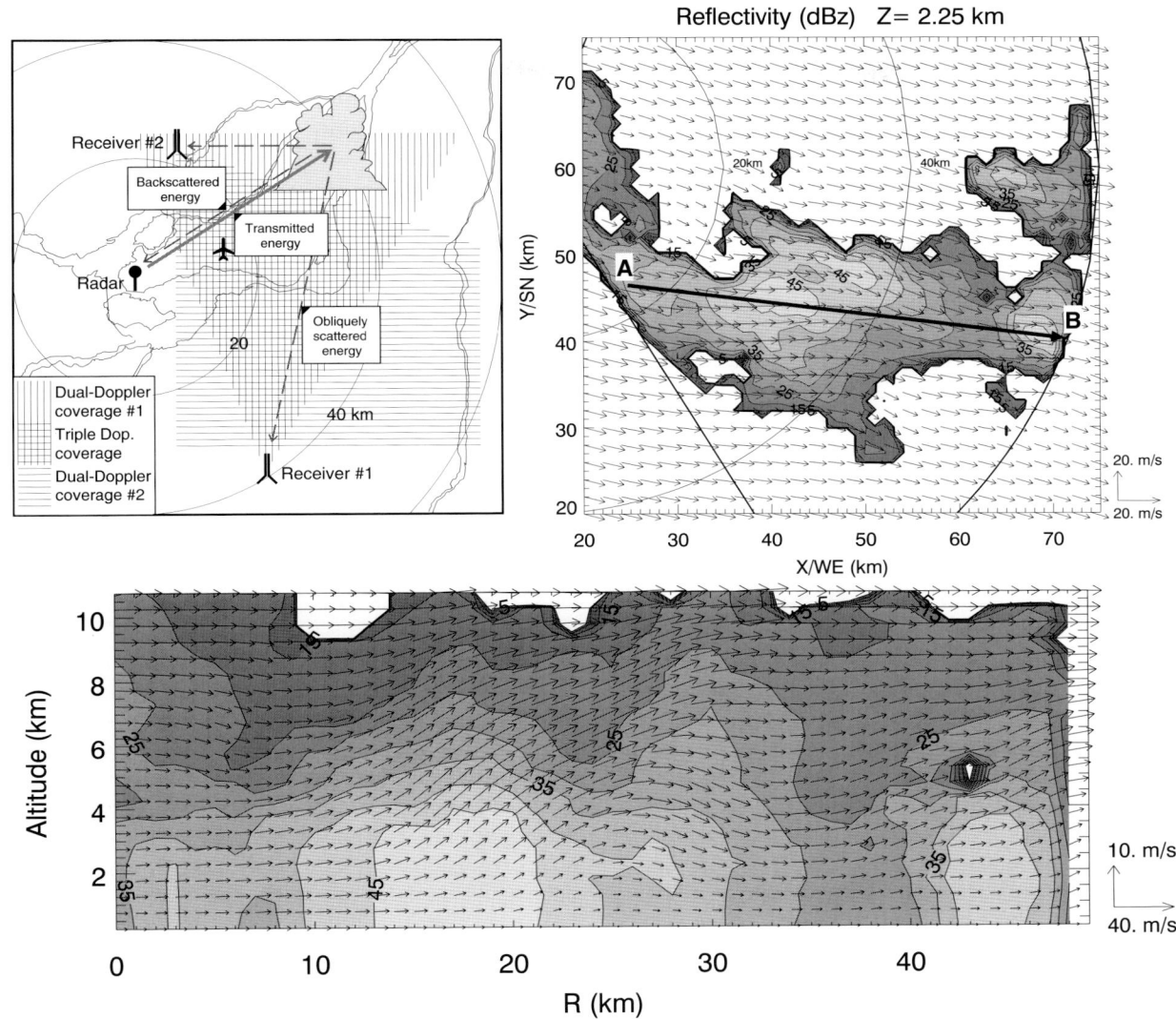

FIGURE 6 (Top left) Configuration and concept behind the Doppler multistatic radar network operated by McGill University. Radar pulses sent by the McGill radar are scattered by meteorological targets like precipitation from a storm. Some of this energy comes back to the main radar site, providing information on the strength and radial velocity of targets, while some other portion is scattered obliquely to additional receivers from which other components of the wind can be obtained. The measurements from all the receivers are then combined to determine the 3-D wind. (Top right) Horizontal section of radar reflectivity and horizontal winds of a convective line within coverage of the multistatic radar network. Winds within the storm are derived from the radar data, while winds outside are set based on a synoptic scale analysis. (Bottom) Vertical section through the convective line along the line A–B. In this image, the winds being plotted are the projection of the 3-D wind on the plane defined by the cross-section. Vertical motion is observed over two intensifying cells (kilometre 18 and 44) while descending motion occurs over a decaying cell (kilometre 29).

layer or with exceptional events like the ash clouds from volcanic eruptions.

Finally, more satellite and ground-based sensors have been deployed to quantify the radiative properties of the atmosphere. Absorption, transmission, and reflection of both visible and infrared radiation are being studied to help understand the complex nature of the greenhouse effect, in particular the contribution from clouds, from which one could better infer the potential effect of humans on the radiative properties of the atmosphere (Harries, this volume).

5. Related Scientific Considerations

Until recently, there has been a relative stagnation in the amount of data used by models, especially above the surface. This is in strong contrast with the constant evolution of models whose number of grid points doubles every 30 months. As a result, model runs have progressively become "data poor", and one of the main limitations of model prediction is the limited accuracy of the initial value and boundary conditions of the parameters. The data from an increasing number of remote sensors should reverse that trend in the next few years, allowing us to obtain a better pay-off of the increased sophistication of models. In a new "data rich" environment, the accuracy of numerical forecasts should be less limited by the accuracy of initial conditions and more by shortcomings in the parameterization of microphysical and subscale processes and by the intrinsic predictability of atmospheric phenomena.

Thanks to progress in technology, building complex instrumentation has never been easier and as inexpensive as now. While in the past sophisticated measurements could only be done with equipment costing multimillions of euros, they are now possible with a new breed of instruments within the reach of the budget of small groups of researchers. This is currently resulting in a spectacular spree in instrument development in the university community, particularly of mobile radars in North America. For example, a significant proportion of the instruments being deployed for the International H_2O Project (IHOP 2002) are still under development, and their number may reach an all-time high for an experiment of this size (Table 1). It is the authors' strongly held belief that while good science can always be done with large instruments, vast research possibilities exist with the data obtained by their lower cost counterparts (e.g. by low-cost radar, see Fig. 4).

An unfortunate result of the possibilities offered by remote sensors is that they capture the interest of measurement-oriented scientists at the expense of laboratories. The decline of the laboratory started some time ago, but is currently accelerating as laboratories often deal with the very thorough study of very specific problems, while remote sensors have the potential to tackle a wider variety of problems, but in less detail. This trend is regrettable because the atmosphere does not lend itself to the controlled experiments needed to solve some meteorological problems like mixing and microphysical processes that only laboratory experimentation can truly help to solve.

D. SOCIETAL IMPACTS

Data dissemination technology may be progressing even faster than data processing. As a result, information from all sources, including data and imagery from new observing platforms, is becoming available not only to every scientist who needs the data for his or her research, but also to anyone who has access to a computer. While a generation ago, a young weather enthusiast would build a hygrometer using a milk carton and hair strand, his children can now have access to the latest satellite or radar imagery via the Internet as well as model prediction prognoses from around the world. As a result, the societal impacts of new observing technologies are broadening their reach.

Weather offices are increasingly dependent on new observing technologies for their forecasts, especially for the very short-term forecasting (0–6 hours) of severe weather events. The *raison d'être* of data from sophisticated instruments in the weather office is to help forecasters in their task of warning the public of dangerous meteorological events that may result in the loss of lives or property. For example, thanks to their capability of collecting data over large areas, the installation of a Doppler radar network in the United States resulted in a marked increase in the probability of detection and the forecasting lead time of severe weather events like flash floods and tornadoes (Polger *et al.*, 1994). With better warnings, it is hoped that the public will be better prepared to face the danger and the disruptions associated with severe weather.

For similar reasons, television became an obvious

TABLE 1. University-owned Remote Sensors Slated for Deployment for the IHOP 2002 Experiment. Entries in Italics Are for Instruments Under Development or Testing. Many More Are Under Development in Government Research Laboratories.

University	Instrument
Desert Research Institute	Mobile dual-channel radiometer
Florida State University	*Mobile dual-polarization C-band Doppler radar*
McGill University	*Mobile dual-polarization X-band Doppler radar*
Texas A&M, Texas Tech., U. Oklahoma	*Two mobile C-band Doppler radars (SMART-Rs)*
University of Alabama	Mobile integrated profiling system (profiler, sodar, ceilometer)
University of Massachusetts	Mobile 95 GHz tornado radar
University of Oklahoma	Two mobile X-band Doppler radars (DOWs)
University of Wyoming	95 GHz airborne cloud radar

medium for using imagery from remote sensors for their weather forecasts. This is because such imagery has both an information and an entertainment value. Weather forecasts are becoming a spectacle thanks to the combination of the entertainment value offered by these technologies and our increasing awareness of weather-related disasters and environmental problems around the world (for example the coverage associated with hurricane Mitch). One has problems imagining the modern TV weather forecast without at least a loop of infrared satellite imagery that so clearly illustrates the movement of weather systems. Furthermore, when precipitation is within the television channel viewing area, radar imagery is often presented. In fact, showing radar imagery is perceived to be so valuable by the media in the United States that many television stations actually own their radar so they can better serve their viewers and boast about it at the same time. The educational value of this exposure to millions of people is difficult to underestimate, as it brings the familiarity with atmospheric phenomena to the level of everyday experience.

Broadcasters are not the only component of the private sector that benefits from remote sensing imagery. In parallel to the weather services, there is a whole private sector in meteorology that specializes in distributing weather forecasts and tailoring them to specialized customers. This private sector is growing rapidly, fuelled by the availability of new data and processing tools. In fact, in the United States where the National Weather Service (NWS) is less protective of its data than in other countries, the private sector tops 1 billion euros in sales each year, which is 150% of the budget of the NWS itself (Rosenfeld, 2000). These specialized forecasts are extremely varied in nature, and many require data from remote sensors. For example, the authors were contacted by a Formula 1 motor racing team to provide guidance on how to interpret and use radar imagery for real-time decision making during the 2000 Montreal race.

One area of activity that benefits particularly from sophisticated weather information is international transportation and commerce, particularly by air. It is creating a need for better operational real-time observing systems that can monitor the environment to improve safety and efficiency and to reduce environmental impacts. Decades ago, the density of air traffic was sufficiently low for airports to afford to be closed for short periods in bad weather. Nowadays, operations must be pushed to the edge of safety. For example, the new Hong Kong airport must operate despite convective wind shear and turbulent airflow over nearby terrain. To tackle this problem, a combination of sophisticated observation systems (radar, wind profiler, anemometers, etc.) and a numerical forecast model was developed. Such examples may become more common as aircraft may soon be able to choose their own flight routes: this new mode of operation would make the system more efficient, provided that good weather information is available on phenomena that may help air traffic, like winds, as well as those that may hinder it, like turbulence, thunderstorms, and icing conditions. Detection of many of these conditions will have to rely on remote sensors.

Finally, one should not underestimate the power of some of the images provided by remote sensors to both illustrate a case and to spark the imagination. One cannot look at a satellite view of the earth and not feel the beauty, the isolation, and the fragility of the planet. The environmental awareness, and the consequent policy making, is certainly enhanced by this everyday spectacle. Similarly, people in general and scientists in particular can use the kind of imagery only possible by remote sensing to illustrate environmental problems and urge policy makers into action. Satellite loops of the Antarctic ozone hole obtained from the Total Ozone Mapping Spectrometer (TOMS) certainly helped to illustrate the pressing nature of the problem and accelerated policy decisions at the international level. This is only because such images, especially when animated, have a credibility that often more accurate but less spectacular *in situ* data do not possess. Intuitively, images are perceived to be more truthful than symbols or lines on a graph. This has its negative counterpart effect: rich imagery can be confounded with knowledge and often leads to superficial decision making.

E. CONCLUSIONS

Thanks to their capabilities of collecting, over large areas, data that are relevant but often difficult to interpret, remote sensors are undergoing a rapid development both as research and operational instruments in atmospheric sciences. With the sophisticated numerical models now available to ingest these data, we have an increasingly better description of the weather phenomena we are observing. The data obtained by these new observing technologies are also being disseminated more efficiently to the general public and scientists alike, who can access them for personal curiosity or professional interest. For a scientist working in the field of remote sensing every new year is the most exciting of his/her entire career.

As more and more parameters are being measured by remote sensors, it is likely that remote-sensing data will have increasing impacts in science and in various aspects of our daily lives.

References

Bösenberg, J., 1998: Ground-based differential absorption lidar for water-vapor and temperature profiling: Methodology. *Appl. Optics.*, **37**, 3845–3860.

Businger, S., S. Chiswell, M. Bevis, J. Duan, R. Anthes, C. Rocken, R. Ware, T. Van Hove, and F. Solheim, 1996: The promise of GPS in atmospheric monitoring. *Bull. Am. Meteorol. Soc.*, **77**, 5–18.

Crook, N. A., 1996: Sensitivity of moist convection forced by boundary layer processes to low-level thermodynamic fields. *Mon. Weath. Rev.*, **124**, 1767–1785.

Fabry, F. and C. Creese, 1999: If fine lines fail, try ground targets. *Preprints, 29th Int. Conf. on Radar Meteorology*, Montreal, QC, Am. Meteorol. Soc., pp. 21–23.

Fabry, F., C. Frush, I. Zawadzki and A. Kilambi, 1997: On the extraction of near-surface index of refraction using radar phase measurements from ground targets. *J. Atmos. Ocean. Tech.*, **14**, 978–987.

Fleming, R. J., 1996: The use of commercial aircraft as platforms for environmental measurements. *Bull. Am. Meteorol Soc.*, **77**, 2229–2242.

Hildebrand, P. H., C. A. Walther, C. L. Frush, J. Testud and F. Baudin, 1994: The ELDORA/ASTRAIA airborne Doppler weather radar: Goals, design, and first field tests. *Proc. IEEE*, **82**, 1873–1890.

Hollingsworth, A., 2000: The use of satellite data in medium-range and extended-range forecasting. Presented at Meteorology at the Millennium, Royal Meterological Society Conference, Cambridge, 10–14 July 2000.

May, P. T., R. G. Strauch, K. P. Moran and W. L. Ecklund, 1990: Temperature sounding by RASS with wind profiling radars. *IEEE Trans. Geosci. Remote Sens.*, **28**, 19–28.

Polger, P. D., B. S. Goldsmith, R. C. Przywarty and J. R. Bocchieri, 1994: National Weather Service performance based on the WSR-88D. *Bull. Am. Meteorol. Soc.*, **75**, 203–214.

Rocken, C., Y.-H. Kuo, W. S. Schreiner, D. Hunt, S. Sokolovskiy, and C. McCormick, 2000: COSMIC system description. *Terrest. Atmos. Ocean. Sci.* **11**, 21–52.

Rosenfeld, J., 2000: Do we still need the National Weather Service? *Sci. Am. Pres.* **11**(1), 28–31.

Serafin, R. J. and J. W. Wilson, 2000: Operational weather radar in the United States: progress and opportunity. *Bull. Am. Meteorol. Soc.*, **81**, 501–518.

Ware, R. H., M. Exner, D. Feng, M. Gorbunov, K. Hardy, B. Herman, H. K. Kuo, T. Meehan, W. Melbourne, C. Rocken, W. Schreiner, S. Sokolovskiy, F. Solheim, X. Zou, R. Anthes, S. Businger and K. Trenberth, 1996: GPS sounding of the atmosphere from low Earth orbit: preliminary results. *Bull. Am. Meteorol. Soc.*, **77**, 19–40.

Ware, R. H., D. W. Fulker, S. A. Stein, D. N. Anderson, S. K. Avery, R. D. Clark, K. K. Droegemeier, J. P. Kuettner, J. B. Minster, and S. Sorooshian, 2000: SuomiNet: A real-time national GPS network for atmospheric research and education. *Bull. Am. Meteorol. Soc.*, **81**, 677–694.

Weckwerth, T. M., V. Wulfmeyer, R. M. Wakimoto, R. M. Hardesty, J. W. Wilson and R. M. Banta, 1999: NCAR-NOAA lower-tropospheric water vapor workshop. *Bull. Am. Meteorol. Soc.*, **80**, 2339–2357.

Whiteman, D. N., S. H. Melfi and R. A. Ferrare, 1992: Raman lidar system for the measurement of water vapor and aerosols in the earth's atmosphere. *Appl. Optics.*, **31**, 3068–3082.

Wulfmeyer, V., 1999: Investigation of turbulent processes in the lower troposphere with water vapor DIAL and radar-RASS. *J. Atmos. Sci.*, **56**, 1055–1076.

Wurman, J., S. Heckman and D. Boccippio, 1993: A bistatic multiple-Doppler radar network. *J. Appl. Meteorol.*, **32**, 1802–1814.

Wurman, J., J. Straka, E. Rasmussen, M. Randall and A. Zahrai, 1995: Design and first results from a portable, pencil-beam, pulsed Doppler radar. Preprints, 27th Conf. On Radar Meteorol., Vail, CO, Am. Meteorol. Soc., pp. 713–716.

Turning Dynamical Ideas into Forecast Practice

A Proposal for a Renewed Graphical Summary of the Synoptic Scale Situation

Alain Joly and Patrick Santurette

Météo-France, Centre National de Recherches Météorologiques and Service Central d'Exploitation Météorologique, Toulouse, France

---

Progress in the understanding of atmospheric dynamics, and in particular, of cyclogenesis has considerably changed scientists' views of midlatitude weather. Yet, it is surprising how little visible impact this better knowledge is having on the preparation of a "subjective forecast" out of raw numerical weather prediction products. The subjective forecast takes the form of a graphical summary representing surface pressure and fronts and of a written bulletin. This graphical language for weather has seldom followed the progress in dynamical meteorology. Since the need for expressing the future weather evolution in a graphical form persists and even grows, a new graphical language is proposed by a group of Météo-France forecasters and scientists. It takes into account the present understanding of cyclogenesis and large-scale dynamics. The actual and potential baroclinic interactions replace the polar front concept. Dynamical meteorology also provides rather precise guidance for structuring and preparing the summary. It has the aim of explaining zones of synoptic weather activity in the initial spirit of the Norwegian method, since this has been so successful in shaping modern operational meteorology. The proposed code is applied to cases taken from the recent FASTEX project and its potential for evolution is also discussed.

A. INTRODUCTION

An atmospheric scientist reasonably well aware of the current understanding of midlatitude cyclogenesis would not undertake to write a review of this phenomenon explaining it in terms of the linear normal mode instability of an extreme "polar front". Yet, this concept is widely implied in the production of forecasts all over the world, and, probably as a result, it is still the favoured explanation in popular books on meteorology and general encyclopaedias.

One can in fact only be bemused by the contrast that exists between the evolution of the scientific understanding of cyclogenesis, spread all along the past century, the parallel evolution of the preparation of the raw atmospheric forecast and the apparent *lack* of evolution of the conceptual embodiment of the final "subjective" forecast – as well as the lack of impact of this evolution in the popular literature. The understanding of cyclogenesis is shared between various baroclinic instability theories favoured by fluid dynamicists and the less linear baroclinic interaction of finite-amplitude precursors preferred by synopticians. The raw forecasts are now prepared independently of any concept, using data assimilation and numerical weather prediction. But, after the forecasters have interpreted the latter in the framework of the former, they can only express their views using the same graphical language as in the mid-1920s, rooted in a completely different conceptual framework.

Part of the reason is that explanations in terms of instabilities have been overwhelmingly favoured by atmospheric research scientists for decades but these explanations never made much of an impression on synopticians, ever since the beginning (see e.g. Petterssen, 1955, last section). The preparation of a forecast is really a synopticians' work. The forecasters on one side and most dynamicists on the other had little influence on each other's methods or practices. This situation has lasted for a long time. It is far removed from the early days in Bergen, when the very same people were preparing the forecasts and carrying on the research.

None-the-less, as will be recalled in this chapter, dynamical meteorology has evolved over the years a framework that can, and indeed should, be the basis for interpreting raw numerical forecasts, assessing them and outlining a final, "Subjective" scenario (Section D). The main purpose of this chapter is to show how this framework has led a mixed group of forecasters and scientists from Météo-France together to define a new graphical code for summarizing the state of the atmosphere on the synoptic scale. This will be done through examples in Section E. The same dynamical framework allows extensions of the approach, either making the summary document much richer or allowing a closer cooperation between the forecasters and the numerical prediction suite (Section F). Prior to this, the importance of graphics and the limits of the existing code will be recalled in Section B. The historical perspective is also important and it is discussed in Section C.

B. THE EXISTING GRAPHICAL SUMMARY MAP: ITS IMPORTANCE AND ITS PROBLEMS

Our society depends more and more on graphics. This is the most efficient way to convey rapidly large quantities of information. As far as the description of natural phenomena is concerned, words are only poor communication tools. Every day, an in-depth study is conducted by central forecasters in monitoring and assessing the most likely evolution of the atmosphere on the synoptic scale. The result is currently translated into the existing summary map and, at the same time, a bulletin. Because of the limitations of the map, the bulletin is mandatory. But this also limits the impact of the central forecasters' conclusions on their colleagues and customers. In order to reach its full potential, this material should be prepared in graphical form for dissemination.

The existing graphical summary of the synoptic situation is a map showing one field, sea-level pressure, and a number of line structures mostly related to fronts (Fig. 1). A few other symbols, such as one for marking tropical cyclones, are available. In France, this document is called *Isofront* when it is based on observations and *Préiso* when it summarizes a forecast. However, the graphical language employed in these maps enjoys quasi-universal recognition. It is normalized by the World Meteorological Organization (WMO publication No. 485, 1992 Edition, II–4–12: Organisation Météorologique Mondiale, 1992). It is employed for international exchanges between weather services, but it also forms the basis of graphical exchanges between meteorologists and "educated" customers such as mariners or aircraft crews. This language is

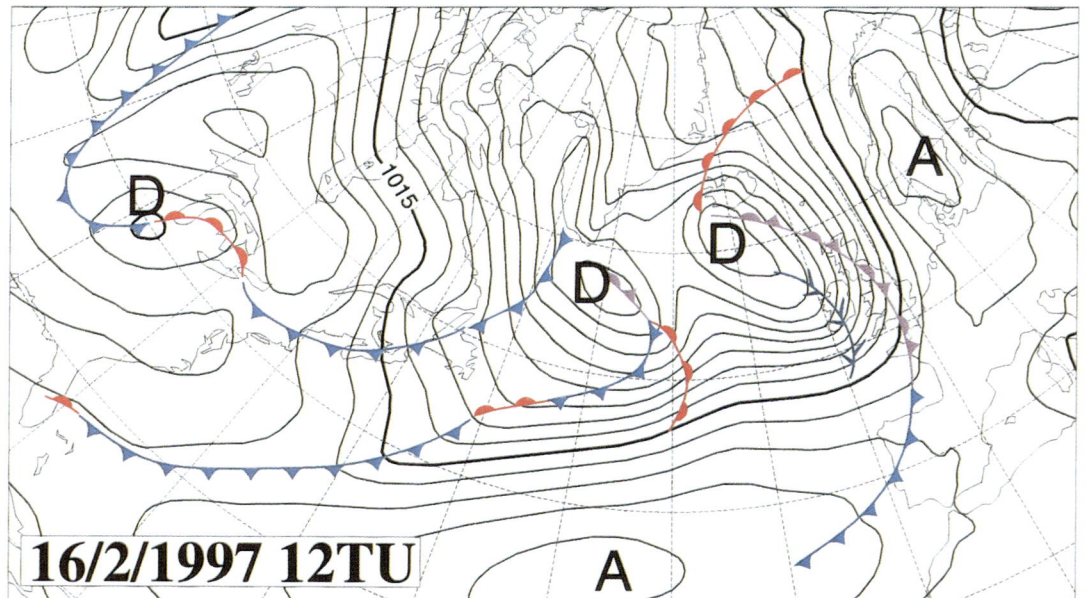

FIGURE 1 An example of a reference synoptic summary map for Europe: these are drawn and printed by the Deutscher Wetterdienst twice daily. While an excellent way to keep the memory of past weather, to be exploited fully, it requires other maps. This figure is an exact redrawing of the DWD on the FASTEX 4D-VAR reanalysis of surface pressure (solid lines, contour interval: 5 mbar, heavier line: 1015 mbar). Note that in this figure and all similar ones the date, the low and high centres are labelled in the French style, with *TU* denoting Universal Time Coordinate, *D* denoting a depression or low-pressure centre, *A* an anticyclone or high-pressure centre.

often familiar to the general public: it is used in newspapers and technical TV bulletins. In many books dealing with meteorology, the "words" of the language are defined with the same 80-year-old vertical and horizontal cross-sections.

So much for the successful side of this graphical summary. This explains why fronts are always involved whenever the topic of meteorology comes up.

At the beginning of the 1990s (and perhaps earlier), French forecasters were meeting the limitations of this approach almost every day. They began to talk about it and discuss possible solutions. The work started from these concerns and of a preliminary proposal discussing what should have been the impact of current knowledge in dynamical meteorology on the graphical depiction of midlatitude weather (Joly, 1993).

The results of the fruitful discussions that took place afterwards first led to an internal definition of a new document (Santurette et al., 1997) and then to developments and tests. This new document is called ANASYG (analyse synoptique graphique, graphical synoptic analysis) when it is based on observations and PRESYG (prévision synoptique graphique, graphical synoptic forecast) when it is based on numerical forecasts.

Many of the questions that were addressed when devising this proposal existed in the minds of many colleagues. Their thoughts have sometimes been published. Most noteworthy is the article written by Mass (1991). It presents a lucid and documented analysis of the position. Some of the deficiencies of the current approach are documented and several lines of attack are proposed. Our own approach can be seen as a mixture of these different possibilities. Some of the criticisms regarding the use of frontal symbols are further addressed by Sanders (1999), also including some proposals.

Here, only some of the most frequent limitations will be mentioned as richer discussions can be found in the above-mentioned articles. On the one hand, having access to satellite imagery and to extremely realistic model runs, meteorologists are perpetually faced with the fantastic variety of shapes and of evolutions that weather can take. On the other hand, they have to describe this variety graphically using a handful of symbols devised during the 1920s and basically confirmed in 1950 when WMO was established under its current status. As a result, a number of symbols are overused or, in other words, the same symbol is used to mark very different atmospheric structures. The writer or preparer of the document knows what phenomena he or she is trying to depict. However, the reader who has access only to the summary map has no chance to interpret correctly the meaning of the message. The graphical language has become ambiguous and confusing.

Figure 2 illustrates some of the shortcomings of the existing graphical summary. The number (1) points to a cold front nearly 3000 km long. Does the presence of this symbol mean that the rainfall activity is the same all along that front? Number (2) points to a front that appears to meander continuously from the Bahamas up to Bergen in Norway. This means that there is, along this long curved line, some significant temperature change from one side to the other. But does it have the same strength all along? More important, a front is also an area of large wind changes. Is it really possible to have a similar large wind shear all along this line? Number (3) points to two structures depicted as a warm front. Can we see from the map whether one is strengthening, or both, or whether one or both are decaying? The pressure field gives no indications, on this summary, that can help to infer tendencies for all the structures that can be seen. This is even more critical for the many lows that can be seen on this map or in Fig. 1: it is very difficult if not impossible to know which

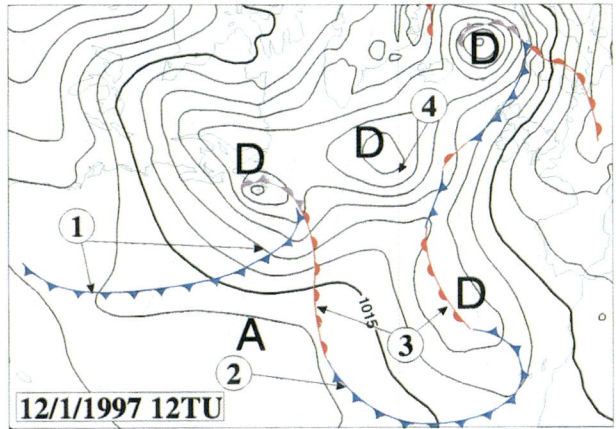

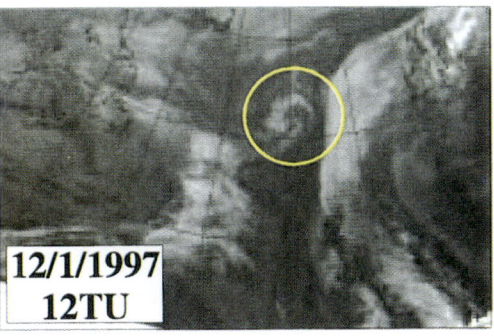

FIGURE 2 Top: redrawing of the DWD synoptic analysis, 12 January 1997 at 12 UTC. The numbers are referred to in the text. They point to serious limitations of the current graphical summary. Bottom: the composite satellite image from NOAA GOES and EUMETSAT METEOSAT at the same time. Note the comma-cloud structure (circled), corresponding to number (4) of the top panel.

lows are developing, and which are dying. Number (4) in Fig. 2 points out one of the worst shortcomings of this map: the satellite image clearly shows an organized cloud system, a comma cloud. The map shows nothing there. Yet, this phenomenon is essential to enable the reader to spot at a glance what can happen next—where the critical areas are in terms of weather; the comma cloud will interact strongly with the front there. Conversely, most of the line marked as a front is barely active and of little practical use. This case will be returned to in Section E.2.

This example reveals an obvious lack of symbols to represent important, real, features of the atmosphere. It is also a good example of an overuse of the frontal symbols. In fact, two of the available symbols are much overused. T_1 is the occluded front mark. The definition referred to is always the original one (Bjerknes and Solberg, 1922), due to Bergeron. However, vertical cross-sections rarely (to say the least) reveal this structure. While it is probably possible to define in dynamical terms an occluded cyclone, the idea of an occluded front seems evasive. The other overused symbol is that of the cold front. Besides actual cold fronts, it is employed for many kinds of line structures on many scales, from a transoceanic cloud band to the border of a mesoscale cold pool. Generally speaking, there is a tendency to mark as a front any area of increased horizontal temperature gradient, whatever the other fields are (Sanders, 1999, among others).

C. HISTORICAL PERSPECTIVE

The Bergen ideas, from which the existing summary map directly comes, remain a model of success. Quite a few aspects of existing practices inherited from V. Bjerknes's group cannot be changed ruthlessly by just claiming that they are old and that they have been designed for a meteorology that could not dream of numerical weather prediction nor of satellites. It is of interest to the present approach to attempt to recover their initial intent.

1. The Original Intent of the Bergen Method

There are two aspects to what shall be called here the "Bergen method", sometimes called the "Norwegian school". One is the use of graphical syntheses in the process of producing a forecast. The other is the meteorological content of these maps, including the summary frontal map. The first aspect is of little use to us: it is the part that has to be carefully rethought. The second aspect, we argue, is worth keeping, worth generalizing, but first, it needs to be dusted off.

Articles written and published by members of the Bergen group around 1920 are an obvious reference for recovering the original intent of the method. Other sources are the international codes (the Utrecht code of 1923, the Zürich code proposed by Bergeron in 1926) put together by several "Commissions" and then endorsed by the "Conference of the Directors of Weather Services" that seems to have been the ruling body of the international organization of meteorology at the time. The most active representatives of the Bergen group in these institutions have been V. Bjerknes and the wonderfully creative T. Bergeron. Their hands are to be seen everywhere in the resulting schemes (see e.g. Giao, 1927). However, the best overview making use of all available documentation, most notably unpublished material such as letters, is the professional historian's book by R. M. Friedman (1989). The plan of the Bergen group is fully exposed.

The Bergen group proposes the polar front assumption (which is not presented a such, but as a fact) as a way to organize the proper use of the very scarce observations available. The idea is that you do not need much information if you already know what you are looking for and what it looks like: the whole idea of a powerful conceptual model lies there. It is not our goal to delve into the details of this, as it has already been analysed by Friedman.

What is worth bringing out is that, while making this analysis, Friedman and, in fact, the Bergen people themselves reveal that the primary weather phenomena that the Bergen method attempts to capture is *cyclogenesis*. The fronts, in the Bergen method, are not central to the analysis of the data for themselves, *they are central to the analysis because fronts are seen as the source and cause of cyclones.* After summarizing his views of the cause of vertical motion in a cyclone, V. Bjerknes (1920) writes, "This view ... gives a clear indication of the lines to be followed in the attempt to predict it. The main point will be to discover the line of demarcation between the two kinds of air ...": things cannot be said more plainly. This is one key idea that time has dimmed somewhat. Just as important, fronts are believed to be the *only* source of cyclones. For this reason, the method makes no allowance for such things as comma clouds or polar lows.

Another idea about the Bergen method is that the surface pressure map with fronts is just a surface map. It is originally meant as a 3-D map. In the same way that the concept of a polar front is very useful to interpret observations in the horizontal, the vertical structure of the fronts and cyclones, resulting from a mixture of observations and assumptions, provides the third dimension. Having the conceptual model in mind, the map can be read in three dimensions. "Thanks to the

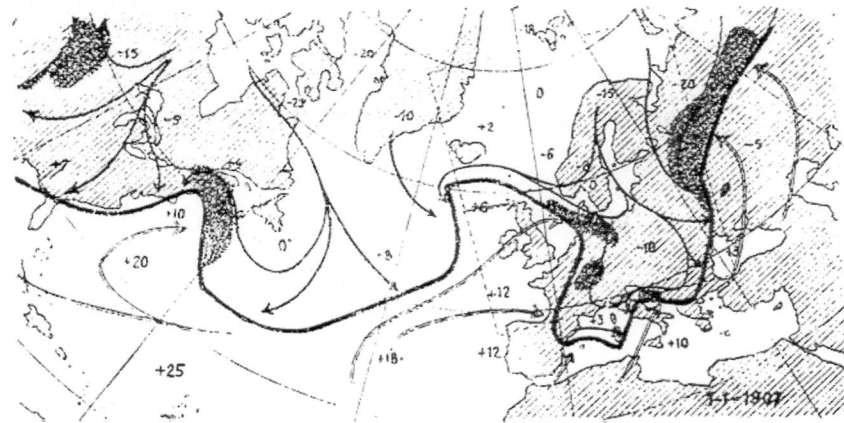

FIGURE 3 One of the first hand-drawn polar front analyses, by H. Solberg in 1920. The various types of fronts are not distinguished by specific symbols, the frontal structure and the activity are represented separately. (From Friedman, 1989.)

surfaces of discontinuity", says an account of V. Bjerknes's lecture of 12 May 1924 to the Royal Meteorological Society (Bjerknes, 1925), "we can simplify the atmosphere and treat it as consisting of nearly barotropic layers, separated from one another by strongly baroclinic layers, or, in extreme cases, by surfaces of discontinuity".

They pay attention to the surface level not because people are living here but because (i) this was the location of most observations and, above all, because (ii) according to their conceptual model, the surface or near-surface levels are driving the atmosphere above: the upper air only follows, especially during cyclogenesis. This is repeated in section 2 of Bjerknes and Holmboe (1944), even though the importance of the upper levels was quite fully grasped at that time: "... the initial impulse for wave formation is always available due to the shear at the polar-front surface. This paper purports to show how the baroclinic air mass *above* the polar-front surface *reacts* to such a wave impulse" (our italics). This initial idea was doomed by the development of radiosounding in the 1930s and later. This led to the third dimension being represented by separate maps.

The authors believe that the two key ideas that were part of the foundation of the Bergen method, namely (i) a synoptic map centred on the cause of cyclogenesis and (ii) showing at the same time the atmosphere in three dimensions, are as important as they were initially. A new approach must not only retain these ideas but most also put them back on the forefront. It will be shown below that it is possible to recover the third dimension while keeping much of the simplicity of a single map.

2. Evolution of the Form of the Graphical Summary

As the early drawing by Solberg shows (Fig. 3), in the original Bergen method:

- there was no distinction made between various types of fronts (they are all represented with a heavy solid line, dashed in the published diagrams);
- the areas of precipitation were represented explicitly (with hatched areas), independently of the frontal structure itself.

This shows, first, that the frontal symbols were not God-given as we know them, and therefore, they can be discussed and changed if needed. To support this, Fig. 4 shows the first map archived by the French weather service and bearing fronts (1929). Note that the symbols are different, but that the various types of fronts have been employed. Note also the separate marking of precipitation.

The second aspect, the separate marks for features that *supplement* the temperature and wind structures, have been kept alive for quite a long time. Figure 5 shows the first map disseminated both inside and, from this time onwards, outside of the French weather service (1949): not only are the active areas shown but a distinction is even made between stratiform rain and convection. Actually, this is still explicitly part of the WMO norm, but is now rarely used. This contributes, like forgetting the initial intents, to impoverishment of the document.

Sometimes one gets the impression that meteorologists show their respect for the initial achievements of the Bergen method by considering that the "letter" of

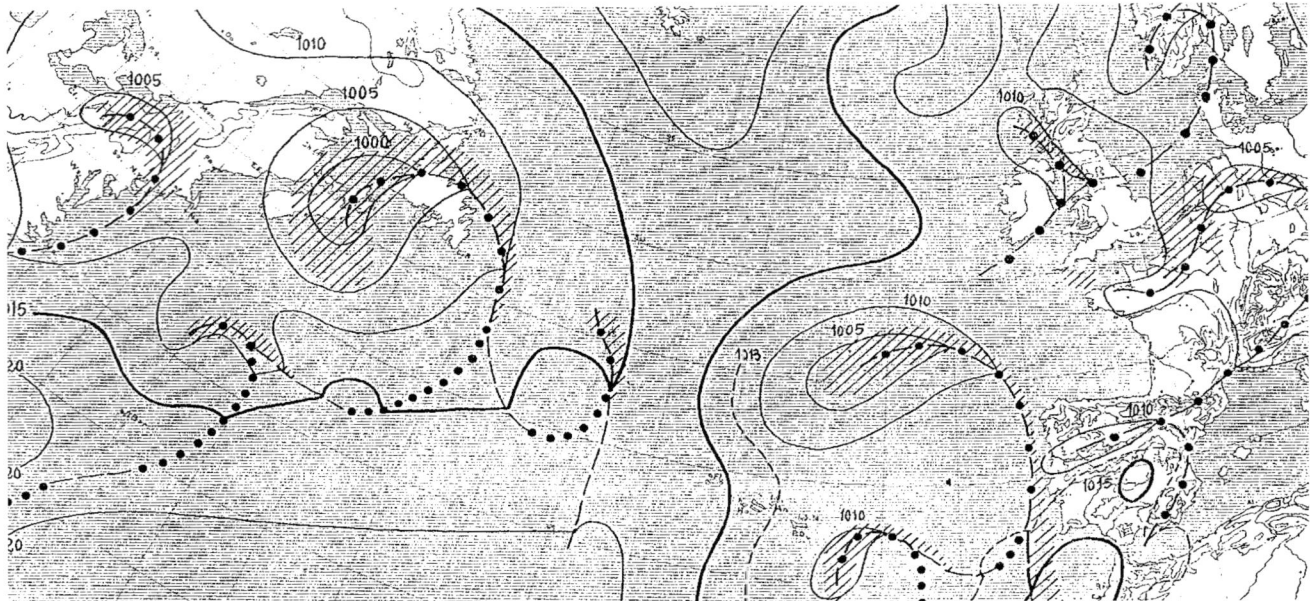

FIGURE 4 An enhancement of the first map bearing fronts printed for internal use at the French meteorological office (Office National Météorologique), showing the situation on 1 July 1929. Fronts were being analysed a few years earlier, but only in the Paris forecast room, and they were not printed: all one can see in the archives are the characteristic breaks in the isobars. Note that the different types of fronts are indicated, but with a different code than the one in use at Bergen at the same time: heavy dots stand for cold fronts, dashes for warm fronts and dash-dots for occlusions. Note also the explicit indication of activity with hatches.

the method must be continually obeyed. The "letter" is the set of frontal symbols on the surface pressure map. This is the wrong kind of obedience. The true heritage of the Bergen method lies not with the current analysis practice but in the initial project of producing a complete, self-contained, but at the same time simple, graphical depiction of synoptic-scale weather and its precursors.

D. RATIONALE OF A DYNAMICALLY GROUNDED SYNOPTIC SUMMARY

It turns out that dynamical ideas arising from the continuing research effort on the synoptic scale since the 1920s can greatly help to place the synoptic-scale graphical summary on a firm dynamical framework.

Helped by our analysis of the successful example from Bergen, the following guidelines can be laid down for a new graphical summary:

- Cyclogenesis remains the primary source of very bad weather at midlatitudes on the synoptic scale: the graphical summary should depict ongoing cyclogenesis as well as allow an assessment of where each case stands in its life cycle. It should also contain information on potential risks of future development or possible alternative evolutions. In short, the graphical summary should convey an idea of what *explains* or causes the cyclogenesis.
- We know now that the key mechanism to depict is, therefore, *baroclinic development*. This term is used as a way to distinguish the finite-amplitude interactions between various kinds of coherent structures in the real atmosphere from linear baroclinic instability. Recent climatological work (Ayrault and Joly, 2000) shows that the baroclinic interaction is the *only* mechanism that explains large developments (deepening larger than 10 mbar). However, this mechanism can take many different forms, from the classical Petterssen Type B paradigm (Petterssen and Smebye, 1971, or, rather, the Sutcliffe, 1947, paradigm) to the direct interaction between a weak low and a barely disturbed jet-stream aloft (Baehr *et al.*, 1999) that the preferred path to explosive cyclogenesis off the European coast.
- None-the-less, cyclones are not the only sources of "weather", i.e. of strong surface winds, deep precipitating clouds and other related features. There are other features of the flow, which may or may not have a clear signature near the surface, that can lead to such activity, such as zones of large changes in the jet-stream. The graphical summary should include indications of most sources of balanced vertical motion. As for cyclogenesis, the

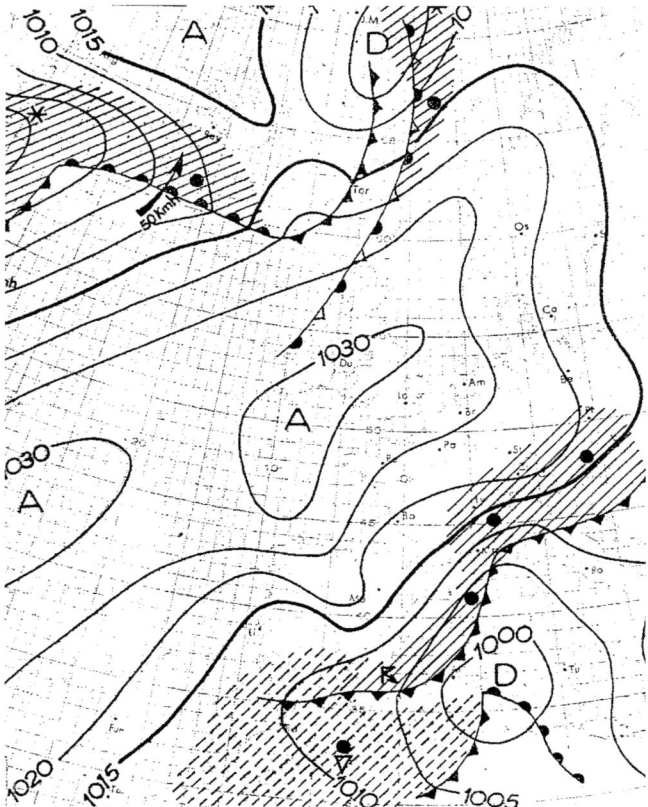

FIGURE 5 The first map bearing a frontal analysis and released to the public by the French meteorological office, showing the situation on 30 April 1949. The frontal symbols are the usual ones but note the distinction made between areas of shower activity (dashed shading) and stratiform rain (hatching). Note the explicit indication of displacements (top left) and the strange single frontal band with special marking (top right area).

graphical summary should contain not just an indicator of the process itself, but also indications of its cause.

- In the interest of completeness, the graphical summary should also enable the identification of a number of known more-or-less coherent features seen in the atmosphere and that can be a potential danger, but may not be explained by the synoptic flow configuration alone. This typically includes the various forms of organization of convection: in clusters, in line, etc. This also covers polar lows, tropical cyclones and others. The important point is that specific symbols should be attached to these structures so that there is no interference with the balanced features of interest. The graphical summary should then allow the identification of the possible interactions between such organized or semi-organized features with the balanced part of the flow.

Furthermore, if this attempt is to be successful, it cannot ignore the important use of the existing "surface map", either in analysis or forecast forms. This imposes an important constraint:

- It should be possible to easily extract from the new summary a simplified form close to the existing one, as defined by the current WMO norm.

The main dynamical foundation on which our proposal rests is the existence, on the synoptic scale, of some kind of balance. This implies that variables like potential temperature θ, wind **V** and others are not independent but that each can be deduced from the other. Using symbols, this property can be seen as arising from the idea of a conceptual model that was introduced in meteorology during the 1910s (independently in at least Britain, France and Norway). The idea is that, under certain circumstances, meteorological fields always combined themselves in a similar way. Once this is accepted, a single field or even a specific symbol will be enough to depict the whole combination. The existence of a balance in the atmosphere is reflected in the works of Charney (1947), Eady (1949) or Sutcliffe (1947), but it was really formalized by Charney (1948) and yields the extremely useful quasi-geostrophic representation of the atmosphere. The structures that will be highlighted as "causes" or potential causes of meteorological activity or cyclogenesis are those that are identified by finite-amplitude quasi-geostrophic theory and its close parent, semi-geostrophic theory (Hoskins, 1975).

The presence of a balance means that the state of the atmosphere can be reduced to two such fields: potential vorticity and moisture. In the approach proposed here, moisture is not treated explicitly, but indirectly. Using the balance assumption, it is possible to retrieve the structure of the other "dry" state variables from the potential vorticity: this is called "inversion" (Kleinschmidt, 1950; Hoskins et al., 1985).

A further reduction is possible. Indeed, most weather depends on what takes place in the troposphere. Another fact of observation, like the existence of a balance, is that potential vorticity in the troposphere can be considered to be, to a good approximation, *uniform*. In this case, the mathematical form of the inversion problem becomes very close to the classical inversion of Laplace operator. The interior part of the fields is then entirely determined by the boundaries, in our case the near-surface or top of the boundary layer and the tropopause (see Hoskins and Bretherton, 1972, for further references, in particular in the Monge–Ampère form of the inversion problem taken assuming semi-geostrophic balance).

Combining the idea of balance with the assumption of uniform tropospheric potential vorticity leads us to a

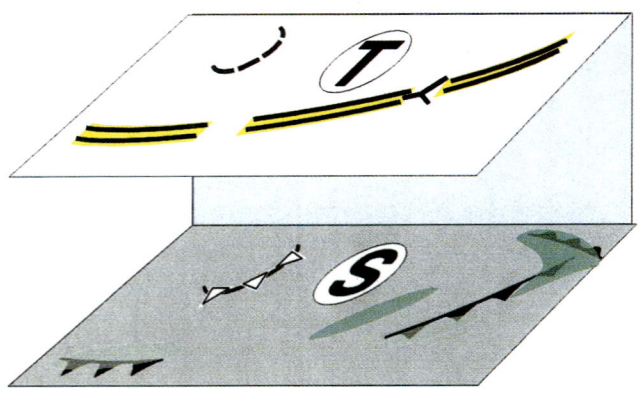

FIGURE 6 A schematic showing the principle of the proposed graphical summary. It is a two-level representation of the troposphere, based on the assumption of uniform potential vorticity: the top of the boundary layer (S) and the tropopause (T). All the significant sources of weather can be represented by a symbol located at one of the two levels, where their amplitude is the largest.

two-level representation of the synoptic scale (Fig. 6). We know, from the dynamical theorems briefly recalled here, that this is the *minimum* information set that can enable us to give a reasonably complete picture of the current weather. This background also explicitly sets the limit of this approach: when large tropospheric anomalies of potential vorticity are present, our description will not be the most complete one.

This background clearly tells us that we cannot do anything simpler. A one-level representation will never enable the formation of a complete picture showing systems and their development, contrary to the initial hopes of the Bergen group: not everything is driven from the surface. Independent events occur at the tropopause and must be indicated as such.

It has long been known, in synoptic practice, that further maps at upper levels are required to supplement the surface map. However, the assumptions of balance and uniform tropospheric potential vorticity tell us that only one more level is really needed. There is more to it: these same theorems tell us *where* these levels are best chosen. The mathematical result is that the amplitude of the departure from a mean or climatological flow of fields like θ, **V**, geopotential ϕ, vorticity ζ, etc. (fields that may be called "geostrophic" for simplicity) is a maximum at the boundaries. Therefore, the best signal-to-noise ratio for picturing these fields will be found at the near-surface and at the tropopause. These theorems give extremely solid and very clear indications as to which levels are important. The tropopause level, for example, is defined by fields on a map of constant potential vorticity value such as 1.5 pvu or 2 pvu (poten-

tial vorticity unit: 1 pvu = 10^{-6} K m^2 kg^{-1} s^{-1}). These theorems are therefore meaningful both for structuring the new summary document and for organizing its preparation.

E. EXAMPLES OF USE OF THE NEW GRAPHICAL FRAMEWORK

The new graphical summary will now be shown using examples rather than by a formal description. The latter can be found in the Appendix and is discussed in Santurette and Joly (2001). A few winter examples will precede a summer case. The winter examples are selected from the well-documented FASTEX cases (Fronts and Atlantic Storm-Track Equipment, Joly *et al.*, 1999). The base maps employed to prepare the graphical summary result from a 4D-VAR reanalysis of all the data.

1. A Large Transoceanic Cloud Band

This first example will be constructed step by step in order to outline the components and the methodology of the graphical summary. This case is introduced by Fig. 7. It has been chosen because the presence of a large transoceanic cloud band on a satellite image is perceived by some people as sufficient indication of the presence of a front. The cloud band is then inevitably drawn as a very long uniform front line (the corresponding classical analysis of this case is Fig. 1). The modern analysis of these situations attributes most of the banded cloud activity to the horizontal changes of the velocity in the jet-stream. The new framework introduces the possibility of representing the jet and associated baroclinicity *per se*, thus leading to a quite different perception of the situation.

The first step is to identify the large-scale regime (in the sense of Vautard, 1990) and then the significant upper-level features. To do this, entirely new symbols are needed. They represent explicitly the jet-stream by a double heavy solid line (⇒) as well as some of its remarkable velocity gradients (such as ◁ for a diffluent zone) and several other upper-level coherent features.

The regime here is clearly zonal, reinforced by a confluence between the subtropical jet-stream out of the Gulf of Mexico and a higher-latitude large-scale trough. This trough is over the Great Lakes area, while the confluence lies south of Newfoundland. Using a map of the wind vectors above some threshold on the 1.5 pvu surface together with its geopotential height, potential temperature and relative vorticity fields, the flow at the tropopause is first represented (Figs 8 and 9). Here, it has been chosen to

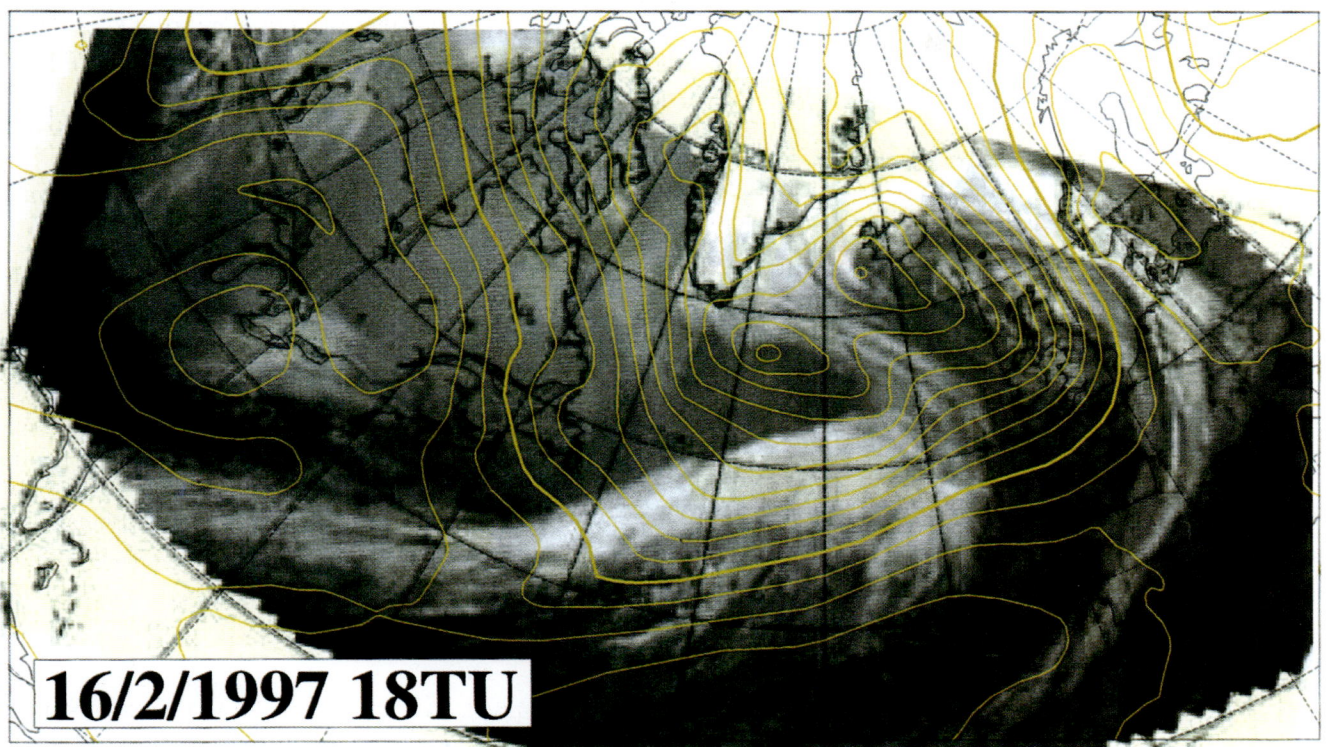

FIGURE 7 Composite infrared satellite image for 16 February 1997 18UTC, with the mean-sea-level surface pressure field from the FASTEX 4D-VAR reanalysis. Contour interval: 5 mbar, heavier line: 1015 mbar. The image mixes NOAA GOES east and EUMETSAT METEOSAT radiances. The compositing has been performed by the Centre de Météorologie Spatiale of Météo-France in Lannion. This case is an example of the presence of a large cloud band over the Atlantic ocean that is too often interpreted as a very long, single front: see Fig. 1.

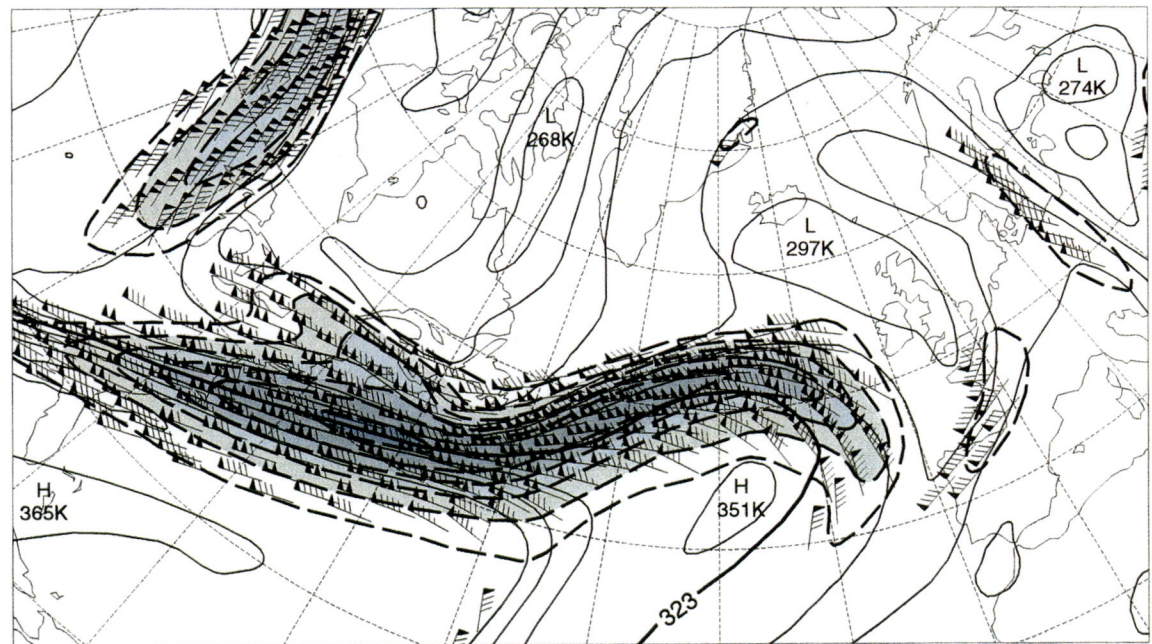

FIGURE 8 Tropopause wind vectors above 40 m s^{-1} (arrow flecks, to be interpreted in kt) on 16 February 1997 at 18UTC. The tropopause surface is defined as the 1.5 pvu potential vorticity surface. Heavy dashed lines: wind velocity, contour interval 10 m s^{-1}. Solid lines: potential temperature θ, contour interval 10 K.

FIGURE 9 The upper-level part of the proposed graphical summary on 16 February 1997 at 18UTC, showing specific new symbols in relation to some of the fields they are summing up. These are the tropopause geopotential height (contour interval: 1000 mgp, heavier line: 8000 mgp) and relative vorticity at the same level (heavy lines, values larger than 10×10^{-5} s^{-1} for readability, contour interval: 5×10^{-5} s^{-1}). The background shading is the wind velocity at the same level, shown in detail in Fig. 8. The tropopause is defined as the 1.5 pvu potential vorticity level.

underline the zonal regime by setting the new jet-stream symbol over only 50 m s^{-1} isotachs on the north Atlantic and by not marking the northernmost inflow branch upstream of the confluence. Its presence is strongly suggested by the northwesterly flow of the trough over Canada and by the confluence symbol. With these choices, the anticyclonic flow over Europe is also only suggested, even when moving the threshold for the jet-stream to 40 m s^{-1}.

The type of 3-D structure to be mentally associated with the jet-stream symbol is shown by Fig. 10 (the corresponding *conceptual model*). The temperature gradient across the zone is visible at all levels and maximum near the surface and near the tropopause. On the other hand, the signature on the wind field is concentrated near the tropopause. The new code allows for drawing further features such as confluent/diffluent zones, secondary branches (which we propose to call *drift*, following the oceanographic usage), etc. This should be done carefully so that the main stream is readily visible. Although the main signature of this symbol will be found, by definition, in the horizontal organization of the maximum wind velocity near the tropopause, one should also keep an eye on the corresponding low-level temperature gradient. This wind-based symbol is in principle also a temperature gradient indicator. The location of the low-level area of largest temperature gradient, denoted by $y(\nabla T_{\max \text{ low lev}})$ for example, is given to first order by semi-geostrophic theory (Hoskins, 1975) as:

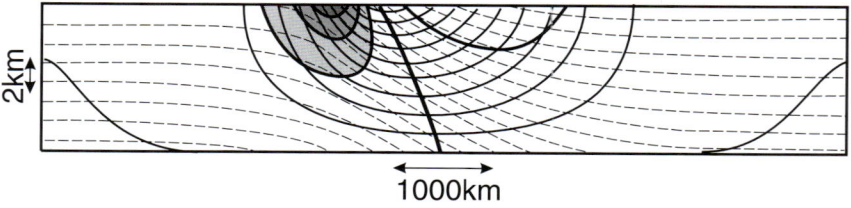

FIGURE 10 Vertical cross-section showing an idealized jet-flow baroclinic zone. The fields are potential temperature θ (thin dashed lines, contour interval 4 K), wind into the section (solid lines, contour interval 6 m s^{-1}, smallest value 0 m s^{-1}) and absolute vorticity (heavy solid lines and shading, contour interval 2×10^{-5} s^{-1}, heaviest line 10×10^{-5} s^{-1}). This flow is the result of an inversion in a uniform potential vorticity fluid.

$$y(\nabla T_{\text{max low lev}}) = y(U_{\text{max}}) - \frac{U_{\text{max}}}{f}$$

in a cross-jet vertical plane with y pointing towards the cold air, U_{max} the maximum velocity of the jet, $y(U_{\text{max}})$ the location of the maximum velocity at the tropopause and f the Coriolis parameter. The stronger the jet, the stronger the warmward displacement of the low-level counterpart of the baroclinic zone with respect to the location of the symbol.

Once the main flow is represented, the same fields can be used to locate significant vortex-like upper-level anomalies. Here, attention must be paid to the time series of the basic fields employed. Only structures with a minimum time persistence should be indicated. The code calls for making a distinction between "actively involved" (solid marking) and "other" (dashed marking) upper-level features. This requires some input from other fields, such as the surface pressure map and the 400 mbar or lower vertical velocity. A more detailed example of upper-level feature drawing is presented below (Section E.2 and Fig. 15).

The next step is to identify the frontal zones and the other synoptic scale line organizations (Fig. 11). The basic fields for this analysis are low level (850 mbar, say) wet-bulb potential temperature θ_w, vorticity and wind vectors and magnitudes.

The proper conceptual model to associate with a frontal symbol is represented by the vertical cross-section of Fig. 12. It must be compared with Fig. 10. Both structures have a noticeable cross-plane temperature gradient. In a real front (see e.g. Thorpe and Clough, 1991, for well-documented, objectively analysed cases from the FRONTS 87 project), the relevant thermal parameter is a moist potential temperature (θ_e, or θ_w). Therefore, the temperature field *cannot* be employed alone in order to distinguish a frontal zone from a jet-flow baroclinic zone. The main differences are in the low-level wind field. Realistic idealized models reveal two distinctive features of the wind field associated with a front: (i) the presence of a well-defined low-level vorticity maximum, and (ii) the presence of at least one low-level jet, which can be either on the cold or warm side or on both. These features are confirmed by observations.

We propose and we insist that frontal symbols should *not* be employed in the presence of a single field signature, such as a strong thermal gradient. The value of the concept of a front is that it is a balanced structure that associates characteristic features in several fields, most importantly vorticity and a moist potential temperature field (dry balanced fronts on the synoptic scale are rare). In this respect, our multiparameter approach to the definition of a front is very different from that of

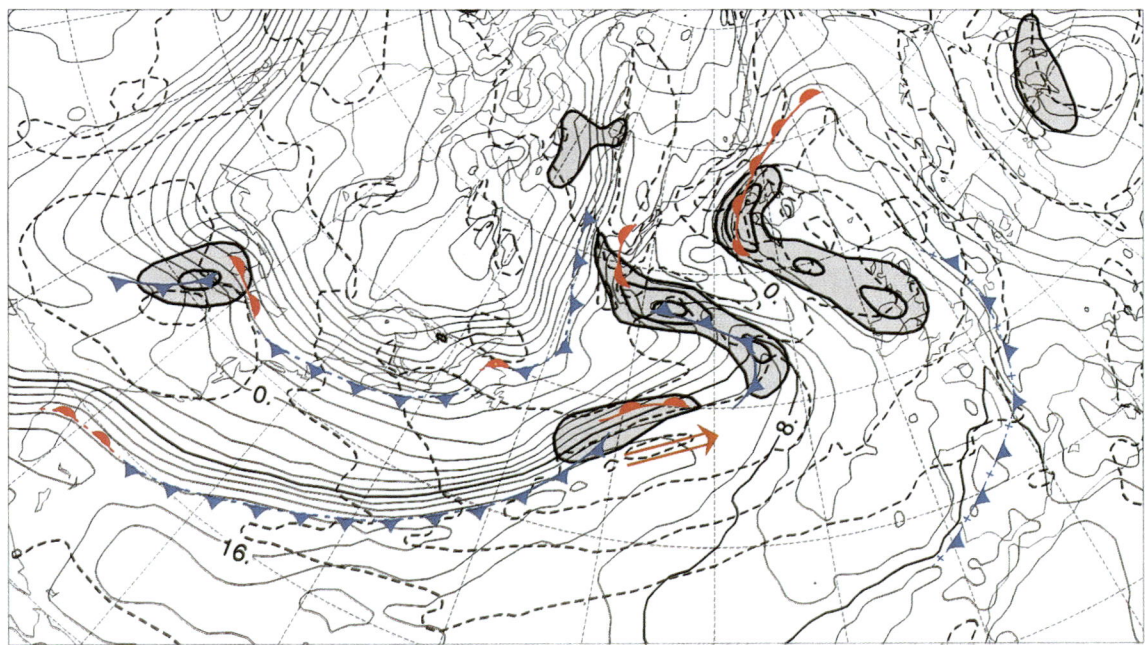

FIGURE 11 The low-level part of the graphical summary on 16 February 1997 at 18UTC, showing the dedicated symbols with the fields on which they are based. These are moist potential temperature θ_w at 850 mbar (thick solid lines, contour interval 2 K), 850 mbar wind velocity (dashed lines, contour interval: 10 m s^{-1} above 10 m s^{-1}) and relative vorticity (heavy solid lines, shading, contour interval 5 × 10^{-5} s^{-1} above 5 × 10^{-5} s^{-1}. The 850 mbar level is taken as being representative of the top of the boundary layer.

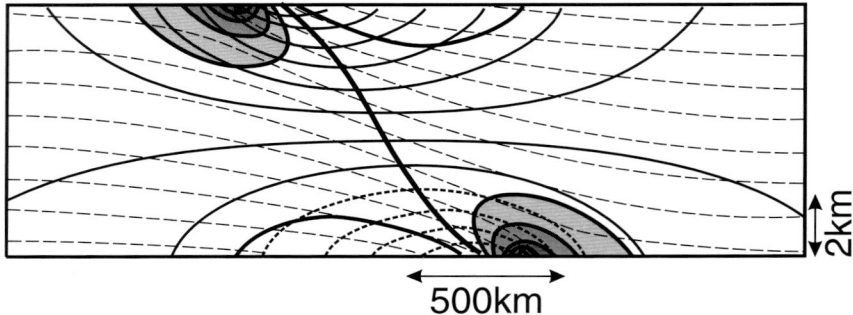

FIGURE 12 Vertical cross-section showing an idealized frontal flow. The fields are the same as in Fig. 10 with the same drawing conventions, except for the wind into the section (solid lines, negative values dotted, contour interval 5 m s^{-1}). Like the jet, this front is the result of an inversion in a uniform potential vorticity flow. The inversion does not take the boundary layer into account: this means that the lowest level in these cross-sections is not the surface but is akin to the top of the boundary layer.

Hewson (1998) and of earlier work cited therein. They use a thermal parameter only, while we insist on the need to attribute the symbol only in places where thermal *and* wind features suitably combine. Of course, in a strictly balanced flow over a uniform aquaplanet, this would not be required. However, the practice of everyday analysis reveals the existence of numerous areas of significant temperature changes that have little or nothing to do with fronts. These are discussed below.

Concluding with taxonomy, frontal flows and jet flows should be viewed as two distinct "species" of the more general "family" of baroclinic zones. We argue that understanding the current weather implies having the means to separate graphically these two types of structures.

To draw a front, the horizontal gradient of θ_w should be reinforced with respect to the background gradient attributable to the broader baroclinic zone represented by the jet-flow symbol. A lower troposphere vertical velocity may also be helpful. The time series of these fields are needed to check whether a front is growing or decaying, as over western Europe in the example shown (Figs 11, 13 and 14).

Areas of significant thermal gradient that have no counterpart in the low-level wind can then be signalled as a pseudo-front. The pseudo-frontal symbols (for example, ▼⋯▼⋯) are introduced to account for areas of nonfrontal large thermal gradients. Pseudo-fronts are either the remnants of a past frontal band or the result of boundary layer processes influenced by changes in surface properties, orography, etc. Sanders (1999) independently introduces a similar concept.

Another use of the pseudo-front symbol is for locating the near-surface trace of the upper-level jet. Although this introduces some redundancy in the full document, it brings it closer to the existing one when the upper-level component is removed.

It is the case in the example shown, with a line that begins over the Great Lakes as the surface trace of the northernmost inflow to the Newfoundland confluence and continues as a boundary-layer phenomenon relating to the southern limit of ice-covered lands and seas. Most notably, there are few symbols to add to the main baroclinic zone. This is because it is already indicated. Note that in its south-west portion, the confluence implies a frontogenetical situation but the resulting vorticity has not yet reached the modest threshold of 5×10^{-5} s^{-1} (roughly 0.5^f) employed here.

Next comes the documentation of active areas. The main fields to be used are vertical velocity at various levels (Fig. 13), precipitation rates and, in the case of analyses, as done here, satellite images (recall Fig. 7). The vertical circulation along the jet-stream and its changes can be set down. It can be checked that it fits the image reasonably well. Notice the specific indication of line convection over Ireland. The complete code (Santurette and Joly, 2001) has several other symbols available to represent these features: see the Appendix for a summary.

The final document is then shown by Fig. 14 (to be compared with Fig. 1). At this step, a last check of the connection between various elements can be carried out. Most of the activity over the Atlantic ocean can be related to the changes in the horizontal wind at upper levels. There are, however, reinforcements of this activity related to the presence of low-level features, a front-low complex near the main diffluent eastern end of the main jet and an incipient low finely located and analysed here (near 50°N, 40°W). The richness of the new document clearly appears, while, at least in the original colour version, the overall readability is not too reduced. Note again how the reintroduction of an independent activity indicator prevents drawing actual fronts all over the place. It can also help to locate fron-

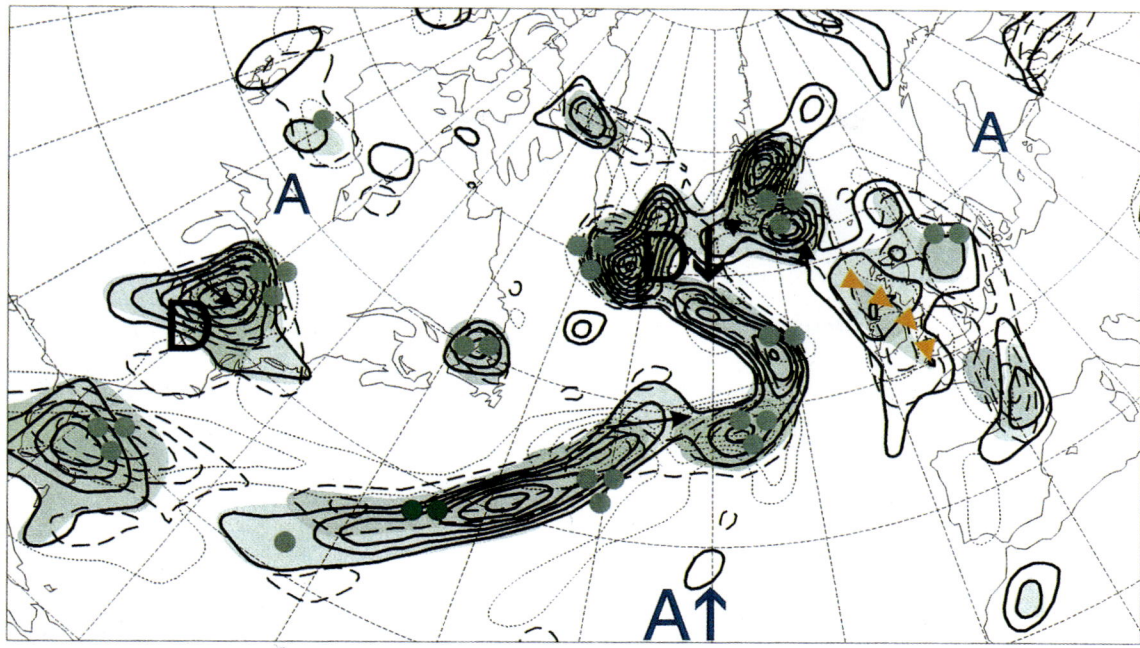

FIGURE 13 An important component of the proposed graphical summary is an explicit mapping of meteorological activity (shading and large dots). One source of information is the vertical velocity field. Contours show the areas of ascent (for readability) at three levels: 800 mbar (heavy solid lines), 600 mbar (dashed lines) and 400 mbar (light dotted lines) for 16 February 1997 at 18 UTC. The field is pressure vertical velocity ω, contour interval 0.1 Pa s^{-1} below 0 Pa s^{-1}. Other sources are the precipitation rate or accumulated precipitation and, as in this case, the satellite image (Fig. 7) and observations, e.g. that used to add a convective line symbol over the Irish Sea.

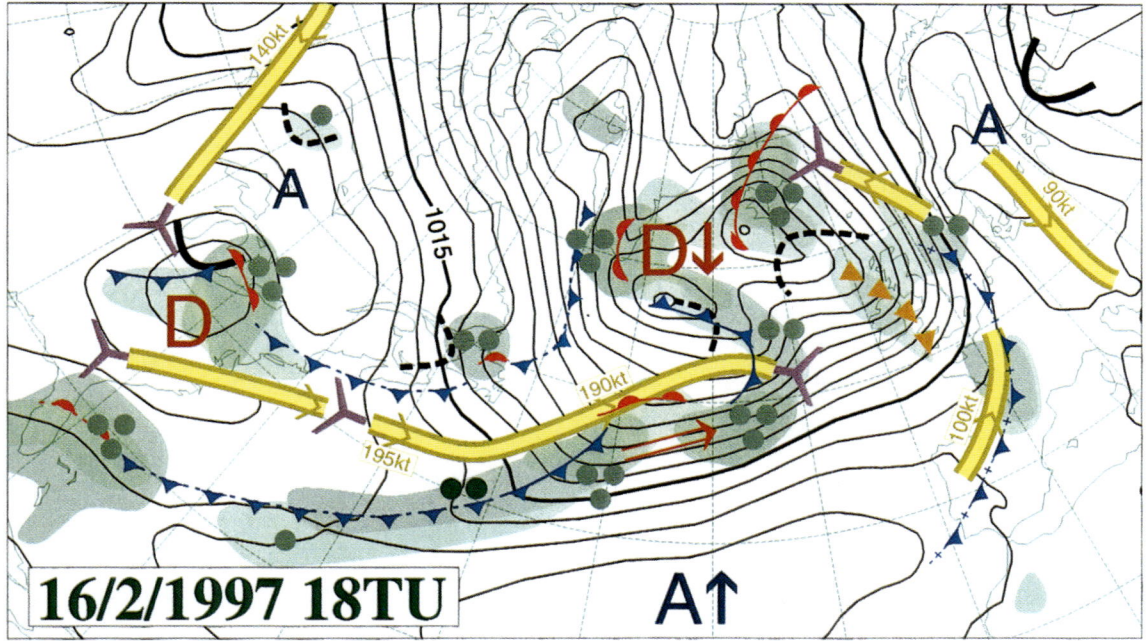

FIGURE 14 Graphical summary of the synoptic situation on 16 February 1997 at 18 UTC using the new framework proposed here and collecting together the information shown in Figs 7–13. This example shows the maxima of the jet-stream velocity: this is one proposed way of indicating the "strength" of the baroclinicity. This summary is to be compared with Fig. 1.

togenetic areas that are not yet strong enough to be termed fronts. The analysis in this map suggests the future possibility of an interaction between the upper-level anomaly in the trough and the Newfoundland confluent zone over the sea that could lead to cyclogenesis. The subsequent evolution is described in detail in Cammas *et al.* (1999).

2. Comma-cloud–Baroclinic-zone Interaction

The comma-cloud case seen in Section B is here considered in the new framework. The satellite picture (Fig. 2) revealed a structure in the cold air which is nearing a meridionally oriented baroclinic zone. The existence of such structures was not anticipated in the Bergen methodology and, as a result, they remain difficult to represent in the existing framework.

The new framework proposed here, on the other hand, makes it easy to clearly identify and signal cloud systems that are driven by upper-level features (Fig. 15). The structure seen in the satellite images does correspond to well-defined, easy-to-track signatures in upper-level fields. This definitely invites the use of the "active upper-level vortex-like" symbols. As a result, the subsequent interaction between this structure and the baroclinic zone, leading to the rapid formation of a cyclone, can be properly anticipated and represented. The activity of convective cloud bands organized by the upper-level vortex can also be represented explicitly.

3. Blocking

A final example from the FASTEX period, shown in Fig. 16, is used here to illustrate how it is proposed to handle blocks. Blocks may move to and fro, but they generally persist for several days in the same broad area. The low-pressure component of a block has a dynamics that is completely different from that of a midlatitude cyclone (Hoskins et al., 1983). It is elusive to try to see fronts in the low of a block. These lows undergo bursts of activity. The most active parts tend to rotate around the low, obviously with the wind, while fronts are remarkably fixed with respect to the low. All these features, which can be extremely intense, require specific marking. The maintenance of blocks depends on their discontinuous "feeding" of warm and cold air by the lows and dynamical highs coming from the diffluent baroclinic zone (Vautard and Legras, 1988).

4. A Summer Situation

Consider now the summer situation of 17 July 1997 (Fig. 17). The new framework diagram has been pre-

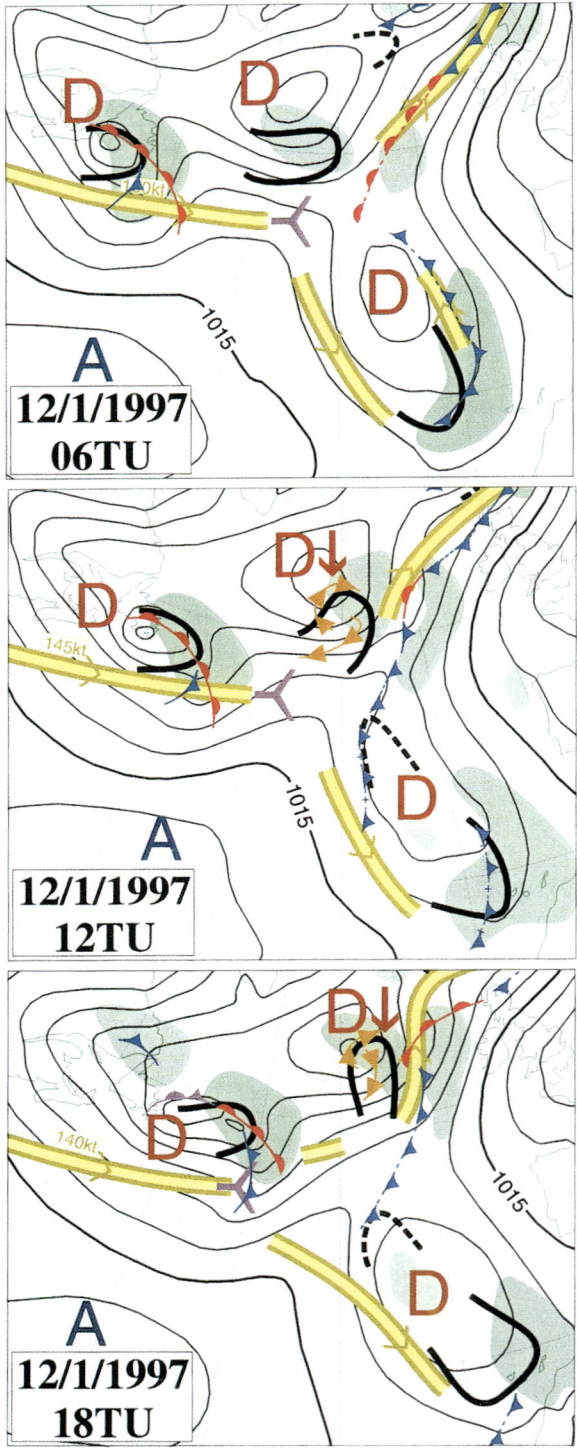

FIGURE 15 Graphical summary of the situation also shown in Fig. 2, showing the ability of the new framework to describe the dynamical and actual weather signatures of the comma cloud.

E. EXAMPLES OF USE OF THE NEW GRAPHICAL FRAMEWORK 97

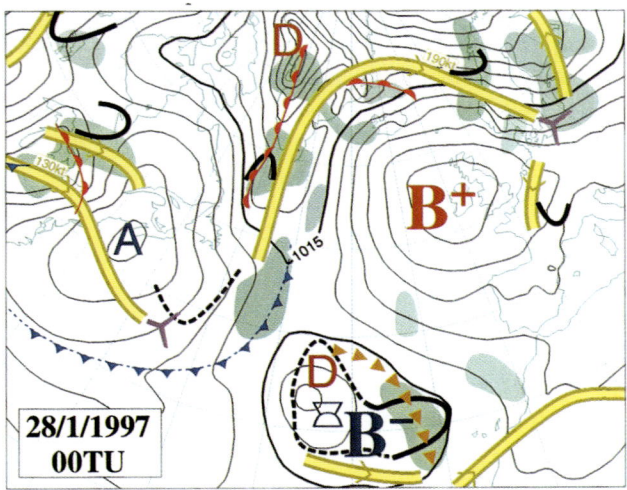

FIGURE 16 Graphical summary of a blocking situation during the FASTEX period. Instead of being marked as an anticyclone and a low, the components of a block are indicated with a large **B**, explicitly showing their association. The weather structures within the low of a block are generally not fronts. The description of these features given here relies on the satellite images. The shape of the upper-level part of the blocking low is also entirely outlined as an indication of possible tilts.

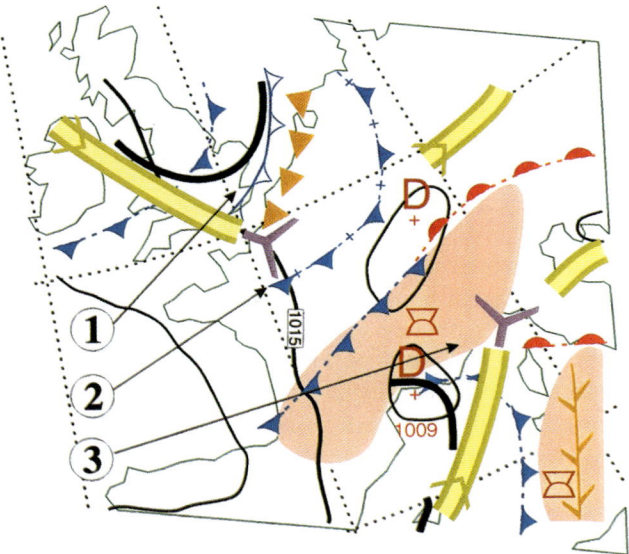

FIGURE 17 Graphical summary of a 24 h summer forecast, for 17 July 1997 at 12UTC. The classical summary, not shown, represented only three or even four cold fronts. The new framework is able to help locate the places of real activity and to describe a wide range of events, including convective outbreaks. Note the split front structure anticipated over northern France and southern England. The features indicated by the numbers (1) to (3) are referred to in the text.

pared entirely with operational products and is based on a forecast rather than analyses and observations. The thresholds employed to mark the upper-level features are also naturally lower than in the winter seasons.

This example shows the amount of information that can be conveyed by the new framework, even in summer. Symbols that are initially required for a proper handling of cyclogenesis turn out, in our experience, to be extremely telling and useful for capturing and understanding the organization, amplitude and persistence of convection,

The traditional representation of this situation is limited to a quite flat, barely useful, pressure field and to three or four cold-front line symbols, all looking the same. This is all that is left to the forecaster, who knows, when drawing the summary, that none of these structures is actually a front. In fact, all the useful information must be written up in words in a separate document, the written bulletin, which needs to be read carefully.

The new graphical summary is immediately useful. First, note that there are, indeed, no actual fronts. There are three areas of interest. In area (1), line-organized convection is associated with a split front. Convection originating from the mid-troposphere is driven by the upper-level circulation. It comprises a well-defined upper-level anomaly and a jet exit, implying that this organization will persist and may become quite active when the convection benefits from the upper-level divergence associated with the transverse circulation resulting from the diffluence. The cold front in area (2) is undergoing frontolysis: no real activity is to be expected in this area, but it is useful to mark it because weak temperature and humidity gradients persist at low levels and influence the actual weather, such as by supporting low clouds. Area (3), on the other hand, indicates explicitly the presence of convection and it also shows several factors that may make it locally intense. There is a broad area of significant vertical shear and there is the more localized zone of diffluence that can help the convective clouds to become organized into a large cluster. Note also the convergence line over the Mediterranean Sea.

Because this new graphical framework reaches the heart of the forecaster's task, the progress of this project within Météo-France has been cautious and slow since it was first outlined in 1993. The current status is as follows. The required drawing and editing tool has been part of Météo-France's SYNERGIE forecasting workstation since the 1996 versions, and has been maintained and upgraded since then. This production tool is used on a regular experimental basis at the forecast laboratory to train the central forecasters to produce the new document. At the same time, this document is useful if it can be fully read and exploited. The only way to achieve this goal is to set up a massive training programme for all Météo-France forecasters. This is a hugely ambitious project that was started in 1998 and which is currently at the stage of finalizing the training material. The training sessions themselves will take place in 2001 and probably 2002.

F. POTENTIAL FOR UPGRADING

The main body of this chapter details our proposal for providing a telling and rich graphical synthesis of the synoptic-scale situation which is, at the same time simple. Tools of this kind are required by weather forecasting to meet at least two goals:

- to summarize and organize the enormous amount of information available from the multiple-source observing system and from forecast models; this summary may be used "as is" as a forecast, or more generally, as a tool for organizing a coherent search and detailed study of the raw material with a view to adapting the forecast locally;
- to enable forecasters to provide changes to the reference numerical scenario that can be more efficiently, convincingly and powerfully conveyed than with a word bulletin.

It takes the form of a two-level representation of the atmosphere which is an extensive overhaul of the existing graphical summary rather than a completely new document. At the same time, it is based on a consistent dynamical framework. The latter is, at this stage, underused. The purely graphical approach discussed so far can be seen as a first step and is amenable, in that same framework, to extensions that will allow an even better match of the main goals listed above. This section briefly presents this potential for extensions which is one of the richer aspects of this proposal.

The general motivation, here, is that numerical weather prediction provides a consistent set of fields. A growing number of applications depend on these fields rather than on the "subjective products". As a result, the in-depth analysis work performed by central forecasters needs not only to be well translated graphically, it now needs to be translated in the semi-objective form of a consistent set of fields.

1. Towards a Dual Representation of Meteorological Structures: Graphical Objects and Field Anomalies

The ANASYG and PRESYG graphical summaries presented in the previous sections translates into one-dimensional symbols three-dimensions structures of related (nonindependent) fields. These complex and intricate structures, which are represented in the original model fields, are reduced to purely graphical objects. The synthetic power of this approach is large, but so is the loss of information. Apart from the jet-stream symbol, which represents an essential feature of the large-scale flow, many of the other symbols, most notably those of confluent/diffluent zones, the surface and upper-level fronts, the upper-level vortices correspond to "anomalies" of a smooth large-scale pattern. The interdependence of the fields is due to the scale of these structures, which is large enough to render them subject to a balance condition.

Instead of reducing these anomalies directly to graphical symbols, an intermediate stage that makes use of the same dynamical framework as outlined in Section D can be exploited. The idea is to model the anomalies in one way or another in terms of one field. The "standard" way would be (with the uniform potential vorticity assumption) to work with potential temperature at the top of the boundary layer and at the tropopause. A more practical way is to use vorticity, because the anomalous features are better defined in this field for which the large-scale component is less masking. This can be done by using time filtering (this way of bringing anomalies up is implemented in Météo-France's SYNERGIE workstation), global-space filtering or probably even better, local-space filtering. The number of parameters defining a given anomaly needs to be kept relatively low, since expert intervention must remain at the heart of the approach.

The result is that the atmospheric state at a given time is split into two smooth 2-D fields (or one smooth 3-D field depending on the assumptions made for potential vorticity) representing the very large-scale flow (that could be potential vorticity or geopotential) and a number of semi-automatically defined anomalies represented by a set of parameters such as a location, an amplitude and several space dimensions. This is still an enormous reduction with respect to a model state trajectory. But there is considerably more information kept than with the purely graphical approach. Instead of having purely graphical representations, we now have some kinds of dynamical subject.

In this extended framework illustrated by Fig. 18, the following possibilities arise:

- Much as in the purely graphical approach, the "objects" are editable by the forecaster, depending on her/his evaluation of the base model trajectory: locations can be changed, amplitudes as well as spatial extensions can be modified. In rare extreme cases, structures can be removed or added.

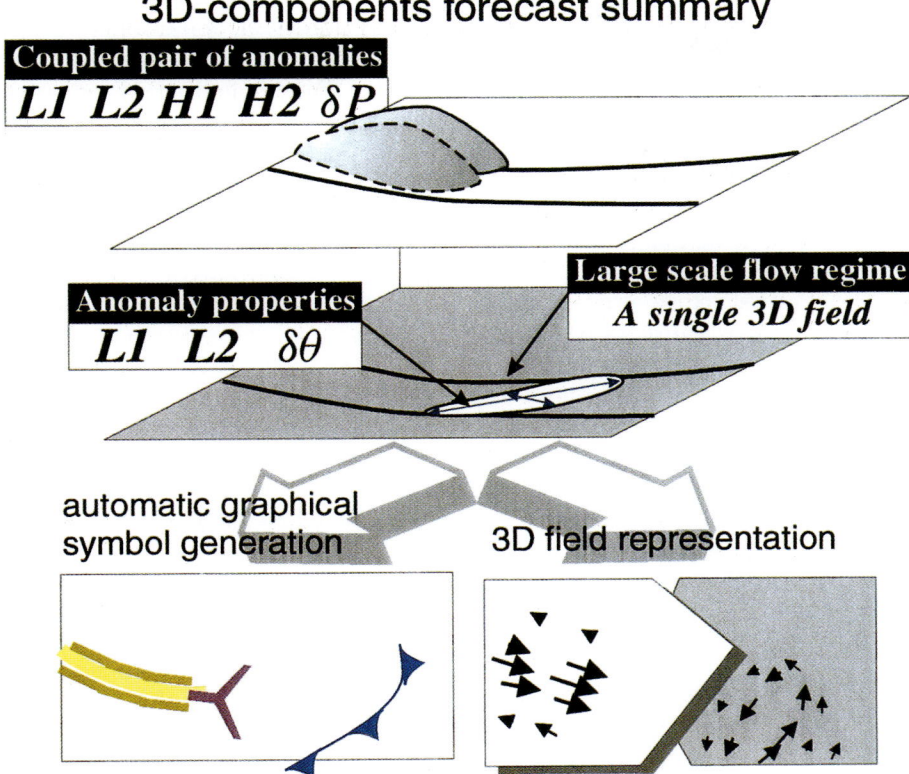

FIGURE 18 A schematic explaining the dual representation of significant weather structures. The same framework as in Fig. 6 is employed, except that now significant features are represented by parameterized anomalies and the background flow is kept as a field. Using two kinds of algorithm, pattern recognition on the one hand, potential vorticity inversion on the other, the information can be delivered either in the form of the graphical summary or as 3D fields.

- Starting from this relatively simple database (one background field, a set of modelled anomalies), pattern-recognition algorithms can be devised that will automatically generate the graphical symbol to represent such an isolated anomaly or coupled subset of anomalies. This will produce a PRESYG as described in previous sections.
- However, now using potential vorticity inversion concepts, the modelled anomalies can be inverted and translated in terms of fields. Superimposed on the background field, a consistent set of 3-D fields influenced by the forecasters becomes available, exactly in line with the graphical summary. A number of quantitative predictions can be derived based on the synoptic-scale fields of wind, temperature, geopotential and, probably, vertical velocity.

The possibility of producing both a graphical summary and a set of fields out of the same compact database is what is called here the dual representation of the state of the atmosphere.

As already hinted, it is possible to extend further the idea by removing the uniform potential vorticity assumption and work with full 3-D potential vorticity features. The technique then becomes parent to the 3-D field modification tool presented by Carroll (1997), except that an automatic pattern-recognition tool produces a graphical summary out of it.

The extension to four dimensions is, on the other hand, nontrivial. This is because of the importance of deformation and of dispersion in the atmosphere: the useful conceptual idea of a coherent structure that propagates while keeping some kind of integrity (an idea that is present, to some extent, in the Type B paradigm of cyclogenesis, or Sutcliffe's development theory) is difficult to make objective. A fluid cannot be approximated by a finite set of semi-solid objects. One possible approach is to link together several 3-D states using automatic feature tracking (Hodges, 1994; Ayrault and Joly, 2000). This can provide a simplified history of the core of the most time-consistent structures.

2. Towards a Subjective Forecast that Looks Like a Numerical Model Output

Probably the best way towards having an expert, the forecaster, influencing a fully featured 4-D model-trajectory evolution of the atmosphere is to have him/her working *upstream* of the last model time integration, rather than, as in all the previous sections, downstream of the most recent run (Fig. 19).

The basis for this ultimate extension is the visionary article of McIntyre (1999). This article also relates this final idea and the ones previously discussed. They all rely on various consequences of the existence of an invertibility principle in large-scale atmospheric dynamics. The purely graphical summary is based on the consequence of inverting a uniform potential vorticity flow. The extension discussed above explicitly performs the inversion to provide a field description of characteristic structures. Inversion can also be performed on man-modified initial conditions from which a fully consistent model integration can be performed and exploited. Chaigne and Arbogast (2000) have developed a variational inversion algorithm and applied it to FASTEX cases in this spirit.

Recently, another mixed working group of Météo-France forecasters and scientists inspired by these views put forward the concept of *reference synoptic trajectory*. The reference synoptic trajectory is a standard model trajectory, with all the time and space consistency as well as the richness of parameters that can be computed out of it, but influenced by the central forecasters.

The sources of information that lead forecasters to decide changes to the raw numerical scenario are, much as with the previously discussed techniques:

- very recent observations that have not yet been included in the assimilation cycle from which the raw forecast results;
- observations that are not assimilated: this is currently the case for most geostationary and radar imagery; although they may ultimately be assimilated, there is always a long gap between the availability of a given remote sensing instrument and its use in a data assimilation system: forecasters can make use of the product much earlier;
- knowledge, hopefully quantitative and objective, of model systematic errors;
- access to numerical forecasts from other centres or from an ensemble run.

Rather than relying only on potential vorticity inversion, a variety of techniques will be tested in order to draw the raw fully objective forecast towards a different scenario (or a small set of scenarios) that the evidence listed above makes more likely. These are:

- Initial state potential vorticity manipulation and inversion. This technique concentrates the work on the initial stages of the forecast, when observations are available.
- Initial state modification by a sensitivity-based state vector resulting from one adjoint-model integration. This provides a relatively objective way of working on later forecast times. The technique can be either constrained by observations (Hello *et al.*, 2000) or by the average amplitude of the linear correction in an attempt to apply it to a given property of the forecast (e.g. to lower the surface pressure by 15% in a given area at the 24 h range).
- To include a small set of 3-D corrections to the base scenario at several key forecast times (analysis, 24 h, 36 h, say), either in the form of potential vorticity anomalies or corrections obtained by inverting these anomalies and then use an algorithm to extract from an ensemble of forecasts the 4-D trajectory closest to the corrected scenario.

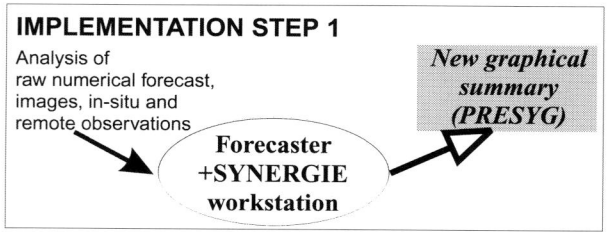

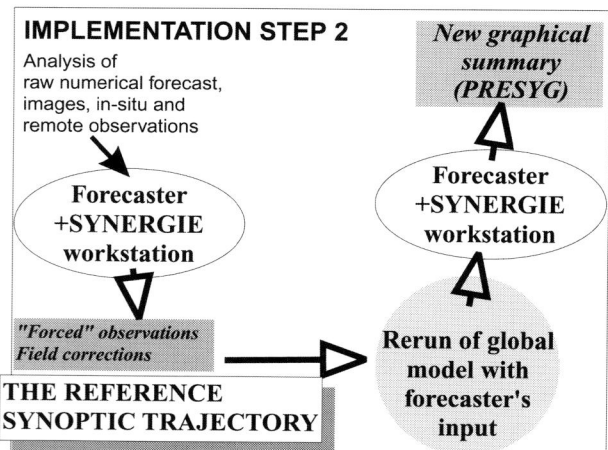

FIGURE 19 The flowchart corresponding to the two stages of implementation currently planned for allowing the critical work of central forecasters to influence forecast production consistently and efficiently. The first stage is close to existing practice, except that more information is conveyed through the new graphical summary. The second stage, which is a development project, includes an attempt at producing a full model trajectory modified by the forecasters.

Then, the rest of the production is linked to this "expertized" model trajectory, including the mesoscale model forecast.

The graphical summary described in the earlier sections remains the ultimate way for the forecasters to convey their views when this man-influenced model run still departs significantly from what they expect. It is likely that the next decade will see the traditional approach of deterministic forecasting rocked on its very basis. A probabilistic approach will slowly prevail at all ranges. However, only a few users are in a position to actually use, mostly in an economical sense, the probabilistic information. Probably the best way to combine the predominantly deterministic trajectory culture and the probabilistic view is by working out one or two alternative scenarios. The proper framework to do that is called "tubing" (Atger, 1999). The graphical framework described here as well as its extensions is extremely suitable for rapidly presenting the essential features of two or three tubes.

G. CONCLUSION

The graphical translation of knowledge is needed more than ever in the present modern, information-driven society. This chapter is about a proposal to represent graphically the various structures involved, at a given time, in making "weather" and its potential for extensions to more quantitative but subjectively influenced forecasts.

This proposal somewhat shakes the current reading of the letter of the Bergen methodology from which many aspects of the subjective forecast practices come. However, we have shown that we stand closer to the initial spirit and intention of the Bergen group than the current practice. The essence of our proposal is to combine this initial intention with bridging the gap between current practices and the state of our theoretical understanding of the atmosphere.

The addition of information on the upper-level evolution in a graphical form is not particularly new. Interactions on the synoptic scale between upper and lower levels in the midlatitude troposphere are so important that many research and review articles in recent books have worked out ways to represent them (see e.g. Bader *et al.*, 1995, that uses lines of little arrows, or several articles in Newton and Holopainen, 1990). The novel aspect here is to promote the idea into operational practice.

In this way, several ideas well known in the atmospheric sciences community will be explicitly employed in everyday forecasting. The expected effect of this change will not just be that the operational and research sides of weather forecast will feel happier; there is here an opportunity to recover a potentially fruitful common working ground and for both groups to talk and make progress together. As soon as the concepts underlying our proposal are put to the test of everyday use, many questions will be raised by forecasters to the dynamicists. From the experience already gathered, it is clear that basic instability ideas do not really help to answer these questions. Potential vorticity ideas, on the other hand, and the possibilities of describing the dynamics in terms of interacting finite-amplitude structures is far better suited to tackle them. Around this framework, several concepts need to be formalized, and these problems really call for the best dynamicists to look at them, e.g.: Can reasonably objective definitions of "anomalies" be devised? Do thresholds exist, in terms of amplitudes, scales, etc. that control the various stages in the development of a finite amplitude cyclone? How can processes like downstream development be diagnosed and included in the proposed framework?

It is hoped that this chapter, following the gathering in Cambridge of many influential atmospheric scientists, will help convince some of them that the building of a finite-amplitude understanding of cyclogenesis and other synoptic-scale sources of weather is one of the challenges set out for the next century.

ACKNOWLEDGEMENTS

The authors thank Professor R. P. Pearce for his extensive revision of the original version of this chapter. The ANASYG/PRESYG working group, which defined the graphical code presented here involved Raymond Beugin, Cécile Guyon, Hubert Brunet, Jean Coiffier, Joel Crémoux, Thierry Lefort, Lionel Mercier, Bernard Roulet and François Lalaurette, together with the authors. Philippe Arbogast also brought important ideas in the very first outline of these proposals. The training of many at Météo-France in these ideas strongly relies on Jean Coiffier and Sylvie Malardel. Gwenaëlle Hello is especially thanked for providing the raw graphics from the 4D-VAR reanalysis of FASTEX used to prepare the sample ANASYGs that form the basis of this article. The concept of *reference synoptic trajectory* is the result of the numerous and fruitful discussions led by Eric Brun under his SYMPOSIUM 2 project. Apart from him and the authors, the colleagues most heavily involved in this part of the project are Etienne Arbogast, Philippe Arbogast, Hubert Brunet, Jean Coiffier, Gwenaëlle Hello, Jean Pailleux, as well as Corinne Mithieux, Claude Pastre and Florence Rabier.

The FASTEX examples are very well-documented cases. An atlas of maps and images, in situ data and operationally analysed fields are available on line to work on these cases at http://www.cnrm.meteo.fr/fastex/. The sample fields shown for these cases and used to prepare the graphical synthesis do not come from this source, however, but from the recent 4D-VAR reanalysis of all the available, corrected, data. Fields from the reanalysis area available upon request to the FASTEX data manager.

APPENDIX. THE GRAPHICAL CODE OF ANASYG AND PRESYG

Tables A1 to A6 present the complete proposed graphical code for reference. Definitions are commented on in Santurette and Joly (2001).

TABLE A1 Symbols for Tropopause and Tropospheric-Deep Features in the New Synoptic-Scale Graphical Summary

Symbol	Structure	Colour
⇒	Upper-level jet, main baroclinic zone	Brown yellow
⇒<	Strong upper-level diffluent zone	Brown (jet) purple (diffluence)
)⇒	Strong upper-level confluent zone	Purple (confluence) brown (jet)
≥ ⇒	Other diffluent or confluent zones, marked by interrupting the jet symbol	
—	Active coherent upper-level anomaly, tropopause front (generally appears as a curved heavy line)	Black
--	Other coherent upper-level structure	Black

TABLE A2 Symbols for Low-Level Frontal Features

Symbol	Structure	Colour	WMO norm[a]
▲▲	Active, possibly strengthening, cold front	Blue	•
⌒⌒	Active, possibly strengthening, warm front	Red	•
▲⌒▲⌒	Occluded front (in the strict original sense)	Purple	•
▼⌒	Quasi-stationary front	Magenta	•
⇒	Strong low-level jet	Red orange	—
△ △	Split cold front, upper-level mark (about 600 mbar)	Blue	×
▲...▲...	Split cold front, surface mark	Blue	—
▲+▲+-	Decaying cold front, cold front under frontolysis	Blue	•
⌒+⌒+-	Decaying warm front	Red	•
▲⌒+▲⌒+-	Decaying occluded front	Purple	•
▼+▼+-	Decaying quasi-stationary front	Magenta	•

[a] Key:
•: symbol exists in WMO norm, proposed usage is similar.
×: close symbol exists in WMO norm, but proposed usage is different.
—: new symbol.

TABLE A3 Symbols for Other Low-Level Line Features

Symbol	Structure	Colour	WMO norm[a]
	Areas of thermal gradient (pseudo-fronts)		
▲··▲···	Pseudo cold front	Blue	—
●··●···	Pseudo warm front	Red	—
◆··◆···	Pseudo quasi-stationary front	Magenta	—
	Areas of wind changes		
---------	Surface trough	Blue	b
←⃪←⃪	Convergence line	Orange	•
	Line organized convection		
ᴠ—ᴠ—	Squall line	Red	—
◄◄◄	Strong line convection, including comma clouds	Orange	—

[a] For key see Table A2.
[b] The "surface trough" symbol, a solid line in the WMO norm, is dashed to avoid conflict with the "upper-level anomaly" symbol.

TABLE A4 Symbols for Convective Zones and Clusters and Remarkable Features

Symbol	Structure	Colour	WMO norm[a]
	Convective zones and clusters[b]		
⌶	Strong convection in warm unstable air mass	Red	×
⌶	Strong convection in a post-frontal area	Blue	—
⌶̄	Cyclone wake with marked subsidence, not favourable to convection	Blue	—
	Tropical cyclones and vortex-like structures		
⊙	Tropical low, tropical convective cluster	Red	•
∮	Tropical storm	Red	•
∫	Tropical cyclone	Red	•
D_p or ⌇	Polar low, strong mediterranean cyclone	Red	—

[a] For key see Table A2.
[b] Generally overlayed on a light-red or a light-blue surface delimiting a zone of convective activity according to the symbol's colour.

TABLE A5 Symbols for Noticeable Features of the Surface Pressure Field

Symbol	Structure	Colour	WMO norm[a]
D↓ D↑	Deepening or filling low	Red	×
A↑ A↓	Rising or weakening high	Blue	×
D_{name}	Former tropical cyclone in storm track	Red	—
d	Thermal low	Red	—
δ	Orographic low	Red	—
B^-, B^+	Coupled pair of blocking high-low system	Blue–red	—

[a] For key see Table A2.

TABLE A6 Symbols for Weather Activity (Vertical Velocity, Precipitations)[a]

Symbol	Structure	Colour	WMO norm[b]
•	Weak activity	Light green	×
••	Moderate activity	Green	×
•••	Strong activity	Dark green	×

[a] The symbols are either:
 • overlayed on a green-coloured area delimiting the active zone, or
 • placed next to an active front, the area implicitly being the whole of the frontal band

[b] For key see Table A2.

References

Atger, F., 1999: Tubing: an alternative to clustering for the classification of ensemble forecasts. *Weath Forecast.*, **14**(5), 741–757.

Ayrault, F. and A. Joly, 2000: Une nouvelle typologie des dépressions météorologiques: classification des phases de maturation. *C.-R. Acad. Sci. Sci. Terre Planètes*, **330**, 167–172.

Bader, M. J., G. S. Forbes, J. R. Grant, R. B. E. Lilley and A. J. Waters, 1995: *Image in Weather Forecasting, a Practical Guide for Interpreting Satellite and Radar Imagery*. Cambridge University Press.

Baehr, Ch., B. Pouponneau, F. Ayrault and A. Joly, 1999: Dynamical characterization and summary of the FASTEX cyclogenesis cases. *Q. J. R. Meteorol. Soc.*, **125**, 3469–3494.

Bjerknes, J. and J. Holmboe, 1944: On the theory of cyclones. *J. Meteorol.*, **1**, 1–22.

Bjerknes, J. and H. Solberg, 1922: Life cycle of cyclones and the polar front theory of atmospheric circulation. *Geofys. Publ.*, **3**, 1.

Bjerknes, V., 1920: The structure of the atmosphere when rain is falling. *Q. J. R. Meteorol. Soc.*, **46**(194), 119–140.

Bjerknes, V., 1925: Polar front meteorology. *Q. J. R. Meteorol. Soc.*, **51**, 261–268.

Cammas, J. P., B. Pouponneau, G. Desroziers, P. Santurette, A. Joly, Ph. Arbogast, I. Mallet, G. Caniaux and P. Mascart, 1999: FASTEX IOP 17 cyclone: introductory synoptic study with field data. *Q. J. R. Meteorol. Soc.*, **125**, 3393–3414.

Carroll, E. B., 1997: A technique for consistent alteration of NWP output fields. *Meteorol. Appl.*, **4**, 171–178.

Chaigne, E. and Ph. Arbogast, 2000: Multiple potential vorticity inversions in two FASTEX cyclones. *Q. J. R. Meteorol. Soc.*, **126**, 1711–1734.

Charney, J. G., 1947: The dynamics of long waves in a baroclinic westerly current. *J. Meteor.*, **4**, 135–162.

Charney, J. G., 1948: On the scale of atmospheric motions. *Geofys. Publ.*, **17**(2), 1–17.

Eady, E. T., 1949: Long-waves and cyclone waves. *Tellus*, **1**(3), 33–52.

Friedman, R. M., 1989: *Appropriating the Weather, Vilhelm Bjerknes and the Construction of a Modern Meteorology*. Cornell University Press.

Giao, A., 1927: La météorologie à Bergen aujourd'hui et demain. Société Belge d'Astronomie, de Météorologie et de Physique du Globe: Bruxelles.

Hello, G., F. Lalaurette and J. N. Thepaut, 2000: Combined use of sensitivity information and observations to improve meteorological forecasts: a feasibility study applied to the 'Christmas Storm' case. *Q. J. R. Meteorol. Soc.*, **126**, 621–647.

Hewson, T. D., 1998: Objective fronts. *Meteorol. Appl.*, **5**, 37–65.

Hodges, K., 1994: A general method for tracking analysis and its application to meteorological data. *Mon. Weath. Rev.*, **122**, 2573–2586.

Hoskins, B. J., 1975: The geostrophic momentum approximation and the semi-geostrophic equations. *J. Atmos. Sci.*, **32**, 233–242.

Hoskins, B. J. and F. P. Bretherton, 1972: Atmospheric frontogenesis models: mathematical formulation and solution. *J. Atmos. Sci.*, **29**, 11–37.

Hoskins, B. J., I. N. James and G. H. White, 1983: The shape, propagation and mean-flow interaction of large-scale weather systems. *J. Atmos. Sci.*, **40**(7), 1595–1612.

Hoskins, B. J., M. E. McIntyre and R. W. Robertson, 1985: On the use and significance of isentropic potential vorticity maps. *Q. J. R. Meteorol. Soc.*, **111**, 877–946.

Joly, A., 1993: L'évolution de la météorologie dynamique et ses conséquences sur la prévision synoptique. In *Actes de l'Atelier de prospective sur l'évolution de la prévision centrale*, p. 41p, Météo-France, Service Central d'Exploitation de la Météorologie, 42 av. G. Coriolis, F-31057 Toulouse cedex 1, France.

Joly, A., K. A. Browning, P. Bessemoulin, J. P. Cammas, G. Caniaux, J. P. Chalon, S. A. Clough, R. Dirks, K. A. Emanuel, L. Eymard, R. Gall, T. D. Hewson, P. H. Hildebrand, D. Jorgensen, F. Lalaurette, R. H. Langland, Y. Lemaitre, P. Mascart, J. A. Moore, P. O. G. Persson, F. Roux, M. A. Shapiro, C. Snyder, Z. Toth and R. M. Wakimoto, 1999: Overview of the field phase of the Fronts and Atlantic Storm-Track Experiment (FASTEX) project. *Q. J. R. Meteorol. Soc.*, **125**, 3131–3164.

Kleinschmidt, E., 1950: Über Aufbau und Enstehung von Zyklonen, I Teil. *Met. Rundsch.*, **3**, 1–6.

Mass, C. F., 1991: Synoptic frontal analysis: time for a reassessment? *Bull. Am. Meteorol. Soc.*, **72**(3), 348–363.

McIntyre, M. E., 1999: Numerical weather prediction: a vision of the future, updated still further. In *The Life Cycles of Extra-tropical Cyclones*, American Meteorological Society, pp. 337–355.

Newton, C. W. and E. O. Holopainen (Eds), 1990: *Extratropical Cyclones, the Erik Palmén Memorial Volume*. American Meteorological Society.

Organisation Météorologique Mondiale, 1992: *Manuel du système mondial le traitement des données*, 1992 edition, Vol. 1, Annexe IV au Règlement technique de l'OMM, Aspects Mondiaux, Supplément N.4, Référence OMM 485. Organisatiodn Météorologique Mondiale, Genève, Suisse.

Petterssen, S., 1955: A general survey of factors influencing development at sea level. *J. Meteorol.*, **12**, 36–42.

Petterssen, S. and S. J. Smebye, 1971: On the development of extratropical cyclones. *Q. J. R. Meteorol. Soc.*, **97**, 457–482.

Sanders, F., 1999: A proposed method of surface map analysis. *Mon. Weath. Rev.*, **127**(6), 945–955.

Santurette, P. and A. Joly, 2001: ANASYG/PRESYG, Météo-France's new graphical summary of the synoptic situation. *Meteorol. Appl.*, **7**, accepted.

Santurette, P., A. Joly, R. Beugin, C. Guyon, H. Brunet, J. Coiffier, J. Crémoux, T. Lefort, L. Mercier, B. Roulet and F. Lalaurette, 1997: *Les ANASYG et PRESYG*. Technical Report, Météo-France, Service Central d'Exploitation de la Météorologie, 42 av. G. Coriolis, F-31057 Toulouse cedex 1, France. Version 4.

Sutcliffe, R. C., 1947: A contribution to the problem of development. *Q. J. R. Meteorol. Soc.*, **73**, 370–383.

Thorpe, A. J. and S. A. Clough, 1991: Mesoscale dynamics of cold fronts—part I: structures described by dropsoundings in fronts 87. *Q. J. R. Meteorol. Soc.*, **117**, 903–941.

Vautard, R., 1990: Multiple weather regimes over the north-Atlantic: analysis of precursors and successors. *Mon. Weath. Rev.*, **118**, 2056–2081.

Vautard, R. and B. Legras, 1988: On the source of midlatitude low-frequency variability. Part II: Nonlinear equilibration of weather regimes. *J. Atmos. Sci.*, **45**(20), 2845–2867.

Weather Forecasting
From Woolly Art to Solid Science

Peter Lynch

Met Éireann, Dublin, Ireland

When the Royal Meteorological Society was founded, weather forecasting was a sorry business: methods were dubious and results unreliable. Worse still, prospects for improvement looked bleak. Around 1900, Vilhelm Bjerknes realized that the development of thermodynamics had completed the jigsaw of atmospheric laws, and that these laws provided a rational basis for precise scientific forecasting. But his was a method without a means: he saw no possibility to put his ideas to practical use. Lewis Fry Richardson was bolder, attempting a direct assault on the equations of motion. His methodology was unimpeachable but his results disastrously wrong. His glorious failure discouraged further investigation along these lines.

Developments in meteorology and numerical analysis in the ensuing decades, and the invention of the radiosonde and digital computer, led to the possibility of a practical implementation of Bjerknes' and Richardson's ideas. In 1950 the first computer forecast was made using a greatly simplified mathematical model. It was so successful that within 10 years computer predictions were used in operational forecasting. Numerical weather prediction has advanced rapidly, with the development of efficient algorithms and refinement of the treatment of physical processes. By Trojan efforts, assimilation techniques have been devised to squeeze benefits from quantitative satellite data. Forecasts with some skill up to a week ahead are now routine.

Progress since 1950 has been spectacular. Advances have been based on sound theoretical principles so that the woolly art of weather forecasting has been ennobled to a precise scientific discipline.

A. THE PREHISTORY OF SCIENTIFIC FORECASTING

At the time the Royal Meteorological Society was founded, weather forecasting was very imprecise and unreliable. Moreover, there was little practical improvement, or indeed much hope of improvement, for the following 50 years. Observations were scarce and irregular, especially for the upper air and over the oceans. The principles of theoretical physics played little or no rôle in practical forecasting: the forecaster used crude techniques of extrapolation, knowledge of local climatology and guesswork based on intuition; forecasting was more an art than a science. The observations of pressure and other variables were plotted in symbolic form on a weather map and lines drawn through points with equal pressure revealed the pattern of weather systems—depressions, anticyclones, troughs and ridges. The forecaster used his experience, memory of similar patterns in the past and a menagerie of empirical rules to produce a forecast map. To a great extent it was assumed that what had been happening up to now would continue for some time. The primary physical process attended to by the forecaster was *advection*, the transport of fluid characteristics and properties by the movement of the fluid itself. But the crucial quality of advection is that it is nonlinear; the human forecaster may extrapolate trends using an assumption of constant wind, but is quite incapable of intuiting the subtleties of complex advective processes.

The difficulties of forecasting during this period are illustrated by the tragic fate of Admiral Robert Fitzroy. Around 1860, Fitzroy started to issue forecasts and storm warnings for coastal regions, which were published in the London *Times*. However, with so few observations and such limited understanding of the atmosphere, he had no real hope of success. The forecasts generated considerable hostility and indeed some unjustified criticism of Fitzroy, which contributed to his untimely death.

1. Vilhelm Bjerknes

The first explicit analysis of the weather prediction problem from a scientific viewpoint was undertaken at the beginning of the last century when the Norwegian

FIGURE 1 A recent painting (from photographs) of the Norwegian scientist Vilhelm Bjerknes (1862–1951), painted by Rolf Groven, 1983, (original in the Geophysical Institute, Bergen).

scientist Vilhelm Bjerknes (Fig. 1) set down a two-step plan for rational forecasting (Bjerknes, 1904):

> "If it is true, as every scientist believes, that subsequent atmospheric states develop from the preceding ones according to physical law, then it is apparent that the necessary and sufficient conditions for the rational solution of forecasting problems are the following:
>
> 1. "A sufficiently accurate knowledge of the state of the atmosphere at the initial time.
>
> 2. A sufficiently accurate knowledge of the laws according to which one state of the atmosphere develops from another."

Bjerknes used the medical terms *diagnostic* and *prognostic* for these two steps (Friedman, 1989). The diagnostic step requires adequate observational data to define the three-dimensional structure of the atmosphere at a particular time. There was a severe shortage of observations, particularly over the seas and for the upper air, but Bjerknes was optimistic:

> "We can hope ... that the time will soon come when either as a daily routine, or for certain designated days, a complete diagnosis of the state of the atmosphere will be available. The first condition for putting forecasting on a rational basis will then be satisfied."

In fact, such designated days, on which upper-air observations were made throughout Europe, were organized around that time by the International Commission for Scientific Aeronautics.

The second or prognostic step was to be taken by assembling a set of equations, one for each dependent variable describing the atmosphere. Bjerknes listed seven basic variables: pressure, temperature, density, humidity and three components of velocity. He then identified seven independent equations: the three hydrodynamic equations of motion, the continuity equation, the equation of state and the equations expressing the first and second laws of thermodynamics. As pointed out by Eliassen (1999), Bjerknes was in error in listing the second law of thermodynamics; he should instead have specified a continuity equation for water substance. The same error was repeated in the inaugural address at Leipzig (Bjerknes, 1914a). While this may seem a minor matter it proves that, while Bjerknes outlined a general philosophical approach, he did not attempt to formulate a detailed procedure, or algorithm, for applying his method. Indeed, he felt that such an approach was completely impractical.

Since Bjerknes knew that an exact analytical integration was beyond human ability, he developed instead a more qualitative, graphical method. His idea was to represent the initial state of the atmosphere by a number of charts giving the distribution of the variables at different levels. Graphical methods based on the fundamental equations could then be applied to construct a new set of charts describing the atmosphere some hours later. This process could be iterated until the desired forecast length was reached. Bjerknes realized that the prognostic procedure could be separated into two stages, a purely hydrodynamic part and a purely thermodynamic part; the hydrodynamics would determine the movement of an airmass over the time interval and thermodynamic considerations could then be used to deduce changes in its state.

In 1912, Bjerknes became the first Director of the new Geophysical Institute in Leipzig. In his inaugural lecture he returned to the theme of scientific forecasting. He observed that "physics ranks among the so-called exact sciences, while one may be tempted to cite meteorology as an example of a radically inexact science". He contrasted the methods of meteorology with those of astronomy, for which predictions of great accuracy are possible, and described the programme of work upon which he had already embarked: *to make meteorology into an exact physics of the atmosphere.* Considerable advances had been made in observational meteorology during the previous decade, so that now the diagnostic component of his two-step programme had become feasible:

> "Now that complete observations from an extensive portion of the free air are being published in a regular series, a mighty problem looms before us and we can no longer disregard it. We must apply the equations of theoretical physics not to ideal cases only, but to the actual existing atmospheric conditions as they are revealed by modern observations. ... The problem of accurate pre-calculation that was solved for astronomy centuries ago must now be attacked in all earnest for meteorology." (Bjerknes, 1914a)

Bjerknes expressed his conviction that the acid test of a science is its utility in forecasting: "There is after all but one problem worth attacking, *viz.*, the precalculation of future conditions." He recognized the complexity of the problem and realized that a rational forecasting procedure might require more time than the atmosphere itself takes to evolve, but concluded that, if only the calculations agreed with the facts, the scientific victory would be won. Meteorology would then have become an exact science, a true physics of the atmosphere. He was convinced that, if the theoretical problems were overcome, practical and applicable methods would soon follow:

> "It may require many years to bore a tunnel through a mountain. Many a labourer may not live to see the cut finished. Nevertheless this will not prevent later comers from riding through the tunnel at express-train speed."

At Leipzig, Bjerknes instigated the publication of a series of weather charts based on the data which were collected during the internationally agreed intensive observation days and compiled and published by Hergessel in Strassbourg. One such publication (Bjerknes, 1914b) was to provide Richardson with the initial conditions for his forecast.

2. Lewis Fry Richardson

Bjerknes saw no possibility to put his ideas to practical use. The English Quaker scientist Lewis Fry Richardson was bolder, attempting a direct solution of the equations of motion. Richardson (Fig. 2) first heard of Bjerknes' plan for rational forecasting in 1913, when he took up employment with the Meteorological Office. In the Preface to his book *Weather Prediction by*

FIGURE 2 Lewis Fry Richardson (1881–1953). (From Richardson's *Collected Works,* reproduced with permission of Bassano and Vandyk Studios.)

Numerical Process (Richardson, 1922, denoted *WPNP*) he writes:

> "The extensive researches of V Bjerknes and his School are pervaded by the idea of using the differential equations for all that they are worth. I read his volumes on *Statics* and *Kinematics* soon after beginning the present study, and they have exercised a considerable influence throughout it."

Richardson's book opens with a discussion of then-current practice in the Meteorological Office. He describes the use of an *Index of Weather Maps*, constructed by classifying old synoptic charts into categories. The Index assisted the forecaster to find previous maps resembling the current one and therewith to deduce the likely development by studying the evolution of these earlier cases:

> "The forecast is based on the supposition that what the atmosphere did then, it will do again now. There is no troublesome calculation, with its possibilities of theoretical or arithmetical error. The past history of the atmosphere is used, so to speak, as a full-scale working model of its present self." (*WPNP*, p.vii)

Bjerknes had contrasted the precision of astronomical prediction with the "radically inexact" methods of weather forecasting. Richardson returned to this theme in his Preface:

> "—the *Nautical Almanac*, that marvel of accurate forecasting, is not based on the principle that astronomical history repeats itself in the aggregate. It would be safe to say that a particular disposition of stars, planets and satellites never occurs twice. Why then should we expect a present weather map to be exactly represented in a catalogue of past weather? ... This alone is sufficient reason for presenting, in this book, a scheme of weather prediction which resembles the process by which the *Nautical Almanac* is produced, in so far as it is founded upon the differential equations and not upon the partial recurrence of phenomena in their ensemble."

Richardson's forecasting scheme amounts to a precise and detailed implementation of the prognostic component of Bjerknes' programme. It is a highly intricate procedure: as Richardson observed, "the scheme is complicated because the atmosphere is complicated". It also involved a phenomenal volume of numerical computation and was quite impractical in the pre-computer era. But Richardson was undaunted:

> "Perhaps some day in the dim future it will be possible to advance the computations faster than the weather advances and at a cost less than the saving to mankind due to the information gained. But that is a dream."

Today, forecasts are prepared routinely on powerful computers running algorithms which are remarkably similar to Richardson's scheme—his dream has indeed come true.

We shall not consider Richardson's life and work in detail, but refer readers to a number of valuable sources. A comprehensive and readable biography has been written by Ashford (1985). The Royal Society Memoir of Gold (1954) provides a more succinct description and the recently published *Collected Papers* of Richardson (Drazin, 1993; Sutherland, 1993) include a biographical essay by Hunt (1993). The article by Chapman (1965) is worthy of attention and some fascinating historical background material may be found in the superlative review article by Platzman (1967).

3. Richardson's Forecast

Richardson began serious work on weather prediction in 1913 when he was appointed Superintendent of Eskdalemuir Observatory, in Scotland. He had had little or no previous experience of meteorology when he took up this position "in the bleak and humid solitude of Eskdalemuir". Perhaps it was this lack of formal training in the subject which enabled him to approach the problem of weather forecasting from such a breathtakingly original and unconventional angle. Richardson's idea was to express the physical principles which govern the behaviour of the atmosphere as a system of mathematical equations and to apply his finite-difference method to solve this system. He had previously used both graphical and numerical methods for solving differential equations and had come to favour the latter:

> "whereas Prof. Bjerknes mostly employs graphs, I have thought it better to proceed by way of numerical tables. The reason for this is that a previous comparison of the two methods, in dealing with differential equations, had convinced me that the arithmetical procedure is the more exact and the more powerful in coping with otherwise awkward equations" (*WPNP*, p.viii)

The basic equations had already been identified by Bjerknes but they had to be simplified using the hydrostatic assumption and transformed to render them amenable to approximate solution. The fundamental idea is that atmospheric pressures, velocities, etc., are tabulated at certain latitudes, longitudes and heights so as to give a general description of the state of the atmosphere at an instant. Then these numbers are processed by an arithmetical method which yields their values after an interval of time Δt. The process can be repeated so as to yield the state of the atmosphere after $2\Delta t$, $3\Delta t$ and so on.

Richardson was not concerned merely with theoretical rigour, but wished to include a fully worked example to demonstrate how his method could be put to use. Using the most complete set of observations available to him, Richardson applied his numerical method and calculated the changes in the pressure and winds at two points in central Europe. The results were something of a calamity: Richardson calculated a change in surface pressure over a 6-h period of 145 hPa, a totally unrealistic value. The calculations themselves are presented in his book, on a set of 23 computer forms. These were completed manually, and the changes in the primary variables over a 6-h period computed. Richardson explains the chief result thus:

> "The rate of rise of surface pressure ... is found on Form P_{XIII} as 145 millibars in 6 hours, whereas observations show that the barometer was nearly steady. This glaring error is examined in detail below ... and is traced to errors in the representation of the initial winds."

Richardson described his forecast as "a fairly correct deduction from a somewhat unnatural initial distribution". He speculated that reasonable results would be obtained if the initial data were smoothed, and discussed several methods of doing this. In fact, the spurious tendencies are due to an imbalance between the pressure and wind fields resulting in large-amplitude high-frequency gravity-wave oscillations. The "cure" is to modify the analysis so as to restore balance; this process is called *initialization*. A numerical model has been constructed, keeping as close as possible to the method of Richardson, except for omission of minor physical processes, and using the same grid discretization and equations as used by him (Lynch, 1999). The results using the initial data which he used were virtually identical to those obtained by him; in particular, a pressure tendency of 145 hPa in 6 h was obtained at the central point. The initial data were then initialized using a digital filter, and the forecast tendencies from the modified data were realistic.

In Table 1 we show the 6 h changes in pressure at each model level. The column marked LFR has the values obtained by Richardson. The column marked MOD has the values generated by the computer model. They are very close to Richardson's values. The column marked DFI is for a forecast from data initialized using a Dolph filter (Lynch, 1997). The initial tendency of surface pressure is reduced from the unrealistic 145 hPa/6 h to a reasonable value of less than 1 hPa/6 h (bottom row, Table 1). These results indicate clearly that Richardson's unrealistic prediction was due to imbalance in the initial data used by him. Fuller details of the forecast reconstruction may be found in Lynch (1999).

The initial response to *Weather Prediction by Numerical Process* was unremarkable, and must have been disappointing to Richardson. It was widely reviewed, with generally favourable comments—Ashford (1985) includes a good coverage of reactions—but the impracticality of the method and the apparently abysmal failure of the solitary example inevitably attracted adverse criticism. The true significance of Richardson's work was not immediately evident; the computational complexity of the process and the disastrous results of the single trial forecast both tended to deter others from following the trail mapped out by him. Despite the understandably cautious initial reaction, Richardson's brilliant and prescient ideas are now universally recognized among meteorologists and his work is the foundation upon which modern forecasting is built.

B. THE BEGINNING OF MODERN NUMERICAL WEATHER PREDICTION

While Richardson's dream appeared unrealizable at the time his book was published, a number of key developments in the ensuing decades set the scene for progress. There were profound developments in the theory of meteorology, which provided crucial understanding of atmospheric dynamics. There were advances in numerical analysis, which enabled the design of stable algorithms. The invention of the radiosonde, and its introduction in a global network, meant that timely observations of the atmosphere in three dimensions were becoming available. And, finally, the development of the digital computer provided a means of attacking the huge computational task involved in weather forecasting.

1. John von Neumann and the Meteorology Project

John von Neumann was one of the leading mathematicians of the twentieth century. He made important contributions in several areas: mathematical logic, functional analysis, abstract algebra, quantum physics, game theory and the theory and application of computers. A brief sketch of his life may be found in Goldstine (1972) and a recent biography has been written by Macrae (1999).

In the mid-1930s von Neumann became interested in turbulent fluid flows. The nonlinear partial differential equations which describe such flows defy analytical

Table 1. Six-hour Changes in Pressure (units: hPa/6 h)

Level	LFR	MOD	DFI
1	48.3	48.5	−0.2
2	77.0	76.7	−2.6
3	103.2	102.1	−3.0
4	126.5	124.5	−3.1
Surface	145.1	145.4	−0.9

LFR, Richardson; MOD, model; DFI, filtered.

assault and even qualitative insight comes hard. Von Neumann saw that progress in hydrodynamics would be greatly accelerated if a means of solving complex equations numerically were available. It was clear that very fast automatic computing machinery was required. He masterminded the design and construction of an electronic computer at the Institute for Advanced Studies. This machine was built between 1946 and 1952 and its design had a profound impact upon the subsequent development of the computer industry. The Electronic Computer Project was "undoubtedly the most influential single undertaking in the history of the computer during this period" (Goldstine, p. 255). The Project comprised four groups: (1) Engineering; (2) Logical design and programming; (3) Mathematical; and (4) Meteorological. The fourth group was directed for the period 1948–1956 by Jule Charney (Fig. 3).

Von Neumann recognized weather forecasting, a problem of both great practical significance and intrinsic scientific interest, as a problem *par excellence* for an automatic computer. Moreover, according to Goldstine (p. 300), Von Neumann

> "knew of the pioneering work ... [of] Lewis F. Richardson.... Richardson failed largely because the Courant condition had not yet been discovered, and because high speed computers did not then exist. But von Neumann knew of both".

Von Neumann had been in Göttingen in the 1920s when Courant, Friedrichs and Lewy were working on the numerical solution of partial differential equations

FIGURE 3 Jule Charney (1917–1981). (From the cover of *EOS*, Vol. 57, August, 1976. © Nora Rosenbaum.)

and he fully appreciated the practical implications of their findings. However, Goldstine's suggestion that the CFL criterion was responsible for Richardson's failure is wide of the mark. This erroneous explanation has also been widely promulgated by others. Von Neumann made estimates of the computational power required to integrate the equations of motion and concluded tentatively that it would be feasible on the IAS computer. A formal proposal was made to the U.S. Navy to solicit financial backing for the establishment of a meteorology project. According to Platzman (1979) this proposal was "perhaps the most visionary prospectus for numerical weather prediction since the publication of Richardson's book a quarter-century earlier". The proposal was successful in attracting support, and the Meteorological Research Project began in July 1946.

A meeting—the Conference on Meteorology—was arranged at the Institute the following month to enlist the support of the meteorological community and many of the leaders of the field attended. Von Neumann had discussed the prospects for numerical weather forecasting with Carl Gustaf Rossby, who arranged for Jule Charney to participate in the Princeton meeting. Charney was at that time already somewhat familiar with Richardson's book. Richardson's forecast was much discussed at the meeting. It was clear that the CFL stability criterion prohibited the use of a long time step such as had been used in *WPNP*.

The initial plan was to integrate the primitive equations; but the existence of high-speed gravity-wave solutions required the use of such a short time step that the volume of computation might exceed the capabilities of the IAS machine. And there was a more fundamental difficulty: the impossibility of accurately calculating the divergence from the observations. Thus, two obstacles loomed before the participants at the meeting: how to avoid the requirement for a prohibitively short time step, and how to avoid using the computed divergence to calculate the pressure tendency. The answers were not apparent; it remained for Charney to find a way forward.

2. The ENIAC Integrations

In his baroclinic instability study, Charney had derived a mathematically tractable equation for the unstable waves "by eliminating from consideration at the outset the meteorologically unimportant acoustic and shearing-gravitational oscillations" (Charney, 1947). The advantages of a filtered system of equations would not be confined to its use in analytical studies. The system could have dramatic consequences for numerical integration. Charney analysed the primitive equations using the technique of *scale analysis*, and was able to simplify them in such a way that the gravity-wave solutions were completely eliminated (Charney, 1948). The resulting equations are known as the *quasi-geostrophic* system. In the special case of horizontal flow with constant static stability, the vertical variation can be separated out and the quasi-geostrophic potential vorticity equation reduces to a form equivalent to the nondivergent barotropic vorticity equation

$$\frac{d(f + \zeta)}{dt} = 0$$

The barotropic equation had, of course, been used by Rossby *et al.* (1939) in their analytical study of atmospheric waves, but nobody seriously believed that it was capable of producing a quantitatively accurate prediction of atmospheric flow.

By early 1950 the Meteorology Group had completed the necessary mathematical analysis and had designed a numerical algorithm for solving the barotropic vorticity equation. The scientific record of this work is the much-cited paper in *Tellus* by Charney *et al.* (1950). Arrangements were made to run the integration on the only computer then available, the Electronic Numerical Integrator and Computer (ENIAC). The story of the mission to Aberdeen was colourfully told by Platzman (1979). Four 24 h forecasts were made, and the results clearly indicated that the large-scale features of the mid-tropospheric flow could be forecast barotropically with a reasonable resemblance to reality. Each 24 h integration took about 24 h of computation; that is, the team were just able to keep pace with the weather.

Addressing the Royal Meteorological Society some years after the ENIAC forecast, Jule Charney said that "… to the extent that my work in weather prediction has been of value, it has been a vindication of the vision of my distinguished predecessor, Lewis F. Richardson …" (Chapman, 1965). It is gratifying that Richardson was made aware of the success in Princeton; Charney sent him copies of several reports, including the paper on the ENIAC integrations. His letter of response is reprinted in Platzman (1968). Richardson wrote that the ENIAC results were "an enormous scientific advance" on the single, and quite wrong, forecast in which his own work had ended.

3. The Barotropic Model

The encouraging initial results of the Princeton team generated widespread interest and raised expectations that operationally useful computer forecasts would soon be a reality. Within 2 years there were research groups at several centres throughout the world. In an interview with Platzman (see Lindzen *et al.*, 1990) Charney remarked: "I think we were all rather surprised … that the predictions were as good as they were." Later, Fjørtoft described how the success of the ENIAC forecasts "had a rather electrifying effect on the world meteorological community" (Taba, 1998, p. 367). In fact, as Platzman observed in his review of the ENIAC integrations (Platzman, 1979), nobody anticipated the enormous practical value of this simple model and the leading role it was to play in operational prediction for many years to come.

Not everyone was convinced that the barotropic equation was useful for forecasting. The attitude in some quarters to its use for prediction seems to have been little short of antagonistic. At a discussion meeting of the Royal Meteorological Society in January 1951 several scientists expressed strong reservations about it. In his opening remarks, Sutcliffe *et al.*, (1951) reviewed the application of the Rossby formula to stationary waves with a somewhat reluctant acknowledgement:

"Although the connection between non-divergent motion of a barotropic fluid and atmospheric flow may seem far-fetched, the correspondence between the computed stationary wavelengths and those of the observed quasi-stationary long waves in the westerlies is found to be so good that some element of reality in the model must be suspected."

Scorer expressed his scepticism more vehemently, dismissing the barotropic model as "quite useless":

"Even supposing that wave theory did describe the actual motion of moving disturbances in the atmosphere there is nothing at all to be gained by applying formulae derived for a barotropic model to obtain a forecast because all it can do is move on disturbances at a constant velocity, and can therefore give no better result than linear extrapolation and is much more trouble." (Sutcliffe *et al.*, 1951)

The richness and power encapsulated in its nonlinear advection were greatly under estimated. Sutcliffe reiterated his doubts in his concluding remarks, saying that "when a tolerably satisfactory solution to the three-dimensional problem emerges it will derive little or nothing from the barotropic model—which is literally sterile". These meteorologists clearly had no confidence in the utility of the single-level approach.

The 1954 Presidential Address to the Royal Meteorological Society, *The Development of Meteorology as an Exact Science*, was delivered by the Director of the Met. Office, Sir Graham Sutton. After briefly considering the methods first described in Richardson's "strange but stimulating book", he expressed the view that automated forecasts of the weather were unlikely in the foreseeable future:

"I think that today few meteorologists would admit to a belief in the possibility (let alone the likelihood) of Richardson's dream coming true. My own view, for what it is worth, is definitely against it."(Sutton, 1954)

The prevalent view in the Met. Office at that time was that, while numerical methods had immediate application to dynamical research, their use in practical forecasting was very remote. This cautious view may well be linked to the notoriously erratic nature of the weather in the vicinity of the British Isles and the paucity of data upstream over the Atlantic Ocean. Despite various dissenting views, evidence rapidly accumulated that even the rudimentary barotropic model was capable of producing forecasts superior to those produced by conventional manual means.

Several baroclinic models were developed in the few years after the ENIAC forecast. They were all based on the quasi-geostrophic system of equations. The Princeton team studied the severe storm of Thanksgiving Day, 1950 using two- and three-level models. After some tuning, they found that the cyclogenesis could be reasonably well simulated (Charney, 1954). Thus, it appeared that the central problem of operational forecasting had been cracked. However, it transpired that the success of the Thanksgiving forecast had been something of a fluke. Shuman (1989) reports that the multi-level models were consistently worse than the simple barotropic equation; and it was the single-level model that was used when regular operations commenced in 1958.

4. Primitive Equation Models

The limitations of the filtered equations were recognized at an early stage. In a forward-looking synopsis in the *Compendium of Meteorology*, Charney (1951) wrote:

> "The outlook for numerical forecasting would indeed be dismal if the quasi-geostrophic approximation represented the upper limit of attainable accuracy, for it is known that it applies only indifferently, if at all, to many of the small-scale but meteorologically significant motions."

He discussed the prospects for using the primitive equations, and argued that if geostrophic initial winds were used, the gravity waves would be acceptably small.

In his 1951 paper "The mechanism of meteorological noise", Karl-Heinz Hinkelmann tackled the issue of suitable initial conditions for primitive equation integrations. Hinkelmann had been convinced from the outset that the best approach was to use these equations. He knew that they would simulate the atmospheric dynamics and energetics more realistically than the filtered equations. Moreover, he felt certain, from his studies of noise, that high-frequency oscillations could be controlled by appropriate initialization. A number of other important studies of initialization, by Charney (1955), Phillips (1960) and others, followed.

The first application of the primitive equations was a success, producing good simulation of development, occlusion and frontal structure (Hinkelmann, 1959). Routine numerical forecasting was introduced in the Deutscher Wetterdienst in 1966; according to Reiser (1986), this was the first ever use of the primitive equations in an operational setting. A six-level primitive equation model was introduced into operations at NMC in June 1966, running on a CDC6600 (Shuman and Hovermale, 1968). There was an immediate improvement in skill: the S_1 score for the 500 hPa 1-day forecast was reduced by about five points.

We have seen that the view in the Met. Office was that single-level models were unequal to the task of forecasting. As a result, barotropic models were never used for forecasting in the UK and, partly for this reason, the first operational model (Bushby and Whitelam, 1961) was not in place until the end of 1965. In 1972 a 10 level primitive equation model (Bushby and Timpson, 1967) was introduced. This model incorporated a sophisticated parameterization of physical processes and with it the first useful forecasts of precipitation were produced. The remarkable progress in computer forecasting since the ENIAC integrations is vividly illustrated by the record of skill of the 36 h 500 hPa forecast produced at NCEP (formerly, NMC, Washington). The forecast skill, expressed as a percentage of an essentially perfect forecast score, is plotted in Fig. 4, for the period 1955–1999. The score has improved steadily over this period. The sophistication of prediction models is closely linked to the available computer power; the introduction of new machines is indicated in the figure. Each introduction of faster computer machinery or major enhancement of the prediction model has been followed by increased forecast accuracy.

C. NUMERICAL WEATHER PREDICTION TODAY

It is no exaggeration to describe the advances made over the past half century as revolutionary. Thanks to this work, meteorology is now firmly established as a quantitative science, and its value and validity are demonstrated daily by the acid test of any science, its ability to predict the future. Operational forecasting today uses guidance from a wide range of models. In most centres a combination of global and local models is used. By way of illustration, we will consider the global model of the European Centre for Medium-range Weather Forecasts and the local model called HIRLAM (High Resolution Limited Area Model) used for short-range forecasting.

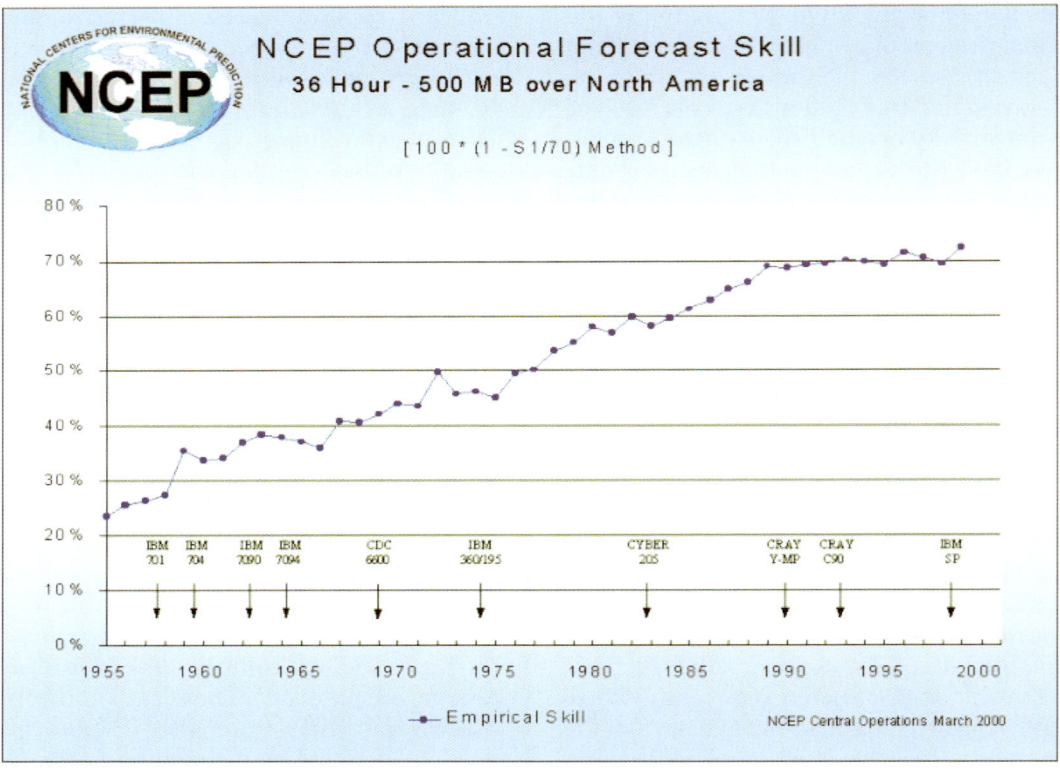

FIGURE 4 Skill of the 36 h 500 hPa forecast produced at NCEP. Forecast skill is expressed as a percentage of an essentially perfect forecast score, for the period 1955–1999. The accuracy of prediction is closely linked to the available computer power; the introduction of new machines is indicated in the figure. (Thanks to Bruce Webster of NCEP for the graphic of S_1 scores.)

1. ECMWF

The European Centre for Medium-range Weather Forecasts celebrated its twenty-fifth anniversary in 2000. This world-leading meteorological centre provides an extensive range of products for medium-range forecasting to the 18 member states and to cooperating states. The ECMWF model is a spectral primitive equation model with a semi-Lagrangian, semi-implicit time scheme and a comprehensive treatment of physical processes. It is coupled interactively to an ocean wave model. Its spatial resolution is about 40 km and there are 60 vertical levels. Initial data for the forecasts are prepared using a four-dimensional variational assimilation scheme, which uses a huge range of conventional and satellite observations over a 12 h time window. A sustained and consolidated research effort has been devoted to exploiting quantitative data from satellites, and now these observations are crucial to forecast quality. Forecasts out to 10 days are made, starting from the mid-day analysis each day. These deterministic 10-day forecasts are the primary products of the Centre, but are complemented by many other forecast products.

The computer system in current use at ECMWF is a Fujitsu VPP5000, with 100 processing elements and a peak computation speed of one Teraflop, that is, one million-million floating-point operations per second, more than nine orders of magnitude faster than the ENIAC. It represents an increase in computer power over the intervening 50 years which is broadly in agreement with Moore's law, an empirical rule governing growth of computers. According to this rule, chip density doubles every 18 months. Over 50 years, this implies an increase of $2^{50/1.5} = 1.08 \times 10^{10}$. The increases in memory size and computational speed from ENIAC to Fujitsu, indicated in Table 2, are in good agreement with this rule.

Table 2 Comparison of ENIAC and Fujitsu VPP5000/100 Computation Speed and Memory Size

	ENIAC	VPP5000	Ratio
Speed	500 flops	1 Teraflops (10^{12} flops)	2,000,000,000 2×10^9
Memory	10 words (40 bytes)	400 Gbytes (4×10^{11} bytes)	10,000,000,000 10^{10}

The chaotic nature of the atmospheric flow is now well understood. It imposes a limit on predictability, as unavoidable errors in the initial state grow rapidly and render the forecast useless after some days. The most successful means of overcoming this obstacle is to run a series, or ensemble, of forecasts, each starting from a slightly different initial state, and to use the combined outputs to deduce probabilistic information about future changes in the atmosphere. An Ensemble Prediction System (EPS) of 50 forecasts is run daily at ECMWF, each forecast having a resolution half that of the deterministic forecast. Probability forecasts for a wide range of weather events are generated and disseminated for use in the operational centres. These have become the key guidance for medium-range prediction. For a recent review, see Buizza *et al.* (2000).

Seasonal forecasts, with a range of 6 months, are also prepared at the European Centre. They are made using a coupled atmosphere/ocean model, and a large number of forecasts are combined in an ensemble each month. These forecast ensembles have demonstrable skill for tropical regions. Recent predictions of the onset of El Niño and La Niña events have been impressive. However, in temperate latitudes, and in particular for the European region, no significant skill has yet been achieved, and they are not yet used in operations for this region. Indeed, seasonal forecasting for middle latitudes is one of the great problems facing us today.

The primary products prepared and disseminated by ECMWF are:

- Forecast to 10 days from 12Z analysis (40 km);
- Fifty-member EPS Forecast to 10 days (80 km);
- Global Ocean Wave Forecast to 10 days (56 km);
- North Atlantic Ocean Forecast to 5 days (28 km);
- Ensemble Seasonal Forecast to 6 months.

The deterministic full-resolution forecasts are the basis for forecasts in the early medium range, 3–5 days. The EPS is most valuable for the range 6–10 days, beyond the threshold of deterministic predictability, where probabilistic forecasting is more appropriate.

Progress over the period of operations at ECMWF is illustrated by Fig. 5. It shows the anomaly correlation (AC) for the 500 hPa height forecast for the Northern Hemisphere, as a function of forecast length. It is generally agreed that the minimum value of AC for useful forecasts is 0.6. The three curves are for different winters. In 1972, forecasts out to $3\frac{1}{2}$ days were providing useful guidance. By 1980 this threshold had reached $5\frac{1}{2}$ days. By 1998, the limit of usefulness by this measure had reached almost 8 days. Thus, the representative range of useful forecasts is now greater than a week! This must be seen as quite remarkable progress over the lifetime of the European Centre.

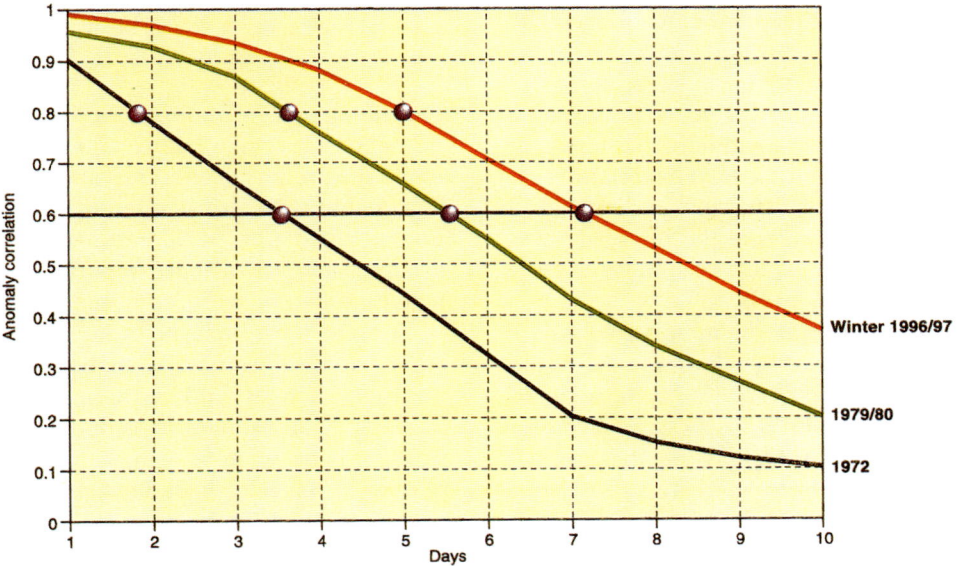

FIGURE 5 The anomaly correlation (AC) for the 500 hPa height forecast for the Northern Hemisphere, as a function of forecast length. The three curves are for different winters: black, 1972, green 1980 and red 1998. Thanks to David Burridge, Director of ECMWF, for the graphic.

2. HIRLAM

Short-range forecasting requires detailed guidance which is updated frequently. Virtually all European weather services run limited-area models (LAMs) with high resolution to provide such forecast guidance. These models permit a free choice of geographical area and spatial resolution, and forecasts can be run as frequently as required. LAMs make available a comprehensive range of outputs, with a high time resolution. Nested grids with successively higher resolution can be used to provide greater local detail.

Eight European countries currently combine their efforts in the HIRLAM (High Resolution Limited Area Modelling) Project. They are Denmark, Finland, Iceland, Ireland, Netherlands, Norway, Spain and Sweden. The HIRLAM analysis and forecasting system is the basis for operational short-range forecasting in these countries. As HIRLAM has a limited geographical domain, appropriate values must be specified on the lateral boundaries. ECMWF carries out special model integrations to generate boundary data for limited area models. The boundary condition project at ECMWF has recently been extended to provide more timely and accurate data.

An outline of the structure of the HIRLAM model and its practical applications will be given here (see Lynch *et al.* (2000) for further details). The forecast model is a hydrostatic primitive equation grid-point model. It includes a comprehensive package of physical processes. The system is coded to run on massively parallel computers. A semi-Lagrangian advection scheme and a digital filtering initialization scheme have recently been introduced. The physics package has been enhanced by a higher-order turbulence scheme and a new condensation and convection scheme. A new package for parameterization of surface processes is close to operational implementation. A variational assimilation system is at an advanced stage of development. The operational implementations of HIRLAM differ in horizontal and vertical resolution, and in details of the physical parameterizations. However, the typical resolution used is 20 km, with about 30 vertical levels.

The HIRLAM model is used in the member countries for a wide range of applications. Perhaps the most important application is to provide timely warning of weather extremes. Huge financial losses can be caused by gales, floods and other anomalous weather events. The warnings which result from this additional guidance can enable great saving of both life and property. Transportation, energy consumption, construction, tourism and agriculture are all sensitive to weather conditions. There are expectations from all these sectors of increasing accuracy and detail of short-range forecasts, as decisions with heavy financial implications must continually be made.

Limited-area models such as HIRLAM are used to generate special guidance for the marine community. Winds predicted by the LAM are used to drive wave models, which predict sea and swell heights and periods. Forecast charts of the sea-state, and other specialized products can be automatically produced and distributed to users. Prediction of road ice is performed by specially designed models which use forecasts of temperature, humidity, precipitation, cloudiness and other parameters to estimate the conditions on the road surface. Trajectories are easily derived from LAM. These are vital for modelling pollution drift, for nuclear fallout, smoke from forest fires and so on. Aviation benefits significantly from LAM guidance, which provides warnings of hazards such as lightning, icing and clear air turbulence. Automatic generation of terminal aerodrome forecasts (TAFs) from LAM and column model outputs enables servicing of a large number of airports from a central forecasting facility.

D. CONCLUSIONS

In 1976, Jule Charney was awarded the Bowie Medal of the American Geophysical Union, "for his outstanding contribution to fundamental geophysics and for unselfish cooperation in research". In his acceptance speech he said:

> "I have seen the growth of meteorology from an immature science dominated by the pragmatics of weather forecasting, with theory a weak and ineffectual handmaiden of observation, to a mature science, in which theory and observation exist on an equal footing, where weather prediction has been transformed from an art to a science...."(AGU, 1976)

Developments in atmospheric dynamics, instrumentation and observing practice and digital computing have made the dreams of Bjerknes and Richardson an everyday reality. Numerical weather prediction models are now at the centre of operational forecasting. Forecast accuracy has grown apace over the half-century of NWP activities, and progress continues on several fronts.

Despite the remarkable advances over the past 50 years, some formidable challenges remain. Sudden weather changes and extremes cause much human hardship and damage to property. Nowcasting is the process of predicting changes over periods of a few hours. Guidance provided by current numerical models occasionally falls short of what is required to take effective action and avert disasters. Greatest value is obtained by a systematic combination of NWP products with conventional observations, radar imagery, satellite imagery and other data. But much remains to be done to develop optimal nowcasting systems, and we may be optimistic that future developments will lead to great improvements in this area.

At the opposite end of the timescale, the chaotic nature of the atmosphere limits the validity of deterministic forecasts. The ensemble prediction technique provides probabilistic guidance, but so far it has proved quite difficult to use in many cases. Interaction between atmosphere and ocean becomes a dominant factor at longer forecast ranges. Although good progress in seasonal forecasting for the tropics has been made, the production of useful long-range forecasts for temperate regions remains to be tackled by future modellers. Another great challenge is the modelling and prediction of climate change, a matter of increasing importance and concern.

Perhaps the most frequently quoted section of Richardson's book is §11.2, "The Speed and Organization of Computing", in which he describes in detail his fantasy about *a Forecast Factory* for carrying out the process of calculating the weather:

"Imagine a large hall like a theatre, except that the circles and galleries go right round through the space usually occupied by the stage. The walls of this chamber are painted to form a map of the globe. The ceiling represents the north polar regions, England is in the gallery, the tropics in the upper circle, Australia on the dress circle and the antarctic in the pit. A myriad computers are at work upon the weather of the part of the map where each sits, but each computer attends only to one equation or part of an equation. The work of each region is coordinated by an official of higher rank. Numerous little "night signs" display the instantaneous values so that neighbouring computers can read them. Each number is thus displayed in three adjacent zones so as to maintain communication to the North and South on the map. From the floor of the pit a tall pillar rises to half the height of the hall. It carries a large pulpit on its top. In this sits the man in charge of the whole theatre; he is surrounded by several assistants and messengers. One of his duties is to maintain a uniform speed of progress in all parts of the globe. In this respect he is like the conductor of an orchestra in which the instruments are slide-rules and calculating machines. But instead of waving a baton he turns a beam of rosy light upon any region that is running ahead of the rest, and a beam of blue light upon those who are behindhand." (*WPNP*, p. 219)

FIGURE 6 An artist's impression of Richardson's *Forecast Factory*, described in §11.2 of his book. Richardson estimated that 64,000 "computers" would be needed to keep pace with the atmosphere. (Thanks to the artist, François Schuiten, and to Jean-Pierre Javelle, Météo-France, for providing the image.)

Richardson estimated that 64,000 people would be needed to keep pace with the atmosphere. Since the ENIAC was about five orders of magnitude faster than human computation, the Forecast Factory would have been comparable in processing power to this early machine. (In fact, Richardson's "staggering figure", although enormous, was a significant underestimate. The correct number of people required was over 200,000; see Lynch, 1993).

An impression of the Forecast Factory by the artist François Schuiten appears in Fig. 6. Richardson's fantasy is an early example of a *massively parallel processor*. Each computer is responsible for a specific gridpoint, receives information required for calculation at that point and passes to neighbouring computers the data required by them. Such message passing and memory distribution are features of modern machines, such as the Fujitsu VPP5000 in use at the European Centre. But, at peak computational power, the Fujitsu is more than 14 orders of magnitude faster than a single human computer, and equivalent in purely number-crunching terms to about *three billion* of Richardson's Forecast Factories.

ACKNOWLEDGEMENTS

It is a pleasure to thank those who helped me to obtain illustrations for this article, or permission to reproduce them. Sigbjorn Grønås provided a copy of the painting of Vilhelm Bjerknes (Fig. 1), the original of which hangs in the Geophysical Institute in Bergen. The photograph of Lewis Fry Richardson (Fig. 2) is from Richardson's *Collected Works*, and reproduced with permission of Bassano and Vandyk Studios. The photograph in Fig. 3 of Jule Charney appeared on the cover of *EOS, Transactions of the American Geophysical Union*, Vol. 57, August, 1976 (©Nora Rosenbaum). Bruce Webster of NCEP drew the graphic of S1 scores which appears in Fig. 4. David Burridge, Director of ECMWF, provided the graph of AC scores (Fig. 5). Finally, thanks to Jean-Pierre Javelle, Météo-France and the artist François Schuiten, Bruxelles, for the "realization" of Richardson's Forecast Factory (Fig. 6).

References

AGU, 1976: Awards presented at the Spring Annual Meeting. *EOS, Trans. Am. Geophys. Union*, **57**, 568–569.

Ashford, O. M., 1985: *Prophet—or Professor? The Life and Work of Lewis Fry Richardson*. Adam Hilger, Bristol and Boston.

Bjerknes, V., 1904: Das Problem der Wettervorhersage, betrachtet vom Standpunkte der Mechanik und der Physik. *Meteor. Zeit.*, **21**, 1–7. (Trans. Y. Mintz: The problem of weather forecasting as a problem in mechanics and physics. Los Angeles, 1954, reprinted in *Shapiro and Grønås, 1999*, pp. 1–4.)

Bjerknes, V., 1914a: Meteorology as an exact science. Translation of *Die Meteorologie als exacte Wissenschaft. Antrittsvorlesung gehalten am 8. Jan., 1913 in der Aula der Universität Leipzig. Mon. Weath. Rev.*, **42**, 11–14.

Bjerknes, V., 1914b: *Veröffentlichungen des Geophysikalischen Instituts der Universität Leipzig. Herausgegeben von dessen Direktor V. Bjerknes. Erste Serie: Synoptische Darstellungen atmosphärischer Zustände. Jahrgang 1910, Heft 3. Zustand der Atmosphäre über Europa am 18., 19. und 20. Mai 1910.* Leipzig, 1914.

Buizza, R., J. Barkmeijer, T. N. Palmer and D. S. Richardson, 2000: Current status and future developments of the ECMWF Ensemble Prediction System. *Meteorol. Applics.*, **7**, 163–175.

Bushby, F. H. and M. S. Timpson, 1967: A 10-level model and frontal rainfall. *Q. J. R. Meteorol. Soc.*, **93**, 1–17.

Bushby, F. H. and C. J. Whitelam, 1961: A three-parameter model of the atmosphere suitable for numerical investigation. *Q. J. R. Meteorol. Soc.*, **87**, 380–392

Chapman, S., 1965: Introduction to Dover edition of *Weather Prediction by Numerical Process*, pp. v–x. (See Richardson, 1922.)

Charney, J. G., 1947: The dynamics of long waves in a baroclinic westerly current. *J. Meteorol.*, **4**, 135–162.

Charney, J. G., 1948: On the scale of atmospheric motions. *Geofys. Publ.*, **17**, No. 2.

Charney, J. G., 1951: Dynamic forecasting by numerical process. In *Compendium of Meteorology*, American Meteorological Society, Boston, pp. 470–482.

Charney, J. G., 1954: Numerical prediction of cyclogenesis. *Proc. Natl. Acad. Sci.*, **40**, 99–110.

Charney, J. G., 1955: The use of the primitive equations of motion in numerical prediction. *Tellus*, **7**, 22–26

Charney, J. G., R. Fjørtoft and J. von Neumann, 1950: Numerical integration of the barotropic vorticity equation. *Tellus*, **2**, 237–254.

Drazin, P. G. (Gen. Ed.), 1993: *Collected Papers of Lewis Fry Richardson*, Vol. I. *Meteorology and Numerical Analysis.* Cambridge University Press.

Eliassen, A., 1999: Vilhelm Bjerknes' early studies of atmospheric motions and their connection with the cyclone model of the Bergen School. In *The Life Cycles of Extratropical Cyclones*, (M. A. Shapiro and S. Grønås, Eds.), American Meteorological Society, Boston, pp 5–13.

Friedman, R. M., 1989: *Appropriating the Weather: Vilhelm Bjerknes and the construction of a Modern Meteorology*. Cornell University Press.

Gold, E., 1954: Lewis Fry Richardson. *Obituary Notices of Fellows of the Royal Society, London*, **9**, 217–235.

Goldstine, H. H., 1972: *The Computer from Pascal to von Neumann*. Reprinted with new Preface, 1993. Princeton University Press.

Hinkelmann, K., 1951: Der Mechanismus des meteorologischen Lärmes. *Tellus*, **3**, 285–296. (Trans.: The mechanism of meteorological noise. NCAR/TN-203+STR, National Center for Atmospheric Research, Boulder, 1983.)

Hinkelmann, K., 1959: Ein numerisches Experiment mit den primitiven Gleichungen. In *The Atmosphere and the Sea in Motion*. Rossby Memorial Volume, Rockerfeller Institute Press, pp. 486–500.

Hunt, J. C. R., 1993: A general introduction to the life and work of L. F. Richardson. In *Collected Papers of Lewis Fry Richardson*, Vol. I (P. G. Drazin, , Gen. Ed.), Cambridge University Press, pp. 1–27.

Lindzen, R.S., E. N. Lorenz and G. W. Platzman, (Eds.), 1990: *The Atmosphere—A Challenge. The Science of Jule Gregory Charney*, American Meteorological Society, Boston.

Lynch, P., 1993: Richardson's forecast factory: the $64,000 question. *Meteorol. Mag.*, **122**, 69–70.

Lynch, P., 1997: The Dolph–Chebyshev window: a simple optimal filter. *Mon. Weath. Rev.*, **125**, 655–660.

Lynch, P., 1999: Richardson's marvellous forecast. In *The Life Cycles of Extratropical Cyclones*, (M. A. Shapiro and S. Grønås, Eds.), American Meteorological Society, Boston, pp 61–73.

Lynch, P., N. Gustafsson, B. Hansen Sass and G. Cats, 2000: Final Report of the HIRLAM-4 Project, 1997–1999. Available from SMHI, S-60176, Norrköping, Sweden.

Macrae, N., 1999: *John von Neumann*, American Mathematical Society. Originally published by Pantheon Books, 1992.

Phillips, N. A., 1960: On the problem of initial data for the primitive equations. *Tellus*, **12**, 121–126.

Platzman, G. W., 1967: A retrospective view of Richardson's book on weather prediction. *Bull. Am. Meteorol. Soc.*, **48**, 514–550.

Platzman, G. W., 1968: Richardson's weather prediction. *Bull. Am. Meteorol. Soc.*, **49**, 496–500.

Platzman, G. W., 1979: The ENIAC computations of 1950—gateway to numerical weather prediction. *Bull. Am. Meteorol. Soc.* **60**, 302–312.

Reiser, H., 1986: Obituary of Karl Heinz Hinkelmann. WMO Bull., **35**, 393–394.

Richardson, L. F., 1922: *Weather Prediction by Numerical Process*. Cambridge University Press. (Reprinted by Dover Publications, New York, 1965, with a new introduction by Sydney Chapman.)

Rossby, C. G. and collaborators, 1939: Relations between variations in the intensity of the zonal circulation of the atmosphere and the displacements of the semipermanent centers of action. *J. Marine Res.*, **2**, 38–55.

Shapiro, M. A. and S. Gronås, (Eds.), 1999: *The Lifecycles of Extratropical Cyclones*. American Meteorological Society, Boston.

Shuman, F. G., 1989: History of numerical weather prediction at the National Meteorological Center. *Weath. Forecast.*, **4**, 286–296.

Shuman, F. G. and J. B. Hovermale, 1968: An operational six-layer primitive equation model. *J. Appl. Meteorol.*, **7**, 525–547.

Sutcliffe, R. C., E. J. Sumner and F. H. Bushby, 1951: Dynamical methods in synoptic meteorology. Discussion in Q. J. R. Meteorol. Soc., **77**, 457–473.

Sutton, O. G., 1954: Presidential Address to the Royal Meteorological Society: The development of meteorology as an exact science. *Q. J. R. Meteorol.Soc.*, **80**, 328–338.

Sutherland, I., (Gen. Ed.), 1993: *Collected Papers of Lewis Fry Richardson*, Vol. II. *Quantitative Psychology and Studies of Conflict*. Cambridge University Press.

Taba, H., 1988: *The "Bulletin" Interviews*. World Meteorological Organization, Geneva, WMO No. 708.

Part 2. Climate Variability and Change

Problems in Quantifying Natural and Anthropogenic Perturbations to the Earth's Energy Balance
K. P. Shine and E. J. Highwood

Changes in Climate and Variability over the Last 1000 Years
P. D. Jones

Climatic Variability over the North Atlantic
J. Hurrell, M. P. Hoerling and C. K. Folland

Prediction and Detection of Anthropogenic Climate Change
J. F. B. Mitchell

Absorption of Solar Radiation in the Atmosphere: Reconciling Models with Measurements
A. Slingo

The Impact of Satellite Observations of the Earth's Spectrum on Climate Research
J. E. Harries

Problems in Quantifying Natural and Anthropogenic Perturbations to the Earth's Energy Balance

Keith P. Shine and Eleanor J. Highwood

Department of Meteorology, University of Reading, UK

A necessary prerequisite both for understanding past climate change and for predicting future climate change is the ability to quantify the size of the perturbations to the earth's energy balance as a result of either natural or anthropogenic mechanisms which act to drive climate change. Over recent decades, the number of climate change mechanisms being actively considered has grown enormously, but for many of these there are considerable uncertainties in their quantification. This chapter reviews the current status of our knowledge of these mechanisms and illustrates the impact of some of the uncertainties using a simple climate model. Particular attention is given to the impact of aerosols on clouds and the role of volcanic eruptions.

A. INTRODUCTION

The equilibrium global-mean surface temperature change, ΔT_s^{eq}, in response to some climate change mechanism, can be written as

$$\Delta T_s^{eq} = \lambda \Delta F \tag{1}$$

where ΔF is the radiative forcing, which measures the perturbation to the surface–troposphere system energy budget (in W m^{-2}) due to a climate change mechanism, such as a change in carbon dioxide concentration. λ is a climate sensitivity (in K (W m^{-2})$^{-1}$) and its inverse gives a measure of the change in net radiation budget at the top of the atmosphere per unit change in surface temperature.

Since the forcing is itself time varying, and a typical timescale for the response of the climate system to an imposed steady forcing is of order decades, we are often more interested in the time evolution of the change in global mean surface temperature ΔT_s. This can be represented by

$$C \frac{d\Delta T_s}{dt} = \Delta F(t) - \frac{\Delta T_s}{\lambda} + \text{``noise''.} \tag{2}$$

Here C represents an effective heat capacity of the system, including the ocean, land and atmosphere, and we will assume this is known. $\Delta F(t)$ is the time-varying radiative forcing. Global-mean temperature can and does vary in the absence of any external forcing mechanism—we refer to processes achieving this as "noise" and consider it no further here, although we acknowledge that our noise is others' music!

Equation (2) will serve to illustrate some of the issues in understanding past and future climate change. There are two particularly severe problems in using it, and these apply as much to attempts to solve it with sophisticated coupled ocean-atmosphere general circulation models (GCMs), as they do to attempts to solve it using simple global-mean models, as we shall do here. These difficulties are often not fully acknowledged in attempts to simulate climate.

First, there are serious gaps in our knowledge of radiative forcing. Most work to date has concentrated on calculating the present-day forcing relative to pre-industrial times. The difficulties in doing this are amplified if, as is required in equation (2), the time variation of ΔF is needed. This chapter will concentrate mostly on some of the issues in evaluating the radiative forcing, considering contributions from both human activity and natural mechanisms.

The second severe problem is the lack of knowledge of the climate sensitivity, λ. The value of λ depends on the black-body response of the system as it warms or cools in response to forcing, and hence alters the planetary radiation balance to re-establish balance. It also depends on a number of feedbacks, whereby the induced change in climate leads to further changes in the radiative properties of the system—such feedbacks include the water vapour feedback, snow/ice albedo feedback and cloud feedbacks. Our knowledge of cloud feedbacks is particularly rudimentary, to the extent that not even the sign of its effect is known. Estimates for the value of λ range from around 0.4 to 1.2 K(W m^{-2})$^{-1}$ (e.g. IPCC, 1990; IPCC, 1996). Climate sensitivity is also sometimes represented as the equilibrium surface temperature change from equation (1) in response to a doubling of carbon dioxide concentrations. Given a radiative forcing of 3.7 W m^{-2} for a doubling CO_2, the uncertainty in λ gives a range in ΔT_s of between about 1.5 and 4.5 K. In recent years there have been considerable efforts in trying to understand why different climate models yield different values of λ (see e.g. Cess et al., 1997). These have helped elucidate the reasons *how* models differ (for example in how the shortwave and thermal infrared fluxes respond to changes in cloudiness), but they have not yet succeeded in reducing the uncertainty.

A further difficulty is that it is not yet clear whether the value of λ is independent of the nature of the mechanism causing the climate change (for example, whether the climate change mechanism acts mostly by changing the solar radiation budget or the thermal infrared radiation budget), whether it is independent of the spatial distribution of the forcing mechanism or indeed, whether it is independent of the size of ΔF. We will assume here that λ is some "universal", albeit poorly quantified, constant. Further discussion of this topic can be found in, for example, IPCC (1995), Hansen et al. (1997) and Shine and Forster (1999). In some circumstances, it can be difficult to separate out whether mechanisms should be classified as a forcing or part of the feedback—one example is the change in mineral dust loading of the atmosphere, which could be due to changes in land use, to changes in circulation, or some combination of the two. We will briefly return to this topic in the Conclusions.

B. ESTIMATES OF THE GLOBAL-MEAN RADIATIVE FORCING

Over the past few decades, and since 1990 in particular, the number of individual radiative forcing mechanisms under consideration has increased dramatically. Table 1 gives a short history of the mechanisms related to human activity discussed in major international assessments. Given these rapid changes, it would be complacent to believe that we have yet identified (let alone, reliably quantified) all the potential climate change mechanisms.

One recent view of the global-mean radiative forcing, relative to pre-industrial times, is given in Fig. 1, from Shine and Forster (1999); similar figures are also given by IPCC (1996) and Hansen et al. (1998). The figure shows not only a central estimate for each forcing mechanism, but also the spread of recent results (via the "error" bars), plus the degree of confidence we have that the actual forcing lies within those bars. This is necessarily subjective, but reflects the difficulties in estimating each forcing given the available data and methods of modelling.

Several important points emerge from Fig. 1. The so-called well-mixed greenhouse gases (henceforth WMGG), shown in the first column, cause a strong positive and, it is believed, relatively certain forcing. Carbon dioxide changes contribute 60% of this; this indicates that the cumulative smaller contributions of many other gases are considerable.

The other forcings are more uncertain. Even today, observations of the vertical profile (and trends therein) of tropospheric ozone are poor; the same is true for aerosol particles, with the added problem that details of the composition of the particles need to be known to estimate forcing. In order to derive radiative forcings, it is necessary to know the geographical distributions of

TABLE 1 An Approximate "History" of the Inclusion of Different Anthropogenic Climate Change Mechanisms in Major International Assessments

Era	Assessment	Developments
Early 1970s	MIT reports (Matthews et al., 1971; SMIC, 1971)	CO_2, aerosols (but what sign?) ~~CH_4~~
Mid-1980s	WMO (1985) Ozone Assessment	CO_2, CH_4, N_2O, CFCs, stratospheric O_3 (but what sign?) ~~aerosols~~
1990	IPCC (1990)	~~Stratospheric O_3~~ tentative direct and indirect aerosols (but what sign?) and tropospheric O_3
1992	IPCC (1992)	Stratospheric O_3, sulphate aerosols
1994	IPCC (1995)	Tropospheric O_3, biomass aerosols, tentative surface albedo
1995	IPCC (1996)	Soot aerosols
1999	IPCC (1999)	Aircraft "cirrus"
2001	IPCC	?

Mechanisms that have been scored-through received little or no discussion in the assessment or, in the case of methane in the MIT reports, was positively dismissed as having no climatic significance. The Third Assessment Report of the Intergovernmental Panel on Climate Change (IPCC) is due out in 2001 and will likely include a discussion of an expanded range of aerosol types.

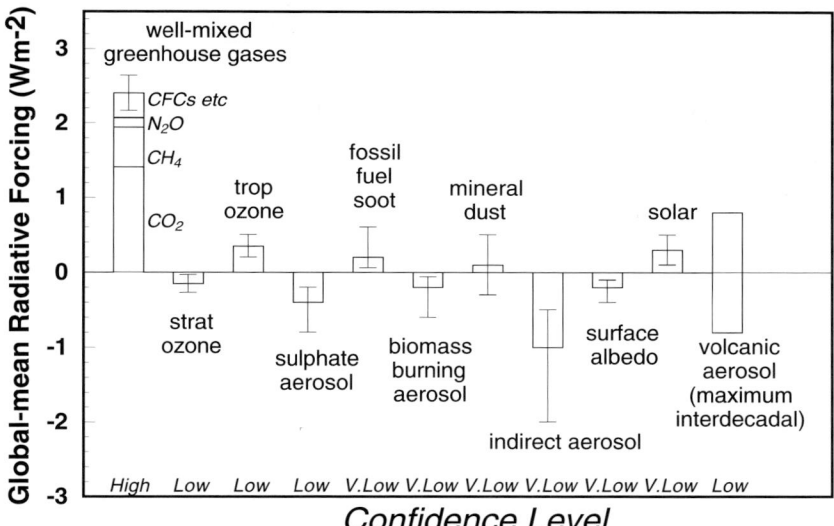

FIGURE 1 Global and annual mean radiative forcing since pre-industrial times for a number of natural and anthropogenic mechanisms. The uncertainty bars indicate, approximately, the range of published results, while the confidence level indicates a subjective confidence level that the true forcing lies within the uncertainty bar, given the difficulties in estimating that forcing. The volcanic forcing is the maximum change between a volcanically active decade and a volcanically quiescent decade, or vice versa. (From Shine and Forster (1999) with permission from Elsevier Science.)

these components in the absence of human activity. This requires heavy reliance on numerical models that, because of the paucity of available data, are difficult to evaluate even for present-day conditions. In the early 1990s there was considerable attention focused on the "cooling" role of sulphate aerosols, because of their enhancement of the scattering of solar radiation. It is now clear that this cooling is, in the global mean at least, largely offset by the warming due to increased concentrations of tropospheric ozone. Further, other components of the aerosol population have now received attention, and these can have quite different effects to sulphate. For example, soot absorbs strongly at visible wavelengths and so causes a positive forcing. The contribution from mineral dust, resulting apparently from changes in land-use, is uncertain even in sign—it absorbs and scatters solar radiation and, because of the relatively large particle sizes, also absorbs significant amounts of thermal infrared radiation.

The strongly negative but highly uncertain indirect aerosol effect results from the impact that aerosols may have on both the optical properties and lifetimes of clouds. The uncertainty is severe—IPCC (1996) declined from giving a central value to the forcing from this source, although Shine and Forster (1999) felt, on the basis of current evidence, that a midrange value of -1 W m^{-2}, with a factor of two uncertainty, was appropriate. The impact of the uncertainty in this forcing will be illustrated in Section C and recent work in this area will be discussed in Section D.

Figure 1 also includes two natural forcings. The first is solar variability. Solar forcings, and possible mechanisms by which solar variations can have an increased impact on climate, have been somewhat "flavour of the day" in recent years. We will not discuss these in detail here, except to note that observational evidence for a proposed link between cosmic rays and cloudiness has been widely questioned (e.g. Kernthaler et al., 1999; Kristjansson and Kristiansen, 2000). Solar variability appears to have a modest, but certainly not a negligible, effect.

The second natural forcing is stratospheric (and mostly sulphate) aerosol resulting from volcanic eruptions. Because the effect of individual volcanic eruptions lasts for only a few years, volcanic eruptions have often been left off figures like Fig. 1. As will be shown in Section C, over the past 100 years there have been periods with either frequent or rare climatically important eruptions; this means that volcanic eruptions are of much greater importance for century-scale climate change than has sometimes been acknowledged. In Fig. 1 volcanic eruptions are shown as either warming or cooling, depending on whether the transition is from a nonvolcanic to a volcanic decade, or vice versa.

C. TIME VARIATION OF RADIATIVE FORCING

Given the difficulties in establishing even the current-day forcings, the derivation of its variation over the past century is even more difficult. It relies on, for example, knowing the time variations of all emissions. Further, for many species there is even less data available with which to evaluate the models than there is for the present day. And yet, as is clear from equation (2), such information is vital if we are to simulate the time variation of climate.

A few attempts have been made to produce such time series (for example, Hansen *et al.*, 1993; IPCC, 1996) and we will illustrate some of the issues using one recent one, Myhre *et al.* (2001). This series produces forcings for the present day which are in general agreement with Fig. 1; it differs from Fig. 1 in that it includes a contribution from fossil fuel organic aerosols (which give a forcing since pre-industrial times of -0.09 W m^{-2}), and excludes the contributions from changes in mineral dust and surface albedo. Figure 2 shows some of the results from this series. First the WMGG forcing shows a near monotonic rise, but with a quite marked flattening in the 1940s. If the indirect forcing is ignored, then the sum of all other anthropogenic forcings yields a curve which, within the uncertainty bars of the individual forcings, is very similar to the WMGG case—this is because of the compensating effects (at least on a global-mean basis) of the various forcings. However, as is expected, the addition of the indirect effect causes an enormous change. Not only is the net forcing notably reduced, but the curve is no longer monotonic, with a relative minimum now occurring in the 1950s.

The volcanic time series, derived from an updated version of Sato *et al.* (1993), illustrates the sporadic nature of climatically significant volcanic eruptions, with an active period between around 1880 and 1910, followed by a quiescent period until 1960, followed by a more active period. This volcanic signal dramatically modulates the anthropogenic curve. Finally, the solar curve is also shown in Fig. 2. One interesting point of note is that its relative impact depends on the degree to which the indirect aerosol effect compensates (in a global-mean sense) the other anthropogenic effects.

In order to illustrate the impact of these uncertainties on climate, it is useful to use the time series in the simple climate model represented by equation (2). We

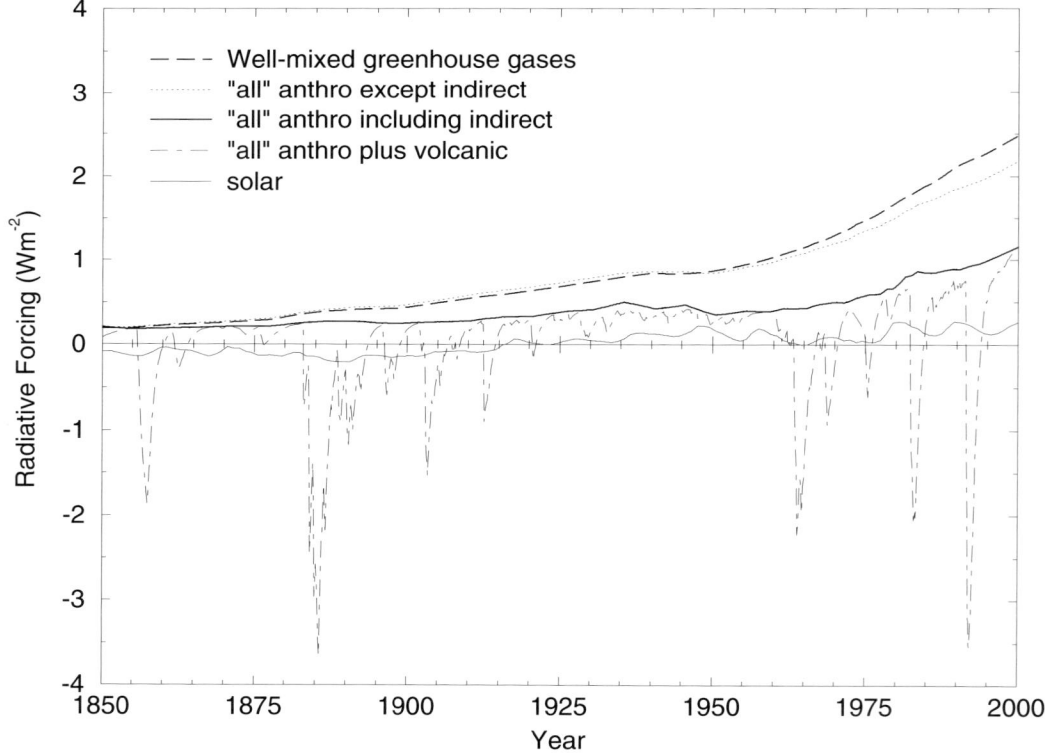

FIGURE 2 A time series of global and annual mean radiative forcing (in W m^{-2}) from 1850 to 1990, from Myhre *et al.* (2001). Various combinations of forcings are included—"all anthro" includes the well-mixed greenhouse gases, tropospheric and stratospheric ozone and sulphate, fossil fuel soot and organics and biomass aerosols. Each forcing has an uncertainty and confidence level associated with it, as indicated in Fig. 1.

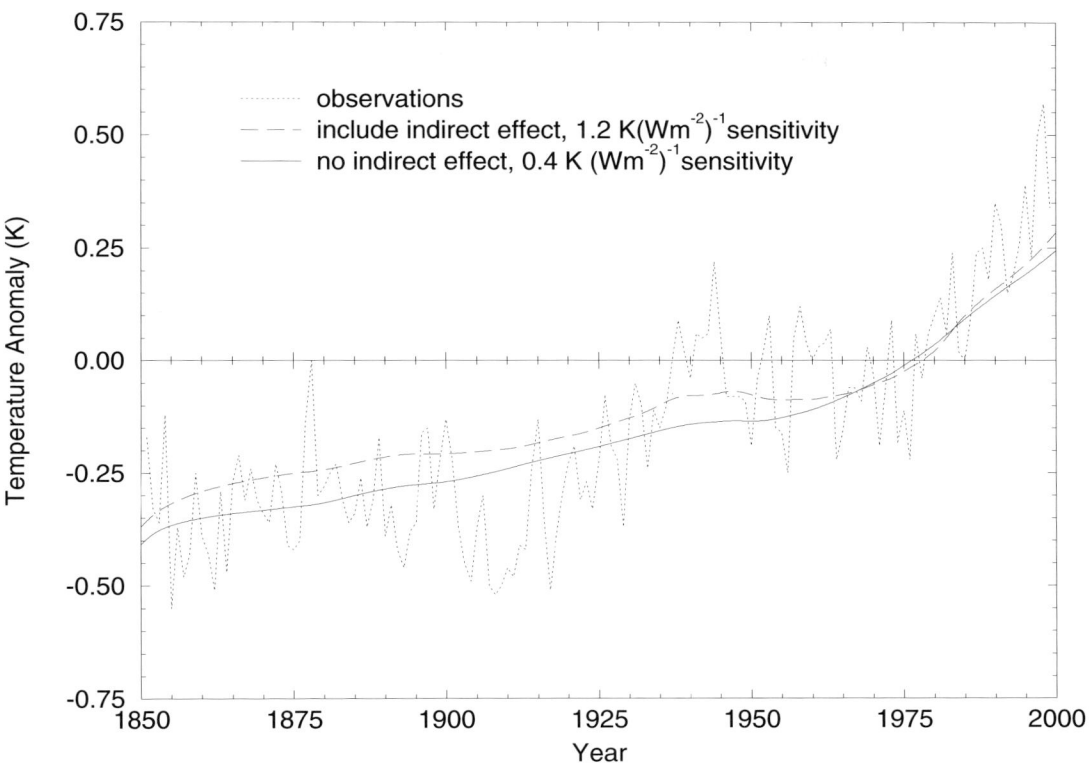

FIGURE 3 Evolution of global mean surface temperature derived using a simple climate model and the anthropogenic forcings of Myhre et al. (2001), as shown in Fig. 2, with and without the indirect aerosol effect and for two different values of the climate sensitivity λ. The observations are from Parker et al. (2000). All temperatures are shown as anomalies from the 1960–1990 mean of each series.

represent the heat capacity of the system using a two-layer ocean, with an upper mixed layer of 100 m depth, coupled to a deep ocean of 1000 m, coupled by a weak diffusion of 1 cm^2 s^{-1}, following Hansen et al. (1981). We use such a simple model in the spirit of it being useful for *illustrating* problems, rather than in the belief that they can *solve* these problems. The limitations of such models are spelt out in, for example, Section 4.7.3 of IPCC (1995). Model parameters, or forcings, can too easily be tuned to match observations for them to be a real guide to the system behaviour. Nevertheless, the model used here can reproduce, to within better than 0.3 K, the warming at time of doubling of CO_2 for a wide range of rates of increase in CO_2 (0.025 to 4%/year) derived using the GFDL Coupled Ocean Atmosphere GCM (see Fig. 6.13 of IPCC, 1996).

The model results are compared with observations of the anomalies of combined land and sea surface temperatures from Parker et al. (2000) relative to the mean for 1961–1990. We do *not* attempt to achieve any "best fit" between models and observations—the purpose of showing the observations is to place the model output in context. Figure 3 shows the impact of the chronic uncertainty in both the forcings and the climate sensitivity. One simulation uses a sensitivity of 0.4 K(W m^{-2})$^{-1}$ and an indirect aerosol forcing of zero; the other uses a sensitivity of 1.2 K(W m^{-2})$^{-1}$, with a present-day indirect aerosol effect of −1 W m^{-2}. The purpose of including or neglecting the indirect aerosol forcing is as a convenient way to illustrate the impact of a 1 W m^{-2} uncertainty in the total forcing; such an uncertainty could, of course, result from errors in other forcings. The overall warming for both cases is almost identical and neither could clearly be rated as "better". This illustrates that there is a large parameter space of radiative forcings and climate sensitivities that generate essentially identical results (as least in terms of the evolution of the global-mean surface temperature).

Next we illustrate the role of volcanic aerosols, which were highlighted in Fig. 2. Many studies have neglected this contribution (e.g. Kelly and Wigley, 1992; Schlesinger and Ramankutty, 1992; Delworth and Knutson, 2000) despite a clear demonstration in many other studies of the potential importance (see e.g. Robock, 2000, and references to many earlier studies therein). Recent demonstrations of the climatic importance of volcanic eruptions include Lindzen and Giannitsis (1998), Rowntree (1998)

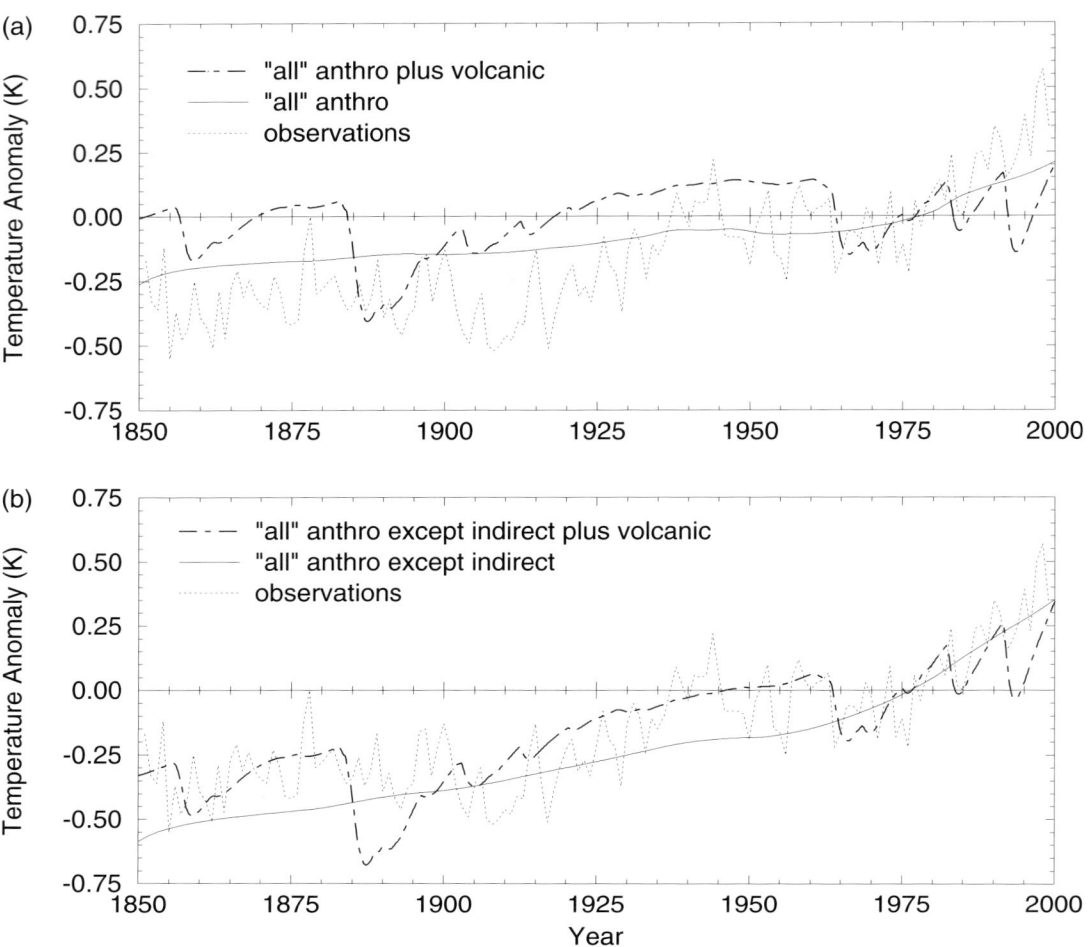

FIGURE 4 Evolution of global mean surface temperature derived using a simple climate model, radiative forcings from Myhre et al. (2001), as shown in Fig. 2, and a climate sensitivity of 0.67 K(W m^{-2})$^{-1}$ for (a) all the anthropogenic forcings in the Myhre et al. series with and without the volcanic aerosol forcing and (b) all anthropogenic forcings except the indirect aerosol effect, again with and without volcanic aerosol forcing. The observations are from Parker et al. and all temperatures are shown as anomalies from the 1960–1990 mean of each series.

and Andronova and Schlesinger (2000). Indeed, some studies have favoured the inclusion of "exotic" forcings related to solar variability, to achieve agreement with observations, while neglecting volcanic forcing which, though there are many uncertainties (Robock, 2000), is arguably much better known. Figure 4a shows, for a midrange climate sensitivity of 0.67 K(W m^{-2}) $^{-1}$, the impact of including volcanic eruptions and the indirect effect. It is apparent that for this particular simple model and for this particular time series, the overall warming during the century is much smaller than observed. Excluding the indirect effect certainly improves agreement (Fig. 4b) but we should not conclude, therefore, that the indirect forcing is indeed small; it may be that other forcings have been incorrectly estimated or even neglected completely. Figure 4 serves to illustrate the tremendous impact of volcanic eruptions on the century timescale, despite the short-term impact of each individual eruption.

It is conventional to plot up such model results as anomalies from some time mean (as in Figs 3 and 4), in order to allow comparison with observations, but this can be somewhat misleading. The climate of the mid-twentieth century certainly was *not* warmer because of volcanoes, as a naïve interpretation of Fig. 4 might imply. This is just a result of taking an anomaly relative to a volcanically perturbed period (1960–1990). Also, the temperature at the end of the twentieth century might appear to have reverted to non-volcanic levels, but again this is not the case. Figure 5 plots up the *absolute* temperature changed from 1850. This makes the impact of volcanic eruptions much clearer. By the end of the "quiescent" 1910–1960 period, the model temperature is indeed back to its nonvolcanic levels.

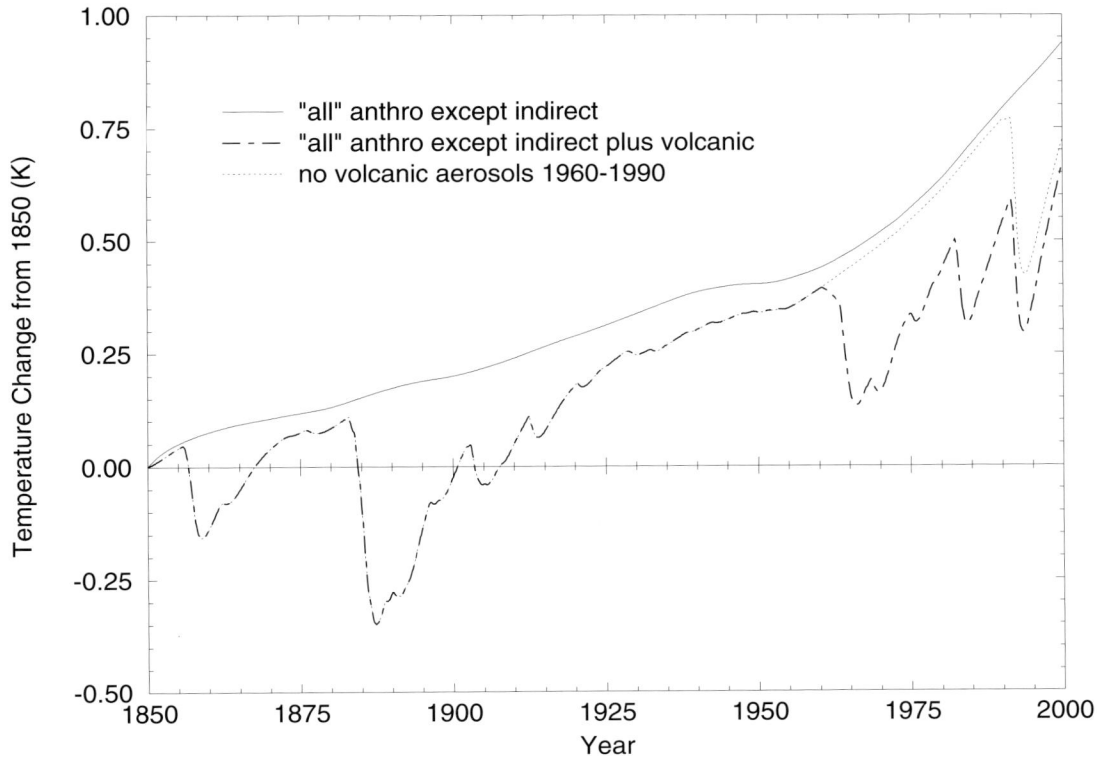

FIGURE 5 Evolution of global mean surface temperature derived using a simple climate model, radiative forcings from Myhre *et al.* (2001), as shown in Fig. 2, and a climate sensitivity of 0.67 K(W m^{-2})$^{-1}$ for all anthropogenic forcings in the Myhre *et al.* series, except the indirect forcing, with and without volcanoes. The dashed line shows the temperatures if there are no volcanic aerosols between 1960 and 1990. Note that, unlike Fig. 4, temperature changes are shown as the absolute change since 1850 rather than as anomalies from a mean.

But while the temperature in 1999 shows a recovery to pre-Pinatubo levels, our simple model appears to indicate that temperatures are still considerably lower than they would have been in the absence of volcanic eruptions. The issue of the degree of "memory" in the climate system between one volcanic eruption and the next depends, of course, on the frequency of eruptions but it also depends on the climate sensitivity, as clearly demonstrated by Lindzen and Giannitsis (1998).

Figure 5 also illustrates an aspect of the short-term impact of aerosols. By "turning off" volcanic aerosols in the 1960–1990 period, it can be seen that the evolution of the 1990s temperatures is almost identical to that with these volcanoes. This illustrates that the overall response in this model is dictated by the frequency of eruptions.

It is again emphasized that the model used here is highly simplified, and cannot represent some of the richness of the dynamical variability and the dynamical response, in particular that resulting from volcanic eruptions (see Robock, 2000). But it does allow the uncertainties in radiative forcings, and the consequences of including or neglecting certain forcings, to be placed in a useful perspective.

D. DEVELOPMENTS IN STUDIES OF THE INDIRECT AEROSOL EFFECT

It is apparent from the above that the uncertainty in the indirect aerosol effect is one of the most severe problems in climate change studies at the present time, and here we briefly elaborate on the issues. These are discussed in more detail in Shine and Forster (1999), Haywood and Boucher (2000) and will also be reviewed in detail in IPCC's Third Assessment Report due out in 2001.

The early studies of the indirect effect concentrated on the impact of sulphate aerosol on cloud properties. Most models derived sulphate mass. To derive a forcing it is necessary to assume relationships between this mass and the number concentration of aerosols or cloud droplets; such relationships were generally based on limited data. There were further concerns. First, sulphate is only one component of the aerosol population. Natural aerosols include sea salt, mineral dust and those derived from natural sulphur emissions. Human activity is also responsible for changes in concentrations of other aerosols. Some of these are shown in Fig. 1 and increased attention is being paid to organic aerosols, emitted directly or formed from hydrocarbons emitted into the atmosphere. The problem is that the indirect aerosol forcing depends crucially on whether the emitted species, such as sulphur, are forming new particles or condensing on to other aerosol particles, of either natural or anthropogenic origin. A recent study by Lohmann et al. (2000) has relaxed many of the limitations of earlier studies by including "background" natural aerosols and anthropogenic organic aerosols, although they recognize many remaining uncertainties. The total forcing is still large and negative; if aerosol components are assumed to mix together, rather than existing as separate entities, the forcing is -1.1 W m^{-2}, but interestingly sulphate causes only 30–40% of the total, with organic aerosols making up the remainder. If verified by other studies, this would constitute something of a paradigm shift in our understanding of the relative importance of different aerosol components, while retaining the large overall importance of the indirect effect.

Two significant developments in studies of the indirect effect can be anticipated in the coming years. First, so far, almost all attention has focused on the impact on liquid water clouds. Clearly, the research needs to be extended to ice processes. The probability of glaciation of a cloud depends quite strongly on the droplet size, and smaller droplets may indicate inhibited probability of glaciation processes (see e.g. Rangno and Hobbs (1994)), especially in midlatitude wintertime clouds. Second, the direct input of ice nuclei into the atmosphere by human activity remains a possibility (see e.g. Szyrmer and Zawadzki, 1997; Haywood and Boucher, 2000); given the complexity of ice cloud processes and the way they interact with solar and thermal infrared radiation, it is difficult to anticipate even the sign of the forcing from such effects.

The second relatively unexplored issue concerns the role that absorbing aerosols have in directly heating the atmosphere, reducing relative humidity and enhancing stability. It has long been recognized that the climatic effect of aerosols changes from a cooling to a warming as aerosol absorption increases. There is a critical value of single scattering albedo* at which the enhanced scattering and absorption by the aerosol balances exactly, giving a radiative forcing of zero (and, by implication from equation (1), no change in surface temperature). Hansen et al. (1997) used a relatively crude GCM and found that when clouds were allowed to respond to climate change, the critical single scattering albedo was higher than when clouds were held fixed. This was because the extra absorption by the aerosols helped to dissipate clouds. Since low clouds act to cool climate, a decrease in such cloud leads to a relative warming effect, enhancing the "clear sky" absorbing effect of the aerosol. More recently, Ackerman et al. (2000) used a large eddy simulation with 200 m horizontal resolution to explore the impact of absorbing aerosols (advected off the Indian subcontinent) on trade wind cumulus in the Indian Ocean. They found a decreased cloud amount as the absorbing aerosol influenced the rate at which the trade wind cumulus was burnt off during the day. The decreased cloud amount has the effect of offsetting the "conventional" indirect effect, and in some cases could even overwhelm it. The fact that aerosol absorption acts in similar ways in two such disparate model types suggests that increased attention needs to be paid to it, to understand the degree to which it may offset the indirect effect at global scales.

E. CONCLUDING REMARKS

The purpose of this short review has been to report on our current understanding of radiative forcing mechanisms and highlight some of the existing uncertainties. Despite the increasing confidence in attribution of some large part of the recent observed climate change to human activity (see e.g. IPCC, 1996), it is evident that our understanding remains fragile. Many forcings are based on chemical model simulations that are difficult to evaluate, and particular problems were highlighted in the case of the aerosol indirect effect. Further, given the rapid rate of developments in this area, we cannot be confident that we have yet identified all relevant climate change mechanisms. Coupled to the uncertainty in the value of the climate sensitivity, it is clear that there is much scope to obtain a good simulation of past climate for the wrong reasons. Certainly there is scope for discriminating between "wrong" and "right" reasons by examining parameters other than global-mean surface temperature change, but progress in this area is difficult given intermodel differences and the limitations of observations of other parameters. One particularly perplexing observation is the apparent decoupling of surface temperature changes from tropospheric temperature changes, particularly in the tropics since the late 1970s (Gaffen et al., 2000).

---

* The single scattering albedo is the ratio of the scattering by a particle to the absorption plus scattering by the particle—thus a non-absorbing aerosol will have a single scattering albedo of 1, with the value decreasing as the aerosol becomes more absorbing.

The impact of absorbing aerosols on clouds, and to some extent, the impact of aerosols on cloud lifetime present a challenge to the simple conceptual model represented by equation (1); should such cloud variations be considered as part of the forcing or part of the feedback? Indeed, is it always possible to unambiguously distinguish between forcing and feedback (see e.g. Haywood and Boucher, 2000)? In terms of the overall climate response, this distinction is not so important, but it certainly does matter if we are to continue to meaningfully compare radiative forcings from different mechanisms. It may be that model comparisons will clearly demonstrate that certain climate change mechanisms have a higher or a lower climate sensitivity than, say, the climate sensitivity for the "classic" CO_2 case. If this could be established, it would allow a more robust means for comparing the sizes of different forcing mechanisms, by accounting for the fact that some mechanisms are more effective than others. Work in this important area is still in its infancy but given that radiative forcing is being implicitly used within the Kyoto Protocol to the United Nations Framework Convention on Climate Change to intercompare emissions of different greenhouse gases, progress in this area is important for societal as well as scientific reasons.

Progress is also necessary on at least three further fronts. First, widespread routine observations of atmospheric composition and climate variables remain crucial and an effective network is likely to need a mixture of observing platforms. Second, much progress has been made in recent years in coordinated campaigns when a wide range of measurements (of, for example, atmospheric composition and radiative fluxes) are made over a limited area from a variety of platforms such as research aircraft, ships and sondes. The aerosol community has been particularly active. Finally, models will need to include mechanisms in a more integrated manner. This is beginning to happen in aerosol studies, but almost all studies of ozone changes treat stratospheric ozone and tropospheric ozone as if they are separate species—the coupling of processes in the upper troposphere and lower stratosphere is likely to be of particular importance. It is also important that a whole hierarchy of models is employed. The state-of-the-art coupled GCMs are a crucial element, but so are simpler, less computer intensive, models that allow a wider range of the uncertain parameter space to be explored.

ACKNOWLEDGEMENTS

EJH is supported by an NERC Postdoctoral Research Fellowship. We would like to thank Gunnar Myhre (University of Oslo) for providing the radiative forcing dataset and for many stimulating conversations on this topic. We also thank Tara O'Doherty for some initial calculations that stimulated some of the work here.

References

Ackerman, A. S., O. B. Toon, D. E. Stevens, A. J. Heymsfield, V. Ramanathan and E. J. Welton, 2000: Reduction of tropical cloudiness by soot. *Science*, **288** 1042–1047.

Andronova, N. G. and M. E. Schlesinger, 2000: Causes of global temperature changes during the 19th and 20th centuries. *Geophys. Res. Letts.*, **27**, 2137–2140

Cess, R. D., M. H. Zhang, G. L. Potter, V. Alekseev, H. W. Barker, S. Bony, R. A. Colman, D. A. Dazlich, A. D. DelGenio, M. Deque, M. R. Dix, V. Dymnikov, M. Esch, L. D. Fowler, J. R. Fraser, V. Galin, W. L. Gates, J. J. Hack, W. J. Ingram, J. T. Kiehl, Y. Kim, H. LeTreut, X. Z. Liang, B. J. McAvaney, V. P. Meleshko, J. J. Morcrette, D. A. Randall, E. Roeckner, M. E. Schlesinger, P. V. Sporyshev, K. E. Taylor, B. Timbal, E. M. Volodin, W. Wang, W. C. Wang, and R. T. Wetherald, 1997: Comparison of the seasonal change in cloud-radiative forcing from atmospheric general circulation models and satellite observations. *J. Geophys. Res. Atmos.*, **102**, 16593–16603.

Delworth, T. L. and T. R. Knutson, 2000: Simulation of early 20th century global warming. *Science*, **287**, 2246–2250.

Gaffen, D. J., B. D. Santer, J. S. Boyle, J. R. Christy, N. E. Graham and R. J. Ross, 2000: Multidecadal changes in the vertical structure of the tropical troposphere. *Science*, **287**, 1242–1245.

Hansen, J., D. Johnson, A. Lacis, S. Lebedeff, P. Lee, D. Rind and G. Russell, 1981: Climate impact of increasing atmospheric carbon-dioxide. *Science*, **213**, 957–966.

Hansen, J., A. Lacis, R. Ruedy, M. Sato and H. Wilson, 1993: How sensitive is the world's climate? *Nat. Geogr. Res. Explor.*, **9**, 142–158.

Hansen, J., M. Sato and R. Ruedy, 1997: Radiative forcing and climate response. *J. Geophys. Res. Atmos.*, **102**, 6831–6864.

Hansen, J. E., M. Sato, A. Lacis, R. Ruedy, I. Tegen and E. Matthews, 1998: Climate forcings in the industrial era. *Proc. Natl. Sci. USA*, **95**, 12753–12758.

Haywood, J. M. and O. Boucher, 2000: Estimates of the direct and indirect radiative forcing due to tropospheric aerosols: a review. *Rev. Geophys*, **38**, 513–543.

IPCC, 1990: *Climate Change: The Intergovernmental Panel on Climate Change Scientific Assessment.* Cambridge University Press, Cambridge, UK.

IPCC, 1992: *Climate Change 1992. The Intergovernmental Panel on Climate Change.* Cambridge University Press, Cambridge, UK.

IPCC, 1995: *Climate Change 1994. The Intergovernmental Panel on Climate Change.* Cambridge University Press, Cambridge, UK.

IPCC, 1996: *Climate Change 1995. The Intergovernmental Panel on Climate Change.* Cambridge University Press, Cambridge, UK.

IPCC, 1999: *Aviation and the Global Atmosphere. The Intergovernmental Panel on Climate Change.* Cambridge University Press, Cambridge, UK.

Kelly, P. M. and T. M. L. Wigley, 1992: Solar-cycle length, greenhouse forcing and global climate. *Nature*, **360**, 328–330.

Kernthaler, S. C., R. Tuomi and J. D. Haigh, 1999: Some doubts concerning link between cosmic ray fluxes and global cloudiness. *Geophys. Res. Lett.*, **26**, 863–865.

Kristjansson, J. E. and J. Kristiansen, 2000: Is there a cosmic ray signal in recent variations in global cloudiness and cloud radiative forcing? *J. Geophys. Res. Atmos.*, **105**, 11851–11863.

Lindzen, R. S. and C. Giannitsis, 1998: On the climate implications of volcanic cooling. *J. Geophys. Res. Atmos.*, **103**, 5929–5941.

Lohmann, U., J. Feichter, J. Penner and R. Leaitch, 2000: Indirect effect of sulfate and carbonaceous aerosols: a mechanistic treatment. *J. Geophys. Res. Atmos.*, **105**, 12193–12206.

Matthews, W. H., W. W. Kellogg and G. D. Robinson, 1971: *Man's Impact on Climate*. The MIT Press, Cambridge, MA.

Myhre, G., A. Myhre and F. Stordal, 2001: Historical evolution of total radiative forcing. *Atmos. Environ*, **35**, 2361–2373.

Parker, D. E., E. B. Horton and L. V. Alexander, 2000: Global and regional climate in 1999. *Weather*, **55**, 188–199.

Rangno, A. L. and P. V. Hobbs, 1994: Ice particle concentrations and precipitation development in small continental cumuliform clouds. *Q. J. R. Meteorol. Soc.*, **120**, 573–601.

Robock, A., 2000: Volcanic eruptions and climate. *Rev. Geophys.*, **38**, 191–219.

Rowntree, P. R., 1998: Global average climate forcing and temperature response since 1750. *Int. J. Climatol.*, **18**, 355–377.

Sato, M., J. E. Hansen, M. P. McCormick and J. B. Pollack, 1993: Stratospheric aerosol optical depths, 1850–1990. *J. Geophys. Res. Atmos.*, **98**, 22987–22994.

Schlesinger, M. E. and N. Ramankutty, 1992: Implications for global warming of intercycle solar irradiance variations. *Nature*, **360**, 330–333.

Shine, K. P. and P. M. D. Forster, 1999: The effect of human activity on radiative forcing of climate change: a review of recent developments. *Global Planet. Change*, **20**, 205–225.

SMIC, 1971: *Inadvertent Climate Modification*. The MIT Press, Cambridge, MA.

Szyrmer, W. and I. Zawadzki, 1997: Biogenic and anthropogenic sources of ice-forming nuclei: a review. *Bull. Am. Meteorol. Soc.*, **78**, 209–228.

WMO, 1985: *Atmospheric Ozone 1985*. Global Ozone Research and Monitoring Project Report No 16. World Meteorological Organization, Geneva, Switzerland..

Changes in Climate and Variability Over the Last 1000 Years

Philip D. Jones

Climatic Research Unit, University of East Anglia, Norwich, UK

The instrumental record of climate extends back in many regions of the world to the mid-nineteenth century. Over the period 1861–2000 average global surface temperatures have warmed by about 0.6°C. The warmest years of this record are in the 1990s with 1998 the warmest year. The warming has not occurred in a linear fashion, but in two phases, early 1920s to the mid-1940s and from the mid-1970s. Patterns of warming during these two periods reveal marked differences, but in both cases few regions exhibit statistically significant warming, but on a global and hemispheric basis the warming is highly statistically significant.

Recent advances in palaeoclimatology have enabled the last 150 years to be placed in the context of the whole millennium, although it is clear that errors of estimates involved in the pre-1850 period are considerably greater than for the instrumental era. The most striking feature of the millennial reconstructions is that the twentieth century was the warmest over the Northern Hemisphere. The 1990s was the warmest decade and 1998 the warmest individual year. Temperatures during the first five centuries of the millennium were warmer than those over the 1500–1900 period, but were at least 0.1°C below the 1961–90 average. The coldest decade, the 1600s, saw average hemispheric temperatures about 0.8°C below 1961–90 levels.

A. INTRODUCTION

Improved understanding of past climate, particularly for the last 1000 years, is not just important from scientific curiosity, but knowledge of past levels of natural variability is vital if we are to say, unequivocally, both whether recent warming is unusual in a long-term context and if it is due to anthropogenic influences through changes in the composition of the atmosphere. The instrumental record of surface air temperatures extends back to the mid-nineteenth century but is far from perfect. Even now it is incomplete, with about a fifth of the earth's surface without data. Using it to assess levels of natural variability is also partly confounded as the period of best data coverage (1951 onwards) is probably inflated by anthropogenic influences.

This chapter will briefly review the instrumental surface air temperature record on a global basis, highlighting the two periods of warming during the twentieth century. Measurements back to the late 1600s/early 1700s are available from Europe and these will be discussed as they clearly provide information for a part of the world before human influences can have had any effect. Palaeoclimatic sources provide evidence for earlier centuries and other regions, clearly less reliably and poorer spatially than instrumental data, but combined together the evidence should be able to place the instrumental period in a much longer-term context. This chapter will conclude with a discussion of recent palaeoclimatic advances and how these relate to perceived wisdom of the course of changes over the millennium for both the Northern Hemisphere and over Europe.

B. THE INSTRUMENTAL PERIOD (1850–)

The surface air temperature record has been extensively analysed over the last 20 years or so (see, e.g., Nicholls *et al.*, 1996; Hansen *et al.*, 1999; Jones *et al.*, 1999a). These and other studies have emphasized that all observations must be taken in a consistent manner (be homogeneous), so that variations are solely due to the vagaries of the weather and climate and not due to non-climatic factors (changes in station locations, observation times, methods of calculating monthly averages and the station environment). The fact that four different groups have analysed land-based records (Jones *et al.*, 1999a; Hansen *et al.*, 1999; Peterson *et al.*, 1998; Vinnikov *et al.*, 1990), used different databases of station temperatures, different approaches to homogeneity

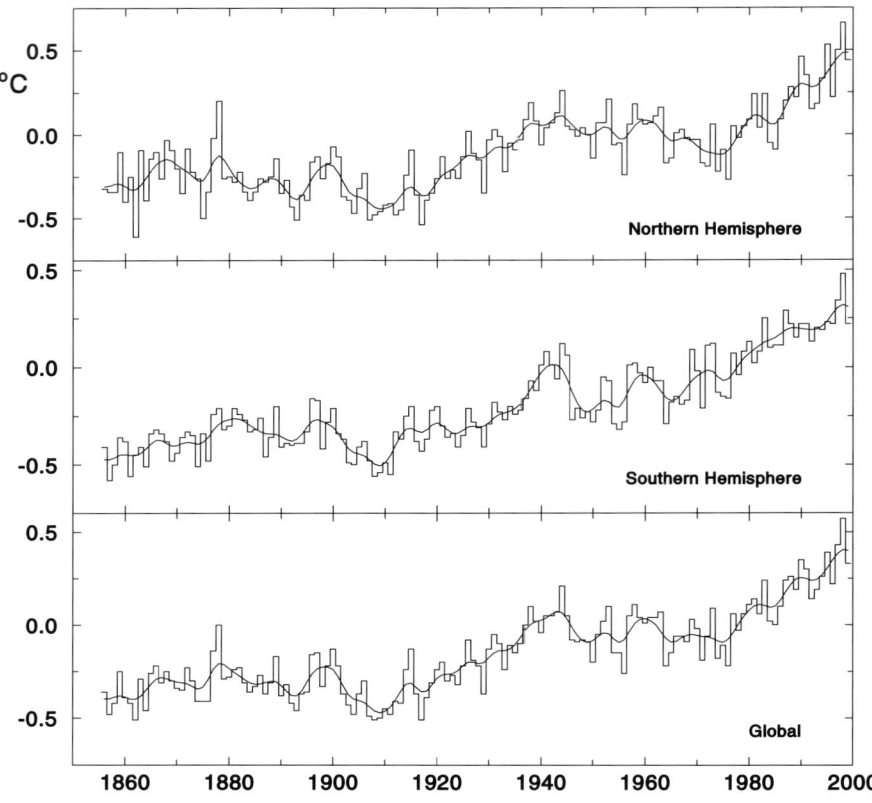

FIGURE 1 Hemispheric and global temperature series on an annual basis, 1855–1999, expressed as anomalies from the 1961–90 base period. The smooth line highlights decadal scale changes.

assessments and different methods of combining the basic data, yet all have produced essentially the same course of change on hemispheric bases, attests to the robustness of the long-term changes that have occurred.

Three groups have analysed ocean temperatures (Parker *et al.*, 1995; Kaplan *et al.*, 1998; Reynolds and Smith, 1994) and despite different methods of homogeneity assessment and data combination arrive at a similar course of change over the last 150 years. Comparison of the land and marine temperature series mutually supports both components attesting that the evidence of a long-term warming since the 1850s of about 0.6°C is beyond question. Two factors, however, remain, as doubts: urbanization over some land regions, although studies (Jones *et al.*, 1990; Peterson *et al.*, 1999) show that effects are an order of magnitude smaller than the long-term warming; and bucket corrections of sea surface temperature measurements before 1940 (Folland and Parker, 1995) as information about the type and use of buckets in the nineteenth century is very limited.

Figure 1 shows annual hemispheric and global temperature series using the dataset of Jones *et al.* (1999a). Both hemispheres show similar amounts of warming over the 1861–2000 period. The warming has clearly occurred in two phases, from about 1920 to 1945 and since 1975. Between these two periods, slight cooling occurred in the NH and slight warming in the SH, but both trends are not statistically significant. Statistical analysis of the two warming periods shows both are highly statistically significant. Sampling errors of estimation of hemispheric/global averages are of the order of ±0.05°C for years since 1951 and roughly twice this before 1900 (Jones *et al.*, 1997).

Spatial patterns of warming are shown for two 25-year periods (1920–44 and 1975–99) by season and year in Fig. 2. Areas where the trend of temperature change is locally statistically significant are relatively few, although there are more in the latter period. The pattern of recent warming covers more of the earth's surface than that of the earlier period this century. The early warming was greatest in the North Atlantic region, particularly in the winter and spring seasons. The recent warming is strongest in midlatitude and some tropical regions, and most significant also in the winter and spring seasons. Cooling is evident in some areas, but only in a few boxes is the trend significant.

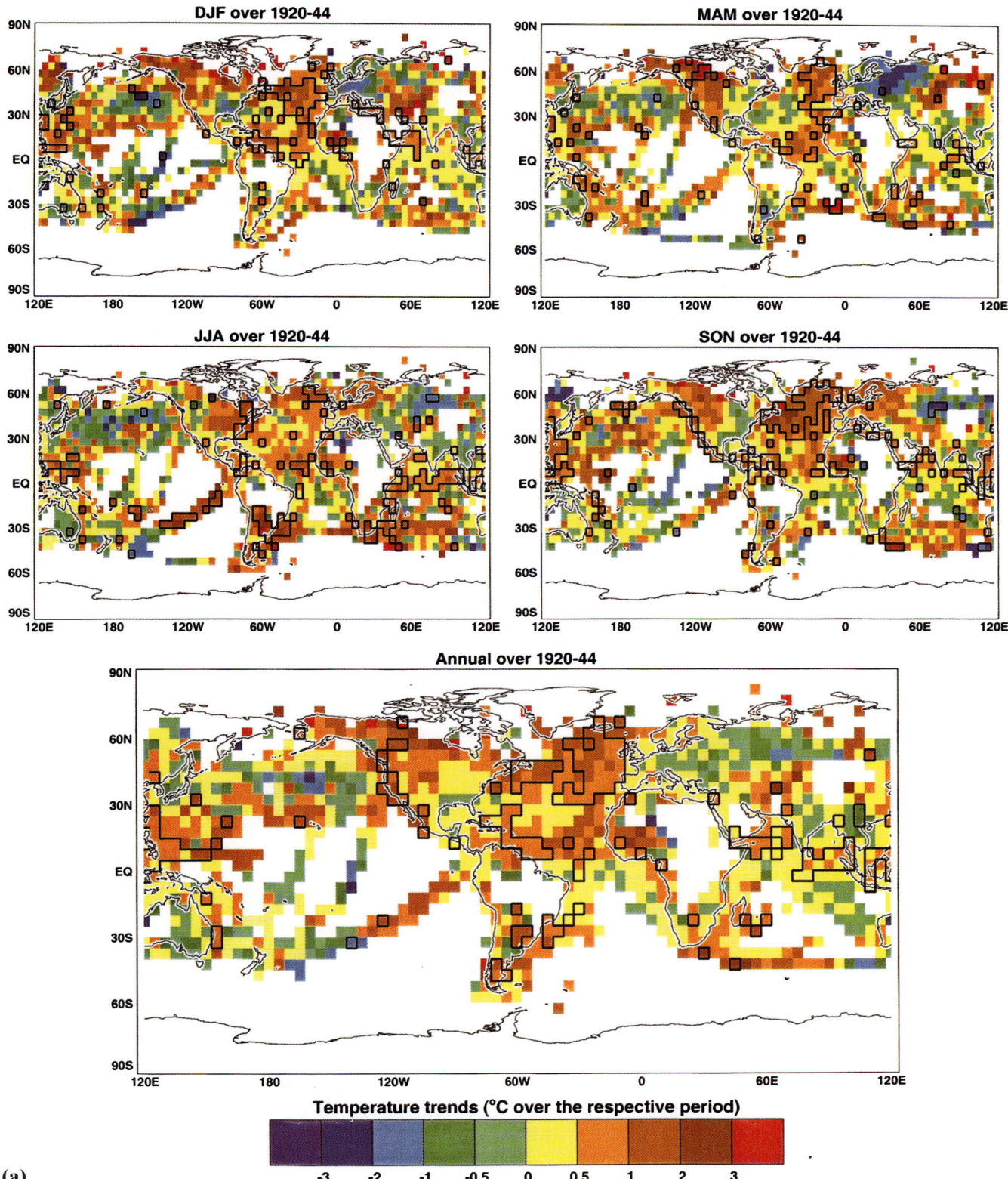

(a)

FIGURE 2 Trend of temperature on a seasonal and annual basis for the two 25-year periods, (a) 1920–44 and (b (overleaf)) 1975–1999. Boxes with significant trends at the 95% level (allowing for autocorrelation) are outlined by heavy black lines. At least 2 (8) months data were required to define a season (year), and at least half the seasons or years were required to calculate a trend.

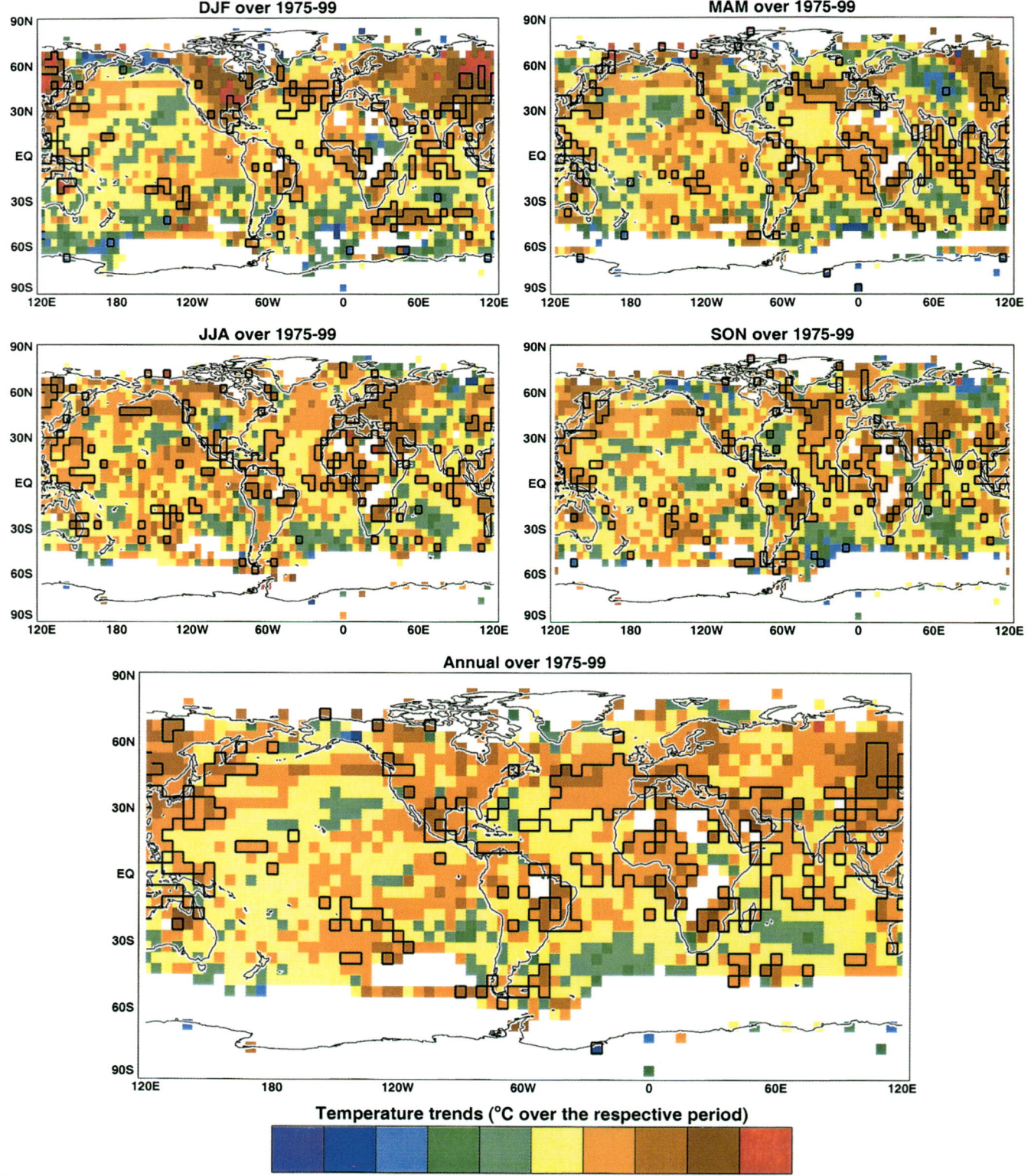

More details of recent changes are discussed in Jones *et al.* (1999a,b) and by Easterling *et al.* (1997). The latter shows that between 1950 and 1993 three quarters of the warming over land areas has occurred for minimum temperatures (nighttimes) and only one quarter for maximum (daytime) temperatures. Jones *et al.* (1999a,b) show that recent increases in temperature have occurred more through reductions in areas affected by extremely cold temperatures than through increases in areas affected by extremely warm temperatures. Warming is not occurring evenly across the temperature spectrum. Even locally, warming of the extremely cold distribution tail occurred at two to three times the change in mean temperatures (Yan *et al.*, 2001a) over much of the twentieth century. So far, the impacts of these changes have largely passed unnoticed, but the lack of killing winter frosts in some regions is beginning to have effects by increasing some insect populations.

C. EARLY EUROPEAN INSTRUMENTAL RECORDS

The invention of the thermometer in Europe in the seventeenth century means that for many parts of the continent there is a potential to develop long daily series back to the early eighteenth century. Realizing this potential, however, requires considerable efforts though, because both observation times were irregular before railways brought national and European times and the exposure of thermometers was nonstandard before the development of "Stevenson-type" screens. Locating all the necessary station history information is difficult as some changes were not considered important enough at the time to have been noted down. A recent European project has digitized and adjusted records for seven sites across Europe (Camuffo *et al.*, 2000). Together with the already available Central England record (Parker *et al.*, 1992) the series have been extensively analysed by Moberg *et al.* (2000), Jones *et al.* (2001) and Yan *et al.* (2001a,b), principally for the assessment of whether there are changes in the frequency of extremes. The conclusions from these studies are not that clear cut. Day-to-day temperature variability has decreased since the 1880s and there is evidence of fewer cold extremes at most sites, although trends are barely significant after accounting for the large year-to-year variability in the derived series. Increases in warm days have occurred recently, but in most of Europe the numbers of warm days per year are only at similar levels compared with some decades in the eighteenth and early nineteenth centuries, especially the 1730s, 1760s and 1820s.

Figure 3 shows annual, winter and summer temperature series for Central England, Fennoscandia and Central Europe. Long-term warming is evident in the annual and particularly in the winter series, but summer temperatures recently are cooler than some decades in the late eighteenth/early nineteenth century, particularly in northern Europe. All these European series clearly show that the coldest period of the last 250 years was the mid-to-late nineteenth century, the time when widespread temperature measurements enabled hemispheric-scale series to begin. The warming since then of about 0.6°C might therefore be accentuated by the cold decades of the 1850s–1880s if the earlier European records are mirrored in other mid-to-high latitude regions of the NH. Early instrumental records outside Europe are very scarce and have not been so extensively analysed as those from Europe, but there is some evidence from North America that the 1820s were not as relatively warm there compared to Europe (see, e.g., Jones and Bradley, 1992). The longest of European records show their coldest periods in the 1690s and the exceptionally cold year of 1740 is apparent in all series that extend back that far.

The early European instrumental data point to considerable decade-to-decade variability of temperatures (the warm 1730s and the cold 1740s), which is large enough to suggest that all recent interdecadal fluctuations could be natural variations. The most clear finding is the relative warming of winters, when compared to summers. This change has reduced the seasonal contrast across Europe, making the climate less continental. Greater warming in winter may be an anthropogenic signal (greenhouse warming throughout the year, but countered by sulphate aerosol cooling in summer), but in most long European series, winters began warming relative to summers well before anthropogenic factors could have been important.

D. THE PRE-INSTRUMENTAL PERIOD (BACK TO 1000)

1. Palaeoclimatic Sources

Over recent decades, several new sources of proxy climate information for the pre-instrumental period have been developed. Sources are generally classified as being of two types. High-frequency sources provide information every year (not necessarily annually, as there is often emphasis on the summer half of the year) and include tree rings (widths and densities), ice cores (accumulation and isotopic composition), corals (growth amounts, isotopic composition and metallic ion ratios), some lacustrine sediments and historical docu-

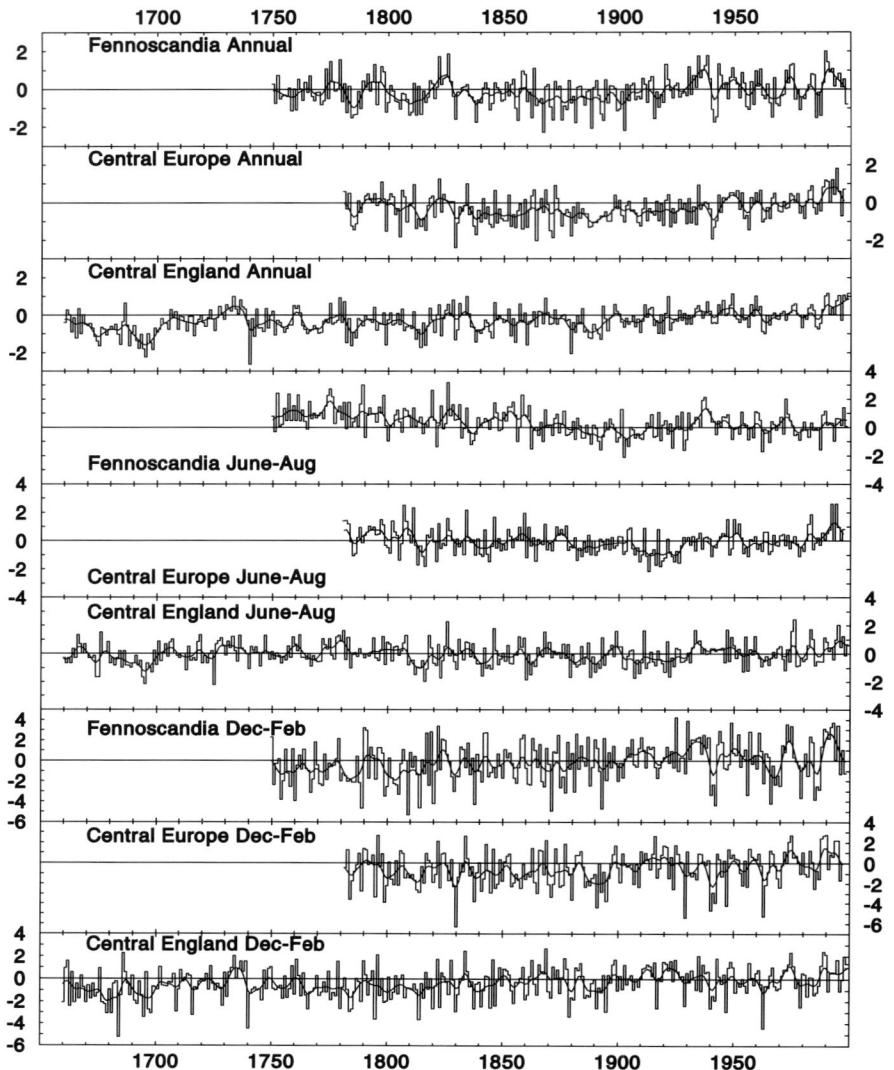

FIGURE 3 Annual, summer (JJA) and winter (DJF) average temperatures for Central England and a Fennoscandian and a Central European average. Data for each series are expressed as anomalies from the 1961–90 base period. Fennoscandia comprises the average of six sites (Uppsala, Stockholm, St. Petersburg, Trondheim, Vardo and Archangel'sk) while Central Europe is based upon five sites (Berlin, Vienna, Kremsmünster, Hohenpeissenberg and Innsbruck). The smooth lines highlights decadal scale variations. Winters are dated by the January.

mentary material (generally from Europe, China, Korea and Japan where written records are plentiful). Less-well resolved sources include some lacustrine and marine sediments, peat bogs, glacial advances/retreats and borehole evidence. Each source is almost its own discipline and some disciplines rarely interact with others. The specifics of all the methods are discussed extensively by Bradley (1999) and the many references therein.

2. Multiproxy Averages from High-frequency Sources

Figure 4 compares five different reconstructions of Northern Hemisphere temperature change for most of the last millennium. The reconstructions (Jones et al., 1998; Mann et al., 1998, 1999; Overpeck et al., 1997; Briffa; and Crowley and Lowery, 2000) are of different seasons, so based on the instrumental record would be expected to differ a little. Here they have all been recalibrated to April to September temperature averages for

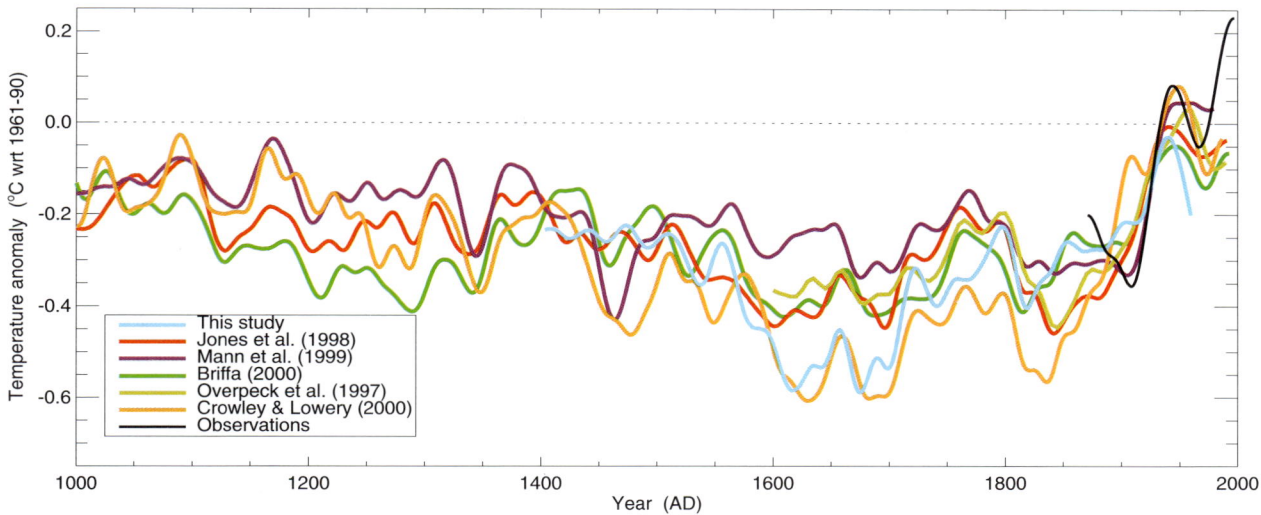

FIGURE 4 Millennial scale reconstructions of 'extended' summer temperature, each calibrated against April to September temperatures for land areas north of 20°N over 1881–1960. Each series has been smoothed with a 50-year Gaussian filter.

land areas north of 20°N for the 1881–1960 period. None is strictly independent of the others, as they contain some common sources, but each has made different assumptions in their averaging.

The most striking feature of the multiproxy averages is the warming over the twentieth century, both for its magnitude and duration. In all the series, the twentieth century was the warmest of the millennium and the warming during it unprecedented. It is likely that the 1990s was the warmest decade and 1998 the warmest year. The coldest year is likely to have been 1601, about 1°C below the 1961–90 average, although this is considerably more uncertain than the warmth of 1998. If standard errors are assigned to these series, as was the case for the instrumental period, errors would be considerably larger and for times before 1600 it is difficult to be confident about any changes that have occurred as there are markedly fewer records for this time.

The various series differ in some respects with regard to the cooler and warmer periods of the millennium, but they have all incorporated orders-of-magnitude more data than available in the early 1970s (e.g. available to Lamb, 1977). The cooler centuries of the millennium were the sixteenth to the ninteenth, the seventeenth being the coldest in Europe and the nineteenth coldest in North America. These regions are still the best studied and it is important to extend our knowledge to other areas, particularly the Southern Hemisphere. At present, for every one long SH record there are about 20 in the NH. Just as with the instrumental record, it is important to gain as much evidence from as many regions as possible, if we are to fully understand how global and hemispheric temperatures have varied over this long time. A more complete understanding of the causes of changes will allow us to determine how much climate can change naturally, enabling us to better distinguish the degree of human influence on surface temperatures during the twentieth century.

3. Low-frequency sources

The two most widely available low-frequency sources for the past millennium are boreholes and glacial advances/retreats. The borehole record for the Northern Hemisphere (Huang et al., 2000) agrees well with instrumental records over the twentieth century but suggests about twice as much long-term warming since 1600. Comparison of boreholes from the British Isles (Jones, 1999) indicates excellent agreement with Manley's (1974) record. Elsewhere, particularly in North America, where the majority of borehole records originate, differences are greater. These may reflect long-term changes in seasonal snow cover (boreholes are insulated from cold air above the snow blanket in winters) but more likely widespread changes in land use. This has been clearly shown for parts of western Canada by Skinner and Majorowicz (1999). For Europe, land-use changes are minimal over this time, probably occurring to a greater extent centuries earlier. Glacial advances and retreats have been extensively studied by Grove (1988) and are commented upon later. Combining a high-frequency tree-ring reconstruction with a simple model of a glacier has been shown to agree with glacial movements for a northern Swedish location over the last 1500 years (Raper *et al.*, 1996), illustrating how the results from different proxies can be reconciled.

E. DISCUSSION AND CONCLUSIONS

1. The Instrumental Period

The world has clearly warmed over the last 150 years, but the warming has not been equally spread across the seasons nor spatially. Warming has occurred in two distinct phases, 1920–45 and since 1975. Taking the standard errors of estimation into account, it can be easily argued that the two warming events are the only significant features of the global and hemispheric temperature series. Warming since 1850 over the NH has been strongest in the non-summer seasons, especially the winter and spring seasons in the two warming phases. The Southern Hemisphere indicates similar warming between the seasons, a result expected because of the dominance of the oceans. Recent warming has been more pronounced during the night (minimum) as opposed to daytime (maximum) and has been greater at colder compared to the warmest temperatures.

Longer European records indicate that the mid-to-late nineteenth century was one of the coldest periods of the long instrumental period. Temperatures were as mild as today in some earlier decades such as the 1730s, 1760s and 1820s. Warming is clear in the long European records during the winter half of the year, but summers today are probably slightly cooler than some experienced in the late eighteenth century. The continentality of European temperatures has reduced over the last 250 years.

2. The Last 1000 Years

High-frequency averages of many proxy climatic sources reveals that the twentieth century was the warmest of the millennium with the sixteenth to nineteenth centuries being cooler than the eleventh to fourteenth. In many respects, the studies encompassed by Fig. 4 are rewriting what has been the 'accepted' course of change for the millennium, although what is 'accepted' clearly varies from researcher to researcher. Reconstructions of past climate have been made for the last 50 years, mostly for locations in Europe, North America and the Far East. Until the recent collection of papers embodied by Fig. 4 they have rarely been combined on a year-by-year basis before. Earlier studies (exemplified by Lamb, 1977) intercompared reconstructions, synthesizing the results in a qualitative manner.

Two major periods of the millennium (the Medieval Warm Epoch (MWE), 900–1200 and the Little Ice Age (LIA), 1450–1850) have been recognized. Neither period has accepted dates, but those given encompass present views. By their very name they suggest a warmer and a cooler period compared to today's prevailing level of temperature. As more evidence has become available from many parts of the world, particularly the Southern Hemisphere, the concepts these terms embody have become increasingly difficult to reconcile (see, e.g., Bradley and Jones, 1993 and Hughes and Diaz, 1994). Colder centuries are clearly evident in the series in Fig. 4, but the long European instrumental series show years in the late eighteenth century and even early nineteenth century as warm as today's levels. Glacial advances have been noted in many alpine regions, particularly during the sixteenth, seventeenth and nineteenth centuries, but the advances are never synchronous across all regions (Grove, 1988).

Evidence for the MWE on a hemispheric basis is even less compelling than for the LIA (Hughes and Diaz, 1994). Warmth clearly occurs in some of the regions with long records, but always at different times in different regions (Crowley and Lowery, 2000). Analyses of the instrumental record indicate that no single region always follows the hemispheric/global average series. Lamb (1977) has argued that Britain and Europe, because of their position in the midlatitude westerlies, should be representative of a larger area. The early 1940s, however, were warm over the globe (see Fig. 1), yet this was one of the coldest five-year periods over Europe (see Fig. 3). Rather than trying to force new reconstructions to conform to outdated terms, it would seem preferable to discuss each series solely on whether it was warm/cold on decade-to-century timescales and integrate them in the forms used in Fig. 4.

In Britain, the two terms seem to have become accepted fact, yet Manley's (1974) record of Central England temperatures clearly does not give much support to the LIA. Manley (1974) was disinclined to use the term and said it was unsatisfactory. One reason for the acceptance of the LIA was the frost fairs of the seventeenth-to-early nineteenth centuries and for the MWE it is a comment by Lamb (1977) that vines were grown near York. The third coldest winter in Manley's series was that of 1962/3 and the Thames did not freeze. A simple explanation of frost fairs is a change in the hydrological regime. London Bridge was rebuilt around 1845 and the weir underneath the old bridge removed. The tidal limit after 1845 was much further upstream at Teddington. Industrial use of the river increased during Victorian times, but it is much easier to freeze fresh water, ponded by a weir. Similarly, it is possible, with considerable effort, to grow vines now in the Vale of York. Monasteries at the time of the MWE may have tried to grow their own vines rather than buy costly supplies from France or even Southern England. England may have exported wine to France in medieval times, but it does so now, albeit in limited quantities. Impressions about the past should be based on hard evidence from diverse sources and not on a few limited pieces of information.

ACKNOWLEDGEMENTS

The author acknowledges the support of the U.S. Department of Energy (grant DE-FG02-98ER62601) and the comments of Keith Briffa and Tim Osborn. Discussions with Ray Bradley, Tom Crowley, Malcolm Hughes and Mike Mann over the last few years have also found their way into some of the text.

References

Bradley, R. S., 1999: *Paleoclimatology: Reconstructing Climates of the Quaternary*. Harcourt/Academic Press, San Diego.

Bradley, R. S. and P. D. Jones, 1993: 'Little Ice Age' summer temperature variations: their nature and relevance to recent global warming trends. *Holocene*, **3**, 367–376.

Briffa, K. R., 2000: Annual climate variability in the Holocene: interpreting the message of ancient trees. *Quater. Sci. Rev.*, **19**, 87–105.

Camuffo, D., G. Demarée, T. D. Davies, P. D. Jones, A. Moberg, J. Martin-Vide, C. Cocheo, M. Maugeri, E. Thoen and H. Bergstrom, 2000: Improved understanding of past climatic variability from early daily European instrumental sources (IMPROVE). *European Climate Science Conference*, Vienna, 19–23 Oct. 1998.

Crowley, T. J. and T. Lowery, 2000: How warm was the Medieval Warm Period? A comment on "Man-made versus Natural Climate Change". *Ambio*, 39, 51–54.

Easterling, D. R., B. Horton, P. D. Jones, T. C. Peterson, T. R. Karl, D. E. Parker, M. J. Salinger, V. Razuvayev, N. Plummer, P. Jamason and C. K. Folland, 1997: Maximum and minimum temperatures for the globe. *Science*, **277**, 364–367.

Folland, C. K. and D. E. Parker, 1995: Correction of instrumental biases in historical sea surface temperatures. *Q. J. R. Meteorol Soc.*, **121**, 319–367.

Grove, J. M., 1988: *The Little Ice Age*. Methuen, New York.

Hansen, J., R. Ruedy, J. Glascoe and M. Sato, 1999: GISS analysis of surface temperature change. *J. Geophys. Res.*, **104**, 30997–31022.

Huang, S., H. N. Pollack and Po-Yu Shen, 2000: Temperature trends over the past five centuries reconstructed from borehole temperatures. *Nature*, **403**, 756–758.

Hughes, M. K. and Diaz, H. F., 1994: Was there a "Medieval Warm Period" and if so, where and when? *Climat. Change*, **26**, 109–142.

Jones, P. D., 1999: Classics in physical geography revisited—Manley's CET series. *Prog. in Phys. Geog.*, **23**, 425–428.

Jones, P. D. and R. S. Bradley, 1992: Climatic variations in the longest instrumental records. In *Climates Since A.D. 1500* (R. S. Bradley and P. D. Jones, Eds), Routledge, London, pp. 246–268.

Jones, P. D., P. Ya, Groisman, M. Coughlan, N. Plummer, W.-C. Wang and T. R. Karl, 1990: Assessment of urbanization effects in time series of surface air temperature over land. *Nature*, **347**, 169–172.

Jones, P. D., T. J. Osborn and K. R. Briffa, 1997: Estimating sampling errors in large-scale temperature averages. *J. Climate*, **10**, 2548–2568.

Jones, P. D., K. R. Briffa, T. P. Barnett and S. F. B. Tett, 1998: High-resolution palaeoclimatic records for the last millennium: interpretation, integration and comparison with General Circulation Model control run temperatures. *Holocene*, **8**, 455–471.

Jones, P. D., M. New, D. E. Parker, S. Martin and I. G. Rigor, 1999a: Surface air temperature and its changes over the past 150 years. *Rev. Geophys.*, **37**, 173–199.

Jones, P. D., E. B. Horton, C. K. Folland, M. Hulme, D. E. Parker and T. A. Basnett, 1999b: The use of indices to identify changes in climatic extremes. *Climat. Change*, **42**, 131–149.

Jones, P.D., K. R. Briffa, T. J. Osborn, A. Moberg and H. Bergström, H., 2001: Relationships between circulation strength and the variability of growing season and cold season climate in northern and central Europe. *Holocene*, in press.

Kaplan, A., Cane, M. A., Kushir, Y. A. and Clement, A. C., 1998: Analyses of global sea surface temperature, 1856–1991. *J. Geophys. Res.*, **103**, 18567–18589.

Lamb, H. H., 1977: *Climate: Present, Past and Future*, Vol. 2, *Climatic History and the Future*. Methuen, London.

Manley, G., 1974: Central England temperatures: monthly means 1659 to 1973. *Q. J. R. Meteorol. Soc.*, **100**, 389–405.

Mann, M. E., R. S. Bradley and M. K. Hughes, 1998: Global-scale temperature patterns and climate forcing over the past six centuries. *Nature*, **392**, 779–787.

Mann, M. E., R. S. Bradley and M. K. Hughes, 1999: Northern Hemisphere temperatures during the past millennium: inferences, uncertainties and limitations. *Geophys. Res. Lett.*, **26**, 759–762.

Moberg, A., P. D. Jones, M. Barriendos, H. Bergström, D. Camuffo, C. Cocheo, T. D. Davies, G. Demarée, J. Martin-Vide, M. Maugeri, R. Rodriguez and T. Verhoeve, 2000: Day-to-day temperature variability trends in 160-to-275-year long European instrumental records. *J. Geophys. Res.*, **106**, 22849–22868.

Nicholls, N., G. V. Gruza, J. Jouzel, T. A. Karl, L. A. Ogallo and D. E. Parker 1996: Observed climate variability and change. In *The IPCC Second Scientific Assessment* (J. T. Houghton *et al.*, Eds), Cambridge University Press, pp. 133–192.

Overpeck, J., K. Hughen, D. Hardly, R. S. Bradley, R. Case, M. Douglas, B. Finney, K. Gajewski, G. Jacoby, A. Jennings, S. Lamoureux, A. Lasca, G. MacDonald, J. Moore, M. Retelle, S. Smith, A. Wolfe and G. Zielinski, 1997: Arctic environmental change in the last four centuries. *Science*, **278**, 1251–256.

Parker, D. E., T. P. Legg and C. K. Folland, 1992: A new daily Central England temperature series, 1772–1991. *Int. J. Climatol.*, **12**, 317–342.

Parker, D. E., C. K. Folland and M. Jackson, 1995: Marine surface temperature observed variations and data requirements. *Climat. Change*, **31**, 559–600.

Peterson, T. C., T. R. Karl, P. F. Jamason, R. Knight and D. R. Easterling, 1998: The first difference method: maximizing station density for the calculation of long-term temperature change. *J. Geophys. Res.*, **103**, 25967–25974.

Peterson, T. C., K. P. Gallo, J. Lawrimore, T. W. Owen, A. Huang and D. A. McKittrick, 1999: Global rural temperature trends. *Geophys. Res. Lett.*, **26**, 329–332.

Raper, S. C. B., K. R. Briffa and T. M. L. Wigley, 1996: Glacier change in northern Sweden from A.D. 500: a simple geometric model of Storglaciären. *J. Glaciol.*, **42**, 341–51.

Reynolds, R. W. and T. M. Smith, 1994: Improved global sea surface temperature analyses using optimum interpolation. *J. Climate*, **7**, 929–948.

Skinner, W. R. and J. A. Majorowicz, 1999: Regional climatic warming and associated twentieth century land-cover changes in northwestern North America. *Climate Res.*, **12**, 39–52.

Vinnikov, K. Ya., P. Ya. Groisman and K. M. Lugina, 1990: Empirical data on contemporary global climate changes (temperature and precipitation). *J. Climate*, **3**, 662–677.

Yan, Z., P. D. Jones, A. Moberg, H. Bergström, T. D. Davies and C. Yang, 2001a: Recent trends in weather and seasonal cycles—an analysis of daily data in Europe and China. *J, Geophys. Res.*, **106**, 5123–5138.

Yan, Z., P. D. Jones, T. D. Davies, A. Moberg, H. Bergström, D. Camuffo, C. Cocheo, M. Mangeri, G. Demarée, T. Verhoeve, M. Barriendos, R. Rodriguez, J. Martin-Vide and C. Yang, 2001b: Extreme temperature trends in Europe and China based on daily observations. *Climat. Change*, in press.

Climatic Variability over the North Atlantic

James W. Hurrell,[1] Martin P. Hoerling[2] and Chris. K. Folland[3]

[1]National Center for Atmospheric Research,[a]
Boulder, CO, USA
[2]Climate Diagnostics Center, NOAA/OAR,
Boulder, CO, USA
[3]Hadley Centre for Climate Prediction and Research,
UK Meteorological Office,
Bracknell, UK

This chapter begins with a broad review of the North Atlantic Oscillation (NAO) and the mechanisms that might influence its phase and amplitude on decadal and longer timescales. New results are presented which suggest an important role for tropical ocean forcing of the unprecedented trend in the wintertime NAO index over the past several decades. We conclude with a brief discussion of a significant recent change in the pattern of the summertime atmospheric circulation over the North Atlantic.

A. INTRODUCTION

The climate of the Atlantic sector and surrounding continents exhibits considerable variability on a wide range of timescales. It is manifested as coherent fluctuations in ocean and land temperature, rainfall and surface pressure with a myriad of impacts on society and the environment. One of the best-known examples of this variability is the North Atlantic Oscillation (NAO). Over the past several decades, an index of the wintertime NAO has exhibited a strong upward trend. This trend is associated with a strengthening of the middle-latitude westerlies and anomalously low (high) surface pressure over the subpolar (subtropical) North Atlantic. Given the large impact of NAO variability on regional temperatures and precipitation, as well as on Northern Hemisphere (NH) temperature changes in general, understanding the physical mechanisms that govern the NAO and its variability on all time scales is of high priority.

In this chapter we present a review of the NAO and the processes that might influence its phase and amplitude, especially on decadal and longer timescales. In addition, some new results that suggest a link between the recent upward trend in the wintertime NAO index and warming of the tropical oceans will be highlighted. Concluding comments are made that focus attention on decadal variability of the extratropical North Atlantic climate during northern summer. The discussion is aimed at a scientifically diverse audience, such as the one that made the "Meteorology at the Millennium" conference a large success.

B. WHAT IS THE NORTH ATLANTIC OSCILLATION AND HOW DOES IT IMPACT REGIONAL CLIMATE?

Monthly mean surface pressures vary markedly about the long-term mean sea-level pressure (SLP) distribution. This variability occurs in well-defined spatial patterns, particularly during boreal winter over the Northern Hemisphere when the atmosphere is dynamically the most active. These recurrent patterns are commonly referred to as "teleconnections" in the meteorological literature, since they result in simultaneous variations in weather and climate over widely separated points on earth. One of the most prominent patterns is the NAO. Meteorologists have noted its

---
[a] The National Center for Atmospheric Research is sponsored by the National Science Foundation.

pronounced influence on the climate of the Atlantic basin for more than two centuries.

The NAO refers to a north–south oscillation in atmospheric mass between the Icelandic subpolar low- and the Azores subtropical high-pressure centres. It is most clearly identified when time-averaged data (monthly or seasonal) are examined, since time averaging reduces the "noise" of small-scale and transient meteorological phenomena not related to large-scale climate variability. The spatial signature and temporal variability of the NAO are usually defined through the regional SLP field, for which some of the longest instrumental records exist, although it is also readily apparent in meteorological data through the lower stratosphere.

The spatial structure and a time series (or index) of more than 100 years of NAO variability are shown in Fig. 1*. Although the NAO is present throughout much of the year, this plot illustrates conditions during northern winter when the NAO accounts for more than one-third of the total variance in SLP over the North Atlantic, far more than any other pattern of variability. Differences of more than 15 hPa in SLP occur across the North Atlantic between the two phases of the NAO in winter. In the so-called positive phase, higher than normal surface pressures south of 55°N combine with a broad region of anomalously low pressure throughout the Arctic (Fig. 1a). Consequently, this phase of the oscillation is associated with stronger-than-average westerly winds across the middle latitudes of the Atlantic onto Europe, with anomalous southerly flow over the eastern United States and anomalous northerly flow across western Greenland, the Canadian Arctic, and the Mediterranean. The easterly trade winds over the subtropical North Atlantic are also enhanced during the positive phase of the oscillation.

This anomalous flow across the North Atlantic during winter transports anomalous warm (and moist) maritime air over much of Europe and far downstream across Asia during the positive phase of the oscillation, while the southward transport of polar air decreases land and sea surface temperatures (SSTs) over the northwest Atlantic. Anomalous temperature variations over North Africa and the Middle East (cooling), as well as North America (warming), associated with the stronger clockwise flow around the subtropical Atlantic high-pressure centre are also notable during high-index NAO winters.

The changes in the mean circulation patterns over the North Atlantic during extreme phases of the NAO are accompanied by changes in the intensity and number of storms, their paths, and their associated weather. The details of changes in storminess differ depending on the analysis method and whether one focuses on surface or upper-air features. Generally, however, positive NAO index winters are associated with a northward shift in the Atlantic storm activity, with enhanced storminess from southern Greenland across Iceland into northern Europe and a modest decrease in activity to the south. The latter is most noticeable from the Azores across the Iberian Peninsula and the Mediterranean. Positive NAO winters are also typified by more intense and frequent storms in the vicinity of Iceland and the Norwegian Sea.

Changes in the mean flow and storminess associated with swings in the NAO change the transport and convergence of atmospheric moisture over the North Atlantic and, thus, the distribution of precipitation. Drier-than-average conditions prevail over much of Greenland and the Canadian Arctic during high NAO index winters, as well as over much of central and southern Europe, the Mediterranean and parts of the Middle East. In contrast, more precipitation than normal falls from Iceland through Scandinavia.

Given this large impact on regional climate, improved understanding of the process or processes that govern variability of the NAO is an important goal, especially in the context of global climate change. At present there is little consensus on the mechanisms that produce NAO variability, especially on decadal and longer timescales. It is quite possible one or more of the mechanisms described below affects this variability.

C. WHAT ARE THE MECHANISMS THAT GOVERN NORTH ATLANTIC OSCILLATION VARIABILITY?

1. Atmospheric processes

Atmospheric general circulation models (AGCMs) provide strong evidence that the basic structure of the NAO results from the internal, nonlinear dynamics of the atmosphere. The observed spatial pattern and amplitude of the NAO are well simulated in AGCMs forced with climatological annual cycles of solar insolation and SST, as well as fixed atmospheric trace-gas composition. The governing dynamical mechanisms are interactions between the time-mean flow and the departures from that flow. Such intrinsic atmospheric variability exhibits little temporal coherence and, indeed, the timescales of observed NAO variability do not differ significantly from this reference. Large changes in the atmospheric circulation over the North Atlantic occur from one winter to the next, and there is also a considerable amount of variability within a given winter

---

* More sophisticated and objective statistical techniques, such as eigenvector analysis, yield time series and spatial patterns of average winter SLP variability very similar to those shown in Fig. 1.

C. WHAT ARE THE MECHANISMS THAT GOVERN NORTH ATLANTIC OSCILLATION VARIABILITY?

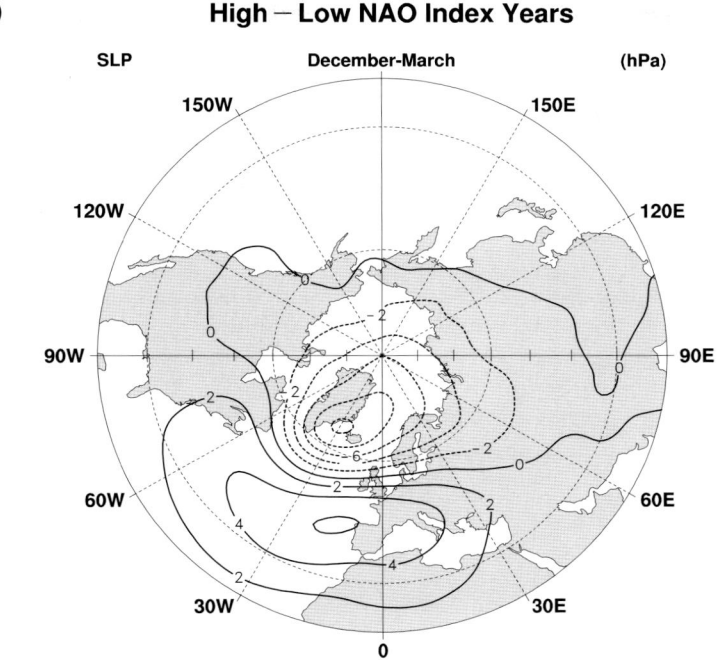

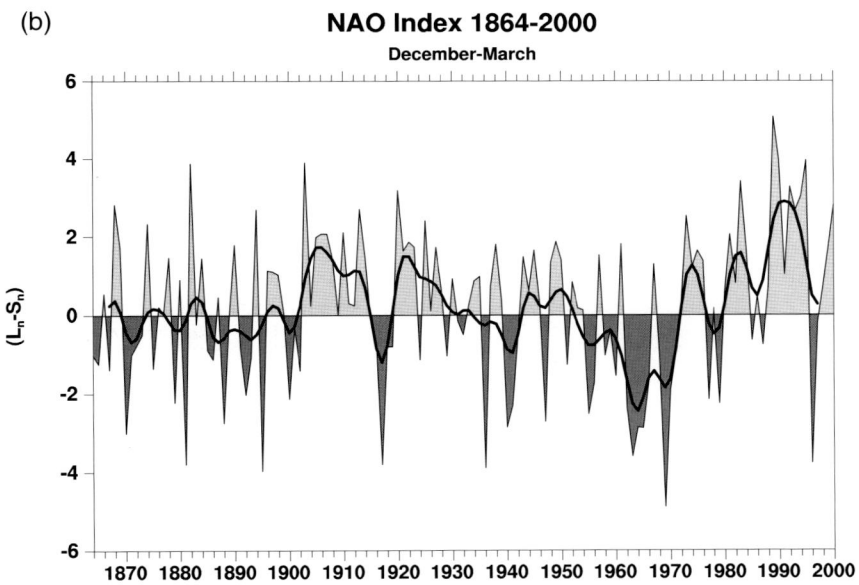

FIGURE 1 (a) Difference in sea-level pressure between winters (December–March) with an NAO index value > 1.0 and those with an index value < −1.0 (high minus low index winters) since 1899. The contour increment is 2 hPa and negative values are dashed. The index (b) is based on the difference of normalized sea-level pressure between Lisbon, Portugal and Stykkisholmur/Reykjavik, Iceland from 1864 through 2000. The heavy solid line represents the index smoothed to remove fluctuations with periods less than 4 years.

season. In this context, any low-frequency NAO variations in the relatively short instrumental record could simply reflect finite sampling of a purely random process.

A possible exception to this interpretation is the strong trend toward the positive index polarity of the NAO over the past 30 years (Fig. 1b). This trend exhibits a high degree of statistical significance relative to the background interannual variability in the observed record; moreover, multicentury AGCM experiments forced with climatological SSTs and fixed atmospheric trace-gas composition do not reproduce interdecadal changes of comparable magnitude.

The equivalent barotropic vertical structure of the NAO, reaching into the stratosphere, suggests that it could be influenced by the strength of the atmospheric circulation in the lower stratosphere. The leading pattern of geopotential height variability in the lower stratosphere is also characterized by a seesaw in mass between the polar cap and the middle latitudes, but with a much more zonally symmetrical (or annular) structure than in the troposphere. When heights over the polar region are lower than normal, heights at nearly all longitudes in middle latitudes are higher than normal. In this phase, the stratospheric westerly winds that encircle the pole are enhanced and the polar vortex is "strong" and anomalously cold. It is this annular mode of variability that has been termed the Arctic Oscillation (AO)*, and the aforementioned conditions describe its positive phase.

During winters when the stratospheric AO is positive, the NAO tends to be in its positive phase. There is a considerable body of evidence to support the notion that variability in the troposphere can drive variability in the stratosphere. New observational and modeling evidence, however, suggests that some stratospheric control of the troposphere may also be occurring.

The atmospheric response to strong tropical volcanic eruptions provides some evidence for a stratospheric influence on the earth's surface climate. Volcanic aerosols act to enhance north–south temperature gradients in the lower stratosphere by absorbing solar radiation in lower latitudes. In the troposphere, the aerosols exert only a very small direct influence. Yet, the observed response following eruptions is not only lower geopotential heights over the pole with stronger stratospheric westerlies, but also a strong, positive NAO-like signal in the tropospheric circulation.

Reductions in stratospheric ozone and increases in greenhouse gas concentrations also appear to enhance the meridional temperature gradient in the lower stratosphere, leading to a stronger polar vortex. It is possible, therefore, that the upward trend in the NAO index in recent decades (Fig. 1b) is associated with trends in either or both of these quantities. Indeed, a decline in the amount of ozone poleward of 40°N has been observed during the last two decades, and the stratospheric polar vortex has become colder and stronger.

2. Ocean Forcing of the Atmosphere

Over the North Atlantic, the leading pattern of SST variability during winter consists of a tripole. It is marked, in one phase, by a cold anomaly in the subpolar North Atlantic, warmer-than-average SSTs in the middle latitudes centred off of Cape Hatteras, and a cold subtropical anomaly between the equator and 30°N. This structure suggests that the SST anomalies are primarily driven by changes in the surface wind and air–sea heat exchanges associated with NAO variations. Indeed, the relationship is strongest when the NAO index leads an index of the SST variability by several weeks. Over longer periods, persistent SST anomalies also appear to be related to persistent anomalous patterns of SLP (including the NAO), although the mechanisms which produce SST changes on decadal and longer timescales remain unclear. Such fluctuations could primarily be the local oceanic response to atmospheric decadal variability. On the other hand, nonlocal dynamical processes in the ocean could also be contributing to the SST variations and, perhaps, the ocean anomalies could be modifying the atmospheric circulation.

A key and long-standing issue in this regard has been the extent to which anomalous extratropical SST and upper ocean heat content anomalies feed back to affect the atmosphere. Most evidence suggests this effect is quite small compared to internal atmospheric variability; for instance, AGCM studies typically exhibit weak responses to extratropical SST anomalies, with sometimes-contradictory results. Yet, some AGCMs, when forced with the time history of observed, global SSTs and sea ice concentrations over the past 50 years or so, show modest skill in reproducing aspects of the observed NAO behaviour, especially its interdecadal fluctuations. One example is given in Fig. 2.

In this plot, the NAO is defined as the spatial structure function corresponding to the first empirical orthogonal function (EOF) of monthly 500 hPa

---

* The signature of the stratospheric AO in winter SLP data, however, looks very much like the anomalies associated with the NAO with centres of action over the Arctic and the Atlantic. The "annular" character of the AO in the troposphere, therefore, reflects the vertically coherent fluctuations throughout the Arctic more than any coordinated behaviour in the middle latitudes outside of the Atlantic basin. That the NAO and AO reflect essentially the same mode of tropospheric variability is emphasized by the similarity of their time series, with differences depending mostly on the details of the analysis procedure.

C. WHAT ARE THE MECHANISMS THAT GOVERN NORTH ATLANTIC OSCILLATION VARIABILITY?

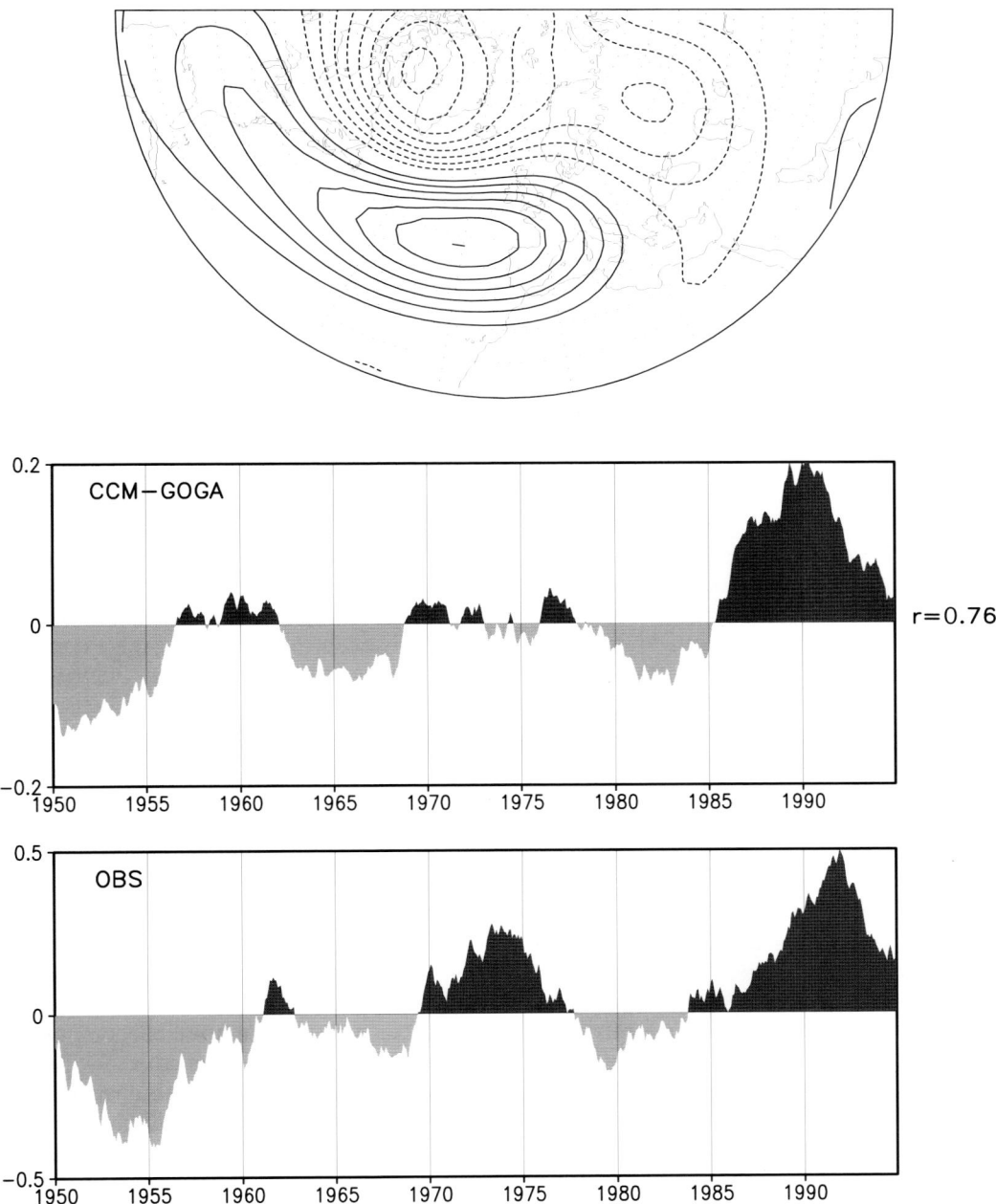

FIGURE 2 Leading empirical orthogonal function of the CCM3 GOGA ensemble-mean monthly 500 hPa geopotential height for the region 100°W–80°E, 20–90°N (top, see text for details). The associated principal component time series (middle), smoothed with a 73-month running mean, correlates with the observed (bottom) at 0.76.

geopotential height anomalies. The principal component time series have been smoothed with a 73-month running mean filter in order to emphasize the low-frequency NAO variations. The model data are from a 12-member ensemble of simulations performed with an AGCM (version 3.6 of the NCAR Community Climate Model, known as CCM3) over 1950–1994. Ensemble experiments are analysed because any individual simulation is dominated by internal atmospheric variability in the extratropics. Since the time history of observed SSTs is specified as a lower-boundary condition at all ocean gridpoints, experiments of this type are often referred to as Global Ocean Global Atmosphere (GOGA) integrations.

The key point in Fig. 2 is that the low-frequency ensemble-mean NAO time series correlates with the observed at 0.76, which is significant at the 5% level, and the overall upward trend is captured. The amplitude of the simulated NAO variability is about one-half that of the observed. This is a consequence of the ensemble averaging, as the simulated NAO in individual members of the ensemble have realistic amplitudes. These results, which have been found in other recent studies as well, suggest that a low-frequency component of the NAO over this period is not merely stochastic climate noise, but a response to variations in global SSTs. They do not necessarily imply, however, that the extratropical ocean is behaving in anything other than a passive manner. It could be, for instance, that long-term changes in tropical SSTs force a remote atmospheric response over the North Atlantic, which in turn drives changes in extratropical SSTs and sea ice. Indeed, additional experiments with CCM3 add merit to this viewpoint.

To isolate the role of changes in tropical SSTs, another ensemble of simulations was performed. In these experiments, the time history of observed SSTs was specified over tropical latitudes (30°S–30°N), but CCM3 was forced by monthly climatological values of SST and sea ice at higher latitudes. These experiments are known as Tropical Ocean Global Atmosphere (TOGA) integrations. Finally, a third set of integrations (Tropical Atlantic Global Atmosphere, or TAGA) were designed to separate the role of tropical Atlantic SST variability by forcing CCM3 with climatological SSTs everywhere outside of the tropical Atlantic. The TOGA and TAGA results shown below are from five-member ensembles.

To illustrate the spatial pattern of the dominant trends in NH wintertime atmospheric circulation over the past several decades, shown are the differences in 500 hPa geopotential heights between 1977–94 and 1950–76 (Fig. 3). The observed changes not only reflect the upward trend in the NAO, but are also characterized by significantly lower heights over the central North Pacific and higher heights over western Canada*. Results from the GOGA ensemble show that CCM3 captures many, but not all, aspects of these hemispheric changes in the wintertime circulation. For instance, over the North Atlantic the height changes associated with the trend in the NAO are well captured (as expected from Fig. 2), but the model extends the negative height anomalies too far to the east.

Perhaps the most striking result in Fig. 3, however, is that the low-frequency changes in 500 hPa heights over the North Atlantic are recoverable from tropical SST forcing alone. The spatial pattern of height change in the TOGA ensemble closely matches that of the GOGA ensemble (the slightly larger amplitudes of the former probably reflect the smaller ensemble size). Tropical (30°S–30°N) SSTs exhibit a warming trend of about 0.2°C since 1950 and this trend is common to all ocean basins. A linear regression of CCM3 500 hPa TOGA ensemble-mean heights onto a low-pass-filtered version of the tropical SST time series projects strongly onto the spatial structure of the leading EOF in Fig. 2, which suggests that, at least for this AGCM, low-frequency variations in the NAO originate from the tropics. Under this scenario, the link between the extratropical Atlantic SSTs and the NAO could be largely diagnostic in the sense that the externally forced NAO variations are themselves contributing to North Atlantic SST variations. The process may be analogous to the atmospheric bridge mechanism that has been invoked for the seasonal variability of North Pacific SSTs driven by teleconnections from the tropical Pacific.

Some recent studies using other AGCMs have suggested variations in tropical Atlantic SSTs significantly modulate NAO variability. The TAGA results in Fig. 3, however, do not suggest a strong influence by the tropical Atlantic. The simulated low-frequency height changes project only weakly onto the spatial structure of the NAO, and the principal component time series of the leading EOF from the TAGA ensemble (not shown) has little correlation with the observed time series (Fig. 2), unlike the GOGA and TOGA results.

The pattern of changes in tropical rainfall between 1977–94 and 1950–76 are shown in Fig. 4 for the TOGA and TAGA ensemble means. The warming trend in tropical SSTs is associated with increases in (TOGA) simulated rainfall (and thus latent heating) over the western and central tropical Pacific, as well as over the

---

* The changes over the Pacific correspond to an enhancement of another prominent mode of atmospheric circulation variability known as the Pacific–North American (PNA) teleconnection pattern.

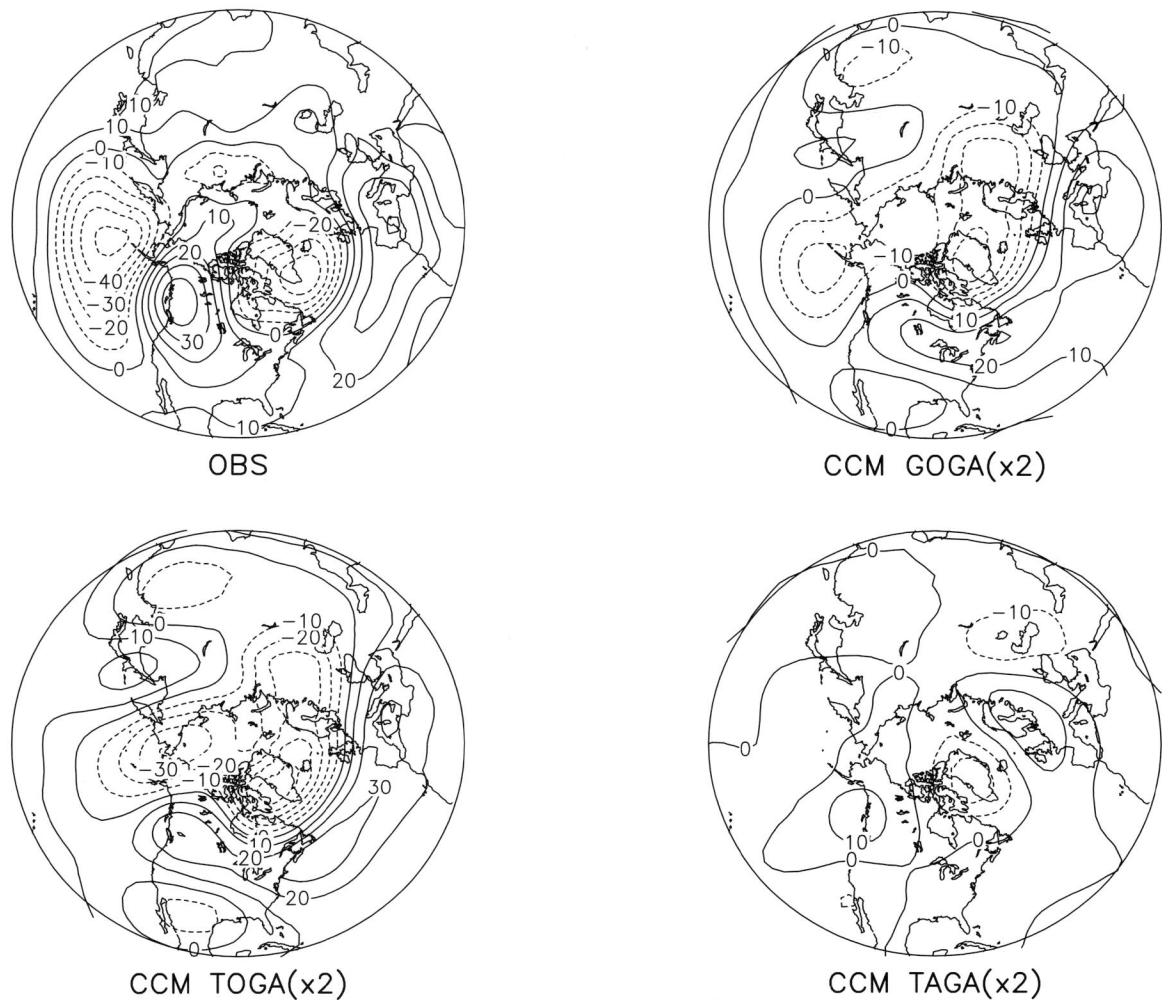

FIGURE 3 Changes in 500 hPa geopotential height (m), between 1977–94 and 1950–76, for the observations and the CCM3 ensemble-mean experiments described in the text. The model changes have been multiplied by a factor of 2, and only winter (December–March) conditions are illustrated. Height decreases are indicated by dashed contours, and the contour increment is 10 m.

tropical Indian Ocean. Over the Caribbean and northern South America, decreases in rainfall are simulated, but these decreases are much weaker in the TAGA ensemble. This suggests that the heating anomalies in these regions are primarily in response to tropic-wide changes in circulation associated with SST forcing outside of the tropical Atlantic. We are currently working to firmly establish the dynamics of this tropical teleconnection and its possible impact on North Atlantic climate variability.

The response of the extratropical North Atlantic atmosphere to changes in tropical and extratropical SST distributions, and the role of land processes and sea ice in producing atmospheric variability, are problems which are currently being addressed by the community. Until these are better understood, it is difficult to evaluate the realism of more complicated scenarios that rely on truly coupled interactions between the atmosphere, ocean, land and sea ice to produce North Atlantic climate variability. It is also difficult to evaluate the extent to which interannual and longer-term variations of the NAO might be predictable.

D. CONCLUDING COMMENTS ON OTHER ASPECTS OF NORTH ATLANTIC CLIMATE VARIABILITY

While the NAO behaviour shown in Fig. 1 explains a substantial portion of the low-frequency climate variability over the North Atlantic, the fact that most NAO variability is concentrated at higher (weekly to

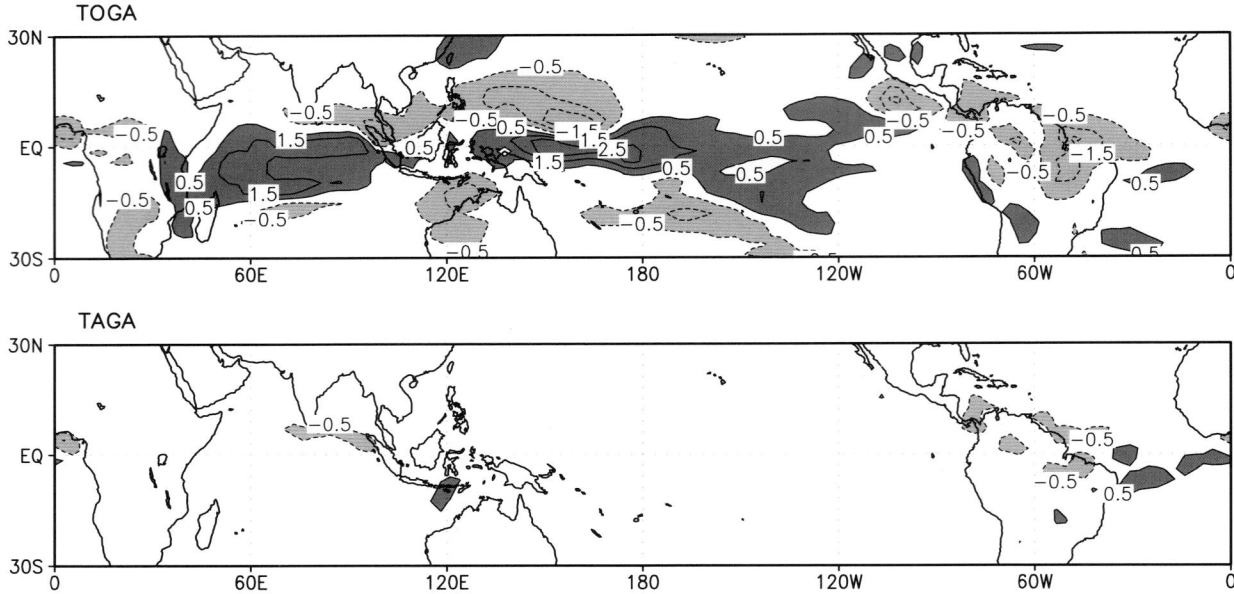

FIGURE 4 Changes in winter (December–March) total precipitation (mm day^{-1}), between 1977–94 and 1950–76, for the TOGA (top) and TAGA (bottom) ensemble means (see text for details). Decreases (increases) are indicated by light (dark) shading, and the contour increment is 0.5 mm day^{-1}.

monthly) frequencies means that there are significant monthly-to-interannual departures from the long-term NAO trend. The winter of 1996, for instance, was characterized by a very negative NAO index (in sharp contrast to the long-term trend toward positive index values), and winter conditions over much of Europe were severe. However, the anomalous anticyclonic circulation was located well to the east of the canonical NAO pattern, with positive SLP anomalies of more than 9 hPa centred over Scandinavia. Such persistent high-pressure anomalies are a typical feature of the North Atlantic climate and are referred to as "blocks". Longitudes near 15°W are particularly favoured locations for the development of blocking highs, which occur when the westerly flow across the middle latitudes of the NH is weak and typified by an exaggerated wave pattern. North Atlantic blocking is most typical during boreal spring and early summer, but it occurs throughout the year.

Over recent decades, for instance, increased anticyclonicity and easterly flows from the relatively warm, summer continent have brought anomalously warm and dry conditions to much of the United Kingdom. This change in circulation is illustrated in Fig. 5, which shows the leading EOF of summer SLP over the North Atlantic. The associated principal component time series is remarkable for its low-frequency variations, including the persistence of high-pressure anomalies over northern Europe during most summers since the late 1960s. Changes in summertime circulation are especially important from the perspective of droughts and heat waves, which have large societal impacts. Indeed, the drying during July–August is particularly evident in England and Wales rainfall, which has exhibited a strong downward trend since about 1960 (not shown). Some of the driest summers over the United Kingdom in recorded history have occurred during this most recent period.

The need to consider the possible links between the summertime climate anomalies noted above to changes in other, more remote, regions of the globe was first illustrated by the work of Chris Folland and colleagues nearly 15 years ago. Their studies noted a strong statistical association between climate variations over the UK region and rainfall over the Sahel region during northern summer. The association is particularly prominent on decadal timescales and it appears to have existed over much of the twentieth century. This intriguing link has not been explained to date. As such, it is also a focus of our current work.

References

Bretherton, C. S. and D. S. Battisti, 2000: An interpretation of the results from atmospheric general circulation models forced by the time history of the observed sea surface temperature distribution. *Geophys. Res. Lett.*, **27**, 767–770.

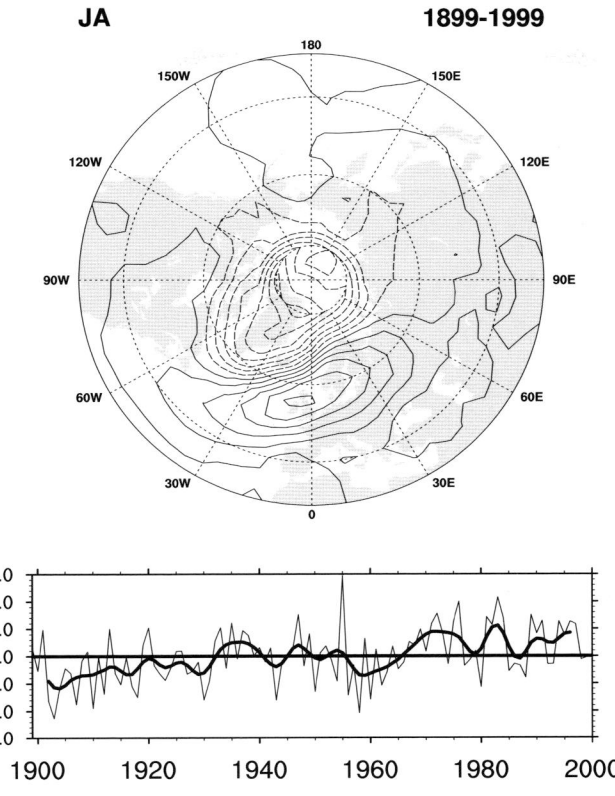

FIGURE 5 The leading empirical orthogonal function of July–August sea-level pressure (top) and the associated principal component time series (bottom) over the domain 60°W–30°E, 30–80°N and period 1899–1999. The contour increment is 0.4 hPa, and the contours correspond to the change in sea level pressure associated with a one standard deviation change of the principal component time series. Negative contours are dashed.

Deser, C., 2000: On the teleconnectivity of the "Arctic Oscillation". *Geophys. Res. Lett.*, **27**, 779–782.

Folland, C. K., D. E. Parker, M. N. Ward and A. W. Colman, 1986: Sahel rainfall, Northern Hemisphere circulation anomalies and worldwide sea temperature changes. Long Range Forecasting and Climate Research Memorandum, No. 7A, UKMO.

Hartmann, D. L., J. M. Wallace, V. Limpasuvan, D. W. J. Thompson and J. R. Holton, 2000: Can ozone depletion and global warming interact to produce rapid climate change? *Proc. Nat. Acad. Sci.*, **97**, 1412–1417.

Hoerling, M. P., J. W. Hurrell, and T. Xu, 2001: Tropical origins for recent North Atlantic climate change. *Science*, **292**, 90–92.

Hurrell, J. W., 1995: Decadal trends in the North Atlantic Oscillation regional temperatures and precipitation. *Science*, **269**, 676–679.

Hurrell, J. W. and H. van Loon, 1997: Decadal variations in climate associated with the North Atlantic Oscillation. *Climat. Change*, **36**, 301–326.

Mehta, V. M., M. J. Suarez, J. Manganello and T. L. Delworth, 2000: Predictability of multiyear to decadal variations in the North Atlantic oscillation and associated Northern Hemisphere climate variations: 1959–1993. *Geophys. Res. Lett.*, **27**, 121–124.

Robinson, W. A., 2000: Review of WETS—The workshop on extratropical SST anomalies. *Bull. Am. Meteorol. Soc.*, **81**, 567–577.

Rodwell, M. J., D. P. Rowell and C. K. Folland, 1999: Oceanic forcing of the wintertime North Atlantic Oscillation and European climate. *Nature*, **398**, 320–323.

Thompson, D. W. J. and J. M. Wallace, 1998: The Arctic Oscillation signature in wintertime geopotential height and temperature fields. *Geophys. Res. Lett.*, **25**, 1297–1300.

Thompson, D. W. J., J. M. Wallace and G. C. Hegerl, 2000: Annular modes in the extratropical circulation. Part II: Trends. *J. Climate*, **13**, 1018–1036.

van Loon, H. and J. C. Rogers, 1978: The seesaw in winter temperatures between Greenland and northern Europe. Part I: General description. *Mon. Weath. Rev.*, **106**, 296–310.

Wright, P. B., J. Kings, D. E. Parker, C. K. Folland and T. A. Basnett, 1999: Changes to the climate of the United Kingdom. *Hadley Centre Internal Note No. 89.*

Wunsch, C., 1999: The interpretation of short climate records, with comments on the North Atlantic and Southern Oscillations. *Bull. Am. Meteorol. Soc.*, **80**, 245–255.

Prediction and Detection of Anthropogenic Climate Change

J. F. B. Mitchell

Hadley Centre for Climate Prediction and Research, UK Meteorological Office, Bracknell, UK

The changes in climate expected with increases in greenhouse gases, derived from model simulations are described, and explained in terms of physical principles. The main sources of uncertainty in these predictions are discussed. The changes in temperature over the last century are compared with climate model simulations which take into account either the natural or the anthropogenic factors which are believed to have influenced recent climate. It is found that the observed changes in temperature over the last century are unlikely to be due to natural factors alone.

A. INTRODUCTION

Atmospheric carbon dioxide concentrations have risen from about 295 ppmv at the beginning of the last century to 367 ppmv at 2000. The twentieth century ended with global mean temperatures higher than at any other time in the instrumental record, and perhaps higher than at any time in the previous millennium. In this chapter we address two questions:

- What sort of climate change would we expect with increases in greenhouse gases?
- Can we detect the effect of greenhouse gases on climate already?

Arrenhius (1896) first noted the possible connection between carbon dioxide concentrations and climate. He estimated that a doubling of carbon dioxide concentrations would lead to a warming of 5.7 K. Subsequent estimates of the equilibrium response to doubling carbon dioxide were based on the energy balance at the earth's surface, ranging from 0.3 K (Newell and Dopplick, 1979) to 10 K (Moller, 1963). This approach is not well posed as the surface and troposphere are strongly thermally coupled, hence the wide range of estimates of the temperature response to doubling CO_2. In 1967, Manabe and Wetherald took this into account using a radiative–convective model, and found an equilibrium warming of 2.3 K on doubling atmospheric CO_2. In 1975, they published the first general circulation model (GCM) calculation of the climatic effects of increasing greenhouse gases.

Many of the mechanisms proposed by Manabe and Wetherald (1975) still stand today. I will review their findings, and then outline how predictions of the climatic effect of increases in greenhouse gases have evolved over the intervening 25 years, using recent work carried out at the Hadley Centre to represent results from typical "state-of-the-art" models. I will then address the question of whether or not we can already detect the influence of past increases in greenhouse gases.

B. AN EARLY GENERAL CIRCULATION MODEL SIMULATION OF THE CLIMATIC EFFECTS OF INCREASES IN GREENHOUSE GASES

In their pioneering simulation, Manabe and Wetherald (1975) used a nine-layer atmosphere with a simple sector (equator to pole over 120° longitude) domain and a horizontal resolution of about 500 km. An idealized land–sea distribution was employed, with no orography, and the surface had no thermal inertia. Sea was treated as dark and wet but had no thermal inertia. Simple parametrizations of radiation, convection, precipitation, snow and sea ice were included. Cloud amounts were prescribed. The model used annually averaged insolation.

On doubling atmospheric carbon dioxide, the domain mean temperature increased by 2.9 K. The troposphere and surface warmed, whereas the stratosphere cooled (Fig. 1), as found by Manabe and Wetherald (1967) using their one-dimensional radiative convective model. The warming is enhanced near the surface in high latitudes due to the reduction of the

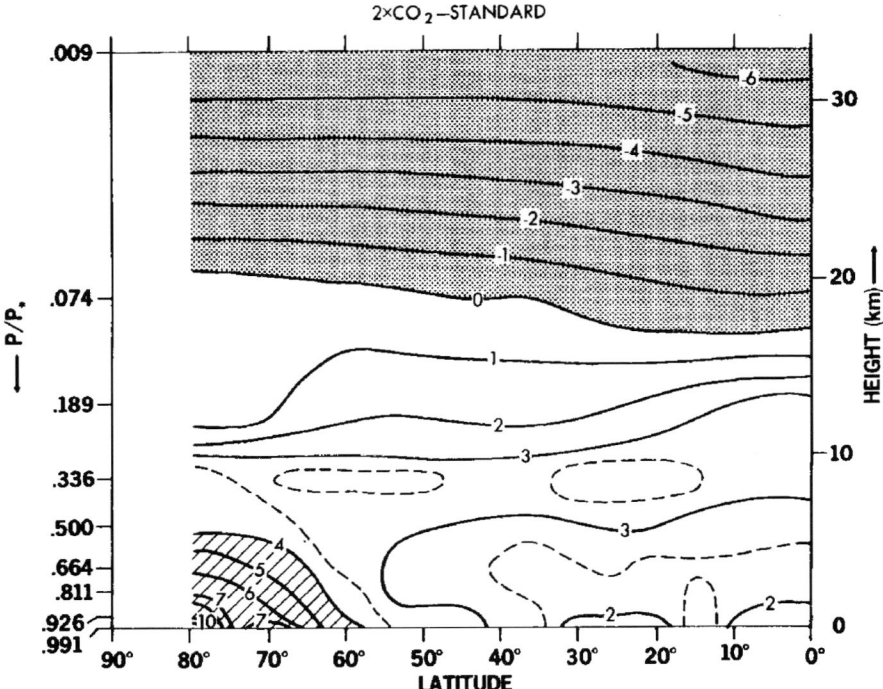

FIGURE 1 Equilibrium changes in zonally averaged atmospheric temperature due to doubling CO_2 in an idealized climate model. Contours every 1 K, dashed contours at 1.5, 2.5 and 3.5 K. (From Manabe and Wetherald, 1975.)

highly reflective snow and sea-ice cover in the warmer climate. In lower latitudes, the tropospheric warming increases with height. The specific humidity in the troposphere increases with warming. In regions of moist ascending air, this leads to increased release of latent heat, enhancing the warming at upper levels. Global mean precipitation and evaporation increase by about 8%. Precipitation increases in the convergence areas of the tropics and high latitudes, but decreases in the subtropics. Relative humidity is largely unchanged, but there are small systematic increases near the model tropopause and in high latitudes, and decreases in the middle troposphere.

C. A RECENT SIMULATION OF THE EFFECTS OF INCREASES IN GREENHOUSE GASES

For comparison, we consider a recent simulation of the effect of increases in greenhouse gases (Mitchell *et al.*, 1998). A global coupled ocean–atmosphere model was used (Gordon *et al.*, 1999) with a 300 km horizontal grid with 19 levels in the atmosphere, and 120 km horizontal grid with 20 levels in the ocean. The model includes realistic geography and orography, the seasonal cycle of insulation, model-generated clouds, as well as complex land surface, boundary-layer and radiative and convective paramerizations. The equilibrium global mean surface temperature increases by 3.3 K when carbon dioxide is doubled in a version of the atmospheric model coupled to a simple mixed-layer ocean. However, the experimental design used in the study with the full ocean model differs in an important way from Manabe and Wetherald's 1975 experiment. Here, greenhouse gases were gradually increased, and the climate has not been allowed to come to a new equilibrium. The greenhouse gases were increased from their pre-industrial value to 1990, and thereafter under the Intergovernmental Panel on Climate Change (IPCC) emissions scenario IS92a (Leggett *et al.*, 1992). The results are shown for a period when the radiative forcing was equivalent to a two-and-a-half-fold increase in CO_2 when the global mean warming was about 3.5 K.

Many of the features that were reported by Manabe and Wetherald (1975) are found in the more recent experiment The tropospheric warming and stratospheric cooling are preserved (Fig. 2). The tropospheric warming is enhanced with height, particularly in low latitudes. The near surface warming is enhanced in high northern latitudes, although not to the same extent as in the original work. This is partly a result of including the seasonal variation of insolation—warming in high latitudes is then found mainly late autumn and winter. In

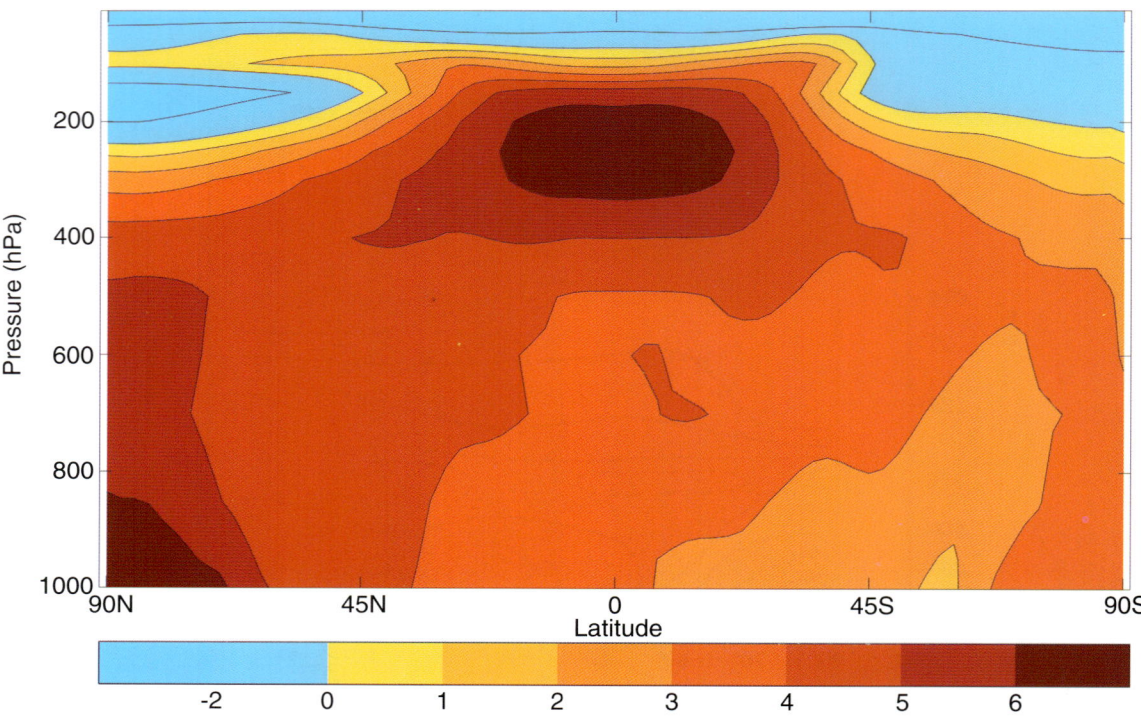

FIGURE 2 Changes in zonally averaged temperature from a pre-industrial control simulation following a gradual increase in greenhouse gases (see text), averaged over 2070–2100. Contours every 1 K above 0 K and at −2 K. (From Mitchell et al., 1998.)

summer, temperatures over Arctic sea ice stay near the ice melting point in both control and perturbed simulations, so there is little warming (see Ingram et al., 1989).

One new feature is the minimum warming at the surface near 50°S. At these latitudes, surface heating is mixed deep into the ocean giving a thermal inertia much larger than that of the continents or most of the remainder of the ocean. Hence this region warms more slowly than elsewhere, and is a long way from equilibrium (see, e.g., Stouffer et al., 1989). Had we held greenhouse gas concentrations fixed at the final values for many centuries, this part of the ocean would have warmed up to give a much more hemispherically symmetrical warming (Stouffer and Manabe, 1999).

The geographical distribution of the surface warming, in addition to the features discussed in the previous section, shows a much more pronounced warming over land than over the sea (Fig. 3). The warming is strongly enhanced over the Arctic due to the thinning and retreat of sea ice, but not noticeably so around Antarctica, where the thermal inertia of the ocean has inhibited warming and hence sea-ice feedbacks. There is also a minimum in the warming in the northern North Atlantic associated with the deeper ocean mixing that occurs in that region. All these features are found in similar experiments carried out using other models (see, e.g., Kattenburg et al., 1996).

Manabe and Wetherald (1975) were unable to investigate ocean circulation feedbacks with their simple sector model. Most coupled models produce a weakening of the North Atlantic thermohaline circulation on gradually increasing greenhouse gases. The degree of weakening varies greatly from model to model, from little change (e. g. Latif et al., 2000) to a total collapse (Manabe and Stouffer, 1994). In the present model, the overturning circulation reduces by 23% by the end of the twenty-first century if greenhouse gases are increased under scenario IS92a (Fig. 4). In an idealized experiment, in which carbon dioxide concentrations were increased rapidly to four times their initial value and held fixed thereafter, the meridional circulation weakens, but does not collapse. The weakening of the circulation is associated with a reduction in the density of the surface water due to increased fluxes of heat and moisture from the atmosphere. The relative importance of these factors varies from model to model, and we still do not understand the reasons why the thermohaline circulation is so much more sensitive in some models than in others.

The changes in annual mean precipitation show the same broad features noted by Manabe and Wetherald

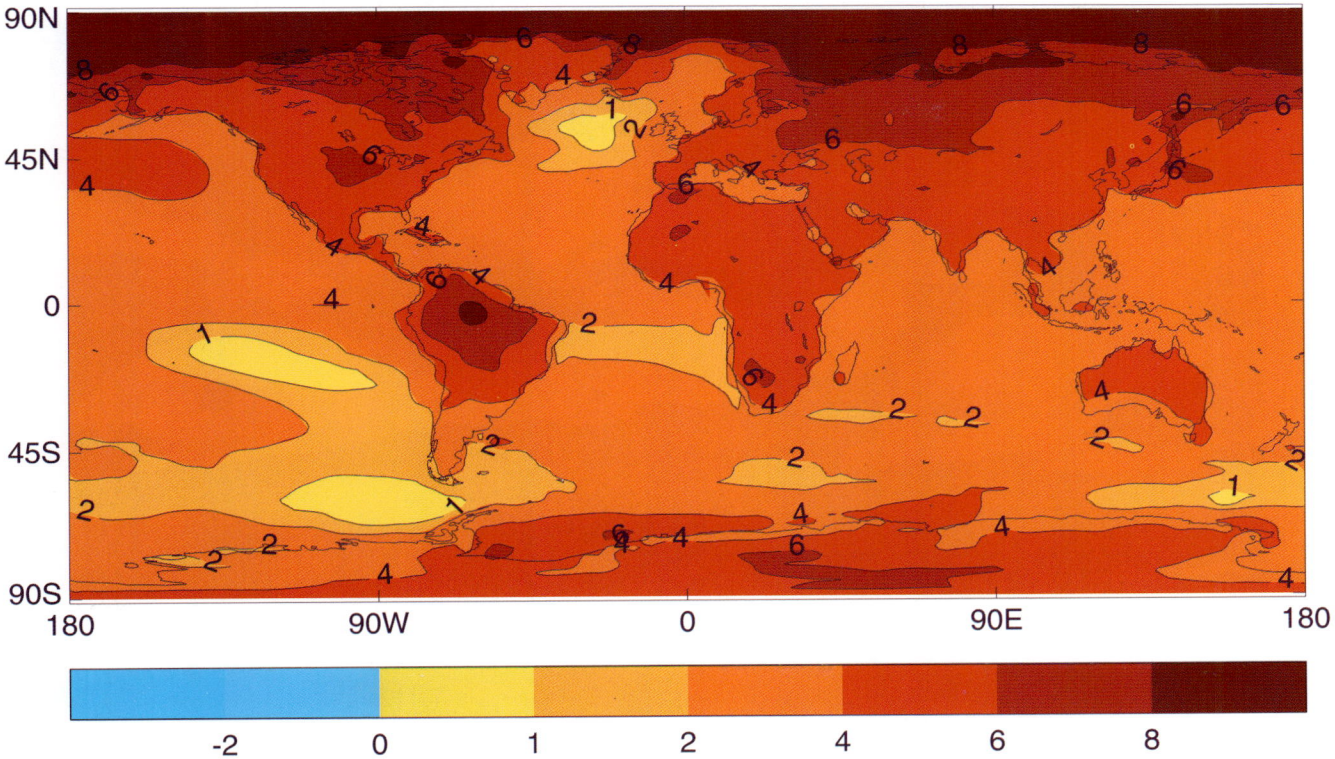

FIGURE 3 As for 2, but for changes in surface air temperature. Contours every 1 K.

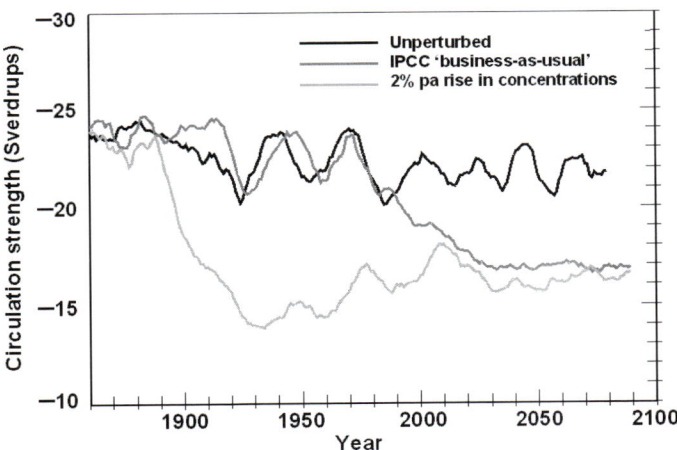

FIGURE 4 Simulated mean meridional overturning ocean circulation (Sverdrups).

(1975)—increases in the convergence zones in low and high latitudes, with areas of decrease in the subtropics (Fig. 5). The simulated changes here vary seasonally—for example, the increases in high latitudes are most pronounced in winter, and there is an enhancement of the southwest Asian summer monsoon. Most recent models also produce a reduction of rainfall over southern Europe in summer. The detailed pattern of simulated precipitation changes varies greatly from model to model, particularly in the tropics.

Recently, attention has been directed to the simulated changes in variability that may accompany increases in greenhouse gases as well as to changes in mean climate. The resolution of current models is

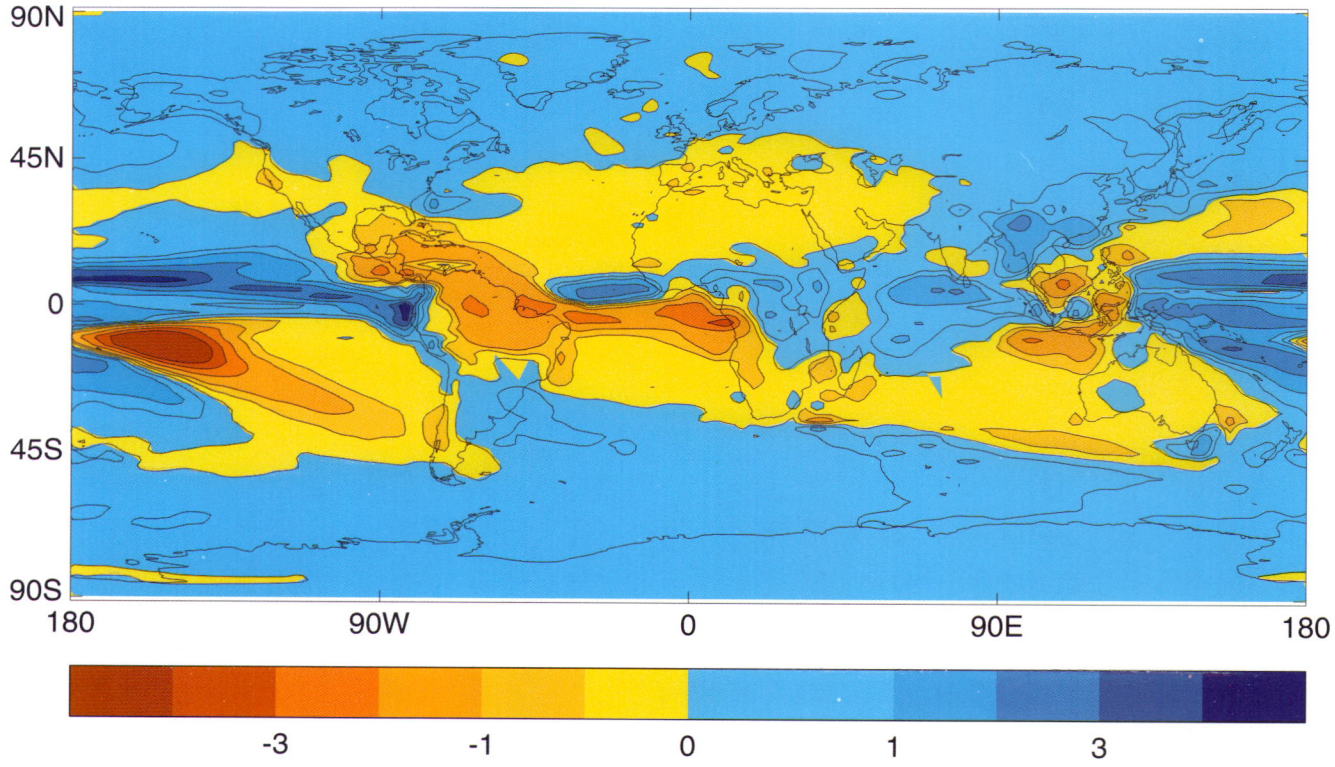

FIGURE 5 As for 3, but for changes in precipitation (mm day^{-1}).

barely adequate to produce a faithful reproduction of modes of variability such as the North Atlantic Oscillation (NAO) and El Niño–Southern Oscillation (ENSO). The current model produces an increase in the NAO index with large (fourfold) increases in CO_2 (McDonald et al., 2000), and increases have been reported in other models (e. g. Ulbright and Christoph, 1999). There is little change in the amplitude or frequency of ENSO in this model (Collins, 2000a), although changes in the character of the simulated behaviour due to increases in greenhouse gases have been reported elsewhere (e.g., Collins, 2000b; Timmermann et al., 1999). However, apart from increased frequency of hot spells, and increased intensity of precipitation events, there is little consensus to be found in changes in variability simulated by different models in response to increases in greenhouse gases.

In the last few years, several centres have started to incorporate chemical and biological factors into their models. The effects of sulphate aerosols have first been represented through a prescribed increase in surface albedo (e.g., Mitchell and Johns, 1997). More recently, simulations have been completed in which sulphur concentrations are determined using an embedded sulphur cycle model, with a full radiative treatment of aerosols, including the direct scattering and indirect cloud effects (e.g., Bengtsson et al., 1999; Johns et al., 2000). The global mean radiative forcing due to increases in greenhouse gases from pre-industrial times to present is about 2.4 W m^{-2}, whereas the negative forcing due to sulphate aerosols used in models is typically 0.6 to 1.0 W m^{-2}. Sulphate aerosol effects produce a cooling, especially in northern mid-latitudes, and weaken the southwest Asian monsoon. In general, the simulated changes are smaller than those due to increases in greenhouse gases, either in simulations of the past, or in scenarios of the future. Cox et al. (2000) have included a model of the carbon cycle in an ocean–atmosphere GCM, allowing the atmospheric concentrations of carbon dioxide to be determined taking into account the simulated changes in climate. They find that the simulated changes in climate enhance the concentrations of atmospheric carbon dioxide, giving a strong positive feedback (see also Cox et al., this volume). Simulations have also been completed in coupled ocean-atmosphere models which include a detailed representation of tropospheric chemistry (C. Johnson, personal communication).

Although much progress has been made since Manabe and Wetherald completed their first GCM doubled CO$_2$ experiment in 1975, particularly in terms of the complexity of the models used, there are some areas where little advance has been made. This is particularly true of climate sensitivity, the equilibrium global mean surface temperature response to doubling atmospheric carbon dioxide. In 1978, the "Charney Report" (National Academy of Sciences, 1979) concluded that the climate sensitivity lay between 1.5 and 4.5 K. This range has remained unchanged in over two decades, and some models have produced even larger values (e. g., Mitchell *et al.*, 1990; Cubasch *et al.*, 2001).

One of the principal reasons for the wide range attributed to climate sensitivity is the uncertainty arising from cloud feedbacks (Cess *et al.*, 1990; Senior and Mitchell, 1993). There is some consistency between the changes in cloud simulated by different models. Most models simulate increases in high cloud and reductions in middle cloud on increasing greenhouse gas concentrations (e.g., Fig. 6) but there are large differences in the details simulated by different models. The net radiative effect of clouds is a balance between strong longwave heating and strong solar cooling. Hence, the simulated changes in cloud feedback are very sensitive to the details of the simulated cloud height and amount and microphysics. Since cloud processes are subgridscale, and there is no accepted theory of cloud formation and dissipation, there is inevitably a degree of arbitrariness in the cloud parametrizations used in climate models. However, there has been some progress in evaluating cloud feedbacks against observations (Senior, personal communication; Webb *et al.*, 2000).

The differences in model response are not just confined to differences in model sensitivity but also differences in the patterns of response. In a recent study, it was found that in general the pattern of a given model's response to increases in greenhouse gases correlated more closely to the pattern in the same model when greenhouse gases and aerosol loadings were increased, than to that produced by a different model in response to greenhouse gases (Table 1). In other words, the pre-

TABLE 1 Pattern Correlation Coefficients of Temperature Change (2021–2050) Mean Relative to (1961–1990) Mean for Simulations with Increases in Greenhouse Gases Only, and Along the Diagonal, Simulations with Increases in Greenhouse Gases Only and Simulations with Increases in Greenhouse Gases and Aerosols. Data from the IPCC Data Distribution Centre

Model	1	2	3	4	5	6	7	8	9
1	**0.96**	0.74	0.65	0.47	0.65	0.72	0.67	0.65	−0.06
2		**0.97**	0.77	0.45	0.72	0.77	0.73	0.80	0.05
3			**0.96**	0.40	0.75	0.72	0.67	0.75	0.10
4				**0.46**	0.40	0.53	0.60	0.53	0.17
5					**0.73**	0.58	0.61	0.69	0.16
6						**0.85**	0.79	0.79	−0.05
7							**0.90**	0.75	0.07
8								**0.89**	−0.03
9									**0.91**

From Cubasch and Fisher-Bruns, 2000.

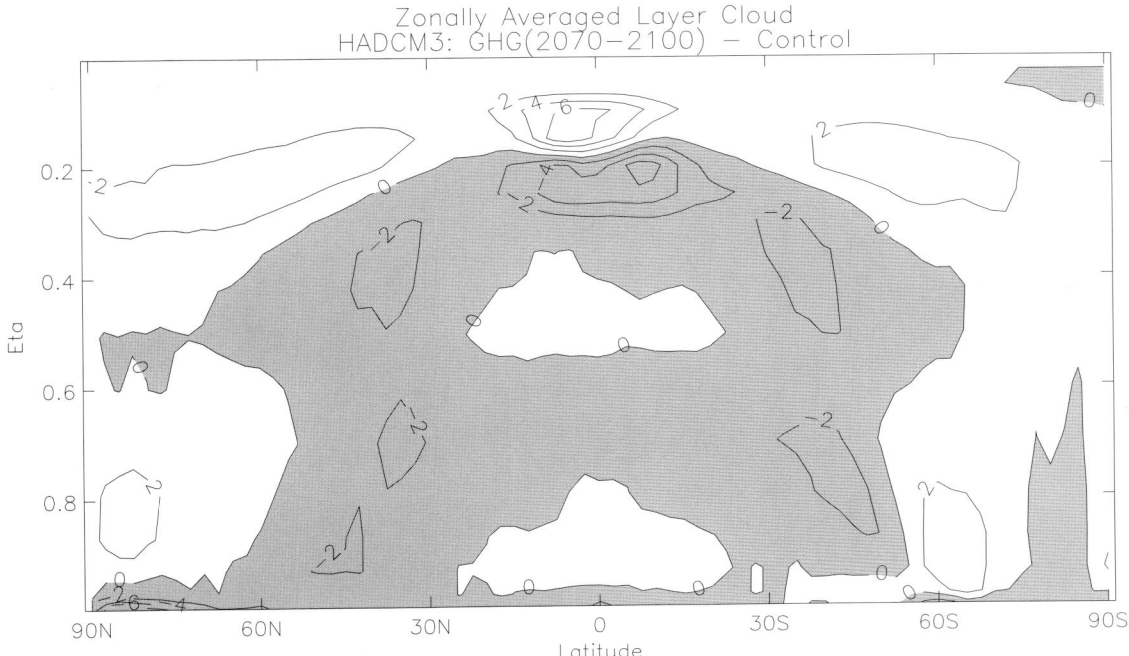

FIGURE 6 As for Fig. 2, but for changes in cloudiness (% of total coverage). (C. Senior, personal communication.)

dicted pattern of future change was more dependent on which model was chosen, than on whether or not sulphate aerosol effects were included. This certainly raises the question—how good are current models at simulating climate change? Traditionally, models have been validated against current climate (e.g., the Atmospheric Model Intercomparison Project, Gates *et al.*, 1999; the Coupled Model Intercomparison Project, Meehl *et al.*, 2000) and past climates (e.g., the Palaeoclimate and Modeling Intercomparison Project, see Joussaume *et al.*, 1999). Here we consider a third approach—validation of simulations of the recent past against the instrumental temperature record.

D. DETECTION AND ATTRIBUTION OF ANTHROPOGENIC CLIMATE CHANGE

By about 1860, the distribution of instrumental measurements of surface air temperature was sufficiently widespread and frequent to justify making estimates of global annual mean surface air temperature. From such measurements, it can be seen that global mean air temperature increased by about 0.3 K during the early part of the twentieth century, then stabilized from about 1940 to 1970, and increased rapidly during the last three decades. Callendar (1938) was aware of the increase in the early part of the twentieth century and having compared it with an estimate of the rise, attributed it to increases in carbon dioxide. However, the subsequent evolution of the temperature record indicates that more than one factor is needed to account for the observed record, and Callendar's conclusion was premature.

The instrumental record is not long enough to show that the recent warming is of human rather than natural origin. Attempts have been made to extend the surface temperature record further back into the past using palaeoclimatic reconstructions (e.g., Mann *et al.*, 1998). If these reconstructions are reliable, then the temperature changes over the last century have certainly been unusual. A similar conclusion is reached from comparing the observed global mean temperature record with time series from long control runs from climate models (Fig. 7) (see also Stouffer *et al.*, 1999). Again, if the model simulation of the amplitude of internal (unforced) climate variability is accurate, then the twentieth-century warming is highly unusual—we have detected a change

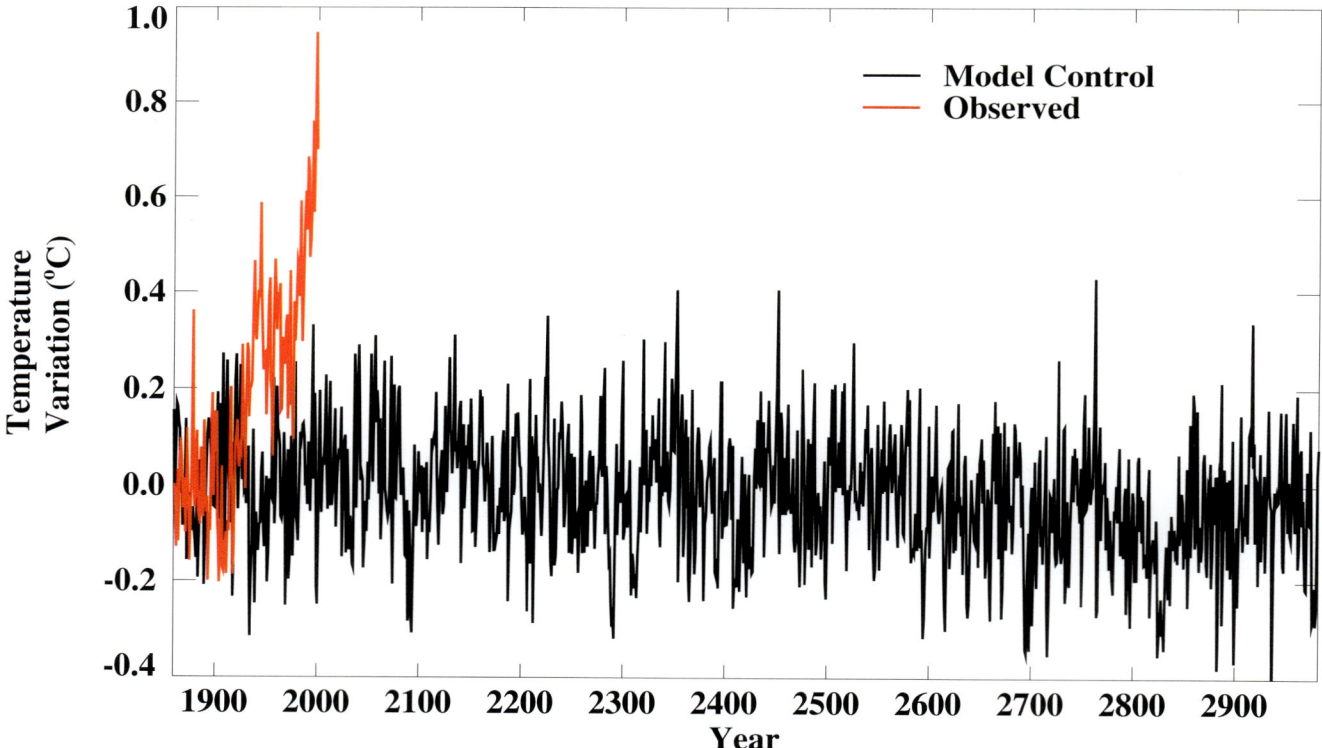

FIGURE 7 Simulated global mean surface air temperatures from a long control simulation (K, black line). The observed changes over the period of the instrumental record are shown for comparison (red line).

in long-term climate. However, from the evidence so far, we cannot tell if it is due to natural or human forcing. This is the problem of attribution.

Strictly speaking, in order to attribute at least part of the recent warming to human activity, we should demonstrate that no other natural forcing agent could explain the recent change. In practice, attribution studies try to demonstrate that the observations are inconsistent with our estimates of the natural effects on climate, but are consistent when estimates of anthropogenic effects are included. The climatic effects of natural and human influences are estimated using climate models.

The main candidates for a natural explanation of twentieth-century temperature changes are increases in solar output and changes in the frequency of volcanic eruptions. The forcing for these agents has to be estimated from proxy data, except for the last 20 years, for which satellite data is available. Increases in well-mixed greenhouse gases (principally carbon dioxide, methane, nitrous oxide, chlorofluorocarbons) are the main anthropogenic factors. There have also been increases in sulphate aerosols (which cool the surface directly through scattering sunlight back to space, and indirectly through increasing the brightness of clouds) and increases in tropospheric ozone, another greenhouse gas. The past concentrations of well-mixed gases are known from measurements taken from the atmosphere or from ice core data, but the ozone and sulphate concentrations have to be estimated from chemical transport models.

When only solar and natural forcing are taken in to account in a recent model study (Fig. 8a), from Tett *et al.*, 2001), the model fails to reproduce the warming over the most recent decades. In fact, an ensemble of four simulations was run, and all four simulations cooled towards the end of the century. The changes in solar forcing changes was small at this time, but there is a strong cooling following the major eruptions of Agung (1963), El Chichon (1983) and Mount Pinatubo (1992). The simulations do, however, reproduce much of the warming in the early twentieth century, when the solar irradiance was probably increasing, and there were few major volcanic eruptions.

In contrast, with the main anthropogenic factors alone (well-mixed greenhouse gases, ozone and direct and indirect sulphate effects) are included, the recent warming is remarkably well simulated (Fig. 8b). On the other hand, none of the ensemble members reproduces the early century warming, indicating that anthropogenic factors and internal climate variability are unlikely to explain global mean temperature changes over the whole of the last century. When both sets of factors are included, the agreement with observations is very good (Stott *et al.*, 2000).

Some cautionary remarks should be made here. First, we could get the right amount of warming if the model sensitivity was too high, and the estimate of forcing was too low (or vice versa). Second, if the real-world unforced variability is sufficiently larger than that of the model, then anthropogenic forcing and unforced variability alone would be sufficient to explain the early century warming. Indeed, if it was sufficiently large, then unforced variability alone would be sufficient to explain all of the recent record. However, the amplitude of the variability in the model used here would have to be at least doubled for this to be the case.

Although ability to reproduce the global mean temperature record may be comforting, we would like to know how well models reproduce the spatial patterns of change. It is much more difficult to judge if the simulated patterns of change are consistent with observations, given the high level of internal variability of temperature on smaller scales. Hence, it is important to carry out formal statistical tests to show that the model and simulated patterns are consistent, allowing for natural internal variability. In addition, statistical procedures allow us to estimate the contributions of different factors to the observed changes, along with uncertainty limits.

Much of the recent detection and attribution work has been based on the optimal detection approach pioneered by Hasselmann (1979, 1993). This can be recast as a form of linear regression (Allen and Tett, 1999). The observations **O** can be expressed as a linear sum of scaled signals $\mathbf{X}_i$ and random unforced variations ε. Thus

$$\mathbf{O} = \Sigma\, \beta_i\, \mathbf{X}_i + \varepsilon$$

The β_i are scaling factors to be estimated through regression. The optimization gives less weight to the modes of variability with greatest amplitude, enhancing the signal-to-noise ratio. Optimization requires the use of the covariance matrix $\varepsilon\varepsilon^T$ estimated from the model's control simulation. The significance levels are estimated from the frequency distributions of the amplitudes of the $\mathbf{X}_i$ in the control simulation (using a different section to that used for optimization). The $\mathbf{X}_i$ are estimated from the model simulations, as for example, in the case of the anthropogenic and natural simulations in Fig. 8. The $\mathbf{X}_i$ may be spatial patterns (e.g., Hegerl *et al.*, 1996), space frequency patterns (North *et al.*, 1995; see also Hegerl and North, 1997) or space–time patterns (Tett *et al.*, 1999, 2001). The analysis is usually carried out in a highly truncated space, so only the large-scale spatial patterns are tested. Significance levels are estimated of the frequency distribution of the amplitude of the $\mathbf{X}_i$ in the control simulation.

The outcome of this highly complex mathematical

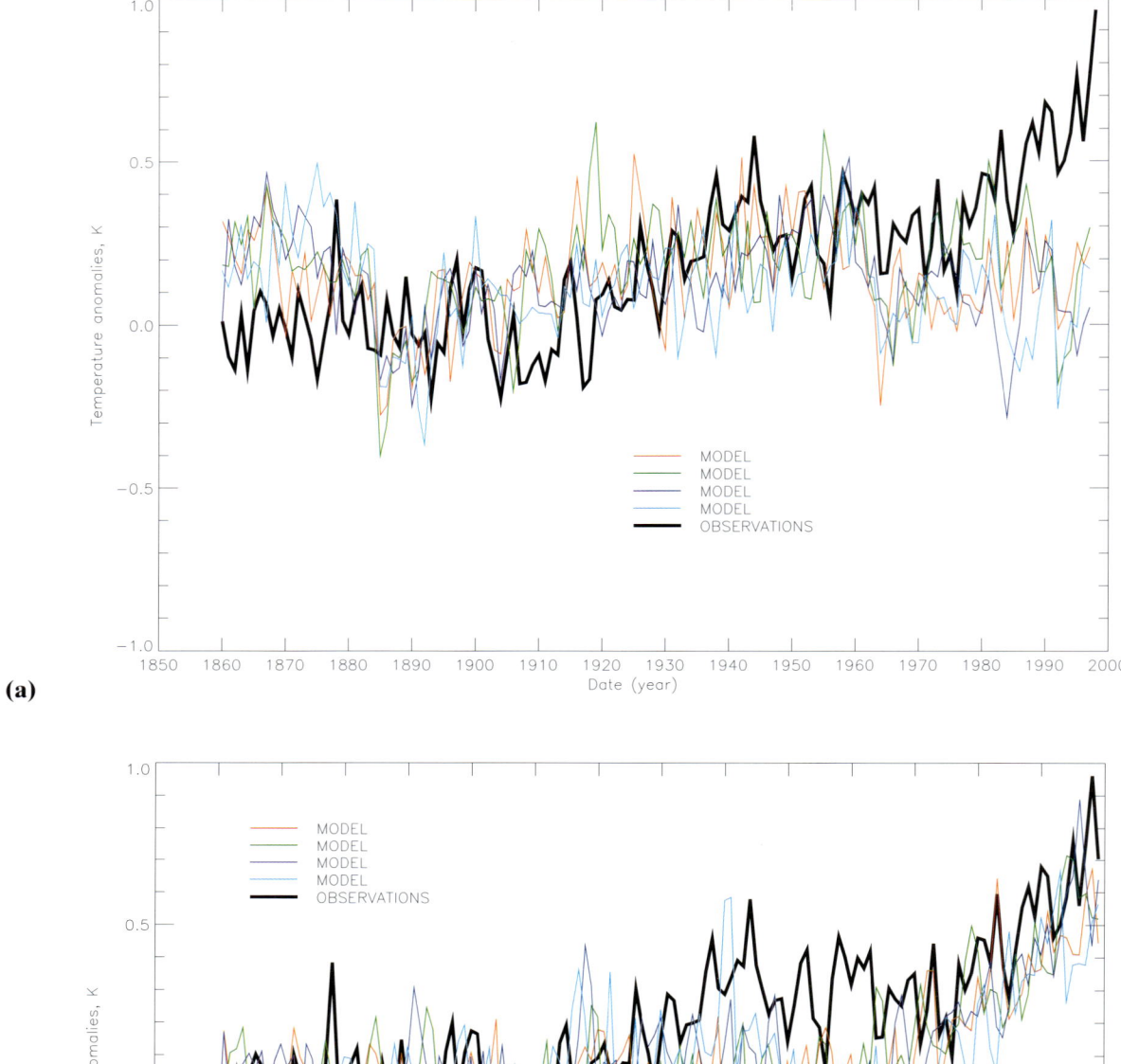

FIGURE 8 Simulations of the response of global mean surface air temperature to (a) natural and (b) anthropogenic factors (K). The thin lines are from individual simulations, started from different parts of the long control simulation shown in Fig. 7. The thick line is the observed change over the same period. (From Tett *et al.*, 2001.)

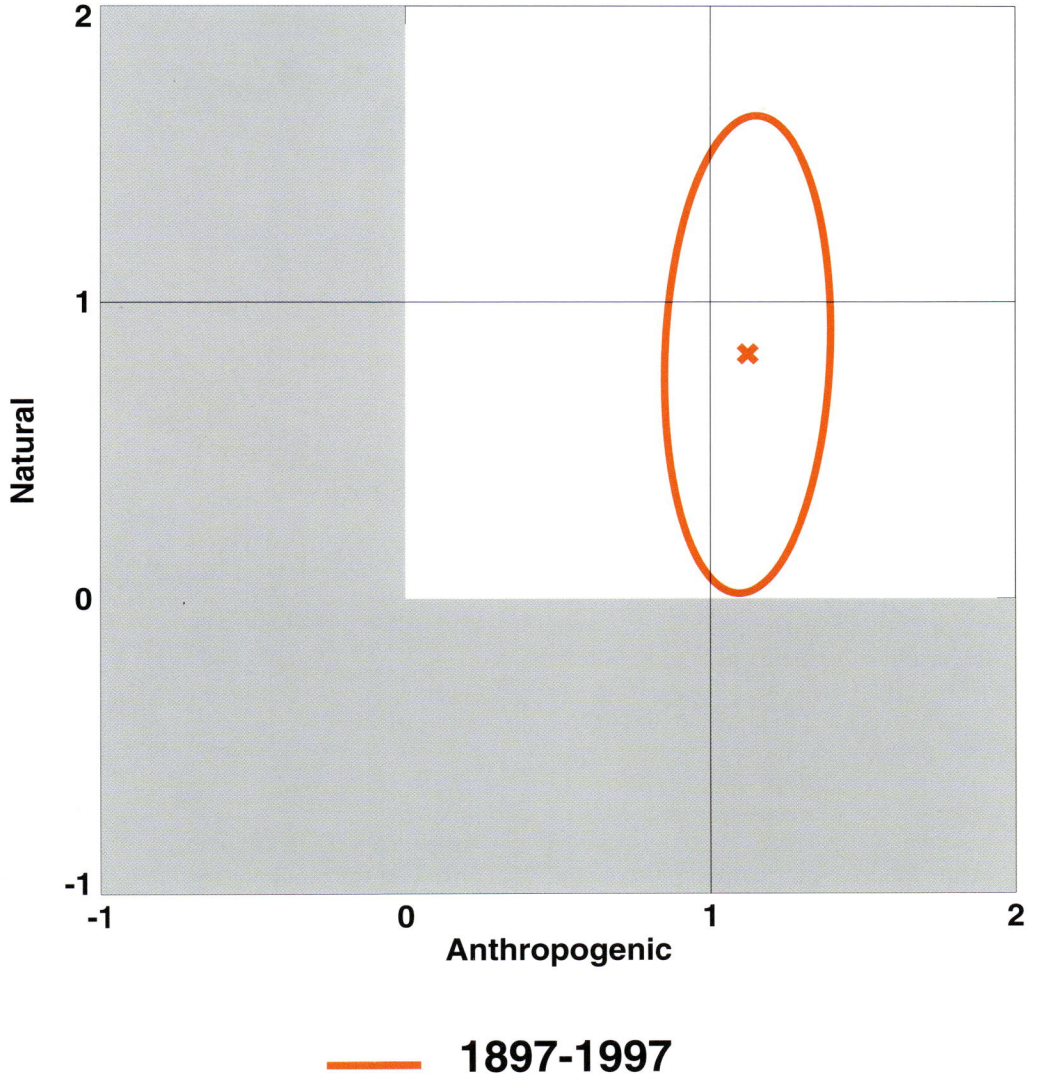

FIGURE 9 Joint probability distribution of the amplitudes of the natural and anthropogenic signals in the observations over the twentieth century. The ellipse outlines the 95% confidence interval: the fact that it encloses the (1,1) point indicates that the model amplitudes are consistent with observations, and the fact that it does not intersect either axis indicates that both the natural and anthropogenic signals are detectable. (P. Stott, based on Tett et al., 2001.)

procedure is the best estimates of the scaling factors and confidence intervals. This allows two tests:

1. If the confidence interval for β_i does not include zero, we have detected signal i.
2. If the confidence interval includes unity, then the amplitude of the model's response is consistent with the amplitude estimated from the observation for signal i.

A further test can be made on the residual (observations minus the best fit) to ensure that it is consistent with the model estimate of unforced variability.

When this procedure is applied to the two ensembles in Figure 8, and the resulting joint probability distribution for the $\beta_{natural}$ and $\beta_{anthropogenic}$ is plotted (Fig. 9), the ellipse denoting the 95% confidence interval includes the point (1,1) indicating that the amplitude of the model responses is consistent with observations. Secondly, the ellipse does not intersect the $\beta_{natural}$ or $\beta_{anthropogenic} = 0$ axes, indicating that both a natural and anthropogenic signal have been detected. The consistency check on the residual variance indicates that it cannot be distinguished from the model's internal variability, so the combined signals provide a statistically consistent explanation of twentieth-century temperature change. If the actual signals are scaled appropriately, then it appears most likely that anthropogenic factors are mainly responsible for the net warming over the last century, with little contribution from natural factors (though these may have contributed significantly in the early part of the century).

The reader is warned that detection results are sensitive to errors in the estimates of forcing, in the simulated response to the warming and in the model estimates of unforced variability (e.g., Barnett *et al.*, 1999; Hegerl *et al.*, 2000). Nevertheless, all studies of this type indicate that the observed warming is unlikely to be entirely natural in origin. The estimates of the contribution from a particular individual component, especially the confidence intervals, are very sensitive to which other components are, or are not, included.

Attempts have been made to use the vertical patterns of temperature change to attribute recent climate change to human activity (Santer *et al.*, 1996; Tett *et al.*, 1996). The pattern of cooling in the stratosphere and warming in the troposphere associated in simulations with increases in CO_2 (Figs 1 and 2), is also seen in the observations. This pattern is hard to attribute to natural forcing—the recent increase in major volcanic eruptions would lead to a warming of the stratosphere and cooling of the troposphere, whereas the increases in solar irradiance would tend to warm the troposphere and leave the stratosphere unaffected (unless the increase in irradiance produced changes in the ozone distribution which lead to stratospheric cooling). Although these physical arguments seem compelling, several factors complicate the interpretation of the vertical temperature record. First, much of the recent cooling in the stratosphere, especially in high latitudes, is likely to be due to reductions in stratospheric ozone. Second, when a rigorous statistical assessment has been made using model simulations, there appear to be shortcomings in the simulated patterns of the response or the model estimates of variability in the upper troposphere and stratosphere (see Tett *et al.*, 2001). Third, the radiosonde temperature record is much shorter than the surface record, with poorer spatial coverage.

Finally, the technique of optimal detection can be used to make estimates of uncertainty in model predictions. Essentially, the components of response (for example, to well-mixed greenhouse gases and to aerosols) can be scaled using the β factors determined by comparing the simulation up to present with observations. Thus, a revised "best" estimate of future warming, along confidence intervals are obtained (Fig. 10). This assumes that there are no changes in the character of the ocean response. The scaling corrects the global

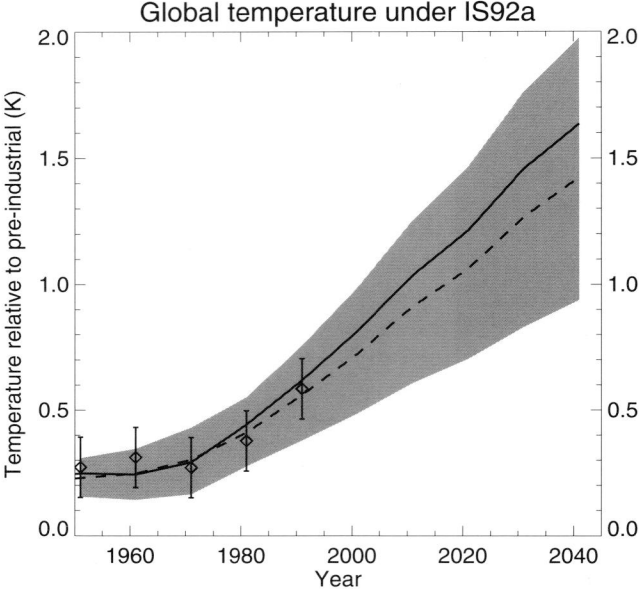

FIGURE 10 Prediction of global mean temperature (K), including estimates of uncertainty, assuming 1% per year increase in atmospheric CO_2, and sulphur emissions as in IPCC scenario IS92a. The solid line denotes the original prediction, the dashed line the revised prediction taking into the observational record, and the shaded area denotes the 5–95% confidence interval. (From Allen *et al.*, 2000.)

mean magnitude of the response, but obviously not errors on the simulated pattern. When this technique is applied across a number of models, it does reduce the discrepancy between predictions from different models (see Allen *et al.*, 2000).

E. SOME CONCLUDING REMARKS

I have tried to indicate some of the progress that has been made in modelling climate change over the last 25 years. A great deal of progress has been made in adding "functionality" to climate models—for example, the inclusion of the ocean dynamics, the addition of chemical and biological processes. Other aspects of climate research have progressed less slowly—in particular the conventionally accepted range for the equilibrium surface temperature response to doubling atmospheric carbon dioxide has remained unchanged for over 20 years.

Many of the ongoing uncertainties are exposed in our attempts to detect and quantify human influence on recent climate change. There are uncertainties in the estimates of radiative forcing due to both natural (solar, volcanic) and anthropogenic (aerosols, changes in land use) factors. Our estimates of the magnitude and patterns of response to a given forcing vary considerably from model to model. There is considerable uncertainty in the magnitude of natural internal variability over multi-decadal timescales.

In the near future, we can expect the resolution of climate models to increase along with computing capacity. This is likely to improve the representation of storm tracks, ocean flow over sills and through channels, and eventually explicit modelling of ocean mesoscale eddies. Increases in computing resources are also likely to allow a more careful exploration of the sources of uncertainty in models. In particular, we would like to know how sensitive model predictions are to the form of the physical parametrizations used and the values chosen for individual parameters. There already exists plans to start to investigate these issues by using idle time on personal computers to run large ensembles of models using different parameter values (Allen, 1999).

Much of the work to date (e.g., in detection and attribution) assumes that the large-scale response of climate is largely linear. Given the inherent non-linearity of atmospheric dynamics, we should not overlook the possibility of non-linear responses in the climate system, such as changes in the frequency of natural climate regimes (e.g., Corti *et al.*, 1999).

Work on detecting climate change over the period of the instrumental temperature record and attributing causes has progressed beyond simple detection to the estimation of the magnitude of the contributions of different factors to the observed change. This will require better estimates of the climate signals, and improved estimates of long-term climate variability.

The last 25 years of modelling climate change has seen climate models grow from basic atmospheric general circulation models to complex earth system models incorporating the main processes relevant to climate change on timescales of up to a century or so. In the next decade or so, I believe more attention needs to be given to validating models and quantifying uncertainty if they are to provide useful and reliable predictions of future climate. This is a challenge worthy of a new millennium.

ACKNOWLEDGEMENTS

I thank Peter Stott for producing Figs 8 and 9, Cath Senior for producing Fig. 6, and to members of the Hadley Centre in general for their help and support over the last decade.

References

Allen, M. R., 1999: Do it yourself climate prediction *Nature*, **401**, 642.

Allen M. R. and S. F. B. Tett, 1999: Checking for model consistency in optimal fingerprinting. *Climate Dynamics*, **15**, 419–434.

Allen, M. R., P. A. Stott, R. Schnur, T. Delworth and J. F. B. Mitchell, 2000: Uncertainty in forecasts of anthropogenic climate change. *Nature*, **407**, 617–620.

Arrhenius, S., 1896: On the influence of carbonic acid in the air upon the temperature of the ground. *Phil. Mag. Ser.* 5, **41**, 237–276

Barnett, T. P., K. Hasselmann, M. Chelliah, T. Delworth, G. Hegerl, P. Jones, E. Rasmussen, E. Roeckner, C. Ropelewski, B. Santer and S. Tett, 1999: Detection and attribution of climate change: A Status report. *Bull. Am. Meteorol. Soc.,* **12**, 2631–2659.

Bengtsson L., E. Roeckner and M. Stendel, 1999: Why is the global warming proceeding much slower than expected. *J. Geophys. Res.*, **104**, 3865–3876.

Callendar, G. S., 1938: The artificial production of CO_2 and its influence on temperature. *Q. J. R. Meteorol. Soc.*, **64**, 233–237

Cess R. D., G. L. Potter, J. P. Blanchet, G. J. Boer, A. D. Del Genio, M. Deque, V. Dymnikov, V. Galin, W.-L. Gates, S. J. Ghan, J. T. Kiehl, A. A. Lacis, H. le Treut, Z.-X. Li, X.-Z. Liang, B. J. Macavaney, V. P. Meleshko, J. F. B. Mitchell, J. J. Morcrette, D. A. Randall, L. Rikus, E. Roeckner, J. F. Royer, U. Schlese, D. A. Scheinin, A. Slingo, A. P. Sokolov, K. E. Taylor, W. M. Washington, R. T. Wetherald, I. Yagai and M.-H. Zhang, 1990: Intercomparison and interpretation of climate feedback processes in 19 atmospheric general circulation models. *J. Geophys. Res*, **95**, 16601–16615.

Collins, M., 2000a: Understanding uncertainties in the response of ENSO to greenhouse warming. *Geophys. Res. Lett.*, **27**, 3509–3512.

Collins, M., 2000b: The El Niño Southern Oscillation in the second Hadley Centre coupled model and its response to greenhouse warming. *J. Climate*, **13**, 1299–1312.

Corti, S., F. Molteni and T. N. Palmer, 1999: Signature of recent climate change in frequencies of natural atmospheric climate regimes. *Nature*, **398**, 799–802.

Cox, P. M., R. A. Betts, C. D. Jones, S. A. Spall, I. J. Totterdell, 2000: Will carbon cycle feedbacks accelerate global warming in the 21st century? *Nature*, **408**, 184–187.

Cubasch, U. and I. Fischer-Bruns, 2000: An intercomparison of scenario simulations performed with different OAGCMs. *Proceedings of the Reg Clim Seminar*, Jevnaker, Norway, 8–9 May 2000.

Cubasch, U., G. A. Meehl, G. Boer, R. Stouffer, M. Dix, A. Noda, C. Senior, S. Raper and K. S. Yap, 2001: Projections of future climate. In *Climatic Change, the Scientific Basis* (J. T. Houghton, Ding Yihui and M. Noguer, eds), Cambridge University Press, to appear.

Gates, W. L., J. S. Boyle, C. Covey, C. G. Dease, C. M. Doutriaux, R. S. Drach, M. Fiorino, P. J. Glecker, J. J. Hnilo, S. M. Marlais, T. J. Phillips, G. L. Potter, B. D. Santer, K. R. Sperber, K. E. Taylor and D. N. Williams, 1999: An overview of the results of the atmospheric model intercomparison project (AMIPI). *Bull. Am. Meteorol. Soc.*, **80**, 29–55.

Gordon C., C. Cooper, C. Senior, H. Banks, J. M. Gregory, T. C. Johns, J. F. B. Mitchell and R. Wood, 1999: Simulation of SST, sea ice extents and ocean heat transports in a coupled model without flux adjustments. *Climate Dynam*, **16**, 147–168.

Hasselmann, K., 1979: On the signal-to-noise problem in atmospheric response studies. In *Meteorology Over the Tropical Oceans* (D. B. Shaw Ed.), Royal Meteorological Society, pp. 251–259.

Hasselmann, K. 1993: Optimal fingerprints for the detection of time dependent climate change. *J. Climate*, **6**, 1957–1971.

Hegerl G. C. and G. R. North, 1997: Statistically optimal methods for detecting anthropogenic climate change. *J. Climate*, **10**, 1125–1133.

Hegerl G. C., H. von Storch, K. Hasselmann, B. D. Santer, U. Cubasch and P. D. Jones, 1996: Detecting greenhouse gas induced climate change with an optimal fingerprint method. *J. Climate*, **9**, 2281–2306.

Hegerl, G. C., P. Stott, M. Allen, J. F. B. Mitchell, S. F. B. Tett and U. Cubasch, 2000: Detection and attribution of climate change: sensitivity of results to climate model differences. *Climate Dynam.*, **16**, 737–754.

Ingram, W. J, C. A. Wilson and J. F. B. Mitchell, 1989: Modelling climate change: an assessment of sea-ice and surface albedo feedbacks. *J. Geophys. Res.*, **94**, 8609–8622.

Johns, T. C., W. J. Ingram, C. E. Johnson, A. Jones, J. F. B. Mitchell, D. L. Roberts, D. M. H. Sexton, D. S. Stevenson, S. F. B. Tett and M. J. Woodage, 2000: Anthropogenic climate change for 1860 to 2100 simulated with the HadCM3 model under updated emissions scenarios. Hadley Centre Tech. Note 22. Available from Meteorological Office, Bracknell RG12 2SY, UK.

Joussaume, S., K. E. Taylor, P. Braconnot, J. F. B. Mitchell, J. E. Kutzbach, S. P. Harrison, I. C. Prentice, A. J. Broccoli, A. Abe-Ouchi, P. J. Bartlein, C. Bonfils, B. Dong, J. Guiot, K. Herterich, C. D. Hewitt, D. Jolly, J. W. Kim, A. Kislov, A. Kitoh, M. F. Loutre, V. Masson, B. McAvaney, N. McFarlane, N. de Noblet, W. R. Peltier, J. Y. Peterschmitt, D. Pollard, D. Rind, J. F. Royer, M. E. Schlesinger, J. Syktus, S. Thompson, P. Valdes, G. Vettoretti, R. S. Webb and U. Wyputta, 1999: Regional climates for 6000 years ago: Results of 18 simulations from the Palaeoclimatic Modelling Intercomparison Project. *Geophys. Res. Lett.*, **26**, 859–862.

Kattenberg, A., F. Giorgi, H. Grassl, G. A. Meehl, J. F. B. Mitchell, R. J. Stouffer, T. Tokioka, A. J. Weaver and T. M. L. Wigley, 1996: Climate Models—Projections of Future Climate. In *Climate Change 1995, the Science of Climate Change*, The Second Assessment Report of the IPCC: Contribution of Working Group I (J. T. Houghton, L. G. Meira Filho, B. A. Callander, N. Harris, A. Kattenberg and K. Maskell, Eds), Cambridge University Press, pp. 285–357.

Latif, M., J. E. Roeckner, U. Mikolajewicz and R. Voss, 2000: Tropical stabilization of the thermohaline circulation in a greenhouse warming simulation. *J. Climate*, **13**, 1809–1813.

Leggett, J., W. J. Pepper and R. J. Stewart, 1992: Emission scenarios for IPCC: an update. In *Climate change 1992, Supplement to IPCC Scientific Assessment* (J. T. Houghton, B. A. Callander and S. K. Varney, Eds), Cambridge University Press, pp. 69–75.

McDonald, R. E., D. R. Cresswell and C. A. Senior, 2000: Changes in storm tracks and the NAO in a warmer climate. In *RegClim, Regional Climate Development Under Global Warming*. General Technical Report No. 2. Presentations from Workshop. 31 May–1 June 1999 at Jevnaker, Norway (T. Iversen and T. Berg, Eds).

Manabe, S. and R. J. Stouffer, 1994: Multiple century response of a coupled ocean-atmosphere to an increase in atmospheric carbon dioxide. *J. Climate*, **7**, 5–23.

Manabe, S. and R. T. Wetherald, 1967: Thermal equilibrium of the atmosphere with a given distribution of relative humidity. *J. Atmos. Sci.*, **24**, 241–259.

Manabe, S. and R. T. Wetherald, 1975: The effect of doubling CO_2 concentrations on the climate of a GCM. *J. Atmos. Sci.*, **32**, 3–15.

Mann M. E., Bradley R. S., Hughes M. K. 1998: Global scale temperature patterns and climate forcing over the past six centuries. *Nature*, **392**, 779–787.

Meehl, G. A., G. J. Boer, C. Covey, M. Latif, and R. J. Stouffer, 2000: The Coupled Model Intercomparison Project (CMIP), *Bull. Am. Meteorol. Soc.*, **81**, 313–318.

Mitchell, J. F. B. and T. C. Johns, 1997: On the modification of greenhouse warming by sulphate aerosols. *J. Climate*, **10**, 245–267.

Mitchell, J. F. B., S. Manabe, V. Meleshko and T. Tokioka, 1990: Equilibrium climate change—and its implications for the future. In *Climate Change, the IPCC Scientific Assessment* (J. T. Houghton, G. J. Jenkins and J. J. Ephraums, Eds), Cambridge University Press, pp. 131–170.

Mitchell, J. F. B., T. C. Johns and C. A. Senior, 1998: *Transient Response to Greenhouse Gases Using Models With and Without Flux-adjustment*. HCTN 2, Hadley Centre, The Met. Office, Bracknell RG12 2SY, UK.

Moller, F, 1963: On the influence of changes in the CO_2 concentration on the radiation balance of the earth's surface and on the climate. *J. Geophys. Res.*, **68**, 3877–3886.

National Academy of Sciences, 1979: *Carbon Dioxide and Climate: a Scientific Assessment*. Report of the ad hoc study group on carbon dioxide and climate. National Academy of Sciences, Washington, DC.

Newell, R. E. and T. G. Dopplick, 1979: Questions concerning the possible influence of anthropogenic CO_2 on atmospheric temperature. *J. Appl. Meteorol.*, **18**, 822–825.

North G. R., K.-Y. Kim, S. S. P. Shen and J. W. Hardin, 1995: Detection of forced climate signals, Part I: Filter theory. *J. Climate*, **8**, 401–408.

Santer, B. D., K. E. Taylor, T. M. L. Wigley, T. C. Johns, P. D. Jones, D. J. Karoly, J. F. B. Mitchell, A. H. Oort, J. E. Penner, V. Ramaswamy, M. D. Schwarzkopf, R. J. Stouffer and S. Tett, 1996: A search for human influences on the thermal structure of the atmosphere. *Nature*, **382**, 39–46.

Senior C. A. and J. F. B. Mitchell, 1993: CO_2 and climate: the impact of cloud parametrizations. *J. Climate*, **6**, 393–418.

Stott, P. A., S. F. B. Tett, G. S. Jones, M. R. Allen, J. F. B. Mitchell and G. J. Jenkins, 2000: External control of twentieth century temperature by natural and anthropogenic forcings. *Science*, **15**, 2133–2137.

Stouffer, R. J. and S. Manabe, 1999: Response of a coupled ocean-atmosphere model to increasing carbon dioxide: sensitivity to the rate of increase. *J. Climate*, pt I, **12**, 224–2237.

Stouffer, R. J., S. Manabe and K. Bryan, 1989: Interhemispheric asymmetry in climate response to a gradual increase of atmospheric CO_2. *Nature*, **342**, 660–662.

Stouffer, R. J., G. C. Hegerl and S. F. B. Tett, 1999: A comparison of surface air temperature variability in three 1000-year coupled ocean-atmosphere model integrations. *J. Climate*, **13**, 513–537.

Tett, S. F. B., J. F. B. Mitchell, D. E. Parker and M. R. Allen, 1996: Human influence on the atmospheric vertical temperature structure: detection and observations. *Science*, **274**, 1170–1173.

Tett, S. F. B., P. A. Stott, M. A. Allen, W. J. Ingram and J. F. B. Mitchell, 1999: Causes of twentieth century temperature change near the earth's surface. *Nature*, **399**, 569–572.

Tett, S. F. B., G. S. Jones, P. A. Stott, D. Hill, J. F. B. Mitchell, M. R. Allen, W. Ingram, T. C. Johns, C. Johnson, A. Jones, D. Sexton and M. Woodage, 2001: Estimation of natural and anthropogenic contributions to 20th century climate temperature change. *J. Geophys. Res.*, submitted.

Timmermann, A., J. Oberhuber, A. Bacher, M. Esch, M. Latif, and E. Roeckner, 1999: Increased El Niño frequency in a climate model forced by future greenhouse warming. *Nature*, **398**, 694–696.

Ulbright, U. and M. Christoph, 1999: A shift of the NAO and increasing storm track activity over Europe due to anthropogenic greenhouse gas forcing. *Climate Dynam.*, **15**, 551–559.

Webb, M. J., C. A. Senior, S. Bony and J.-J. Morcrette, 2000: Assessment of clouds and cloud radiative forcing simulated by three GCM's. *Climate Dynam.*, submitted.

Absorption of Solar Radiation in the Atmosphere:
Reconciling Models with Measurements
A. Slingo
Hadley Centre for Climate Prediction and Research, Met Office, Bracknell, UK

A. INTRODUCTION

The absorption of solar radiation provides the energy source which has enabled life to evolve and prosper on the earth and which also drives the general circulations of the atmosphere and oceans. While about half of the solar radiation incident from space is transmitted through the atmosphere and is absorbed by the land surface and upper layers of the ocean, a significant fraction (about one-fifth) is absorbed within the atmosphere itself. The oft-repeated "fact" that the atmosphere is heated from below by the solar radiation absorbed at the surface is thus a gross simplification of reality.

In this chapter, I will consider the observations and modelling that have contributed to our current understanding of the absorption of solar radiation in the atmosphere, and highlight some of the remaining barriers and uncertainties. One of the themes of this chapter is to note the variety of sources of observations and the interplay between observations, theory and numerical modelling.

B. THE GLOBAL RADIATION BALANCE

Figure 1 summarizes our current understanding of the global heat balance. The amount of solar radiation absorbed in the atmosphere is estimated to be 67 W m^{-2}, with most of the absorption coming from atmospheric gases and only a small contribution from clouds. As will be seen later, some estimates produce a value as high as 93 W m^{-2}, so there is a significant uncertainty in this basic quantity of the system. The higher value is consistent with claims that the cloud contribution is much larger than previously thought, although this assertion is still a matter for vigorous debate.

The figure illustrates the strong coupling between the radiation budget and the hydrological cycle. Evapotranspiration from the surface provides the source of atmospheric moisture, with precipitation acting as the sink. Evapotranspiration cools the surface, but the atmosphere is warmed by the condensation of water vapour to form clouds and its removal by precipitation. The cooling of the surface and warming of the atmosphere by the hydrological cycle (together with the smaller contribution from sensible heat transfer) balances the net heating of the surface and net cooling of the atmosphere from the transfer of solar and terrestrial radiation. The value of 78 W m^{-2} for the global average latent heat released by precipitation is equivalent to a mean precipitation rate of 2.7 mm day^{-1}, although some estimates exceed 3 mm day^{-1} (Kiehl and Trenberth, 1997). If the solar absorption is increased to 93 W m^{-2}, adjustments need to be made to the other terms contributing to the atmospheric net radiative cooling in Fig. 1 to prevent the required precipitation rate from becoming unreasonably small. This requirement shows that the solar absorption cannot be considered in isolation from the net longwave cooling and hydrological cycle.

If we divide the global average column of water vapour of about 24 mm liquid water equivalent by the global average precipitation rate of 2.7 mm day^{-1} we obtain a figure of about 9 days for the mean timescale of the hydrological cycle. Any reduction in the net radiative cooling of the atmosphere (e.g. through enhanced solar absorption) reduces the requirement for latent heating, so the precipitation rate drops and the hydrological cycle slows down.

Figure 1 results from a fusion of satellite observations of the terms at the top of the diagram, plus surface observations and a significant amount of numerical modelling in between. The first such diagram was produced by Dines (1917), and was based on a very limited number of surface-based observations combined with a great deal of physical reasoning. Dines' work is discussed by Hunt *et al.* (1986), who provide a valuable discussion of presatellite estimates of the earth's radiation budget. One of the most uncertain components of the budget was the shortwave albedo, or reflectivity, of the earth, which Dines assumed to be 0.5, based on Hann (1903). The value of the mean incoming solar radiation

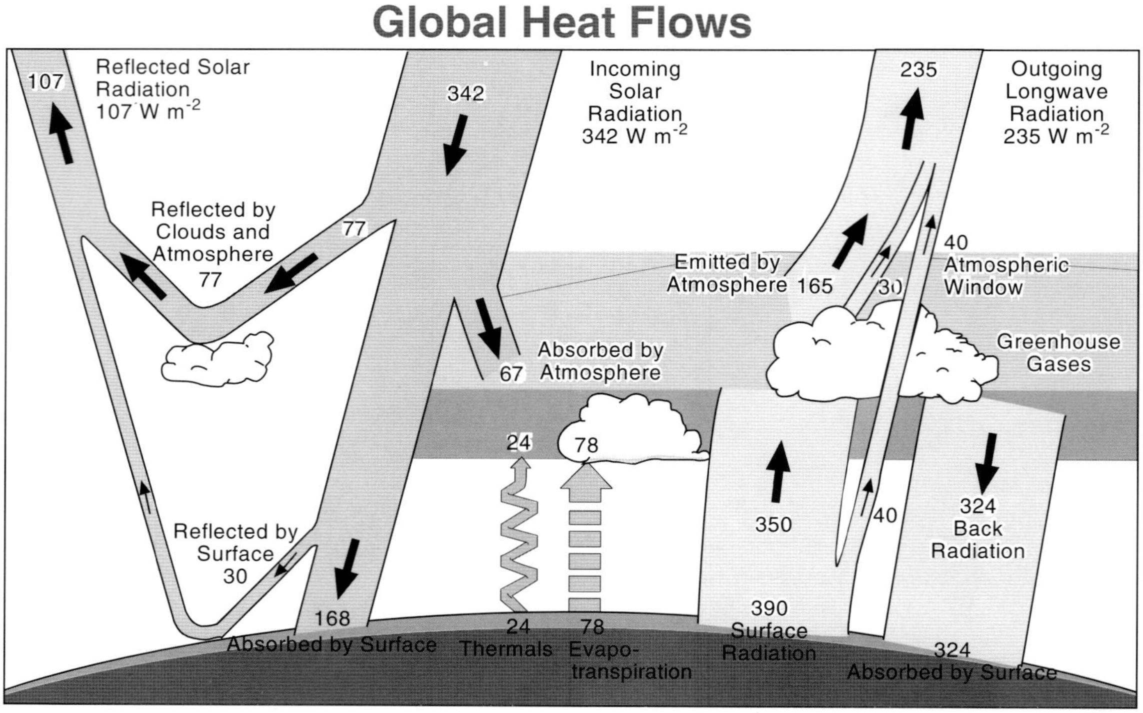

Global Heat Flows

FIGURE 1 Schematic of the global heat balance, from Kiehl and Trenberth (1997).

was also uncertain, but if we take the modern value from Fig. 1, then Dines' value for the reflected solar radiation comes out to be 172 W m^{-2}, much larger than the modern value of 107 W m^{-2}. His estimate of 142 W m^{-2} for the absorption at the surface is lower than in Fig. 1, but is consistent with the higher atmospheric absorption quoted above. Dines' own estimates of this quantity of 29 W m^{-2} is certainly too low, as indeed he mentions in his paper. The outgoing longwave radiation comes out to be 172 W m^{-2}, which is much lower than the modern value of 235 W m^{-2}. His numbers for the longwave budget are a long way from current estimates, but the sum of sensible and latent heat fluxes from the surface ("thermals" and "evapotranspiration" in Fig. 1) was estimated to be 95 W m^{-2}, close to the value of 102 W m^{-2} in the figure.

Dines' work provides a landmark in our understanding of the earth's radiation budget which is as important as the classic paper by Arrhenius (1896) on the influence of changes in carbon dioxide concentrations on climate. Both works combine sparse observations with incisive reasoning, something which is less evident in the data-rich environment we find ourselves in today.

Subsequent work on the radiation balance is summarized by Hunt *et al.* (1986). One might imagine that no further advances would be made in estimating the planetary albedo until the satellite era. In fact, valuable information on the earth's albedo came from observations of earthshine, the illumination by the earth of the non-sunlit portion of the moon. If assumptions are made of the reflectivity of the lunar surface and its dependence on solar zenith angle, then measurements of the intensity of earthshine can be exploited to estimate the earth's albedo and its variation in time. The most detailed study of this phenomenon was made by the French astronomer Andre Danjon, who used observations over much of the lunar cycle to derive values of the earth's albedo ranging from 0.29 to 0.39, much closer to the modern value of about 0.30 than that from Dines.

C. SATELLITE MEASUREMENTS

The arrival of the satellite era provided the first direct estimates of the earth's albedo, beginning with Explorer 7 in 1959 (House *et al.*, 1986). This was followed by estimated from the Nimbus series of experimental satellites, notably the highly successful radiation budget instruments on Nimbus 7. The best estimates to date have come from the Earth Radiation Budget Experiment (ERBE) in 1985–90 (see Harrison *et al.*, 1990), followed by the ScaRaB instruments in 1994–95 and 1998–99 (Kandel *et al.*, 1998) and the CERES instruments on the TRMM and TERRA satellites (Wielicki *et al.*, 1996).

ERBE estimates of the shortwave albedo for the December to February (DJF) season are shown in Fig. 2a. The corresponding distribution from an inte-

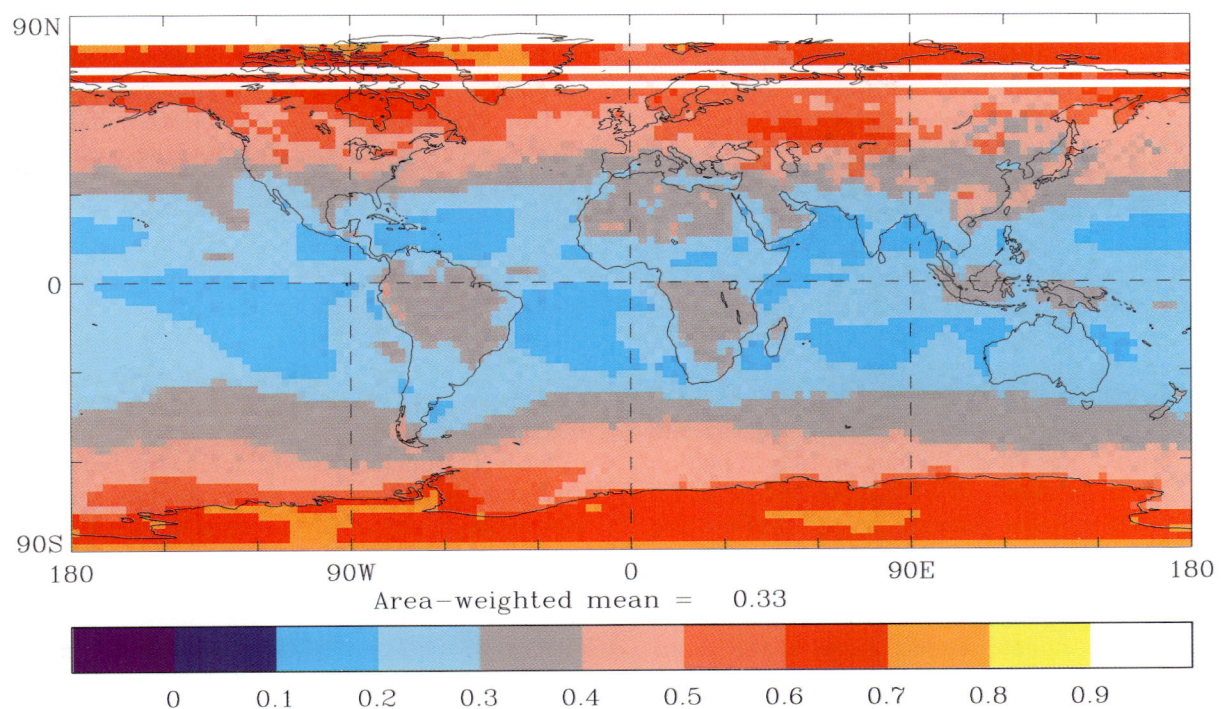
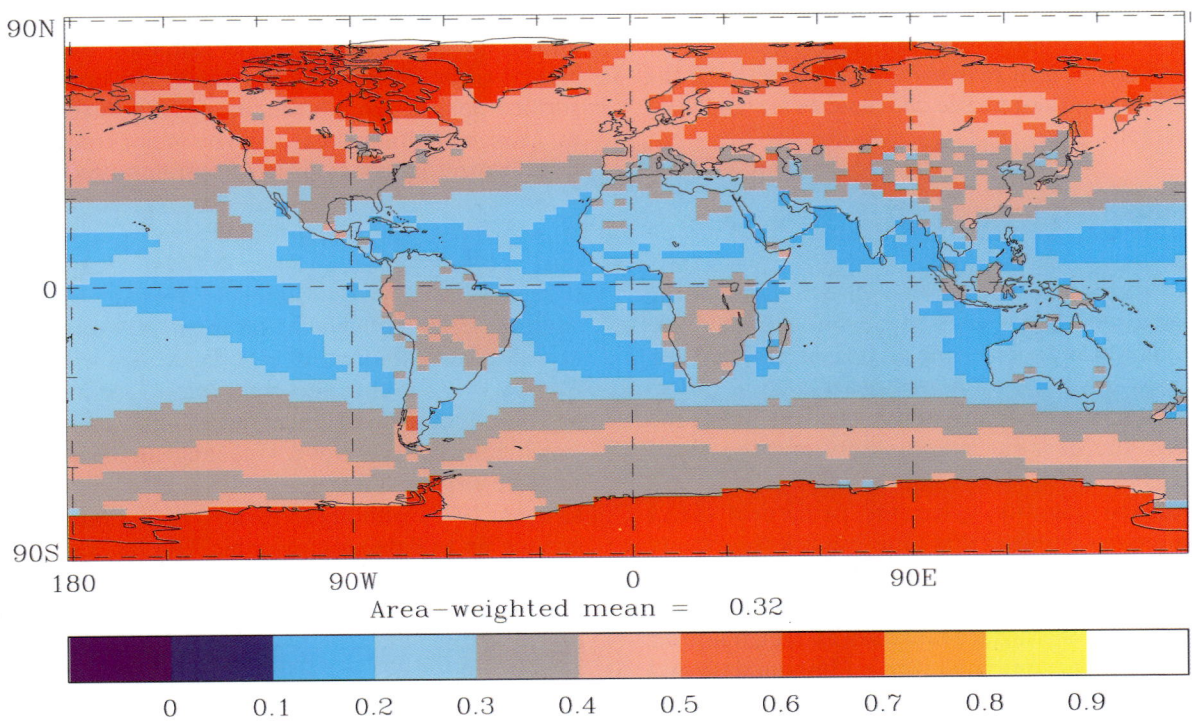

FIGURE 2 Mean shortwave albedo for December to February from (a) ERBE data for 1985–90 and (b) version HadAM3 of the Hadley Centre climate model.

gration of the atmospheric component of the Hadley Centre climate model (version HadAM3) is shown in Fig. 2b. The model results come from an integration using the observed sea surface temperatures and sea-ice distributions (for details, see Pope et al., 2000). The most obvious features in these figures are the high albedo regions at high latitudes, due to areas of permanent land ice and seasonable areas of sea ice and snow-covered land. The lowest albedos occur over the cloud-free oceans, while intermediate albedos are associated with the bright desert regions and clouds, the latter being most obvious in the Southern Hemisphere depression belt and regions of tropical convection.

The strong surface albedo features in Fig. 2 make it difficult to distinguish the contribution of clouds to the shortwave albedo. Recognizing the need for a separate measure of the impact of clouds in the radiation budget, ERBE provided an archive of the reflected shortwave radiation for clear-sky scenes. Subtracting this from the all-sky data provided the "shortwave cloud radiative forcing" diagnostic, which has been tremendously useful in radiation budget research and in the evaluation of climate models. Maps of the shortwave cloud radiative forcing from ERBE and HadAM3 are shown in Fig. 3. The cloud contribution is now clearly visible, being greatest in the southern hemisphere (where the illumination and length of day are greatest in this season), as well as over South America, southern Africa, the maritime continent and in the North Pacific and North Atlantic storm tracks. The model distribution is in reasonable agreement with the observed, although systematic deficiencies are apparent in some regions.

D. ABSORPTION BY THE SURFACE

The first estimates of the global distribution of the surface radiation budget were made with simple parametrizations which depend only on parameters that can be calculated easily or estimated from surface observations, such as the incoming radiation at the top of the atmosphere, the near-surface humidity and the cloud cover (Budyko, 1974). The accuracy of these estimates was difficult to quantify, given the paucity of direct measurements available at that time, but it was certainly much larger than the 5–10 W m^{-2} attributed to the ERBE monthly means. Satellite data have been used to make estimates of the surface radiation budget, but to do this they have to be combined with additional data and modelling. The advantage of this approach is that global products are produced, but the disadvantage is that the products are strongly influenced by the modelling component and their accuracy may be compromised by the assumptions made in the modelling. Using these estimates to evaluate climate models can lead to confusion if the assumptions made in deriving the estimates (e.g. with regard to cloud vertical structure and overlaps) are incompatible with those made in the climate models.

Direct measurements of the surface radiation budget provide a check on the accuracy of these estimates and have been gathered and quality controlled in the Global Energy Balance Archive, GEBA (Garratt, 1994; Wild et al., 1995). The archive includes data from several hundred stations employing relatively low-cost pyranometers, which may be contrasted with the global coverage at the top of the atmosphere from a single, but expensive satellite instrument such as the scanning radiometers used in ERBE (Fig. 2). One problem with the surface measurements is that they are made almost exclusively from the land surface, so the oceans are represented by only a few island stations (Fig. 5). The inherent difficulty in estimating or measuring the global surface radiation budget is reflected in the significant range of values for the shortwave flux absorbed by the surface, which ranges from the value of 168 W m^{-2} in Fig. 1 to 142 W m^{-2} (Wild et al., 1995)

E. ABSORPTION BY THE ATMOSPHERE

Combining ERBE data with either the indirect estimates or the direct measurements of the surface fluxes allows the absorption of solar radiation by the atmosphere to be derived as a residual. Depending on the surface dataset employed, this results in the 67–93 W m^{-2} range mentioned earlier. The major contribution to this absorption comes from atmospheric gases, notably water vapour, ozone and carbon dioxide (Goody and Yung, 1989). Absorbing aerosols, such as wind-blown dust and carbonaceous aerosols from combustion contribute significant absorption in regions close to the sources. Their global contribution is very uncertain, but estimates of about 5 W m^{-2} have been made (e.g. Cusack et al., 1998). The contribution from clouds will be dealt with in more detail later.

Climate models show a wide range in the atmospheric solar absorption (Wild et al., 1995, 1998; Garratt, 1994; Li et al., 1997), although the most recent models, which include up to date spectroscopic data for water vapour, plus aerosols, are well within the 67–93 W m^{-2} range. For example, HadAM3 absorbs about 77 W m^{-2} in the global mean. Nevertheless, comparisons with a combination of ERBE and GEBA data suggest that models have problems in representing the geographical distribution of the absorption. Wild et al. (1998) compared the top of atmosphere, surface and atmospheric absorption derived from these data with the ECHAM3 climate

E. ABSORPTION BY THE ATMOSPHERE

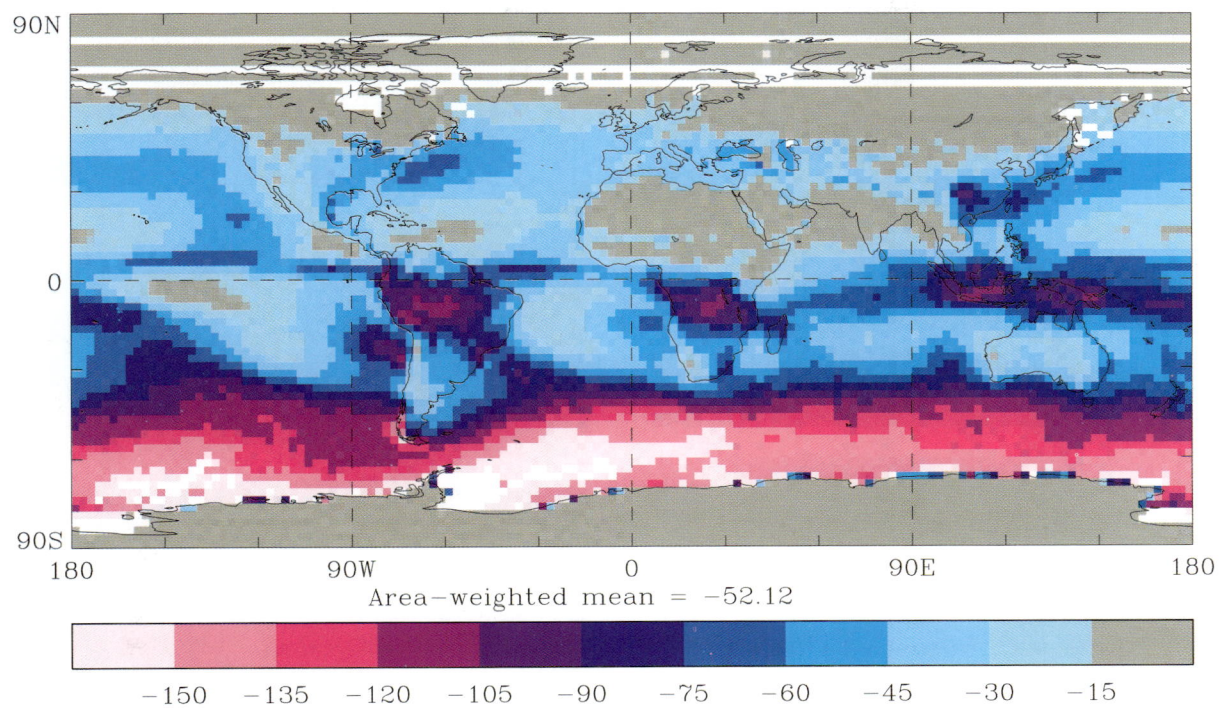

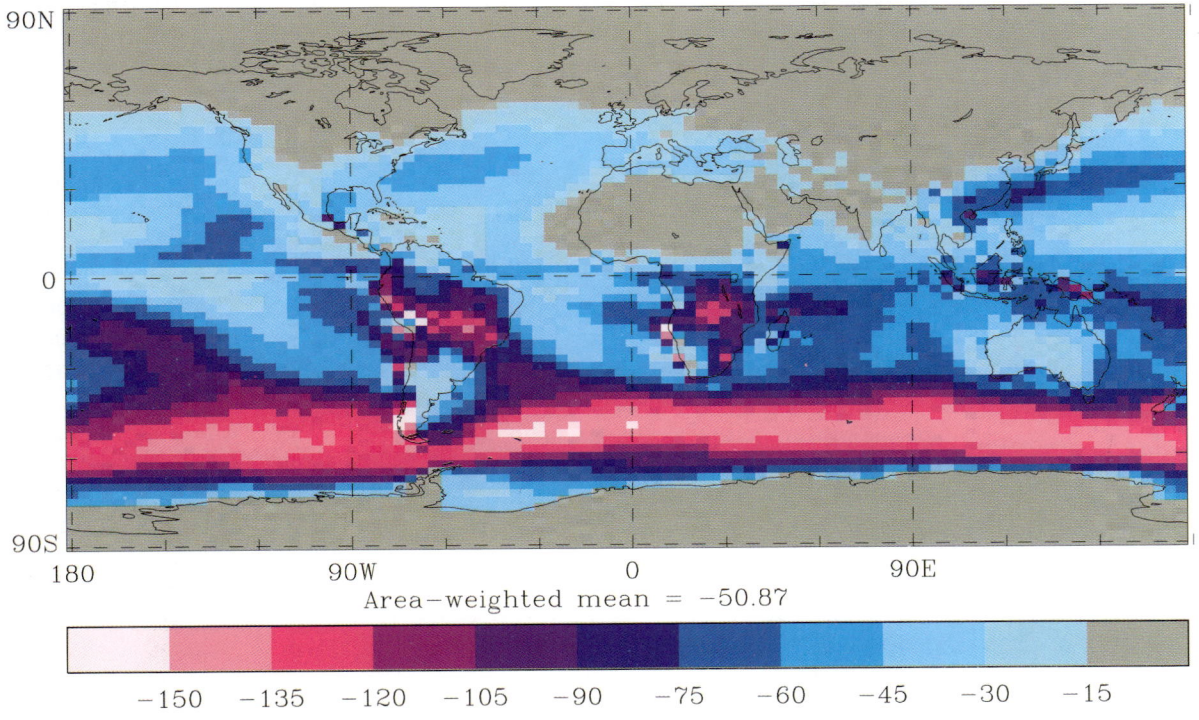

FIGURE 3 As Fig. 2, but for shortwave cloud radiative forcing (W m^{-2}).

FIGURE 4 (Upper) Eppley pyranometer, as used to make measurements of the incoming shortwave radiation at surface sites in GEBA and, in this case, airborne measurements from the Met. Office C-130 research aircraft. (Right) The scanning radiometers used in ERBE. (Photograph courtesy of NASA Langley Research Center.)

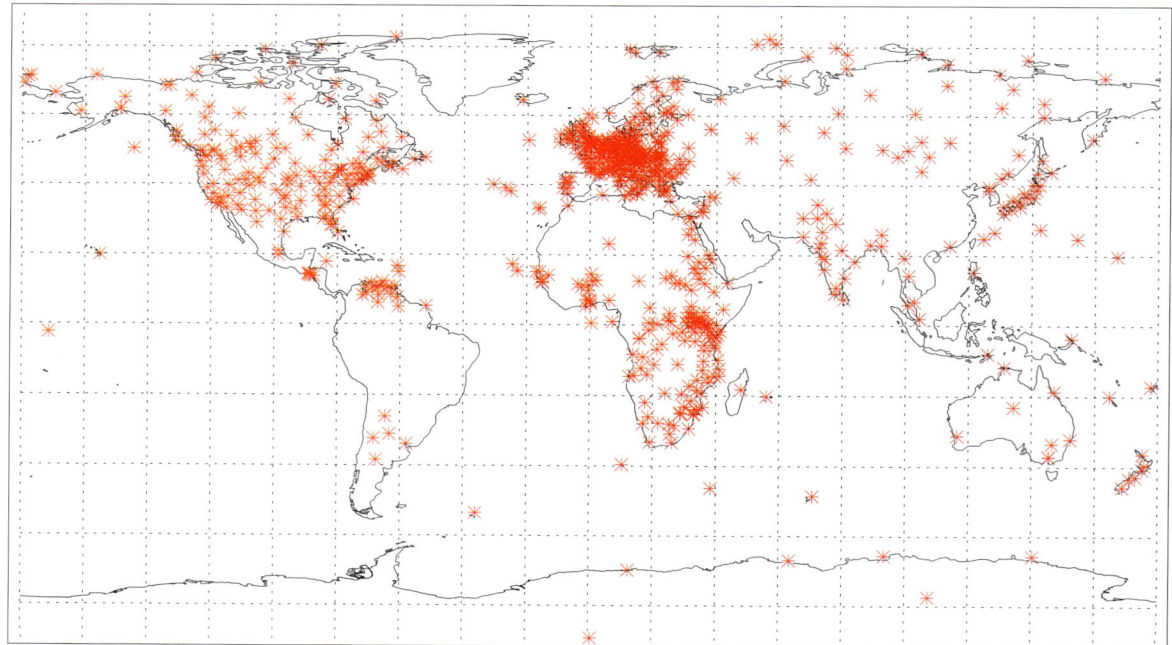

FIGURE 5 Global distribution of stations in the GEBA archive.

E. ABSORPTION BY THE ATMOSPHERE

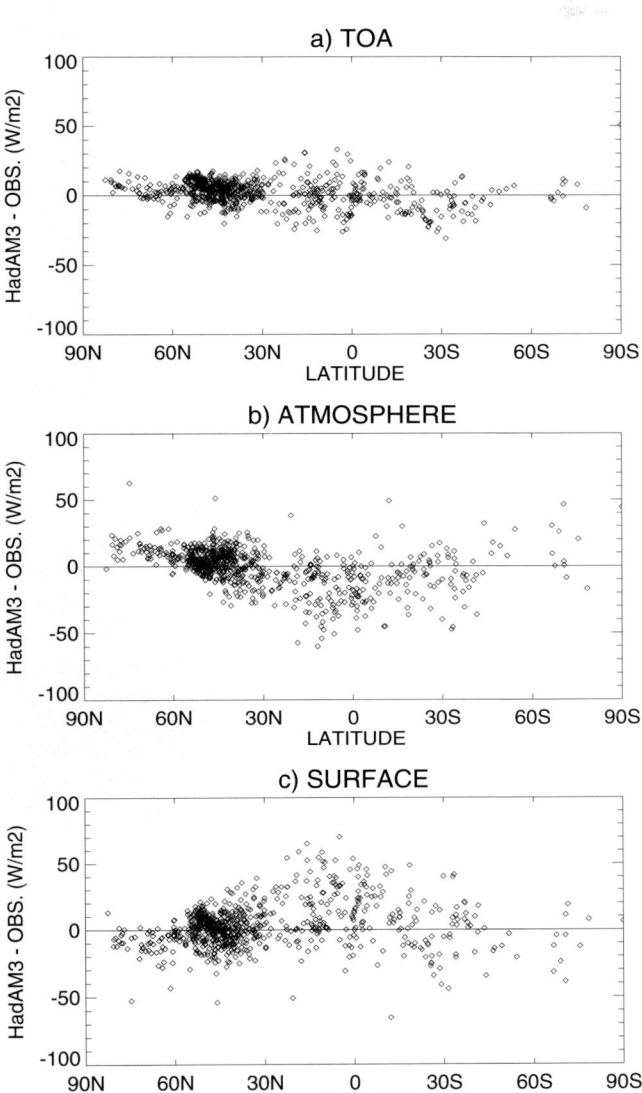

FIGURE 6 Difference between the top of atmosphere, surface and atmospheric shortwave absorption from version HadAM3 of the Hadley Centre climate model and observations from ERBE (TOA) and GEBA (surface). The atmospheric absorption is calculated as the difference between the TOA and surface fluxes. The ERBE data were conditionally sampled over the GEBA sites.

model and found that the model preferentially underestimates the atmospheric absorption at low latitudes. A similar analysis is shown in Fig. 6 for HadAM3. One possible source of the missing absorption comes from the fact that ECHAM3 and HadAM3 include very simple aerosol climatologies, whereas the absorption from both wind-blown dust and aerosols from biomass burning can locally be very large in equatorial regions (e.g. Schwartz and Buseck, 2000). Evidence in favour of this explanation is shown by Wild (1999), who noted that the annual variation in the latitude of greatest disagreement over Africa coincided with the dry season, when biomass burning is at its peak. This suggestion needs further work to quantify the link between biomass aerosols and the atmospheric radiation budget, since it implies that models should include schemes to predict the distribution of such aerosols if they are to produce realistic surface and atmospheric radiation budgets.

Another possible source of the disagreement between ERBE, GEBA and the climate models is errors in modelling the absorption of solar radiation by clouds. While it has long been recognized that clouds absorb some of the incident solar radiation, the amount is generally believed to be small in comparison with that reflected or transmitted (see Stephens and Tsay, 1990). The net effect of clouds on the absorption in the atmospheric column as a whole is small, because the additional absorption both by the cloud (and by the overlying atmosphere of the radiation reflected from the cloud top) is roughly balanced by the reduction in the absorption by water vapour in the atmosphere below the cloud. Extending the definition of the shortwave cloud radiative forcing to the surface, most radiation and climate models produce a ratio R of the forcing at the surface to that at the top of the atmosphere in the range 1.0 to 1.2, with a global mean of about 1.1, so the forcing seen at the top of the atmosphere is enhanced by about 10% by cloud absorption.

This view has been broadly supported by the large number of measurements by research aircraft made since 1949. However, a minority of the measurements have produced larger cloud absorptions than predicted theoretically, a result which has been referred to as the "anomalous absorption paradox" (Wiscombe et al., 1984; Stephens and Tsay, 1990). This is important, because any disagreement could be due to a flaw in the application of radiative transfer to clouds, which is fundamental to a large number of applications in remote sensing and climate modelling. Nevertheless, the problem was generally believed to be not so serious as to challenge the validity of these applications. In 1995, this conventional wisdom was challenged by a series of papers in the journal *Science* (Cess et al., 1995; Ramanathan et al., 1995; Pilewskie and Valero, 1995), asserting that the cloud absorption was *several times* higher than previously believed. The shortwave cloud forcing ratio R was found to be about 1.5, indicating a 50% enhancement from absorption of the forcing at the top of the atmosphere, as opposed to about 10% in the models.

The *Science* papers provoked a vigorous debate

which at times has been acrimonious. Several papers appeared which identified inconsistencies or problems with the analysis methods, notably with regard to the interpretation of the satellite and research aircraft data. One theme which emerged was that proper account may not have been taken of the leakage of photons between clear and cloudy columns because of the three-dimensional nature of the radiative transfer, particularly in partially cloudy conditions. However, the methods were vigorously defended by the original authors. There is no doubt that efficient methods for dealing with 3-D effects are needed in the models and there has been a large increase in the amount of work in this area over the last few years. Other authors noted that the models may underestimate the absorption in cloud-free regions because they ignore the influence of weak water vapour lines, dimers, trace gases and aerosols, and this could be misinterpreted as indicating the need for additional cloud absorption. The assumption in models that clouds are composed of pure water or ice has also been challenged, since it is known that aerosols within or on the surface of cloud drops can enhance the cloud absorption. Whether this effect is sufficiently important globally to explain the discrepancy is not known, although it seems unlikely.

Without new measurements, it was clear that the question as to how much solar radiation is absorbed by clouds was not going to be resolved. Under the Atmospheric Radiation Measurement (ARM) programme, an aircraft campaign was therefore conducted over the Southern Great Plains surface site, with the acronym ARESE (ARM Enhanced Shortwave Experiment: Valero *et al.*, 1997). One case of enhanced cloud absorption was obtained under complete cloud cover, but further analysis of the data revealed inconsistencies and the experiment was inconclusive. A second experiment was conducted in early 2000 and at the time of writing the results are still being analysed.

F. DISCUSSION

Estimates of the amount of solar radiation absorbed by the atmosphere in the global mean cover a significant range (67–93 W m^{-2}), depending on which datasets are used. Some climate models produce values below this range, but the more modern models produce values in the middle of the range, although comparisons with a combination of ERBE and GEBA data suggest that models still underestimate the absorption by up to about 10 W m^{-2} globally. Whether the remaining absorption is due to gases, aerosols, clouds or perhaps a combination of these factors is not yet known. There is some evidence that clouds absorb more solar radiation than predicted by models, but as yet there is no clear evidence of excessive absorption.

What needs to be done to reach a consensus on how much solar radiation is absorbed by the atmosphere? The combination of a few, expensive instruments in space and a large number of inexpensive instruments at the surface provides a powerful resource for defining the global distributions at the top of the atmosphere and at the surface, and hence within the atmosphere. But the datasets need to be contemporaneous to be most useful. More accurate surface measurements at the limited number of sites contributing to the Baseline Surface Radiation Network (BSRN) will help to constrain both the surface data and the satellite-based surface radiation budget climatologies (Ohmura *et al.*, 1998). More surface data over the oceans are required urgently. Detailed analyses at selected locations are made possible by the highly instrumented surface sites such as in the ARM programme. Research aircraft campaigns are vital to provide *in situ* sampling of the radiation and other fields within and around clouds. Spectrally resolved data will be essential in order to identify the regions of the shortwave spectrum where the models are deficient and to determine the source of the missing absorption. An important development over the last few years has been powerful methods for computing three-dimensional radiative transfer. Combining these methods with the output from cloud-resolving models should enable the 3-D radiation fields associated with realistic cloud structures to be explored theoretically for the first time. A combination of these new computational tools and the satellite, surface and airborne measurements should be able to resolve the problem and enable improve formulations to be developed for the climate models.

ACKNOWLEDGEMENT

I am very grateful to my colleagues at the Meteorological Research Flight for the photograph of the pyranometer shown in Fig. 4 and to Martin Wild (ETHZ, Zurich) for providing Figs 5 and 6.

References

Arrhenius, S., 1896: On the influence of carbonic acid in the air upon the temperature of the ground. *Phil. Mag. Ser. 5*, **41**, 237–276.

Budyko, M. I., 1974: *Climate and Life*. International Geophysics Series, Vol. 18. Academic Press, New York.

Cess, R. D. and 19 others, 1995: Absorption of solar radiation by clouds: observations versus models. *Science*, **267**, 496–499.

Cusack, S., A. Slingo, J. M. Edwards and M. Wild, 1998: The radiative impact of a simple aerosol climatology on the Hadley Centre atmospheric GCM. *Q. J. R. Meteorol. Soc.*, **124**, 2517–2526.

Dines, W. H., 1917: The heat balance of the atmosphere. *Q. J. R. Meteorol. Soc.*, **43**, 151–158.

Garratt, J. R., 1994: Incoming shortwave fluxes at the surface—a comparison of GCM results with observations. *J. Climate*, **7**, 72–80.

Goody, R. M. and Y. L. Yung, 1989: *Atmospheric Radiation*. Second edition. Oxford University Press.

Hann, J., 1903: *Handbook of Climatology*. Part 1. *General Climatology*. Translated from the 1901 German edition. Macmillan, New York.

Harrison, E. F., P. Minnis, B. R. Barkstrom, V. Ramanathan, R. D. Cess and G. G. Gibson, 1990: Seasonal variation of cloud radiative forcing derived from the Earth Radiation Budget Experiment. *J. Geophys. Res.*, **95**, 18687–18703.

House, F. B., A. Gruber, G. E. Hunt and A. T. Mecherikunnel, 1986: History of satellite missions and measurements of the earth radiation budget (1957–1984). *Rev. Geophys.*, **24**, 357–377.

Hunt, G. E., R. Kandel and A. T. Mecherikunnel, 1986: A history of presatellite investigations of the earth's radiation budget. *Rev. Geophys.*, **24**, 351–356.

Kandel, R. and ten others, 1998: The ScaRaB earth radiation budget dataset. *Bull. Am. Meteorol. Soc.*, **79**, 765–783.

Kiehl, J. T. and K. E. Trenberth, 1997: Earth's annual global mean energy budget. *Bull. Am. Meteorol. Soc.*, **78**, 197–208.

Li, Z., L. Moreau and A. Arking, 1997: On solar energy disposition: a perspective from observation and modelling. *Bull. Am. Meteorol. Soc.*, **78**, 53–70.

Ohmura, A. and ten others, 1998: Baseline Surface Radiation Network (BSRN/WCRP): new precision radiometry for climate research. *Bull. Am. Meteorol. Soc.*, **79**, 2115–2136.

Pilewskie, P. and F. P. J. Valero, 1995: Direct observations of excess solar absorption by clouds. *Science*, **267**, 1626–1629.

Pope, V. D., M. L. Gallani, P. R. Rowntree and R. A. Stratton, 2000: The impact of new physical parametrizations in the Hadley Centre climate model: HadAM3. *Climate Dynam.*, **16**, 123–146.

Ramanathan, V., B. Subasilar, G. J. Zhang, W. Conant, R. D. Cess, J. T. Kiehl, H. Grassl and L. Shi, 1995: Warm pool heat budget and shortwave cloud forcing: a missing physics? *Science*, **267**, 499–503.

Schwartz, S. E. and P. R. Buseck, 2000: Absorbing phenomena. *Science*, **288**, 989–990.

Stephens, G. L. and S.-C. Tsay, 1990: On the cloud absorption anomaly. *Q. J. R. Meteorol. Soc.*, **116**, 671–704.

Valero, F. P. J., R. D. Cess, M. Zhang, S. K. Pope, A. Bucholtz, B. Bush and J. Vitko, 1997: Absorption of solar radiation by the cloudy atmosphere: interpretations of collocated aircraft measurements. *J Geophys. Res.*, **102**, 29917–29927.

Wielicki, B. A., B. R. Barkstrom, E. F. Harrison, R. B. Lee, G. L. Smith and J. E. Cooper, 1996: Clouds and the earth's radiant energy system (CERES): an Earth Observing System experiment. *Bull. Am. Meteorol. Soc.*, **77**, 853–868.

Wild, M., A. Ohmura, H. Gilgen and E. Roeckner, 1995: Validation of general circulation model radiative fluxes using surface observations. *J. Climate*, **8**, 1309–1324.

Wild, M., A. Ohmura, H. Gilgen, E. Roeckner, M. Giorgetta and J.-J. Morcrette, 1998: The disposition of radiative energy in the global climate system: GCM-calculated versus observational estimates. *Climate Dynam.*, **14**, 853–869.

Wild, M., 1999: Discrepancies between model-calculated and observed shortwave atmospheric absorption in areas with high aerosol loadings. *J. Geophys. Res.*, **104**, 27361–27371.

Wiscombe, W. J., R. M. Welch and W. D. Hall, 1984: The effects of very large drops on cloud absorption, Part I: parcel models. *J. Atmos. Sci.*, **41**, 1336–1355.

The Impact of Satellite Observations of the Earth's Spectrum on Climate Research

J. E. Harries

Blackett Laboratory, Imperial College, London, UK

This chapter discusses the use of measurements of the spectrum of outgoing longwave radiation, measured from space, as a mechanism for detecting and attributing a cause to climate change. Observations are available from two experiments, one flown by NASA in 1970 (the InfraRed Interferometer Spectrometer, IRIS), and the other flown by Japan's NASDA in 1997 (the Interferometric Monitor for Greenhouse gases, IMG). Both experiments employed a Fourier transform spectrometer, and both were well calibrated using onboard reference targets. The chapter discusses the accuracy and precision of these experiments, and illustrates the difference between average spectra from both. A simulation of the difference spectrum allows an interpretation of the main features of the difference spectrum. It is shown that the observed features are consistent with an increase of CO_2, average temperature and CH_4 in the period 1970 to 1997. A preliminary analysis of cloudy-sky data indicates that a significant decrease in outgoing radiation between the two periods has been caused by changes in cloud properties. The chapter concludes with the suggestion that in future, small, sensitive spectrally resolving instruments with the measurement capability of IRIS should be flown as passenger instruments on a regular basis.

A. THE CLIMATE CHANGE DETECTION AND ATTRIBUTION PROBLEM

The problem of climate change has received a very considerable amount of attention, especially over the past decade (see the series of IPCC reports (IPCC, 1990, 1996). A central part of the problem is to not only identify measurable changes in the earth's climate, but also to be able to attribute the changes to specific causes and processes. The nature of the climate change detection and attribution problem lies in the fact that the system under study is one of great complexity, including many processes which are chaotic or nonlinear in nature, making the precise isolation of a particular cause and effect difficult. There are many individual problems we face if we wish to solve this detection and attribution problem, which include:

- How to detect climate change?
- How to distinguish "signal" from "noise"?
- How to attribute cause(s) of climate change?
- What parameters to use in objective, quantitative assessment of climate change?
- What parameters will allow the most direct test of models?

Until now, perhaps the biggest effort has been in the direction of analysing the record of surface and atmospheric temperature measurements around the world ($T(x,y,z,t)$, where x and y are horizontal dimensions, z is height, and t is time). This approach has a number of advantages and disadvantages:

Advantages:
- long length of record (since 1860);
- accuracy—temperature is measurable to a few tenths of a degree with quite simple apparatus.

Disadvantages:
- poor sampling and errors, especially in the pre-satellite data taken using a variety of techniques on ships and at the land surface;
- lack of specificity—T is sensitive to many processes, not all due to forced climate change, which might make attribution difficult.

Figure 1 illustrates the now well-known result of analysing the globally averaged surface temperature since modern records began in 1860 (Hadley Centre, 1997). The figure shows annual and smoothed running mean data, which appear to show an unambiguous increase between about 1920 and the present time, with

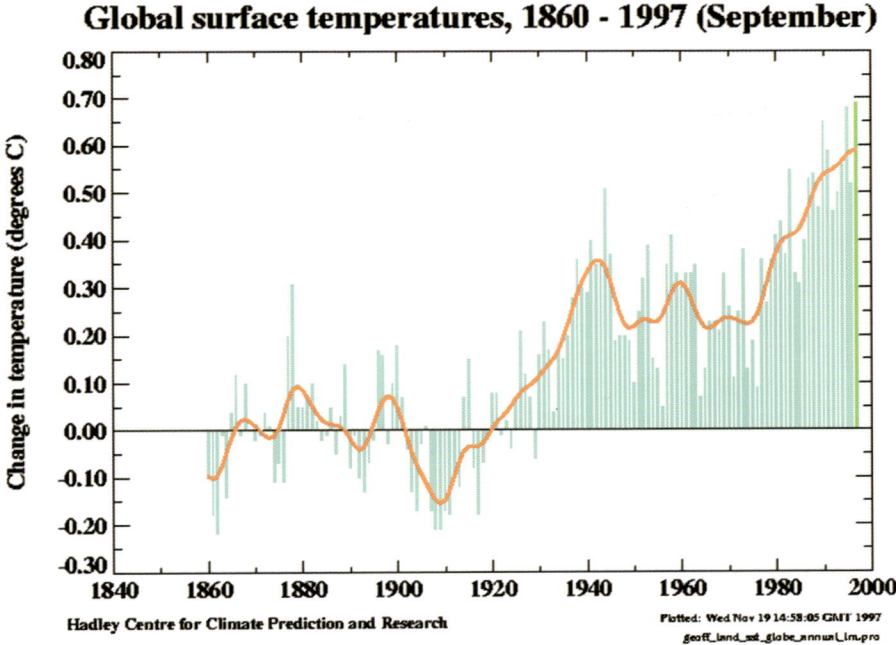

FIGURE 1 Global surface temperatures, 1860–1997 (Hadley Centre, 1997).

concern arising over the ever larger increases in recent years. Several workers, including colleagues at the UK Hadley Centre, have shown that such a trend is consistent with coupled ocean–atmosphere model predictions based on known increases in greenhouse gases, and increases in industrial aerosols. Despite this excellent work, in several centres, we are faced at this level of analysis with the difficulty of discriminating between anthropogenically induced changes and natural fluctuations. It can be said that, at present, our models are simply not capable of providing us with that discrimination. One particular problem is the modelling of various feedback responses to an anthropogenic 'forcing' of the climate, including the response of the global humidity and cloud field, of the ocean circulation and heat transport, and of the land surface vegetation and albedo.

There have also been a number of attempts to use alternative parameters, and to examine whether a specific detection of climate change, and attribution of mechanisms, could be achieved. These parameters include rainfall, cloud, and humidity, amongst others, have been used, so far with similar, somewhat nonspecific results. An unambiguous detection of climate change with an attribution to a clear cause (whether anthropogenic in origin or not) has not yet been achieved. The IPCC have, very reasonably, concluded that the "balance of evidence suggests that there is a discernible human influence on global climate". However, we still seek incontrovertible evidence of this: of course, it is possible that such evidence is at present simply not possible, given the nature of natural variability and of any anthropogenic signal at the present time.

In this chapter we present new results which illustrate the use of measurements of the outgoing longwave radiation (OLR) spectrum as a means of simultaneous detection and attribution of changes in the climate, whether caused by anthropogenic or natural causes.

B. SATELLITE SPECTRAL SIGNATURES OF CLIMATE CHANGE

The spectrum of outgoing infrared energy leaving the earth carries an imprint of the greenhouse effect of the earth in operation, at least in a complementary sense. Energy that is trapped by the earth is not emitted, while in more transparent parts of the spectrum, energy corresponding to higher temperatures deeper in the atmosphere or at the surface does get emitted to space. Figure 2 illustrates these principles. The figure shows calculations* of the outgoing spectrum in the IR for two model atmospheres, the top curve a tropical atmosphere case, and the bottom curve a subarctic winter case. Both are for cloud-free atmospheres. A line-by-line program was used to calculate the spectra (Edwards, 1992).

---

*Thanks to Dr Helen Brindley of Imperial College for these results

The spectrum changes in specific locations if certain key parameters change: for example, if the surface or lower atmosphere temperature increases (decreases), then the radiance in the window region between 800 and 1000 cm^{-1} increases (decreases). If the amount of CO_2 increases (decreases) then the outgoing radiance in the CO_2 band at 600-750 cm^{-1} will decrease (increase). If the amount of H_2O or CH_4 increases (decreases) then the outgoing radiance in bands of these gases (see Fig. 2) will decrease (increase). Not only is there an observable change if anthropogenic forcing occurs, but that change is directly attributable to a cause because of the information available in the resolved spectrum.

There are, of course, both advantages and disadvantages associated with this approach. These include:

Advantages:
- detection and attribution becomes more direct;
- sampling is good, errors self-consistent (long-term, global measurements by well-calibrated satellite sensors).

Disadvantages:
- short duration of record (since only 1970);
- status of global climate models may yet not be sufficiently sophisticated to be able to interpret observations of real changes, if, for example, cloud changes are involved.

Direct observations of the spectrally resolved outgoing longwave radiation (the IR spectrum), or OLR(v), represent direct observations of the external effect of the greenhouse effect in operation: measurement of a difference at two different times would show changes in the greenhouse effect, that is a direct measure and attribution of climate change. In the next section we will briefly review several pieces of work which have recognized this and which have studied different aspects of the issue.

C. RECENT WORK ON SPECTRAL SIGNATURES

Early work included studies by Charlock (1984) and Kiehl (1986) who examined the change in the atmospheric spectrum, especially around the 15 μm CO_2

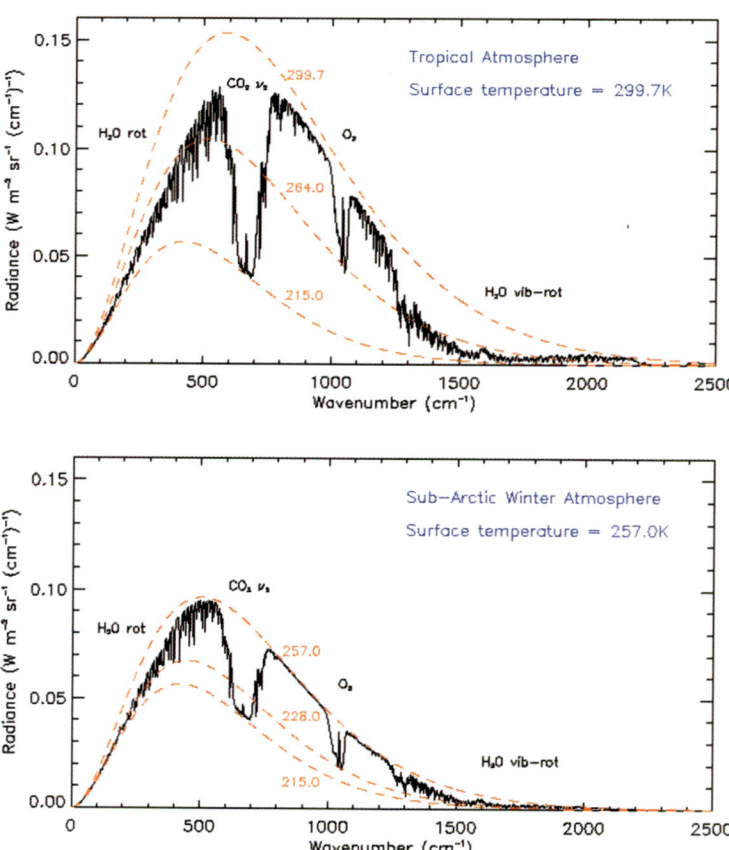

FIGURE 2 Calculations of outgoing longwave radiation spectra for (top) a tropical model atmosphere, and (bottom) a subarctic winter model atmosphere (Brindley, 1998).

band, which might result from anthropogenically induced increases in CO_2. Doherty and Newell (1984) and Clough et al. (1992) illustrated how different parts of the resolved spectrum contribute to the cooling of the Earth to space. Others (e.g. Sinha and Harries, 1997) discussed specifically the importance of the far IR pure rotation band of H_2O to this cooling process. More recently, there have been some detailed studies of the resolved spectrum and its sensitivity to a changing climate. These include:

Theoretical studies:
- Slingo and Webb (1996) used the climate change scenario predictions of the Hadley Centre climate model as input to a calculation of the outgoing spectrum, and studied the changes that would arise between the present day and about the year 2100. They provide detailed results showing the change in global average outgoing radiance which result from growth in CO_2 and other gases.
- Brindley (1998), and Brindley and Harries (1998) report sensitivity studies, in which the responses of the outgoing spectrum to specific changes in atmospheric and surface temperatures, humidity, and other atmospheric constituents, are presented. These provide quantitative estimates of the sensitivity of the $OLR(\nu)$ to specific changes in humidity and other parameters.

Studies of observed spectra:
- Goody et al. (1996) and Iacono et al. (1996) studied the observations obtained using the IRIS spectrometer on the Nimbus 4 satellite in 1970. This mission provided the first high-quality measurements of $OLR(\nu)$ globally from space, and was an experiment ahead of its time. For many years few people recognized the significance of the dataset obtained by IRIS. These studies by Clough, Goody and co-workers analysed the noise, variability and statistics of the spectra.

D. IRIS AND IMG: CHANGE OF $OLR(\nu)$ BETWEEN 1970 AND 1997

1. Previous Broad Band Measurements

Previously, measurements of the spectrally resolved $OLR(\nu)$ leaving the earth have been very rare. We will examine the two notable exceptions below. Far more frequent measurements have been made of the broad band, or spectrally integrated OLR, principally through the series of studies organized by NASA, with scientists at NASA Langley Research Center in the vanguard. Programmes like ERBE and ERBS have flown on a number of polar orbiting satellites (Bess et al., 1989), and recently the CERES experiment has flown on the new NASA EOS programme (Wielicki et al., 1995). Using these OLR data, studies have been carried out of the regional variation of the earth's energy budget, as well as seasonal and other temporal variations: individual studies have examined the OLR signal of phenomena like the El Niño and the Madden–Julian Oscillation (e.g., Soden, 1997; Vincent et al., 1998).

Such studies have been a powerful tool in studies of a number of important processes. However, as we have already seen, if measurements can be made of the resolved $OLR(\nu)$, then in principle much more information becomes available than in the spectrally integrated case. As a "halfway house", some studies have taken band-pass radiance data from meteorological satellite sensors such as HIRS and SSU, which though integrated spectrally are more specific to spectral bands of individual atmospheric components such as CO_2 and H_2O. Thus, Geer et al. (1999) have used HIRS channel 12 measurements (which are tuned to the 6 μm water vapour band) to examine evidence for long-term trends in mid and upper tropospheric water vapour, with some success. Other workers, such as Bates et al. (1996) have also exploited HIRS spectral data in an earlier study.

2. Spectrally Resolved Data

However, spectrally resolved measurements of $OLR(\nu)$, if available, do provide more information. The cost is in complexity of the instruments required to be flown on the satellite, and an increase in data rate due to the large spectral range involved. Nevertheless, we now know that suitable instruments can be developed which are relatively small, and are very efficient in terms of use of onboard resources, data rates, and so on. We shall see that it is now possible to think of a long-term monitoring of the Earth using small, sensitive spectrally resolving instruments.

We have investigated the use of data from two such experiments. The first of these, the InfraRed Interferometer Spectrometer, IRIS, flew on Nimbus 3 and Nimbus 4 in the late 1960s/early 1970s. Only the data from the Nimbus 4 experiment are still available, but these show that the IRIS (which was actually being developed and tested for later planetary missions) was a fine instrument, actually somewhat ahead of its time, in the sense that only in recent years has much serious analysis of its results been attempted. IRIS obtained measurements of $OLR(\nu)$ between 400 and 1600 cm^{-1} at a spectral resolution of 2.8 cm^{-1} between April 1970 and January 1971. Some 700,000 calibrated spectra were taken over all regions of the earth, with a radiometric noise level of 2.3×10^{-4} W m^{-2} (cm^{-1})$^{-1}$ sr^{-1} (Hanel et al., 1971).

In 1996–7, the Japanese Space Agency, NASDA, flew another Fourier spectrometer, called the Interferometric Monitor for Greenhouse Gases, IMG, on the ADEOS 1 spacecraft. This instrument operated between October 1996 and July 1997, and obtained 67,700 calibrated spectra at a spectral resolution of between 0.1 and 0.25 cm^{-1}, and with a radiometric noise level of 5×10^{-4} W m^{-2} (cm^{-1})$^{-1}$ sr^{-1}. IMG covered the rather wider spectral range between 700 and 3030 cm^{-1} (IMG, 1999).

IRIS operated almost continuously during its operational lifetime, but IMG was operated more infrequently, due to power sharing on ADEOS 1 with other instruments. Also, the two instruments had quite different spatial fields of view (100 × 100 km^2 for IRIS, and 8 × 8 km^2 for IMG).

Once we recognized the availability of well-calibrated observations of OLR(ν) at these two times, it became obvious that the possibility existed of forming average spectra in each case for the same month and the same geographical area, and determining whether or not a difference existed between the 1997 and 1970 spectra, and how any observed differences related to the changing climate between the two periods. The key question was: *are these measurements of OLR(ν) in 1970 and in 1997 accurate enough to allow a quantitative assessment of change in the greenhouse effect between these two years?* As we will show below, our conclusion is in the affirmative: a significant difference does occur for clear-sky spectra, and this is consistent with known increases in CO_2, CH_4 and surface temperature.

E. ANALYSIS OF IMG AND IRIS DATA

1. Accuracy

The quoted radiometric accuracy is 0.0002 and 0.0005 W m^{-2} cm^{-1} sr^{-1} for IRIS and IMG, respectively. The magnitude of the signal due to the known changes in temperature, and atmospheric composition between 1970 and 1997 was calculated from model studies to be roughly in the range (depending on wavenumber) 0 to ≈ 0.0025 W m^{-2} cm^{-1} sr^{-1}. Thus a reasonable signal-to-noise ratio for single difference spectra of up to 10:1 is in prospect.

The random error, however, may be reduced further from the figures quoted to insignificance by averaging many spectra (up to hundreds). If the noise is purely random, we would expect the random noise to be reduced in proportion to $N^{1/2}$, where N is the number of spectra averaged. A more serious problem in the differencing of absolute IRIS and IMG spectra would be the presence of different systematic (multiplicative) errors.

Both instruments have the benefit of absolute calibration, and to check for any such systematic errors, we have also differenced normalized spectra, in which each average absolute spectrum is normalized by division by the radiance value at a selected wavenumber in the centre of the window region. We find that within an uncertainty of < 0.5 K this normalization does not change the result obtained with absolute differencing, thereby giving confidence that there are no significant systematic uncertainties that would invalidate our results.

There would be a more serious problem of absolute calibration if a systematic error occurred in either or both datasets which was not spectrally grey. However, while the random uncertainty in the calibrated spectra is expected to rapidly increase as we approach the edge of the pass band of each instrument, there is no reason to expect that systematic errors (e.g. due to change of transmission of filters, degradation of reflecting or transmitting surfaces) should behave in this way. The most likely such effects would exhibit a slow dependence on wavenumber.

Overall, therefore, we expect the systematic uncertainty in the difference spectrum to be within about 0.5 K, as indicated by the normalization test. This is borne out by the results obtained below. In what follows, we shall assume no more than that the absolute difference spectra might be uncertain in absolute level to about 0.5 K. We shall also take it, however, that the random noise in the difference spectra is much smaller than this, at most about 0.1 K. Both these assumptions, about random and systematic uncertainty, are conservative.

2. Validation

We have, in addition, attempted to validate the results obtained in different regions, using the following approach:

- Are observed differences self-consistent?
- To test the absolute calibration, can we simulate absolute spectra given reanalysis data?
- To test the absolute calibration, can we simulate absolute difference spectra given re-analysis data?
- By analysing global and regional data, can we confirm that observations follow known regional/seasonal differences?

As we will show below, we have been able to closely simulate both the IRIS and IMG absolute spectra and the (IMG-IRIS) difference spectra. Furthermore, we have shown that in simulations of the north Atlantic, the central Pacific, the Indian Ocean, and the Sahara, close agreement occurs between the observed difference spectra and those deduced from reanalysis data using line-by-line codes.

3. Results

This chapter will report only a part of the data analysed so far, and principally cloud-free scenes only (see comment below on clouds). We will consider two cases, a global average, and an average for the central equatorial Pacific (defined by the box: 130°W, 180°W, 10°N, 10°S), both for cloud-free cases. Figure 3 shows in panel (a) the globally averaged IMG and IRIS spectra for the April, May, June period in both cases. Panel (b) shows the two averages, for the same season, in this case for the central Pacific area. Panel (c) then shows the difference spectra IMG-IRIS for the global and central Pacific cases.

The general form of the difference spectra shows some general features which appear in all the cases analysed so far. These are: a negative signal in the high frequency wing of the CO_2 band between 700 and about 750 cm^{-1} (consistent with increasing CO_2); a positive difference in the atmospheric window (consistent with a surface warming from 1970 to 1997); a small negative difference in the water vapour v_2 band from 1200 cm^{-1} and above; and a strong negative signal in the Q-branch peak of the CH_4 band centred at 1304 cm^{-1} (consistent

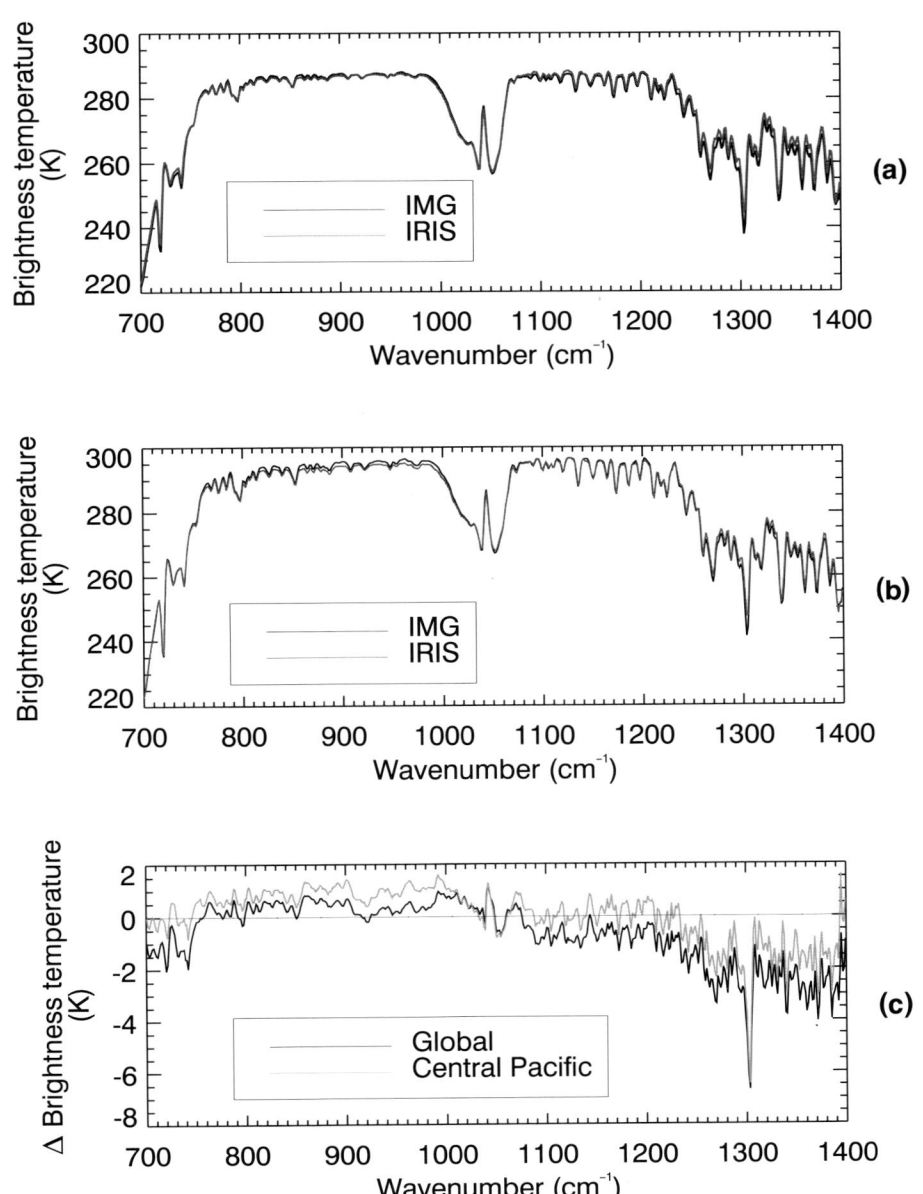

FIGURE 3 Observations of OLR spectra from IMG and IRIS for (a) globe and (b) central pacific; frame (c) shows the difference between IRIS and IMG for the two cases.

with an increase in tropospheric methane and a cooling in the stratosphere). The ozone signal centred at 1050 cm^{-1} indicates a weak signal, as expected, from the known cooling of the stratosphere. Small negative signals in the difference spectra at about 850 and 920 cm^{-1} are due to increase in tropospheric fluorocarbon concentrations between 1970 and 1997.

Figure 4 shows the corresponding simulation of the difference spectrum, in the central Pacific, for April–June, for comparison with Fig. 3. Panel (a) shows

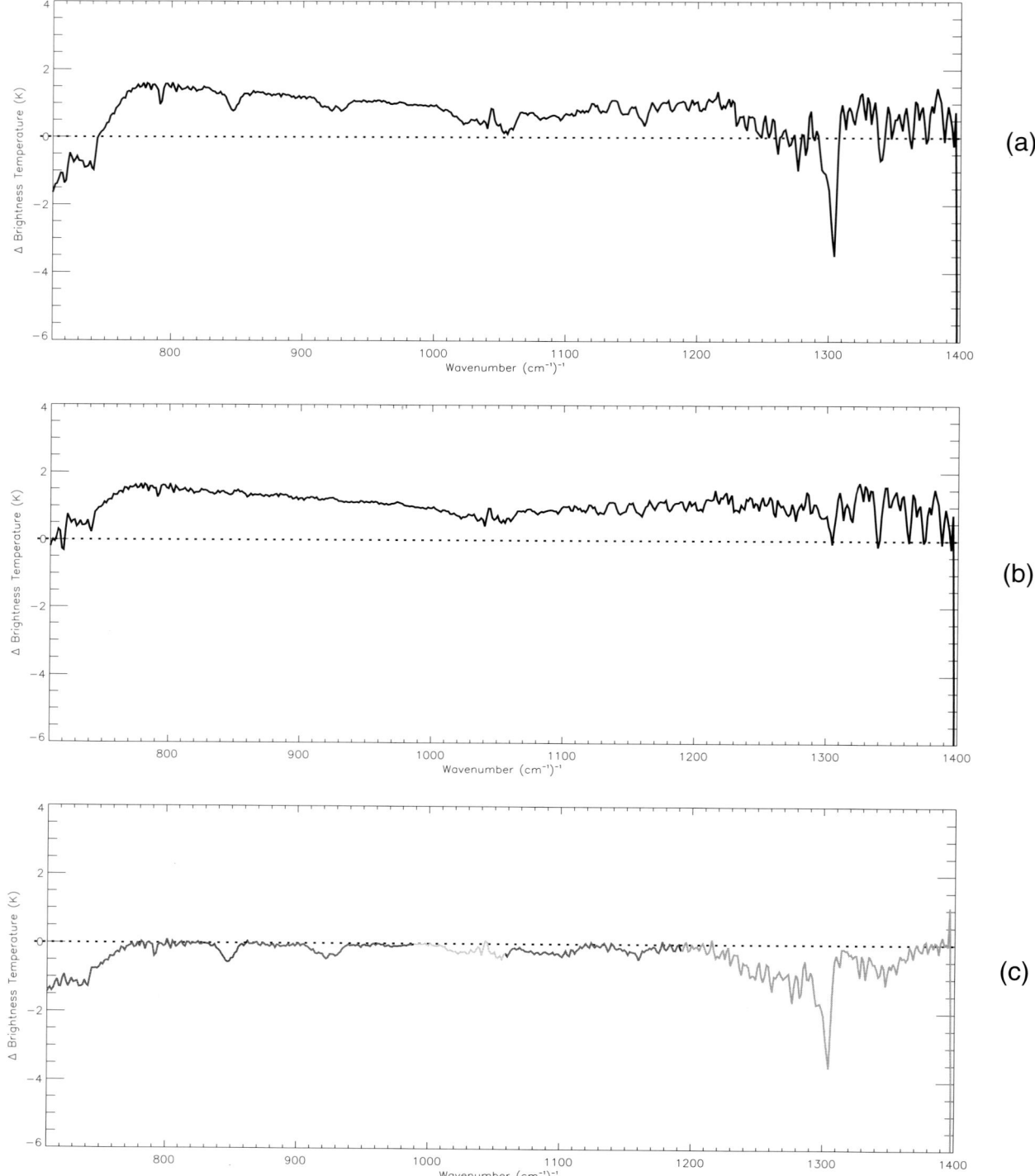

FIGURE 4 Line-by-line model simulations for central Pacific of difference spectrum 1997–1970.

the total calculated spectrum, using NCEP atmospheric parameters, and changes in composition from IPCC reports. Panel (b) shows the effect of changes in temperature and humidity only; and panel (c) shows the effect of the remaining trace gases only, except O_3, which was not changed. Thus, panel (c) shows the effects of CO_2, F-11 and F-12, HNO_3 and CH_4.

This analysis concludes with the comparison of the green curve in Fig. 3c with the top curve in Fig. 4a. The curves are extremely similar in all respects, the only significant differences being:

- In the experimental result, the difference in the window on the high-frequency side of the O_3 band at 1050 cm^{-1} is consistently lower than the difference on the low frequency side. This might be due to the effect of uncleared thin ice clouds in the observations (Smith *et al.*, 1998; Bantges, 2000).
- The magnitude of the negative signal in the CH_4 Q-branch in the experimental data (c. −6.5 K) exceeds that in the simulation (c. −3.5 K). This may be due in part to uncertainty in our knowledge of upper tropospheric methane concentrations at the two times. Little data exists on the methane distribution and trends in the upper troposphere.

We have carried out similar analyses for other regions of the globe: the central north Atlantic; the Sahara; the east and west Pacific; and the Indian Ocean. In each case, while the measured difference spectrum may show small departures from those shown in Fig. 1, the agreement with the simulated spectral differences is good. This gives us confidence that the differences between the 1970 IRIS and the 1997 IMG spectra that we observe are real.

Because we do not have a continuous data set in time, inter-annual variations might have nothing to do with long-term trends and may cause the observations. While we cannot, with the available evidence, refute this, there are two things that are consistent with this result not being due to interannual variability, and for it being evidence of longer-term trends.

- While the F-11 and F-12 features are right at the noise level of the observed difference spectra (about 0.5 K), the appearance of these features, especially the stronger F-11(CFCl$_3$) band at 850 cm^{-1}, can only be due to the long-term trend between 1970 and 1997: inter-annual variability in these well mixed gases is very small (c. 10 pptv^{-1}yr ; IPCC, 1996).
- Again, in the case of methane, inter-annual variability of methane concentrations or of mid-upper tropospheric temperature could not possibly cause the signal observed, which is unambiguously large and negative. Inter-annual variability of methane, which like the FCCs is well mixed, is very small and would not lead to anything like such a large signal (the long-term change between 1970 and 1997 is c. 300 ppb, while year to year changes are c.10 ppbv; IPCC, 1996). This is a definitive detection of the effect of a significant change in methane concentration, and a coincident change in temperatures, between 1970 and 1997.

4. Cloudy scenes

We have performed only an initial analysis of cloudy and all-sky (i.e. clear + cloudy) data. This indicates that the difference between global all-sky spectra in IMG and IRIS is heavily affected by cloud. Thus, while the typical magnitude of difference signal (see Fig. 3) for the clear sky data in the atmospheric window is of order +1 K, in the all-sky case differences across the whole window region from 700–1200 cm^{-1} are between −1 K and −2 K. For the cloudy scenes only, the window differences amount to as much as −4 K across the whole window. If confirmed, these initial results would indicate a significant global warming between 1970 and 1997 due to an increase in amount or height of cloud. In the Pacific case, it is quite possible that this could be due to the difference in ENSO phase between the two years (1970 was not an ENSO year, while in 1997 an ENSO event did take place). However, for the global average, it is doubtful that even a strong ENSO could give rise to a globally averaged depression of OLR across the whole window of several degrees.

This preliminary result is indicative, therefore, of a significant effect of cloud changes on the OLR spectrum between 1970 and 1997. Much more work is required.

F. SUMMARY

The climate change detection and attribution problem will require a combination of all possible methods available to us. We believe we have shown the power of using the outgoing thermal spectrum, and its trends, as one of those methods.

Using IRIS and IMG, we have observed global and regional change of OLR(v) between 1970 and 1997 which is consistent with increased concentrations of CO_2, on average a higher surface temperature, and increased concentration of CH_4. It is not possible to say unambiguously that this result is evidence of long-term climate change, because of the noncontinuous nature of the data, and also the possibility of inter-annual variability. However, inter-annual variability would not be sufficient to explain the observed spectral differences in the

F-11, F-12, and especially the CH_4 bands. These signals can only be accounted for on the basis of the magnitude of change which has occurred in the period 1970 to 1997. On the other hand, inter-annual variability (especially in the Pacific, with ENSO events) could account for the observed central Pacific window difference signal.

Simulations of the expected difference signals in $OLR(v)$ based on NCEP re-analysis atmospheric data agree well with observation, but require greater enhancement of CH_4 than measurements have indicated. There remains an uncertainty over an effect which causes a depression of the window difference signal on the high-frequency side of the ozone band compared with the low-frequency side. This may be due to the effect of uncleared thin ice clouds in the observations (Smith *et al.*, 1998; Bantges, 2000).

An initial analysis of cloudy scenes indicates that cloud effects between 1970 and 1997 data seem to be significant, working in the sense of increasing global warming (reducing the OLR) between 1970 and 1997. This result has yet to be confirmed, but would suggest an important contribution from the cloud–radiation feedback to observed climate change if it is true.

An increasing number of satellite instruments in the near future (AIRS, IASI, MODIS, SEVIRI) will provide more information on $OLR(v)$ in coming years, though accuracy needs to be high in order to make an impact. Moreover, the "leverage" between IRIS and IMG can only be improved at one end of the intervening period, of course. For many years, we will continue to rely on this extraordinary data set from IRIS to provide the other end of the "lever".

We believe that the case is very strong for flying an IRIS-like FTS spectrometer regularly on missions of opportunity: continuous measurement are not as important as regular measurements, perhaps one year of observations between one and three years apart. IRIS is a small, well-designed instrument that will not represent a large incremental cost on most missions, and should be carried as a passenger regularly.

ACKNOWLEDGEMENTS

The excellent work by Dr Helen Brindley and Ms Pretty Sagoo on this project is gratefully acknowledged by the author. Valuable discussions with Professor Richard Goody and Dr William Smith are also warmly acknowledged. Dr Richard Bantges advised on the spectral effects of cirrus clouds.

References

Bantges, R. J., 2000: Cirrus cloud radiative properties in the thermal IR, Ph.D. thesis, University of London.

Bates, J. J., X. Wu and D. L. Jackson, 1996: Interannual variability of upper tropospheric water vapour band brightness temperatures. *J. Climate*, **9**, 427–438.

Bess, T, D., G. L. Smith and T. P. Charlock, 1989: A ten year data set of outgoing longwave radiation from Nimbus 6 and 7. *Bull. Am. Met. Soc.*, **70**, 480–489.

Brindley, H. and J. E. Harries, 1998: The impact of far IR absorption on clear sky greenhouse forcing, *J. Quant. Spec. Rad. Trans.*, **60**, 151–180.

Brindley, H., 1998: An investigation into the impact of greenhouse gas forcing on the terrestrial radiation field: sensitivity studies at high spectral resolution. Ph.D. thesis, University of London.

Charlock, T. P., 1984: CO_2 induced climatic change and spectral variation in the outgoing terrestrial IR radiation. *Tellus*. **36B**, 139–148.

Clough, S., M. Iacono and J.-L. Moncet, 1992: Line-by-line calculations of atmospheric fluxes and cooling rates: Application to water vapor. *J. Geophys. Res.*, **97**, 15761–15785.

Doherty G. M. and Newell, R. E, 1984: Radiative effects of changing atmospheric water vapour. *Tellus*, **36B**, 149–162.

Edwards, D., 1992: GENLN2: A general line by line atmospheric transmittance and radiance model. Version 3.0. Description and users guide. NCAR/TN-367+STR, Boulder, Colorado.

Geer A. J., J. E. Harries and H. E. Brindley, 1999: Spatial patterns of climate variability in upper tropospheric water vapour radiances from satellite data and climate model simulations. *J. Climate*, **12**, 1940–1955.

Goody, R., R. Haskins, W. Abdou and L. Chen, 1996: Detection of climate forcing using emission spectra. *Earth Observ. Remote Sens.*, **13**, 713–722.

Hadley Centre, 1997, *Climate Change and its Impacts, a Global Perspective*. UK Met. Office, Bracknell, UK

Hanel, R., B. Schlachman, D. Rogers and D. Vanous, 1971: Nimbus 4 Michelson interferometer. *Appl. Optics*, **10**, 1376–1382.

Iacono, M. and S. Clough, 1996: Application of infrared interferometer spectrometer clear sky spectral radiance to investigations of climate variability. *J. Geophys. Res.*, **101**, 29439–29460.

IMG mission operation and verification committee, 1999: *Interferometric Monitor for Greenhouse Gases. IMG Project Technical Report* (H. Kobayashi, ed.) Komae-shi, Tokyo.

Intergovernmental Panel on Climate Change, 1990: *Climatic Change: The IPCC Scientific Assessment* (J. Houghton, G. Jenkins and J. Ephraums, Eds) Cambridge University Press, Cambridge.

Intergovernmental Panel on Climate Change, 1996: *Climate Change 1995: The Science of Climate Change* (J. Houghton, *et al.* Ed) Cambridge University Press, Cambridge.

Kiehl, J. T., 1986: Changes in the radiative balance of the atmosphere due to increases in CO_2 and trace gases. *Adv. Space Res.*, **6**, 55–60.

Sinha, A. and J. E. Harries, 1997: The earth's clear sky radiation budget and water vapour absorption in the far IR. *J. Climate*, **10**, 1601–1614.

Slingo, A. and M. Webb, 1996: The spectral signature of global warming. *Q. J. R. Meteorol. Soc.*, **123**, 293–307.

Smith, W. L., S. Ackerman, H. Revercomb, H. Huang, D. H. DeSlover, W. Feltz, L. Gumley and A. Collard, 1998: Infrared spectral absorption of nearly invisible cirrus. *Geophys. Res. Lett.*, **25**, 1137–1140.

Soden, B. J., 1997: Variation in the tropical greenhouse effect during El Niño, *J Climate*, **10**, 1050–1055.

Vincent, D. G., A. Fink, J. M. Schrage and P. Speth, 1998: High and low frequency intraseasonal variance of OLR on annual and ENSO timescales. *J. Climate*, **11** 968–985.

Wielicki, B. A., R. D. Ces, M. D. King, D. Randall and E. F. Harrison, 1995: Mission to planet Earth: role of clouds and radiation in climate. *Bull. Am. Meteorol. Soc.*, **76**, 2125–2153.

Part 3. The Atmosphere and Oceans

Atlantic Air–Sea Interaction Revisited
M. J. Rodwell

The Monsoon as a Self-regulating Coupled Ocean–Atmosphere System
P. J. Webster, C. Clark, G. Cherikova, J. Fasullo, W. Han, J. Loschnigg and K. Sahami

Laminar Flow in the Ocean Ekman Layer
J. T. H. Woods

Ocean Observations and the Climate Forecast Problem
C. Wunsch

Atlantic Air–Sea Interaction Revisited

M. J. Rodwell

Hadley Centre for Climate Prediction and Research, Met Office, Bracknell, UK

Bjerknes published a landmark paper on Atlantic air-sea interaction in 1964. Here the topic is revisited with the benefit of better datasets and statistical techniques and with atmosphere and coupled ocean–atmosphere models. It is found that the UK Met. Office's HadCM3 coupled model represents well the aspects of seasonal timescale atmosphere-to-ocean forcing identified by Bjerknes. In agreement with other studies, it is shown that the strongest forcing is in winter and the oceanic thermal anomalies created in winter persist and decay into the following spring and summer. Part of the extra-tropical thermal anomaly appears to be preserved below the shallow summertime oceanic mixed layer and re-emerges in the following autumn. Analysis of 1948–1998 observational data suggests some atmospheric seasonal predictability based on a knowledge of preceding North Atlantic sea surface temperatures (SSTs); in particular, predictability of the winter "North Atlantic Oscillation" based on previous May or June SSTs with a correlation skill of 0.45. The forcing mechanisms that lead to the seasonal predictability will be essential ingredients in the understanding of decadal variability. Analysis of each simulation period of the ocean–atmosphere model does not identify such forcing mechanisms and may imply a poor representation of decadal variability in the coupled model. On the other hand, the observed predictability for each season is at the top end of, but not outside, the range of predictability suggested by individual simulations of the UK Met. Office's HadAM2b atmosphere-only model. This is the same model that was used by Rodwell *et al.* (1999) to investigate the *potential* predictability of the winter NAO. The present results therefore offer observational evidence that tend to confirm the level of potential predictability found in that study. By using the most predictable patterns from the observations and the model, potential predictability can be increased further to give a correlation skill of 0.63. Ultimately achievable real predictability for the winter NAO may therefore lie in the range 0.45 to 0.63.

A. INTRODUCTION

In his observational study, Bjerknes (1964) (hereafter B) demonstrated convincingly the role of atmospheric forcing of interannual variations in North Atlantic surface temperatures (B, Fig. 14) through the action of surface heat fluxes (particularly latent heat fluxes). At longer timescales, he suggested that sea-surface temperature (SST) anomaly patterns reflected wind-stress-forced decadal changes in the intensity of the subtropical oceanic gyre circulation (B, Fig. 21). These hypothesized timescale-dependent SST responses have since been demonstrated in observations (Kushnir, 1994) and in models (Visbeck *et al.*, 1998). The ocean's role in these mechanisms is "passive" in that it is simply responding to forcing by the atmosphere (Hasselmann, 1976; Frankignoul and Hasselmann, 1977). However, for the global conservation of meridional heat transport, Bjerknes expected that North Atlantic decadal variability would in turn feedback onto the atmospheric circulation. This "active" role for North Atlantic SSTs has been an integral component of ocean–atmosphere coupled mode hypotheses (Kushnir, 1994; Latif and Barnett, 1994; Grötzner *et al.*, 1998) and underlines the importance of studying Atlantic air–sea interaction with coupled ocean–atmosphere models.

The North Atlantic Oscillation (NAO) is the dominant mode of interannual mean sea-level pressure (MSLP) and 500 hPa geopotential height (Z500) variability in winter and spring in the North Atlantic/European (NAE) region (defined here as the region 10°N–80°N, 90°W–45°E). The NAO (see e.g. Wallace and Gutzler, 1981) incorporates a north–south dipole in pressure and one index of the NAO can be defined as the mean sea-level pressure difference between Ponta Delgada (Azores) and Stykkisholmur (Iceland). High index values are manifested with stronger westerly flow across the Atlantic, an intensification and northward shift of the storm track (Rogers, 1990) and, for example, warmer and wetter winters in northern Europe (Walker and Bliss, 1932; Hurrell, 1995; Hurrell and van Loon, 1997). Figure 1 (upper thick curve) shows the observed December–February NAO index from the 50-year period 1947/48 to 1996/97. Variability of the NAO index occurs on all timescales from synoptic to multi-decal. Here, one can see multidecadal variability with a fall of the index to the 1960s and then a rise into the

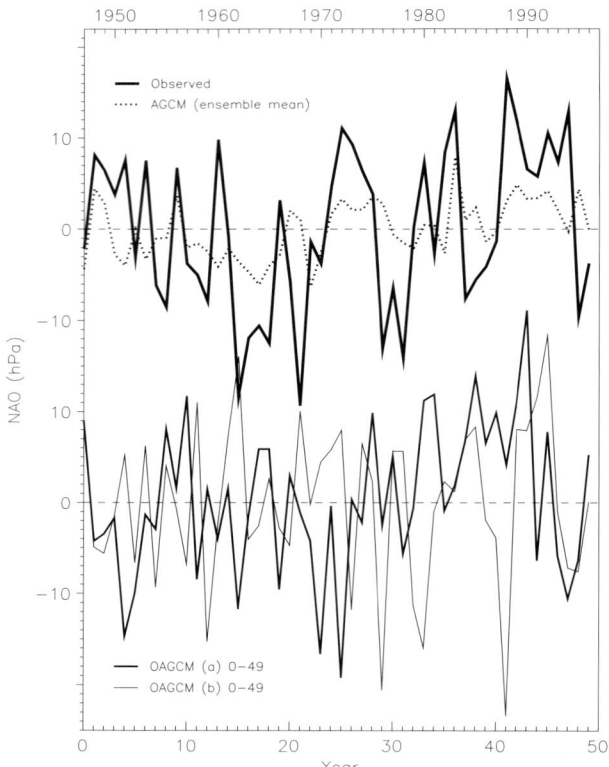

FIGURE 1 December–February North Atlantic Oscillation (NAO). Upper time series are from 50 years of observations (1947/48–1996/97) (thick curve) and the ensemble mean of six atmospheric general circulation model (AGCM) simulations forced with GISST3.0 SSTs for the same period. The lower curves show the first 50 years of two 100-year periods from a coupled ocean–atmosphere general circulation model (OAGCM) simulation with no anthropogenic forcings applied. The NAO index is defined here as the difference in mean sea-level pressure between Ponta Delgada (25.7°W, 37.8°N) in the Azores and Stykkisholmur (22.7°W, 65.1°N) in Iceland. Modelled pressures are first interpolated to these locations. The year corresponds to the December of each season.

1990s. An important question at present is the extent to which the recent multidecadal rise can be attributed to anthropogenic effects (Osborn et al., 1999). "Decadal variability" can also be seen with, for example, two oscillations between about 1972 and 1988 (i.e. an eight-year period). Higher frequency variability is also evident. Overall, for the last 100 years, the power spectrum of the NAO index is weakly red and this may indeed suggest a partly active rather than purely passive role for the ocean.

The possibility that the ocean may play an active role has motivated many past studies. Some of these studies have demonstrated a degree of oceanic forcing of the mean or transient extratropical circulation over the North Atlantic (Namias, 1964; Ratcliffe and Murray, 1970; Namias, 1973; Palmer and Sun, 1985; Rodwell et al., 1999; Bresch and Davies, 2000; Mehta et al., 2000; Latif et al., 2000; Robertson et al., 2000). The SSTs thought to be actively involved have been in the NAE region and elsewhere. Others show a nonexistent or weak (Saravanan, 1998) response.

Analysis of variance suggests that most of the interannual variability in the NAE region is internal to the atmosphere (Davies et al., 1997; Rowell, 1998). A consequence of this result could be that the NAE climatologies of atmospheric models may look quite reasonable even if they do not represent well the active role of the ocean. This may be one reason for the mixed conclusions presented above.

If progress is to be made in the quantification of ocean-forced atmospheric variance and in the understanding of predictability, some form of "model predictability" must be compared with predictability based on observations. However, predictability is extremely difficult to determine from observational data alone, since there is only one realization available, in which there are many processes occurring at the same time. Here a test is developed that attempts to compare models with observations. One model that is compared with the observations is the atmospheric model used by Rodwell et al. (1999) (hereafter R). The winter NAO index from the ensemble mean of six simulations of this model (dotted curve in Fig. 1) had a correlation with the observed NAO index of 0.41. This correlation is significant at the 1% level when tested against an AR(1) model. One question that the present study aims to address is whether this level of potential predictability is reasonable based on observational evidence and how it translates to real predictability when the SSTs themselves need to be predicted.

In Section B, the data sources and models are described. Section C explains the main analysis method used. Section D investigates atmosphere-to-ocean forcing in the winter and serves to demonstrate the analysis method on a topic which is already quite well understood. Section E investigates ocean-to-atmosphere forcing and seasonal atmospheric predictability with the same analysis method. Section F attempts to quantify potential predictability of the atmosphere (when SSTs are required to be known in advance). Section G gives conclusions and a discussion.

B. DATA AND MODELS

The observational data used is monthly and seasonal mean 500 hPa geopotential height (Z500) data for 1948–1998 and monthly mean surface latent heat flux (LHF) data for 1958–1998 from NCEP reanalyses (Kalnay et al., 1996), seasonal average mean sea-level pressure (MSLP) data for 1948–1994 from GMSLP2.1f (Basnett and Parker, 1997) and monthly mean SSTs for 1948–1998 from GISST3.0 (Rayner et al., 1998). Atmospheric seasonal predictability results have been checked by looking at MSLP data.

The atmospheric general circulation model (AGCM) used here is the UK Met. Office's HadAM2b model (Hall et al., 1995). Data is from the period 1947–1997 of a six-member ensemble of 130-year atmospheric simulations forced with "observed" GISST3.0 SSTs and sea-ice extents (Rayner et al., 1998). Seasonal mean Z500 and MSLP and monthly mean SSTs are used.

The coupled ocean–atmosphere general circulation model (OAGCM) is the UK Met. Office's HadCM3 model. This is the first to not use flux adjustment and incorporates a more recent version of the atmospheric model than HadAM2b. Data comes from a ~1000-year period of a control simulation of this model. Two 100-year periods have been used to provide four 50-year simulation segments with which to compare with the observations. For demonstrative purposes, Fig. 1 (lower curves) shows the OAGCM's NAO index from the first 50 years of the two 100-year segments. The period partly depicted by the thicker curve was chosen because it displayed similar multidecadal trends to those seen in the recent record. Since the model simulation was made under nonanthropogenic conditions, it is clear that anthropogenic forcing is not *necessarily* required to achieve the trends observed over the last 50 years. This result is different from that given by Osborn et al. (1999) based on an earlier (but not necessarily poorer in this context) version of the coupled model (HadCM2). The second 100-year period (partly depicted by the thin curve) was chosen at random from the remaining data. Monthly and seasonal mean Z500, and monthly mean LHF and SST data are used.

To keep comparisons as consistent as possible, all data are first gridded onto the atmospheric grid of the AGCM/OAGCM (3.75° longitude × 2.5° latitude).

C. THE ANALYSIS METHOD

The analysis method that is developed and used here is based on singular value decomposition (SVD). SVD (Bretherton et al., 1992) identifies pairs of patterns in two fields that "maximally covary". It is analogous to empirical orthogonal function (EOF) analysis which operates on a single field of data. This study is only interested in the first pair of patterns which are chosen to maximize the covariance, σ_1, of their time coefficients. Of importance in this investigation is the first squared covariance fraction: $SCF = \sigma_1/\Sigma_k \sigma_k$. The higher this fraction, the more dominant the first mode is at explaining joint variability. Area weighting is employed throughout.

In the context of this study, SVD analysis alone is not enough. First, the SVD procedure will produce statistically covarying patterns even if there is no physical connection between the fields (i.e. no "cause" and "effect"). Second, even if there is a physical connection between the two fields, SVD will not readily differentiate between cause and effect.

The problem of differentiating between cause and effect is overcome by applying a time lag to the data before the SVD analysis is performed. The procedure is called here lagged SVD (LSVD). The use of a lag leads to an indication of cause and effect but other tests may have to be applied to confirm this (see later).

The question of whether there is a physical connection between the two fields is addressed in a similar way to Czaja and Frankignoul (1999) by assuming as a null hypothesis that there is no physical connection. If this is the case, then temporarily shuffling the data of the two fields and applying the LSVD analysis should not lead to a statistically different outcome. Here, the combined action of shuffling and applying the LSVD is performed 100 times to produce a "Monte Carlo"-style test (upper shaded region in Fig.2). The probability density function (pdf) of the SCF that arises is compared to the true SCF, giving a significance level (SL) for a physical connection. The *smaller* the SL, the *more* physically significant are the patterns. The method of shuffling is the same as that in Czaja and Frankignoul (1999), namely the years of the 2 fields are randomly shuffled with the constraint that there must be a minimum separation of 2 years between the data from the two fields. Another method of shuffling that preserves the autocorrelation of the individual fields has also been tested and leads to the same conclusions.

The analysis is taken a stage further by attempting to determine the usefulness of any forecast system based on the LSVD technique applied to a single observed or modelled realization. The time series that accompany the patterns are likely to correlate well because of the nature of the SVD procedure. Hence, it would be wrong to use the correlation between these time series to quantify predictability. Instead, two "cross-validated" time series are constructed by projecting in turn the data for each particular year onto the patterns obtained from the LSVD analysis applied to the remaining 49 years of data. The time series based on the leading field is known here as the "predicted time series". The time series based on the lagging field is the "real time series". For display purposes, it is the time series that are standardized (so that they have a standard deviation of 1) and the patterns (which are from the full 50-year LSVD analysis) that show the magnitude

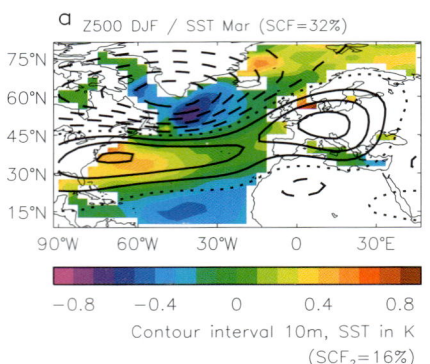
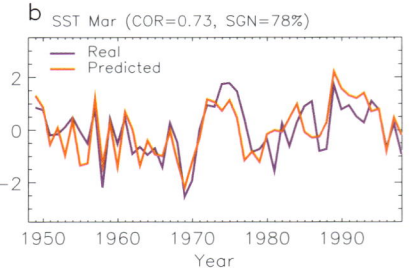
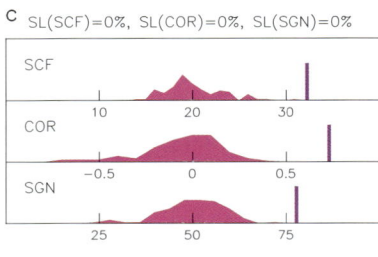

FIGURE 2 Results from a lagged singular value decomposition (LSVD) analysis of December–January (DJF) mean NCEP 500 hPa geopotential height (Z500) leading March GISST3.0 sea-surface temperature (SST) for the period 1948–1998. (a) The patterns of SST (shaded, in K) and Z500 (contoured, with interval 10 m) for one standard deviation of the first LSVD mode. The squared covariance fraction (SCF) for the mode is quoted in the title. (b) The standardized, cross-validated "real" and "predicted" time series for the SST pattern. The correlation between the two time series (COR) and the percentage of time that the correct sign is predicted (SGN) are given in the title. The year corresponds to the month of March. (c) The estimated probability density functions of SCF, COR and SGN derived from the LSVD analysis applied to temporally shuffled data. The significance levels (SLs) of the true values are also given. The squared covariance fraction for the second mode, SCF_2, is given at the bottom of the figure.

corresponding to one standard deviation in the time series. The correlation between the two time series (COR) and the percentage of time that the predicted time series has the same sign as the real time series (SGN) are used to assess the skill of the forecast. In a similar way to the Monte Carlo test of the SCF, Monte Carlo tests of the COR and SGN are performed by applying the above procedure to the temporally shuffled datasets.

By applying the procedure to observations and a multiple of model simulations, it is possible to check whether or not the observational SCF, COR and SGN results fall within the envelope of model results and hence whether the model ocean (atmosphere) is correctly sensitive to atmospheric circulation (SSTs).

Statistical tests often provoke as many questions as they answer. Further tests and modifications to the above method have been made although, for brevity, some of these are not mentioned here.

D. ATMOSPHERIC FORCING OF NORTH ATLANTIC SEA SURFACE TEMPERATURES

First, the lagged SVD technique is applied to observed seasonal mean Z500 and monthly mean SST in the NAE region with Z500 leading SST. Figure 2a shows the DJF mean Z500 pattern (contoured) that leads the March mean SST pattern (coloured). The upper pdf in Fig. 2c corresponds to the Monte Carlo SCFs. It is clear from this pdf that the case for a physical link is strong in winter. The patterns identified are an NAO-like westerly enhancement over the North Atlantic and a "tripole" in SSTs. The patterns look remarkably similar to Bjerknes' interannual anomaly plot (B, Fig. 14). The SST pattern is also very similar to SSTs induced by high-frequency NAO-forcing in Visbeck et al. (1998).

The patterns and SCFs from a similar analysis of the four OAGCM simulation periods (not shown) are in good agreement with those of the observations and are all significant at the 0% level (based on 100 shufflings).

Latent heat fluxes are generally the dominant term in the surface heat flux (Sverdrup, 1942) with sensible heat fluxes generally having the same sign. The role of LHFs is highlighted in terms of lagged correlations with the NAO in Fig 3a (see caption for essential details). The strongly negative correlations seen for winter months at lag 0 represent the LHF forcing of the "tripole" pattern by the "NAO". This forcing is very much as described by B. Weakly negative correlations at lags of +1 and −1 month represent the same forcing but rely on the weak monthly autocorrelation of the NAO. At lags of around 4–6 months after January, February and March, positive correlations are seen. These represent the damping through LHFs of the SST anomalies created in the previous winter.

A very similar set of lagged correlations is found with the OAGCM data (not shown). This suggests that the model is realistically capturing the physics of this forcing.

Figure 3b shows an autocorrelation plot of SSTs. See figure caption for essential details. Correlations at lag 0 are necessarily equal to one and there is some form of symmetry to the plot. Here, the SST anomalies created

D. ATMOSPHERIC FORCING OF NORTH ATLANTIC SEA SURFACE TEMPERATURES 189

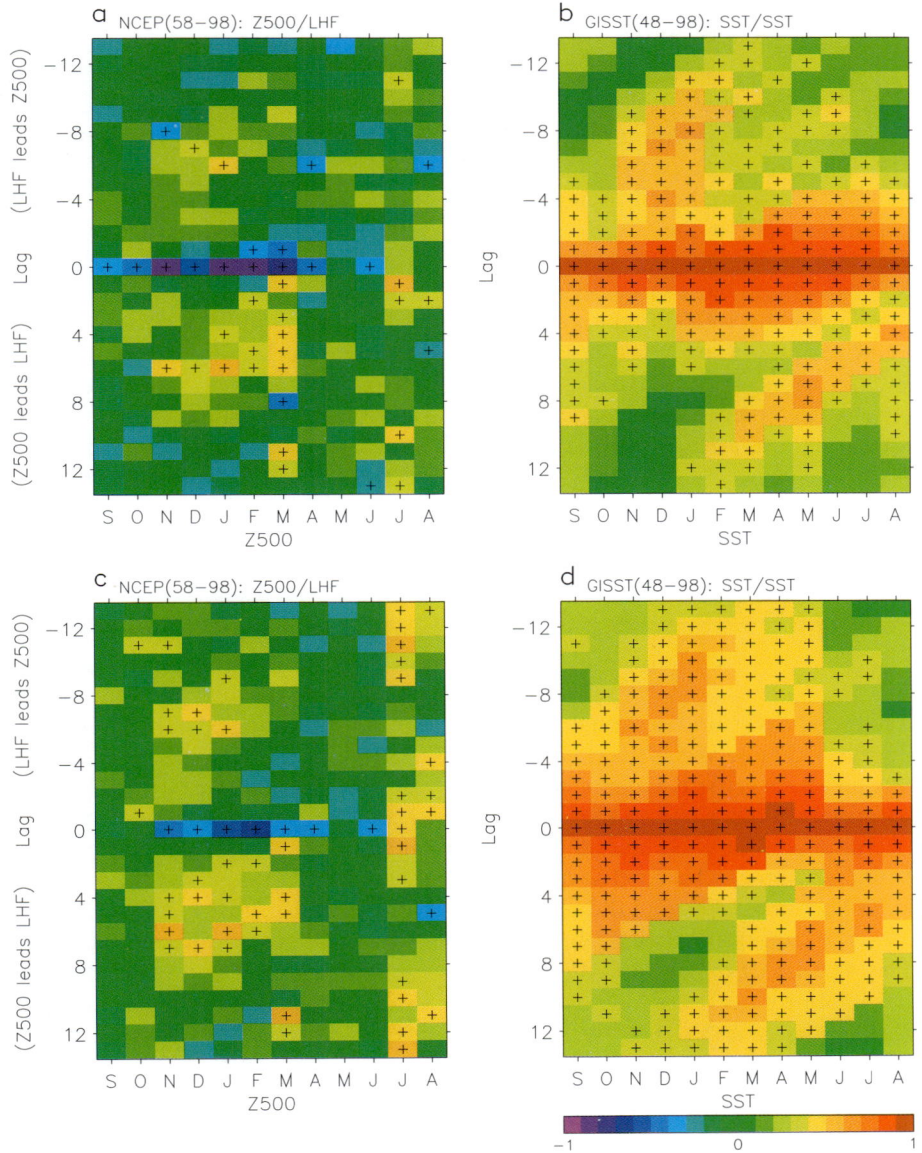

FIGURE 3 Lagged correlation plots. (a) Monthly NCEP Z500 data is projected onto the Z500 pattern in Plate 18a and monthly NCEP surface latent heat flux (LHF) data is projected onto the SST pattern in Plate 18a. Correlations between these time series are plotted. The square corresponding to February Z500 and a lag of −4, for example, represents the correlation between February Z500 and the previous October LHFs. (b) Monthly GISST3.0 SSTs are projected onto the SST pattern in Plate 18a. (Auto) correlations between the time series are plotted. (c) As (a) but using the Z500 and SST patterns in Plate 20d. (d) As (b) but using the SST pattern in Plate 20d. Crosses indicate correlations that are significant at the 5% level when tested against a 1000-member Monte Carlo AR(1) process.

by the strong winter and early spring forcing are seen to persist and decay until around June. Importantly, in November, the correlations restrengthen. This is consistent with the idea that thermal anomalies can get submerged and preserved in May or June below the shallow summertime mixed layer and then re-emerge as the mixed layer deepens in the following winter (Alexander and Deser, 1995; Watanabe and Kimoto, 2000). The symmetry of the plot leads to an equivalent re-emergence signal being seen at negative lags.

Again, the pattern of correlations is similar for the OAGCM simulations (not shown) although SSTs would appear to be a little more persistent than in the observations.

Although it has been demonstrated statistically and physically that the patterns in Fig. 4a are connected with a causal relationship, an important question remains: How useful is the knowledge of this link for prediction purposes?

Figure 2b shows the standardized, cross-validated "real" and "predicted" time series based on the observational data. They have a correlation of 0.73, the correct sign of the tripole pattern is predicted 78% of the time and (for example) the peak in the tripole in the mid-1970s is well predicted.

The mechanisms highlighted here for ocean-to-atmosphere forcing are generally well understood and results with the present technique are consistent with preceding papers on this topic, such as B and Daly (1978). Now the sense is reversed to investigate the less well understood situation with ocean leading atmosphere.

E. NORTH ATLANTIC SEA SURFACE TEMPERATURE FORCING OF THE ATMOSPHERE

1. Observational Evidence

As before, monthly mean SSTs and seasonal mean Z500 data are used in the SVD analysis. This time, it is the SSTs that lead. The aim here is to investigate *instantaneous* ocean-to-atmosphere forcing and so there is an implicit assumption being made that SST anomalies in the month(s) prior to the season can be used as a reasonable prediction for SST anomalies during the season. In other words, there is a temporal "oceanic bridge" that can be utilized by the LSVD analysis. Although this assumed SST predictability need not be based on simple persistence of anomalies, Fig. 3b suggests that, at least for the NAO-forced tripole pattern, there is appreciable persistence in the case of spring and summer, although less so for autumn and winter. For winter, consistent with the re-emergence hypothesis, February-observed SSTs actually correlate better with the preceding May SSTs than they do with the preceding November SSTs in Fig. 3b.

In addition to an oceanic bridge, there is the possibility of a temporal "atmospheric bridge" due to persistence in the atmosphere. If there was appreciable atmospheric persistence, then atmosphere-to-ocean forcing (e.g. in November) could conceivably produce a signal in the (November SST leading DJF Z500) LSVD results. Hence the LSVD technique at short lags may emphasize, but not isolate, one direction of forcing. Below, checks are made for persistence in the atmosphere, longer lags are also used in the LSVD analysis, and conclusions are verified with the analysis of LHF data.

Figure 4 (left-hand panels) shows the patterns from the LSVD analysis of monthly mean SST leading seasonal mean Z500. The significance levels (right-hand panels, upper pdfs) suggest that these patterns do represent some physical connection between SSTs and Z500. Wintertime is perhaps the least clear case (SL(SCF) ≈ 13%) but it will be argued that here too there is a physical linkage.

For all seasons except summer, the Z500 pattern projects strongly onto the NAO. In summer, the area-weighted signal over North Africa may be an important part of the pattern. The magnitudes of the Z500 patterns are broadly in agreement with estimates of Frankignoul (1985) and Palmer and Sun (1985). In each season, the SST patterns look something like the NAO-forced tripole pattern but there are clear differences between the SST patterns leading (Fig. 4d) and lagging (Fig. 2a) the winter atmosphere. In particular the warm SST region at around 45°N in Fig. 4d extends further north and east than in Fig. 2a—extending along, and to the south of, the westerly jet and north Atlantic storm track. This pattern, which is very similar to Bjerknes' Fig. 21 of long-term trends from low to high NAO index and the low-frequency results of Visbeck et al. (1998), may reflect changes in the strength of the oceanic gyres.

The correlations between "real" and "predicted" cross-validated time series (Fig. 4 centre panels b, e, h, k) and the percentages of time that the correct sign of the patterns is predicted are quite substantial in all seasons except winter. The pdfs of COR and SGN (right-hand panels, lower two pdfs) also emphasize the importance of a physical linkage. Although winter appears to be less predictable based on November SSTs, the 1970s peak, for example, is clearly "forecast".

a. SON

The patterns for SON remain very similar when using SSTs at all lags to around 4 or 5 months, although significance levels and COR values deteriorate gradually as would be expected. For July SST leading SON Z500, the significance level is 5% and COR = 0.32. There appears to be very little atmospheric persistence between SON and previous months. For example, the correlation between the strength of the Z500 pattern (Fig 4a) in August and SON is −0.03. Detrending the initial data made practically no difference to the patterns, significance levels, COR or SGN for this season. Hence predictability for this season is likely to be due to ocean-to-atmosphere forcing. Note that the patterns agree well with those of Czaja and Frankignoul (1999) with "SST ASO leading Z500 OND".

b. DJF

Atmospheric persistence between November and DJF (0.31) is higher than the predictability from the

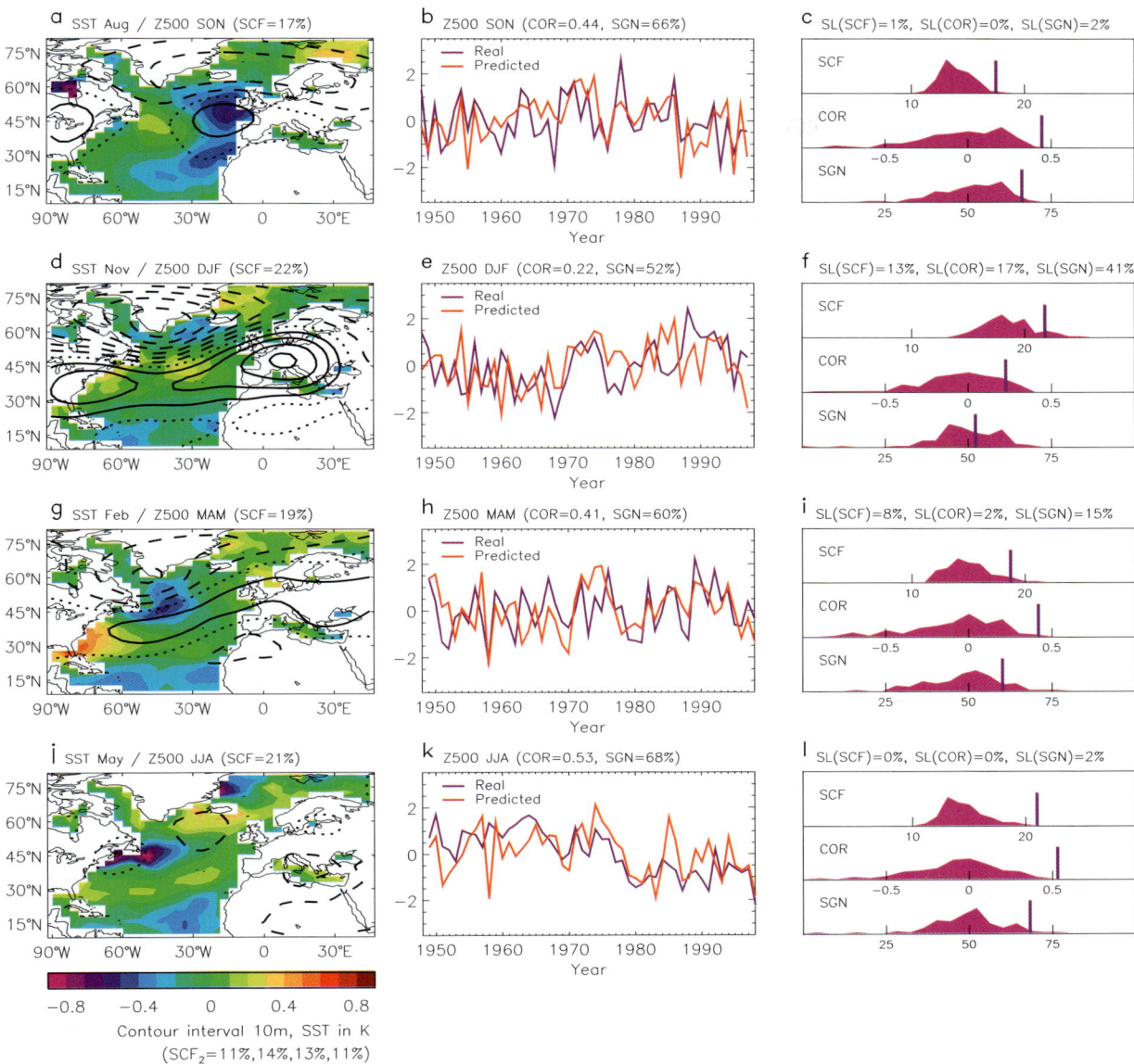

FIGURE 4 As Plate 18 but for the LSVD analysis of monthly mean SST leading seasonal mean Z500. (a–c) August SST leading SON Z500; (d–f) November SST leading DJF Z500; (g–i) February SST leading MAM Z500; (j–l) May SST leading JJA Z500. The years in the time series plots correspond to the first month of the Z500 season.

LSVD analysis. However, there is no atmospheric persistence between October and DJF, whilst LSVD predictability increases slightly to 0.28 and the patterns remain very similar. As with SON, detrending the initial data made no practical difference in this season. Hence it would appear that ocean-to-atmosphere forcing is the dominant reason for the suggested predictability.

The SST autocorrelations shown in Fig. 3b suggested that there could be better winter predictability based on SSTs from around May. The SST pattern that was used in that lagged-correlation analysis was the response to atmospheric forcing. Clearly it would be better here to use the SST pattern for ocean leading atmosphere (Fig. 4d). Fig. 3c, d shows the lagged-correlations results when the patterns are taken from Fig. 4d Figure 3d again shows high correlations between winter SSTs and those around seven months earlier. Figure 3d shows the lagged correlations between Z500 and LHFs. Since the patterns used from Fig. 4d are somewhat similar to those in Fig. 2, it is the *differences* between Fig. 3a and Fig. 3c that are highly relevant for suggesting cause and effect. For example, there are still negative correlations between Z500 and LHFs at zero lag in winter (Fig. 3c) but these are somewhat weaker than in Fig. 3a, since

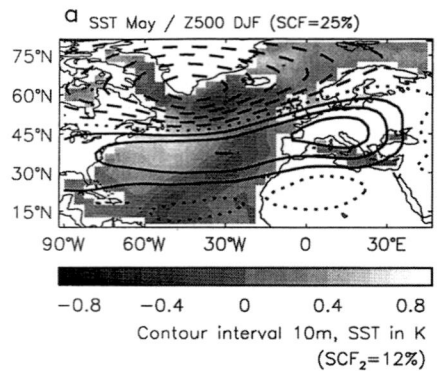 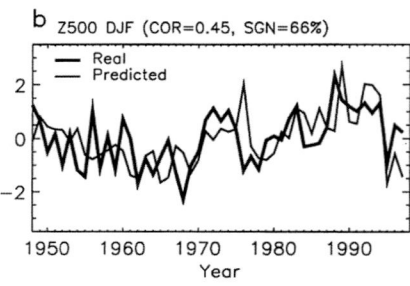 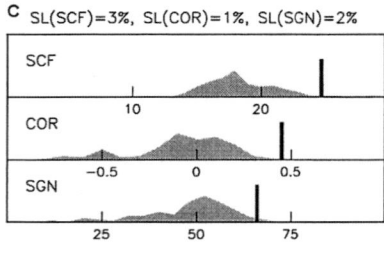

FIGURE 5 As Fig. 4d–f but with May SST leading DJF Z500.

the focus of the analysis has changed from atmosphere-leading to ocean-leading. In the same way, there are now quite strong positive correlations of up to 0.42 at lags of 6 or 7 months before winter. Positive correlations are suggestive of ocean-to-atmosphere forcing, although they may also arise if the patterns in Fig. 4d had occurred by chance. A stronger test is whether the LSVD analysis can itself identify winter predictability based on the previous May SSTs.

Figure 5 shows the analysis results for May SSTs leading DJF Z500. The SST pattern again shows strong meridional gradients along the North Atlantic storm track. The patterns are more physically significant (3%) than at lag −1 (13%). The correlation between "real" and "predicted" time series is 0.45 and the correct sign of the NAO pattern is predicted 66% of the time. The result using June SSTs is quite similar, with COR = 0.35. With SSTs in July and August the correlation is not significant, although the patterns that arise are very similar for all lags up to around 9 months. Since the May results are not very sensitive to detrending of the initial data (COR dropped from 0.45 to 0.34 but SGN increased slightly from 66% to 67%) and, at such a long lead time, atmospheric persistence is not important, the predictability is very likely to come from the oceanic bridge and ocean-to-atmosphere forcing.

These results imply that the submergence/re-emergence hypothesis could play a potentially important role in long-range forecasts for winter. If this is the case, then similar or improved winter forecasts may be possible based on a knowledge of subsurface temperatures in summer and autumn.

Figure 3c now implies more strongly a role for (instantaneous) LHFs in this ocean-to-atmosphere forcing. The fact that negative correlations between Z500 and LHFs are seen at lag zero is because atmospheric forcing of the ocean masks the signal of ocean-to-atmosphere forcing. As with R and Palmer and Sun (1985), the magnitude of the LHF anomalies is not sufficient to account for the atmospheric temperature anomalies associated with the Z500 pattern but the implication is that they may provide a "trigger" for further mean or transient circulation changes.

Another consequence of these results is that whilst Atlantic air–sea interactions and the submergence and re-emergence of thermal anomalies may lead to a reddening of the NAO spectrum, it is difficult to see how there can be an overall positive feedback between the NAO and the tripole SSTs, since opposing LHFs are involved in the two directions of forcing.

c. MAM

Secular trends again have practically no impact on the results for this season. There is quite strong atmospheric persistence between MAM and the preceding 3 months. For longer lags, the oceanic bridge fails (Fig. 4b) due, presumably, to the strong wintertime forcing of SSTs. Hence for this season, the LSVD results are harder to interpret and it is difficult to quantify the active role of the ocean. One view might be that the higher atmospheric persistence is itself at least partly the result of ocean-to-atmosphere forcing although this is clearly speculative.

The patterns in Fig. 4b correspond well to the "SST FMA/Z500 MAM" patterns given by Czaja and Frankignoul (1999).

d. JJA

The atmospheric pattern for this season (Fig. 4j) shows considerable persistence which outweighs the LSVD predictability. A large part of this "persistence" appears to be due to secular trends. When trends are removed from the initial data before the LSVD analysis is performed, there is a small drop in COR from 0.53 to 0.47 and the pattern no longer has a strong signal over North Africa. The atmospheric persistence of the new pattern is reduced considerably so it would appear that there may well be ocean-to-atmosphere forcing in this season but it is difficult to quantify this with any certainty.

e. Summary of Observational Evidence

Determining predictability from observational data is as difficult as it is essential. The above analysis appears to have gone some way to demonstrating that the North Atlantic (10°–80°N) does play an active as well as passive role in Atlantic air–sea interaction. Figure 6 summarizes the COR values for the observations with SSTs leading by 1 month (thick solid curve) and with May SSTs leading DJF Z500 (square box). The values shown for autumn and winter in particular, would appear to be almost purely due to ocean-to-atmosphere forcing.

Seasonal predictability of the NAO with a correlation skill of around 0.45 may be useful but one should bare in mind that the NAO only accounts for perhaps 35% of seasonal mean variance (based on EOF1 of DJF Z500 in the NAE region). There may, of course, be additional predictability of other (higher) modes of variability and the NAO may additionally be influenced by other SST regions. In addition, the seasonal mean SSTs may be more predictable with a knowledge of subsurface conditions and oceanic dynamics. Hence, these estimates of predictability should perhaps be viewed as lower bounds for ultimately achievable predictability. In the next sections, ocean-to-atmosphere forcing is assessed in the two models and seasonal predictability is re-evaluated.

2. Model Results

a. OAGCM

The significance levels for the first SCF from the observations and from the four simulation periods of the OAGCM are given in Table 1. It is clear that the OAGCM does not give physically significant patterns when the ocean leads the atmosphere. This is equally the case for the simulation periods with high multi-decadal variability of the NAO index (as seen in the observational record) as it is for the other periods. One partial explanation could be that there is weaker atmospheric persistence in the OAGCM (for example the correlation between the time series of EOF1 of MAM Z500 and the time series of the same pattern of February Z500 is 0.41 in the observations but only 0.32, 0.32, 0.25, 0.20 in the four OAGCM simulation periods, respectively). However, the observational results, particularly for SON and DJF, did not rely on atmospheric persistence. In addition, since the magnitude of interannual variability in the OAGCM is quite reasonable, it is not likely that this could be unduly masking the ocean-to-atmosphere signal. Hence it would appear that ocean-to-atmosphere forcing in the NAE region is too weak or not well represented in the OAGCM.

Figure 6 summarizes the COR values for the four OAGCM periods (thin solid lines). These values are all less than those of the observations and not significant when tested against an AR(1) model.

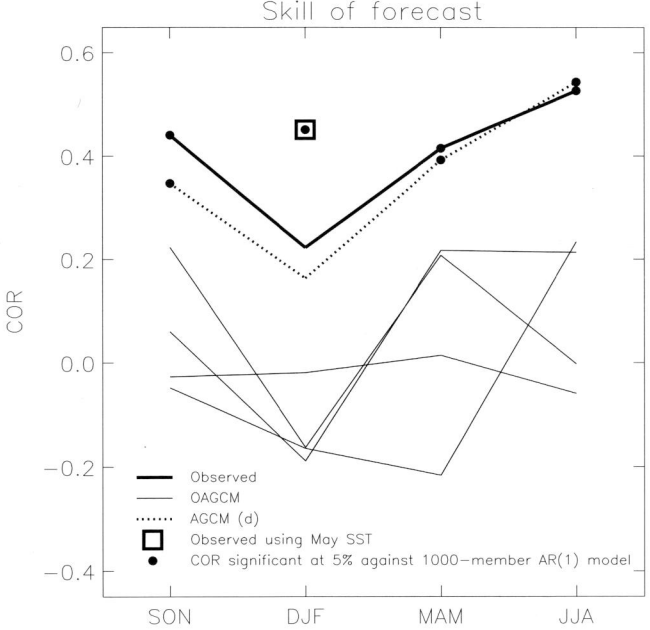

FIGURE 6 The correlation (COR) results from the analysis shown in Fig. 4 applied to NCEP/GISST observations, and for similar analyses applied to the four OAGCM simulation periods and the AGCM(d) simulation. Also plotted is the COR value for when the analysis is applied to May SSTs leading DJF Z500 in the observations.

TABLE 1 Significance Levels of the First Lagged SVD Mode in the NAE Region of Monthly SST Leading Seasonal Z500

Data source	Period	SST/Z500 significance level (%)			
		Aug/SON	Nov/DJF	Feb/MAM	May/JJA
NCEP/GISST	1948–1998	1	13	8	0
(detrended)		0	13	2	5
OAGCM(a)	0–50	45	61	32	56
OAGCM(a)	50–100	18	81	16	40
OAGCM(b)	0–50	69	80	66	17
OAGCM(b)	50–100	28	90	22	20
AGCM(a)	1947–1997	48	15	3	21
AGCM(b)	1947–1997	43	1	22	2
AGCM(c)	1947–1997	64	8	50	6
AGCM(d)	1947–1997	6	43	0	0
AGCM(e)	1947–1997	12	18	64	6
AGCM(f)	1947–1997	89	98	74	55
AGCM(ens6)	1947–1997	2	0	2	1

Significance based on 100 Monte Carlo reorderings.

In addition to the implied poor seasonal predictability based on this model, there may also be implications for its representation of natural low-frequency variability and thus for its use in isolating anthropogenic climate change signals.

b. AGCM

Clearly for an AGCM, it is not necessary to apply a lag to determine cause and effect in the SVD analysis. However, a lag is required to investigate real predictability (where SSTs are not known in advance) and for consistency with the analysis of observations and the OAGCM simulations.

Table 1 also shows the significance levels of the first SCF fraction of SST leading Z500 for the six AGCM simulations. Generally these values are higher than those of the observations, although, with the exception of SON, at least one AGCM simulation matches the observed significance level in each season. The AGCM(d) simulation in particular produces values quite similar to those of the observations. Hence the AGCM would appear to be representing quite well North Atlantic ocean-to-atmosphere forcing. By defining one set of numbers as the four observed significance levels (Table 1) and another set of numbers as the 24 AGCM significance levels, the null hypothesis of no difference in means can be rejected at (only) the 11% level assuming equal variances. Either there is still some room for model improvement or there is a degree of "luck" in the observational results.

Figure 6 shows the COR values from the AGCM(d) simulation. COR results from other AGCM simulations are all poorer except in the winter case where there is a COR value as high as 0.46 from simulation AGCM(b). This latter value matches the observed COR value for May SST leading DJF Z500. It is clear that, again, the observational results are at the top end of, but not outside, the range of predictability estimates based on the AGCM. Note that since the AGCM only represents ocean-to-atmosphere forcing, the agreements in autumn and winter may be the clearest test of the model, since here ocean-to-atmosphere forcing has been most successfully isolated in the observations.

The ensemble mean of the six AGCM simulations should retain the ocean-to-atmosphere forced signal but have reduced internal variability compared to any single AGCM simulation. By applying the LSVD analysis to the ensemble mean, it may be possible to get a better indication of the model's predictable patterns and, by comparing the resulting time series with those of the observations, an estimate of potential predictability.

The significance levels for this analysis, given in Table 1, show that physically linked patterns have been identified in all seasons. For autumn, winter and spring the patterns from the individual simulations are quite consistent with each other and these patterns arise again in the analysis of the ensemble mean. For winter, spring and summer, the patterns from the ensemble mean (not shown here) look remarkably similar to those from the observations.

As with the observations (Fig. 4d), the winter SST and Z500 patterns display the NAO dipole and enhanced meridional SST gradients exist beneath the strengthened westerly jet.

The spring SST and Z500 patterns are similar to those of the observations (Fig. 4g) although the positive height anomaly region is split into two centres.

In summer, the observed North African response (Fig. 4j) is clearly seen in the ensemble mean results but here it tends to extend over the tropical and subtropical North Atlantic—closely matching the underlying cool SST anomalies. There are also similarities in summer with the observations further north over Newfoundland and northern Europe.

For autumn, the first and second SVD modes from the *individual* AGCM simulations are not well separated. This could be due to model deficiencies and may explain the poor correspondence between the SON patterns from the ensemble mean and the observations.

Since it appears that the AGCM should indeed be responding to North Atlantic SSTs and since the AGCM patterns from individual simulations (for winter in particular) look so similar to those of the observations, it may be possible to estimate a scaling for the ensemble mean NAO curve in Fig. 1 (upper left dotted curve) that allows for model systematic error. The standard deviations of the (unstandardized) winter Z500 time series are consistently lower for the six AGCM simulations than the standard deviation of the observational time series. The values suggest that the ensemble mean NAO curve in Fig. 1 should be scaled by a factor of ~1.5 to allow for systematic error.

F. POTENTIAL SEASONAL PREDICTABILITY BASED ON THE ATMOSPHERE GENERAL CIRCULATION MODEL

An estimate of potential predictability based on the AGCM ensemble can be made by correlating the "real" time series of the observations (Fig. 4 centre panels) and that of the AGCM ensemble mean. The correlations skills are: SON 0.31, DJF 0.45, MAM 0.39, JJA 0.59 all of which are significant at the 3% level or better when tested against an AR(1) model. These values represent potential predictability, since the AGCM time series has implicit knowledge of the seasonal mean SSTs.

The observational data suggests that a higher estimate of winter predictability may be gained by using May or June SSTs. The LSVD analysis has also been applied to the AGCM ensemble mean data for May SST leading DJF atmosphere and indeed predictability is enhanced.

The use of May SSTs may also lead to a better estimate of the most predictable patterns for the observational data. Figure 7a,b shows the LSVD patterns obtained from the observations and the ensemble mean with May SSTs leading DJF MSLP, respectively. MSLP has been used here for comparison with R. It is clear that the patterns are quite similar, although the MSLP contour in Fig. 7b is half that in Fig 7a. In both cases, enhanced meridional SST gradients are seen below a strengthened westerly jet. Figure 7 shows the observed and AGCM cross-validated MSLP time series. The correlation between these, which again represents potential predictability, is 0.63. The sign of the winter "NAO" is correctly simulated 67% of the time. This value for winter is higher than the potential predictability of the two-station NAO index presented in R (0.41) presumably because the impact of systematic model errors is reduced by not prescribing in advance the exact nature of the NAO and not insisting that the NAO has to be exactly the same in the observations and the model.

Hence there would appear to be some genuine skill for all seasons, although it should be remembered that the SON Z500 patterns are rather different between the observations and the model.

Potential predictability should perhaps be considered more as an upper bound for ultimately achievable predictability since SSTs need to be known in advance. Hence, for example, the results of this study would suggest that the ultimate predictability of the winter NAO based on correlation skill would be somewhere between 0.45 (as obtained in section E.1.b) and 0.63.

G. CONCLUSIONS AND DISCUSSION

Quantifying the effects of atmosphere-to-ocean forcing and ocean-to-atmosphere forcing in the observational record is not trivial but it is important if the representation of air–sea interaction in models is to be validated. A correct representation of these interactions is essential for the understanding of seasonal predictability and for the realistic simulation of longer-timescale climate variability. Here, a lagged singular value decomposition (SVD) approach has been used to compare Atlantic air–sea interactions in observations, an atmosphere model and a coupled ocean-atmosphere model.

Analysis of observed seasonal mean 500 hPa geopotential height and the sea-surface temperatures for the *following* month shows strong links in winter between the North Atlantic Oscillation (NAO) and a "tripole" of sea-surface temperature (SST) anomalies in the North Atlantic (Fig. 3b). A knowledge of the seasonal mean NAO can be used to predict the evolution of this tripole in SSTs. For winter the correlation between the real and predicted tripole time series is 0.73 with interannual and multi-annual variability well predicted. Since the time series are "cross-validated" this correlation represents genuine hindcast skill.

As indicated by Bjerknes, atmosphere-forced surface latent heat fluxes (LHFs) play a major role in this linkage (Fig. 4a). Mixed layer thermal anomalies created in winter appear to be partly submerged and preserved below the shallow summertime mixed layer and to re-emerge in the falling autumn as the mixed layer deepens (Fig. 3b) (Alexander and Deser, 1995; Watanabe and Kimoto, 2000).

When the same analysis is applied to each of four periods of the ocean–atmosphere model, very similar results are obtained. This suggests that the model represents quite well the mechanisms of atmosphere-to-ocean forcing.

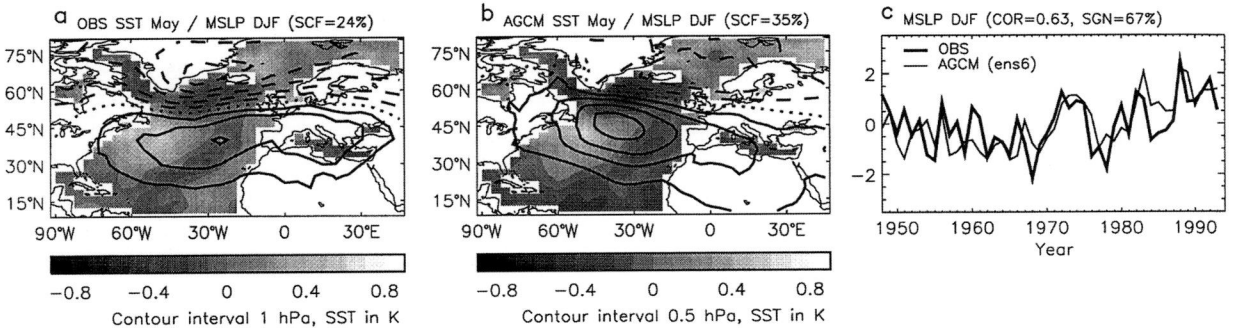

FIGURE 7 (a) The first SVD mode of observed (GISST3.0) May SST leading observed (GMSLP2.1f) DJF MSLP for the winters 1948/49–1993/94. Magnitudes correspond to one standard deviation. (b) As (a) but using AGCM ensemble mean data. (c) The standardized, cross-validated time series of the observed and AGCM MSLP patterns.

Analysis of observed seasonal mean 500 hPa geopotential height and the sea-surface temperatures for the *preceding* month also shows significant links in spring, summer and autumn. Figure 6 (thick curve) shows the (cross-validated) correlations between real and predicted atmospheric time series for each season based on SSTs from the preceding months. The link in winter is weaker but for this season there appears to be significant real predictability based on May North Atlantic SSTs (the correlation skill is 0.45). The latter result is consistent with the submergence/re-emergence hypothesis.

The SST pattern that precedes the seasonal mean Z500 NAO pattern is somewhat different from that which is forced by the NAO. This is particularly the case in winter where the preceding SSTs agree well with SST changes associated with fluctuations in the gyre circulation.

The role of atmospheric persistence between the predictor month and the prediction season, which could lead to misleading results has been accounted for, particularly for the autumn and winter situations. This has been done by using longer lags in the LSVD analysis and by directly quantifying the level of atmospheric persistence. For spring, the lack of SST persistence through the preceding winter makes it more difficult to discount the role of atmospheric persistence, although the identified spring predictability may be useful regardless of whether it is based on ocean-to-atmosphere forcing or on atmospheric persistence. For summer, trends in the observational data result in a high degree of perceived month-to-month atmospheric persistence. However, predictability remains when these secular trends are removed and this may well mean that ocean-to-atmosphere forcing has been partially isolated. See below for further evidence of this latter conclusion.

In the light of the LSVD results, the negative (positive) lagged correlations between the NAO and tripole-like latent heat fluxes (Fig. 7a) at lag 0 (lag ~ seven months before winter) can be interpreted as surface heat flux forcing of SSTs (NAO). These opposing surface latent fluxes probably imply little possibility for an instantaneous positive feedback. However, the re-emergence phenomenon may be enough to redden the NAO spectrum and allow for a modified Latif and Barnett (1994)-type coupled mode.

The coupled ocean–atmosphere model used here does not appear to reproduce the observed North Atlantic forcing of the atmosphere. This is the case even when considering the simulation periods of the model that show multidecadal trends in the winter NAO which are as strong as those seen in the observations. If coupled models do not adequately represent Atlantic air—sea interaction, they may underestimate seasonal predictability and not correctly simulate decadal variability.

The atmospheric model used here (which is different from the atmospheric component of the coupled model) reproduces better the observed ocean-to-atmosphere links. Real seasonal predictability based on the observations is at the top end, but not outside, the levels of real seasonal predictability suggested by the *individual* simulations of the atmospheric model. This suggests that the potential predictability skill of the *ensemble mean* winter NAO, which could be as high as 0.63, may be reasonable. At the very least, if offers observational evidence that tends to confirm the potential predictability of the (Azores minus Iceland) NAO index given in Rodwell *et al.* (1999). Scaling arguments suggest that the ensemble mean winter NAO index from a perfect model could be about 1.5 times the magnitude of the ensemble mean NAO index shown in Fig. 1.

Potential predictability for autumn is difficult to quantify, since the atmospheric patterns are rather different for the observations and the AGCM ensemble mean. For spring, a correlation skill of 0.39 was found. For summer, the correlation was 0.59. This latter result implies that ocean-to-atmosphere forcing may well have been isolated in the observational record even before detrending of the initial data.

Clearly all these numerical results could be sensitive to data errors and sampling uncertainties.

This analysis has focused on coupling mechanisms of the first covarying (SVD) mode of ocean surface temperatures and seasonal mean 500 hPa geopotential height. This mode typically has a squared covariance fraction of ~20%. The first EOF of winter NAE Z500 accounts for 35% of the variance and looks very similar to the Z500 pattern of the SVD analysis (Fig. 2a). If it was possible to predict the NAO several months ahead with a skill correlation within the range 0.45 to 0.63, this may be of some (limited) use. Scope for improvement in predictability may come from: (a) coupled model improvements, if subsurface information can be used to better predict the seasonal mean SSTs; (b) using "non-standard" seasons such as November–January—a search through all such seasons was not made here because it may have had an impact of the estimation of significance levels; (c) the use of higher-order LSVD modes which may also possess some physical basis; (d) using other predictors; (e) considering nonlinearity in ocean–atmosphere interaction (this analysis has focused only on linearly covarying modes).

ACKNOWLEDGEMENTS

The author would like to thank the following for useful discussions in connection with this chapter: Chris Folland, Chris Gordon, Dave Rowell, David Sexton, Claire Cooper, Briony Horton, Nick Rayner, Peili Wu.

References

Alexander, M. A. and C. Deser, 1995: A mechanism for the recurrence of wintertime midlatitude SST anomalies. *J. Phys. Oceanogr.*, **25**, 122–137.

Basnett, T. A. and D. E. Parker, 1997: Development of the global mean sea level pressure data set GM–SLP2. Climate Research Technical Note 79, UK Meteorological Office, Bracknell.

Bjerknes, J., 1964: Atlantic air–sea interaction. *Adv. Geophys*, **10**, 1–82.

Bresch, D. and H. Davies, 2000: Covariation of the mid-tropospheric flow and the sea surface temperature of the North Atlantic: A statistical analysis. *Theor. Appli. Climatol.*, **65** (3–4), 197–214.

Bretherton, C. S., C. Smith and J. M. Wallace, 1992: An intercomparison of methods for finding coupled patterns in climate data. *J. Climate*, **5**, 541–560.

Czaja, A. and C. Frankignoul, 1999: Influence of North Atlantic SST on the atmospheric circulation. *Geophys. Res. Lett.*, **26**(19), 2969–2972.

Daly, A. W., 1978: The response of North Atlantic sea surface temperature to atmospheric forcing processes. *Q. J. R. Meteorol. Soc.*, **104**, 363–382.

Davies, J. R., D. P. Rowell and C. K. Folland, 1997: North Atlantic and European seasonal predictability using an ensemble of multi-decadal AGCM simulations. *Int. J. Climatol.*, **17**, 1263–1284.

Frankignoul, C. and K. Hasselmann, 1977: Stochastic climate models, Part II: Application to sea-surface temperature anomalies and thermocline variability. *Tellus*, **29**, 289–305.

Frankignoul, C., 1985: Sea surface temperature anomalies, planetary waves and air-sea feedback in the middle latitudes. *Rev. Geophys.*, **23**, 357–390.

Grötzner, A. M., M. Latif and T. P. Barnett, 1998: A decadal climate cycle in the North Atlantic ocean as simulated by the ECHO coupled GCM. *J. Climate*, **11**, 831–847.

Hall, C. D., R. A. Stratton and M. L. Gallani, 1995: Climate simulations with the Unified Model: AMIP runs. Climate Research Technical Note 61, UK Meteorological Office, Bracknell.

Hasselmann, K., 1976: Stochastic climate models. Part I: Theory. *Tellus*, **28**, 473–485.

Hurrell, J. W. and H. van Loon, 1997: Decadal variations in climate associated with the North Atlantic Oscillation. *Climate Change*, **36**, 301–326.

Hurrell, J., 1995: Decadal trends in the North Atlantic Oscillation: regional temperatures and precipitation. *Science*, **269**(5224), 676–679.

Kalnay, E., M. Kanamitsu, R. Kistler, W. Collins, D. Deaven, L. Gandin, M. Iredell, S. Saha, G. White, J. Woollen, Y. Zhu, M. Chelliah, W. Ebisuzaki, W. Higgins, J. Janowiak, K. C. Mo, C. Ropelewski, J. Wang, A. Leetmaa, R. Reynolds, R. Jenne and D. Joseph, 1996: The NCEP/NCAR 40-Year Reanalysis Project. *Bull. Am. Meteorol. Soc.*, **77**(3), 437–471.

Kushnir, Y., 1994: Interdecadal variations in North Atlantic sea surface temperature and associated atmospheric conditions. *J. Climate*, **7**, 141–157.

Latif, M. and T. P. Barnett, 1994: Causes of decadal variability over the North Pacific and North America. *Science*, **266**, 634–637.

Latif, M., K. Arpe and E. Roeckner, 2000: Oceanic control of decadal North Atlantic sea level pressure variability in winter. *Geophys. Res. Lett.*, **27**(5), 727–730.

Mehta, V., M. Suarez, J. Manganello and T. Delworth, 2000: Oceanic influence on the North Atlantic Oscillation and associated Northern Hemisphere climate variations: 1959–1993. *Geophys. Res. Lett.*, **27**(1), 121–124.

Namias, J., 1964: Seasonal persistence and recurrence of European blocking during 1958–1960. *Tellus*, **16**, 394–407.

Namias, J., 1973: Hurricane Agnes—an event shaped by large-scale air-sea systems generated during antecedent months. *Q. J. R. Meteorol. Soc.*, **99**, 506–519.

Osborn, T. J., K. R. Briffa, S. F. B. Tett, P. D. Jones and R. Trigo, 1999: Evaluation of the North Atlantic Oscillation as simulated by a coupled climate model. *Climate Dynam.*, **15**, 685–702.

Palmer, T. N. and Z. Sun, 1985: A modelling and observational study of the relationship between sea surface temperature in the northwest Atlantic and the atmospheric general circulation. *Q. J. R. Meteorol. Soc.*, **111**, 947–975.

Ratcliffe, R. A. S. and R. Murray, 1970: New lag associations between North Atlantic sea temperatures and European pressure, applied to long-range weather forecasting. *Q. J. R. Meteorol. Soc.*, **96**, 226–246.

Rayner, N. A., E. B. Horton, D. E. Parker and C. K. Folland, 1998: Versions 2.3b and 3.0 of the global sea-ice and sea surface temperature (GISST) data set. Hadley Centre internal note 85, Hadley Centre, Meteorological Office, Bracknell, UK.

Robertson, A., C. Mechoso and Y.-J. Kim, 2000: The influence of Atlantic sea surface temperature anomalies on the North Atlantic Oscillation. *J. Climate*, **13**, 122–138.

Rodwell, M. J., D. P. Rowell and C. K. Folland, 1999: Oceanic forcing of the winter North Atlantic Oscillation and European climate. *Nature*, **398**, 320–323.

Rogers, J. C., 1990: Patterns of low-frequency monthly sea level pressure variability (1899–1986) and associated wave cyclone frequencies. *J. Climate*, **3**, 1364–1379.

Rowell, D. P., 1998: Assessing potential seasonal predictability with an ensemble of multidecadal GCM simulations. *J. Clim.*, **11**, 109–120.

Saravanan, R., 1998: Atmospheric low-frequency variability and its relationship to midlatitude SST variability: Studies using the NCAR climate system model. *J. Climate*, **11**, 1386–1404.

Sverdrup, H., 1942: *Oceanography for Meteorologists*. Prentice-Hall, Englewood Cliffs, NJ.

Visbeck, M., H. Cullen, G. Krahmann and N. Naik, 1998: An ocean model's response to North Atlantic Oscillation-like wind forcing. *Geophys. Res. Lett.*, **25**(24), 4521–4524.

Walker, G. T. and E. W. Bliss, 1932: World weather V. *Mem. R. Meteorol. Soc.*, **4**, 53–84.

Wallace, J. M. and D. S. Gutzler, 1981: Teleconnections in the geopotential height field during the northern hemisphere winter. *Mon. Weath. Rev.*, **109**, 784–812.

Watanabe, M. and M. Kimoto, 2000: On the persistence of decadal SST anomalies in the North Atlantic. *J. Climate*, **13**, 3017–3028.

The Monsoon as a Self-regulating Coupled Ocean–Atmosphere System

Peter J. Webster,[1] Christina Clark,[1] Galina Cherikova,[1] John Fasullo,[1] Weiqing Han,[1] Johannes Loschnigg[2] and Kamran Sahami[1]

[1]*Program in Atmospheric and Oceanic Science, University of Colorado: Boulder, Boulder, CO, USA*
[2]*International Pacific Research Center, University of Hawaii, Honolulu, HI, USA*

Observational studies have shown that the Asian–Australasian monsoon system exhibits variability over a wide-range of space and timescales. These variations range from intraseasonal (20–40 days), annual, biennial (about 2 years), longer-term interannual (3 to 5 years) and interdecadal. However, the amplitude of the interannual variability of the South Asian monsoon (at least as described by Indian precipitation) its summer pluvial phase is smaller than variability exhibited in other climate systems of the tropics. For example, drought or flood rarely extend to multiple years, with rainfall oscillating biennially from slightly above average to slightly below average precipitation.

We argue that variability of the monsoon is regulated by negative feedbacks between the ocean and the atmosphere. The annual cycle of the heat balance of the Indian Ocean is such that there is an ocean heat transport from the summer hemisphere to the winter hemisphere resulting principally from wind-driven Ekman transport. Given the configuration of the low-level monsoon winds, the Ekman transport is in the opposite sense to the lower tropospheric divergent wind. The cross-equatorial ocean heat transport is large with amplitudes varying between +2 PW (northward) in winter and −2 PW (southward) in summer. Thus, the wind-induced heat transport works to cool the summer hemisphere upper ocean while warming the winter hemisphere. Similar regulation occurs on interannual timescales. For example, during anomalously strong Northern Hemisphere monsoon summers (typically a La Niña), strong winds induce a stronger than average southward flux of heat. When the monsoon is weak (typically an El Niño), the wind-driven ocean heat flux is reduced. In this manner, the monsoon regulates itself by reducing summer hemisphere sea-surface temperatures during strong monsoon years and increasing it during weak years. In this manner, the monsoon is self regulating.

It is noted, however, that the ocean heat transport theory of monsoon regulation does not necessarily insist that heat anomalies persist from one year to the next. Furthermore, the theory does not include the Indian Ocean dipole as a dynamic entity. Finally, we develop a more general theory in which the slow dynamics of the dipole are integral components of a sequence of processes that regulate the monsoon, thus minimizing radical year-to-year departures of the monsoon from climatology.

A. INTRODUCTION

The majority of the population of the planet resides in the monsoon regions. The livelihood and well-being of these monsoon societies depends on the variations of the monsoon and the establishment of a symbiotic relationship between agricultural practices and climate. Whereas the summer rains recur each year, they do so with sufficient variability to create periods of relative drought and flood throughout the region. Hence, forecasting the variations of the monsoon from year-to-year, and their impact on agricultural and water resources are a high priority for humankind and are a central issue in the quest for global sustainable development. With accurate forecasts and sufficient lead time for their utilization, the impact of variability of the monsoon on agricultural practices, water management, etc., could be optimized. Yet, despite the critical need for accurate and timely forecasts, our abilities to predict variability have not changed substantially over the last few decades. For example:

(i) For over 100 years the India Meteorological Department has used models based on empirical relationships between monsoon and worldwide climate predictors with only moderate success. These endeavours have been extended for both the South Asian region and in other monsoon regions to predict interannual variability in different monsoon regions (see reviews by Hastenrath, 1986a,b, 1994). Many of these schemes use a measure of El Niño-Southern Oscillation (ENSO) as a major predictor for monsoon variability (e.g., Shukla and Paolina, 1983; Rasmussen and Carpenter, 1983; Shukla, 1987a,b). Whereas there are periods of extremely high association between ENSO and monsoon variability, there are decades where there appears to be little or no association at all (e.g., Torrence and Webster, 1999). The association has been particularly weak during the last 15 years.

(ii) Numerical prediction of monsoon seasonal to interannual variability is in its infancy and severely handicapped by the inability of models to simulate either the mean monsoon structure or its year-to-year variability. Model deficiencies have been clearly demonstrated in the Atmospheric Model Intercomparison Program (AMIP: Sperber and Palmer, 1996; Gadgil and Sajani, 1998) where the results of 39 atmospheric general circulation models were compared.

In summary, a combination of modelling problems and empirical nonstationarity has plagued monsoon prediction. Empirical forecasts have to contend with the spectre of statistical nonstationarity while numerical models currently lack simulation fidelity.

As noted above, most empirical forecast schemes have concentrated on taking advantage of an ENSO–monsoon rainfall relationship. A recent series of empirical studies have shown that the relationships between Indian Ocean SST and Indian rainfall are stronger than portrayed in earlier studies. For example, Sadhuram (1997), Hazzallah and Sadourny (1997) and Clark et al. (2000a) found that correlations as high as +0.8 occurred between equatorial Indian Ocean SSTs in the winter prior to the monsoon wet season and Indian precipitation. A combined Indian Ocean SST index generated by Clark et al. (2000a) has retained an overall correlation of 0.68 for the period 1945 to 1994, after the removal of ENSO influence. Between 1977 and 1995 the correlation has risen to 0.84 again after the removal of ENSO influence. In addition, the so-called Indian Ocean dipole (Webster et al., 1999; Saji et al., 1999; Yu and Rienecker 1999, 2000) appears to have strong coupled ocean–atmosphere signatures across the basin, especially in the equinoctial periods, exerting a very significant influence on the autumnal solstitial rainfall over East Africa (Latif et al., 1999, Clark et al., 2000b).

These empirical relationships found between the Indian Ocean SST and monsoon variability are important because they suggest that there are basin-wide relationships that may be, to some extent, independent of ENSO and, thus, inherent to the Indian Ocean–monsoon system. Some structural independence of basin-wide modes is evident in Fig. 1a,b which describes the persistence SST both in space (Fig. 1a) and in time along the equator (Fig. 1b). The Pacific Ocean possesses a strong persistence minimum in the boreal spring. This is the Pacific Ocean "predictability barrier" of Webster and Yang (1992) and underlines the rapid changes that often occur in the Pacific Ocean during the boreal spring. The pattern in the Indian Ocean is quite different. Strong persistence occurs from the end of the boreal summer until the late spring of the following year. This behaviour supports the contention of Meehl (1994a,b, 1997) that there is biennial component in the Indian Ocean SST and monsoon rainfall. Meehl (1994a) also noted that strong Indian monsoons were generally followed 6 months later by strong North Australian summer monsoon so that the entire Asian–Australian monsoon system followed a biennial pattern. This related behaviour can be seen in Fig. 2, which plots both the Indian and North Australian rainfall. The South Asian monsoon (at least as described by Indian precipitation) exhibits a smaller range of variability during its summer pluvial phase than variability exhibited by the North Australian monsoon. In fact, the South Asian monsoon has a variability that is generally smaller than most heavy rainfall regions of the tropics. For example, drought, or flood, in South Asia rarely extend to multiple years, with rainfall oscillating biennially from slightly above average to slightly below average precipitation.

Given the problems in the predictions of monsoons, and in the light of the recent discoveries regarding the structure of the Indian Ocean and the monsoon, a number of questions need to be addressed:

(i) What factors determine the phase and the amplitude of the monsoon annual cycle?
(ii) What factors control the interannual rainfall variability of the South Asian monsoon so that the persistent multiyear anomalies and large excursions from long-term means are rare?
(iii) To what extent is the monsoon a coupled ocean–atmosphere phenomena? For example, do the correlations between Indian Ocean SST and monsoon rainfall indicate the existence of coupled modes?
(iv) What are the relationships, if any, between the

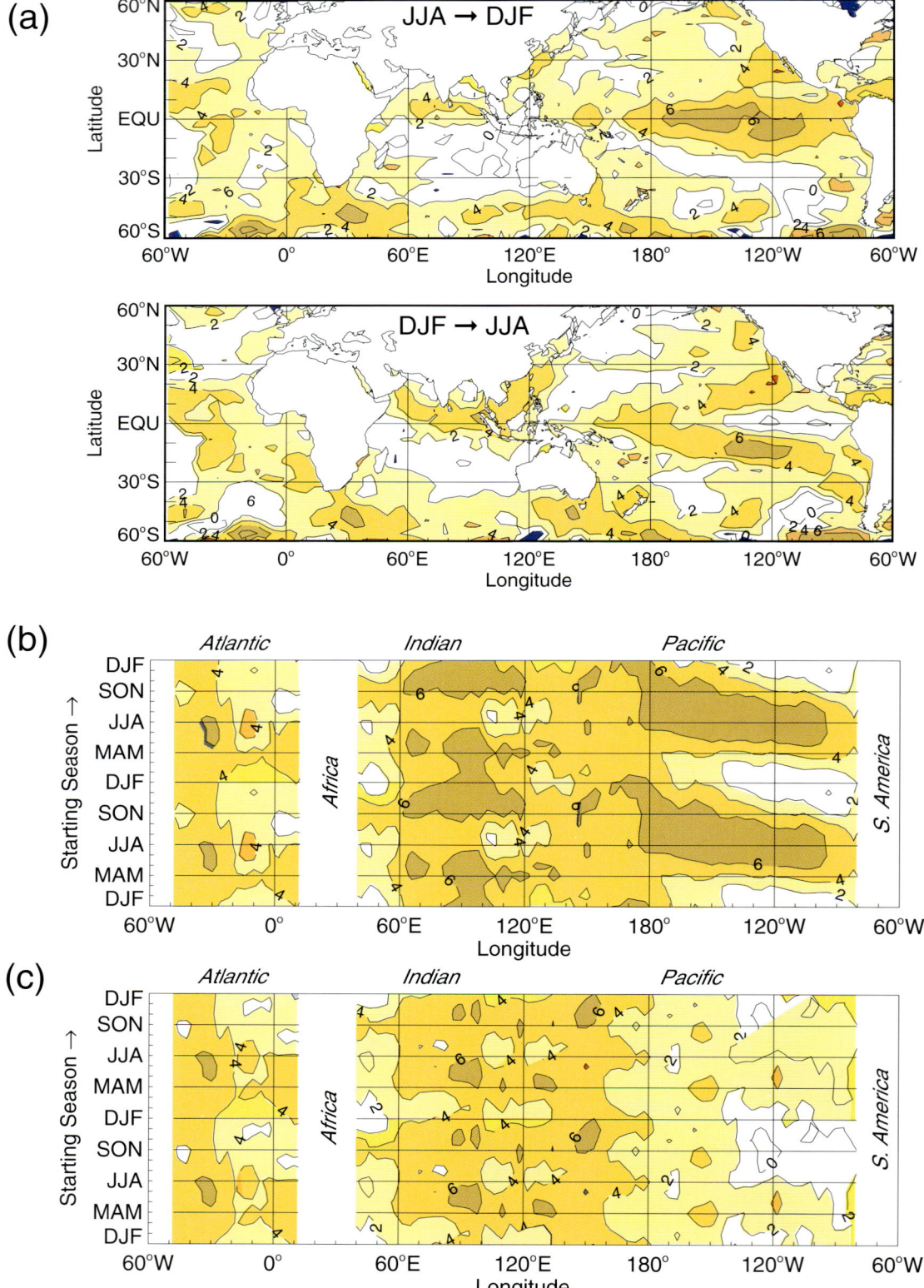

FIGURE 1 (a) Persistence of the SST over two 6-month periods: summer to winter and winter to summer. Extremely strong persistence can be seen in the eastern Pacific Ocean and the North Indian Ocean between summer and winter. However, persistence decreases substantially in the Pacific Ocean between winter and summer. It is maintained, however, in the North Indian Ocean. (b) Time sequence of the climatological persistence of SST along the equator for a 2-year cycle. (c) Persistence with Niño-3 influence removed. In the Figure, a value of 4 (for example) means that 40% of the variance over a six month period can be explained by the SST anomaly at the beginning of the six month period. (Torrence and Webster, 1999.)

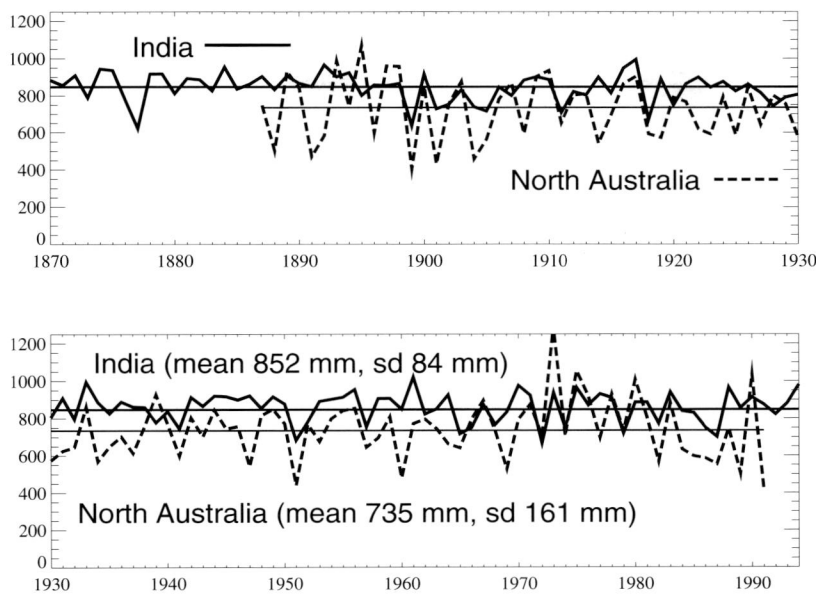

FIGURE 2 Time sequence of mean annual precipitation for India and North Australia. All-India rainfall data from Parthasarathy *et al.* (1992, 1994) and Australian data from Lavery *et al.* (1997).

bienniality of the monsoon, the Indian Ocean dipole, and longer-term interannual variability?

(v) To what degree is variability of the South Asian monsoon independent of outside influences such as ENSO? Is the different character of the persistence of SST shown in Fig. 1 indicative of relative independence between the two basins?

The purpose of this study is to make preliminary progress in answering these questions.

The organization of the chapter is based on the proposition that in order to understand interannual variability of a phenomenon, it is first necessary to understand its annual cycle. In the next section we will investigate aspects of the annual cycle viewed from a coupled ocean–atmosphere perspective. It will be argued that the ocean and the atmosphere, in concert, act to limit the seasonal extremes of the monsoon. In Section C the interannual variability of the monsoon system will be considered, again from the view of a coupled system. It will be shown that similar coupled physical processes regulate monsoon interannual variability. In Section D, a more general theory of coupled ocean–atmosphere regulation of the monsoon is proposed. Here, it will be shown that the bienniality of the monsoon and the Indian Ocean dipole are integral parts of a suite of processes and phenomena that regulate the interannual variability of the monsoon.

B. REGULATION OF THE MONSOON ANNUAL CYCLE

1. The Climatological Annual Cycle

Fig. 3 shows the annual cycle of the mean SST, near-surface wind vectors, and outgoing longwave radiation (OLR) for the four seasons. Some pertinent characteristics may be listed briefly as follows:

(i) In general, the warmest SSTs occur in the boreal spring where 29°C surface water covers most of the Indian Ocean north of 10°S. Except in the very far north of the basin, the SST during winter is about 28°C. With the coming of summer and the quickening of the monsoon, the SST of the North Indian Ocean cools especially in the Arabian Sea. The autumn temperatures are similar to spring but cooler in general by about a degree. In all seasons, the maximum SST gradient occurs south of the equator.

(ii) The lower tropospheric flow of the Asian summer monsoon is much stronger than its wintertime counterpart and possesses a concentrated cross-equatorial flow in the western Indian Ocean compared to the broader, but weaker, reverse flow during the boreal winter.

(iii) During the summer, maxima in convection (low

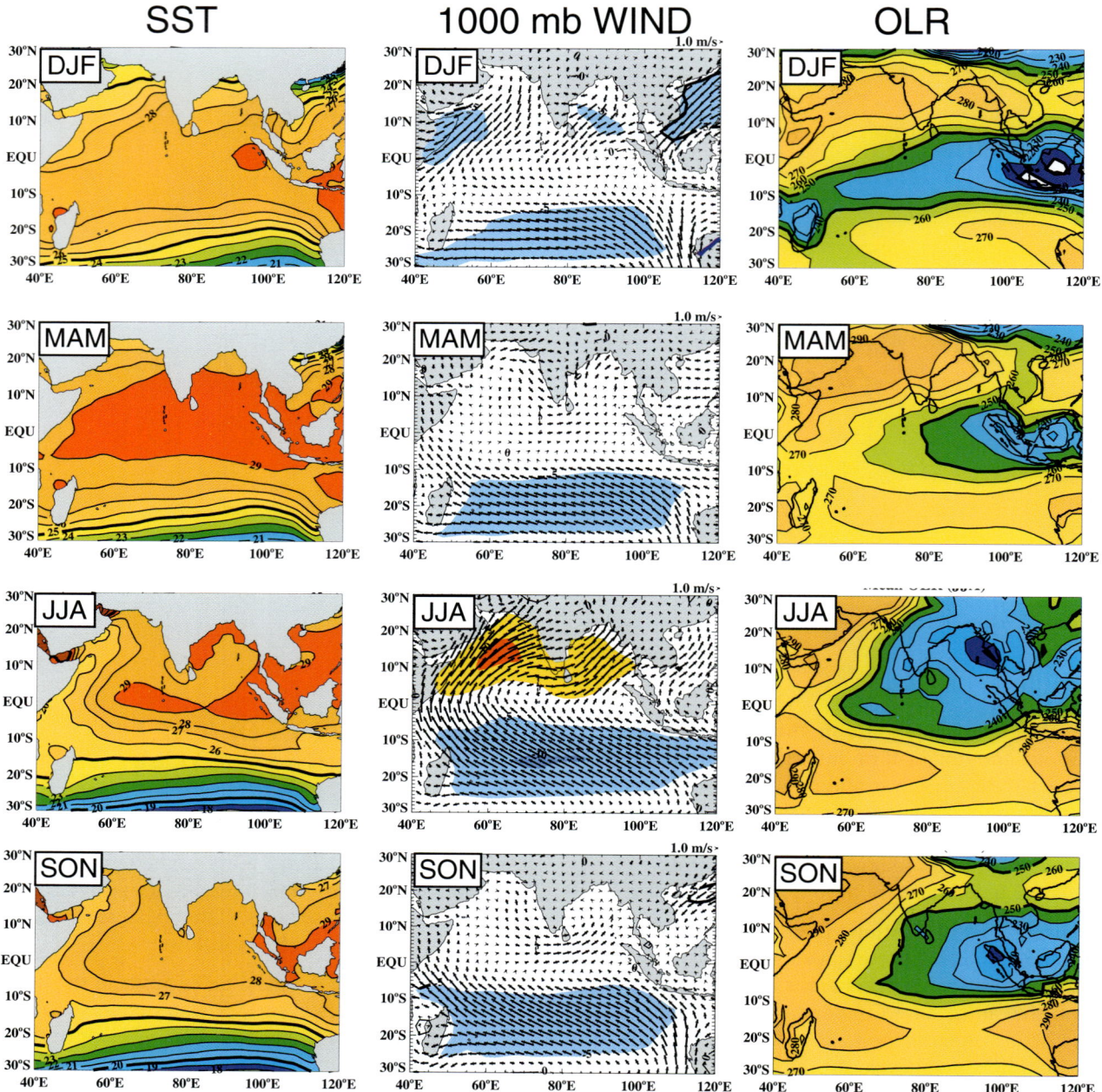

FIGURE 3 Mean annual climatology of SST (°C), near-surface wind vectors (m s^{-1}), and outgoing longwave radiation (W m^{-2}). Data is plotted for the four seasons March, April and May (MAM), June, July and August (JJA), September, October and November (SON), and December, January and February (DJF).

OLR) occurs over south Asia with weaker extensions over equatorial north Africa and in the near-equatorial Southern Hemisphere. During the boreal winter, an elongated band extends across the Indian Ocean and North Australia culminating in a broad maximum over Indonesia and North Australia. Convection over Asia is located farther poleward than its Southern Hemisphere counterpart, and the circulations are not symmetrical between the seasons.

Perhaps the most interesting aspect of these climatologies is the lack of variation of the SST between seasons. Most notably, the SST appears to change little during the boreal spring over much of the North Indian Ocean despite weak winds and high insolation. Similarly, the

TABLE 1 The Components of the Surface Heat Balance for the North Indian Ocean

		Flux (W m^{-2})					dT/dt (K yr^{-1})
		S	LW	LH	SH	NET	
(a)	Entire North Indian Ocean						
	DJF	125	−57	−103	−4	15	2.2
	MAM	223	−49	−82	−2	90	13.4
	JJA	190	−39	−117	−0	34	5.1
	SON	187	−48	−88	−3	50	7.5
	Annual	181	−48	−98	−2	47	7.1
(b)	Equator to 10°N						
	DJF	181	−49	−96	−4	24	
	MAM	202	−47	−88	−3	65	
	JJA	178	−42	−116	−2	17	
	SON	186	−46	−87	−4	50	
(c)	North of 10°N						
	DJF	174	−61	−105	−5	3	
	MAM	232	−48	−74	−1	112	
	JJA	191	−35	−113	0	44	
	SON	179	−47	−88	−2	42	

The net solar radiation, net longwave radiation, latent heat flux, sensible heat flux, and the net flux at the surface are denoted by S, LW, LH, SH, and NET, respectively. Data from COADS (Oberhuber, 1988). Heating rates of a 50 m layer for the entire North Indian Ocean are shown in the right-hand column.

winter SST in the North Indian Ocean is only a degree or so cooler than in the fall or the spring despite significant reductions in surface heating. These aspects of the annual cycle will be addressed in the next section.

2. Processes Determining the Annual Cycle of the Monsoon

One of the major outstanding problems in climate is how the amplitude and phase of the annual cycle of the tropics, such as the cycle described above for the Indian Ocean, adopts the form observed. This problem has been the subject of intensive discussion, especially with respect to the western Pacific Ocean following the introduction of the "thermostat hypothesis" by Ramanathan and Collins (1991). We are particularly interested in how the regulation of the Indian Ocean SST occurs because of the close relationship between the sign and magnitude of SST anomalies with the vigour of the ensuing monsoon (e.g., Clark et al., 2000a).

Determining the surface heat balance of the tropical oceans is difficult, and large differences may occur between estimates. This should not be surprising as the surface heat balances are the relatively small sums of large terms. Furthermore, these large terms are obtained from empirical rules some of which are not known accurately (Godfrey et al., 1995). A result of the TOGA Coupled Ocean–Atmosphere Response Experiment (TOGA COARE: Webster and Lukas, 1992; TOGA COARE, 1994), the surface heat flux into the western Pacific Ocean warm pool is estimated to be between 10 and 20 W m^{-2} (Godfrey et al., 1998).

Unfortunately, the surface heat balance is less well known in the Indian Ocean. Table 1 provides estimates of annual cycle of surface flux for the North Indian Ocean during the boreal spring and early summer using COADS data (Oberhuber, 1988). Estimates are listed for the entire North Indian Ocean and for regions north of 10°N and between the equator and 10°N. The daily mean solar radiation into the North Indian Ocean is >200 W m^{-2} during the spring months and 181 W m^{-2} over the entire year compared to an annual average of about 145 W m^{-2} for the western Pacific (Webster et al., 1998). The annual net surface flux averaged over the entire North Indian Ocean is about +50 W m^{-2} or at least a factor of two or three larger than in the western Pacific warm pool (Godfrey et al., 1998). Earlier estimates for the Indian Ocean by Hastenrath and Lamb (1978), Hsiung et al. (1989), Oberhuber (1988), and Hastenrath and Greischar (1993) were of order +50 and +70 W m^{-2} on an annual basis. Despite uncertainties, it is clear that over the year there is a much stronger flux of heat into the North Indian Ocean than into the western Pacific Ocean. Furthermore, there is a far larger seasonality in this flux than in the western Pacific Ocean, with maximum values occurring in spring and early summer.

It is a simple matter to estimate the changes in SST resulting just from the observed net fluxes (Table 1). The simplest way is to assume that the net flux is spread through an upper ocean layer of some defined depth for which the heating rate may be written:

$$dT/dt = -(dF_{net}/dz)/\rho C_p$$

Heating rates, assuming a 50 m surface layer, are listed in Table 1. Alternatively, one may use a sophisticated one-dimensional mixed-layer model to make the same computation. Webster *et al.* (1998) used the model of Kantha and Clayson (1994) finding results quite similar to those listed in Table 1. Both sets of calculations suggest a very different behaviour in the evolution of SST than is observed in nature. Figure 3 shows a rather gradual change in SST from one season to another, whereas the calculations listed in Table 1 suggest that the Indian Ocean would be continually warming at a large annual rate $>7°C\ yr^{-1}$. If this were the case, the observed cyclic equilibrium noted in Fig. 3 would not be achieved. On the other hand, using flux values for the Pacific Ocean also suggests a net warming but at a rate of order $1-2°C\ yr^{-1}$.

Figure 4 shows a schematic diagram of the basic combination of processes used in studies of SST regulation. The first combination of processes (Set 1 in Fig. 4) is *regulation through local thermodynamic adjustment*. Such a state was defined by Ramanathan and Collins (1991) as the "natural thermostat" in which an SST warming produces cloud, which, in turn, reduces surface solar radiation flux and thus regulates SST. The second combination of processes (Sets 1 and 2) describes *regulation by a combination of atmospheric dynamics and transports in addition to negative radiational feedbacks*. Here an invigorated atmospheric circulation is assumed to respond to increased SST gradients which enhance evaporative cooling of the ocean surface (Fu *et al.*, 1992; Stephens and Slingo, 1992; Wallace, 1992; Hartmann and Michelson, 1993). The third combination of processes (Sets 1, 2 and 3) is *horizontal or vertical advection of heat or changes in the storage of heat by oceanic dynamics in addition to local thermodynamic adjustment and cooling by an invigorated atmosphere*. Within this third combination, the lateral transport of heat by the ocean is important.

It is clear that the Ramanathan and Collins (1991) mechanism cannot regulate SST in the Indian Ocean. The warmest SSTs on the planet during spring and early summer are not associated with convection (Fig. 3) so that there is no cloud shielding to mitigate the solar radiation flux at the surface. Furthermore, the second class of theories for regulation of the SST in the Pacific Ocean (Fu *et al.*, 1992; Stephens and Slingo, 1992; Wallace, 1992; Hartmann and Michelson, 1993) also cannot solve the problem; winds remain light during the boreal spring and early summer, and evaporation is relatively low (Fig. 3).

It is clear that the ocean must play an extremely important dynamic role in achieving the cyclic equilibrium of the mean annual cycle. As summarized by Godfrey *et al.* (1995, p. 12):

> "... on an annual average there is positive heat flux into the Indian Ocean, nearly everywhere north of 15°S. The integral of the net heat influx into the Indian Ocean over the area north of 15°S ranges beween $0.5-1.0 \times 10^{15}$ W, depending on the climatology. Thus, on the annual mean, there must be a net inflow of cold water (into the North Indian Ocean), and a corresponding removal of warmed water, to carry this heat influx southward, out of the tropical Indian Ocean".

Oceanic heat transports of heat represent the only means that allows the large net annual surface heating of the North Indian Ocean to be removed without raising the SST substantially. It should be noted that Godfrey *et al.* (1995) were referring to annually averaged ocean heat transports. In the next section we will show that the annual average is made up of seasonal swings with amplitudes of ±2 PW.

3. Role of Ocean Dynamics in the Annual Heat Balance of the Indian Ocean

To understand the role of dynamical transports by the ocean, Loschnigg and Webster (2000) used the McCreary *et al.* (1993) dynamic upper ocean model, which incorporates full upper ocean dynamics and thermodynamics. Although there is little subsurface data in the Indian Ocean, the model has been shown to replicate the surface structure of the Indian Ocean, as well as its annual cycle, when the ocean model is run in stand-alone mode with prescribed atmospheric forcing (Loschnigg and Webster, 2000). The instantaneous

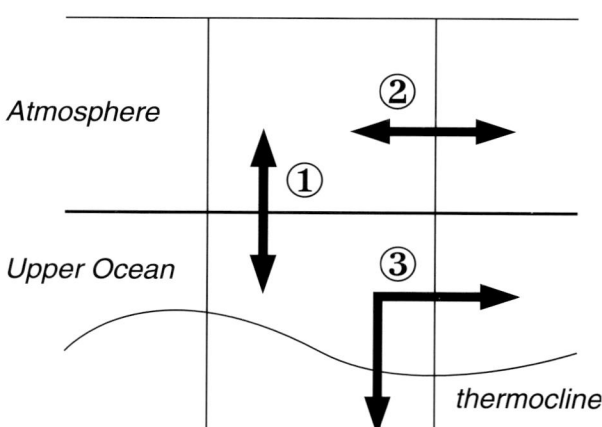

FIGURE 4 Schematic of the basic combinations of processes considered in SST regulation. Process 1 describes a local heat balance through thermodynamic factors. Process 2 uses heat transports by the atmosphere. Process 3 utilizes ocean transports of heat (Webster *et al.*, 1998).

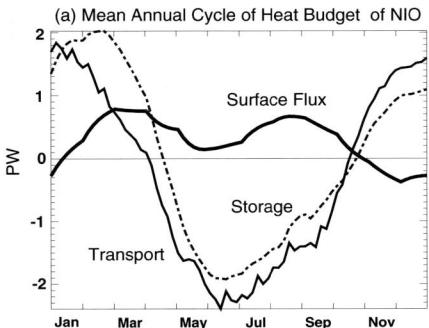
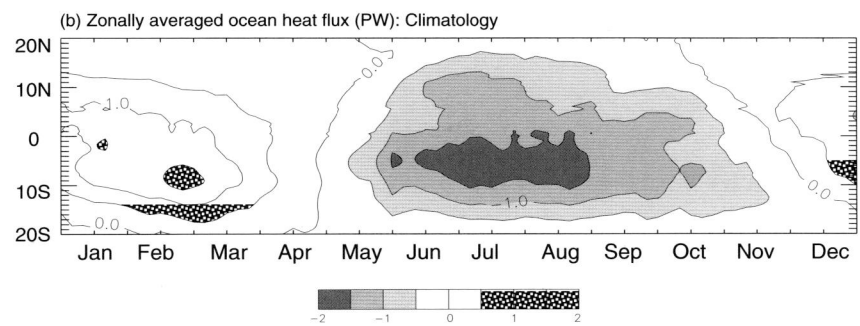

FIGURE 5 (a) Annual cycle of meridional heat transport across the equator, changes in heat storage, and net heat flux into the North Indian Ocean. Positive surface heat flux indicates a heating of the ocean. Positive heat transport is to the north. (b) Annual cycle of meridional ocean heat flux as a function of latitude and longitude. Units are PW. (Loschnigg and Webster, 2000.)

northward heat flux between two positions along a line of constant latitude is defined as

$$F_Q = \rho\omega c\omega \int_x \int_z H_i v_i T_i \, dxdz \qquad (1)$$

where H_i, T_i and v_i are the depth, temperature and meridional velocity component of the ith layer of the model. The storage in a volume is defined by

$$S_Q = \rho\omega C\omega \int_x \int_y \int_z H_i T_i \, dxdydz \qquad (2)$$

Figure 5a shows the annual cycle of meridional heat transport and heat storage changes in the Indian Ocean forced by the annual cycle of climatological winds and heating. The net surface flux into the North Indian Ocean appears to have a semi-annual variation. This is due to the combination of net solar heating and cooling by evaporation. Evaporative cooling is largest in the summer, associated with the strong monsoon winds. There is also a second maximum in winter associated with the winter monsoon. Solar heating also has two minima: in summer where cloudiness has increased with the onset of the monsoon and in winter. Together these two complicated heating mechanisms provide the double maximum evident in Fig. 5a. The major components of heat flow are between the transport and storage terms. A very strong southward flux of heat is evident during the spring and early summer, and a reverse flux is evident during the winter, when the north Indian Ocean is losing considerable amounts of heat by both evaporation, and vertical turbulence mixing which entrains colder water from below the thermocline. The net heat flux across the equator is made up of opposing flows in the upper and lower layer of the model and may be thought of as a seasonally reversing meridional ocean circulation (McCreary et al., 1993). The magnitude of the net southward flux across the equator during the spring and summer months more than makes up for the excess surface flux into the north Indian Ocean, with peak summer magnitudes of meridional heat transport reaching values of 2 PW. Whereas the changes in heat storage had been anticipated by earlier studies (e.g., Vonder Haar and Oort, 1973), the strong role of advection was not considered.

Figure 5b displays a latitude–time section of the annual cycle of the climatological meridional oceanic heat flux averaged across the basin. The year is divided into a period of northward heat flux in winter and spring and a southward, slightly stronger, heat flux between late spring and early fall. Maximum transport occurs at all seasons close to 10°S near the zone of maximum SST gradient.

4. Regulation of the Annual Cycle of the Monsoon: an Ocean–Atmosphere Feedback System

From the calculations above, it is clear that without ocean transport across the equator and changes in the heat storage of the North Indian Ocean, the cross-equatorial buoyancy gradient during early summer would be very large. Yet the processes that accomplish the cross-equatorial transport of heat in the ocean are essentially wind driven (Mohanty et al., 1996; McCreary et al., 1993; Godfrey et al., 1995). In turn, the atmospheric circulation is driven by surface fluxes and heating gradients associated with the buoyancy gradient and atmosphere–land interaction. Thus, the annual cycle in the Indian Ocean is a coupled phenomenon resulting from ocean–land–atmosphere interactions and balanced, to a large extent, by cross-equatorial oceanic transports.

The form of interaction between the atmospheric monsoon flow and the ocean transport results from a coupled ocean–atmosphere feedback. A critical

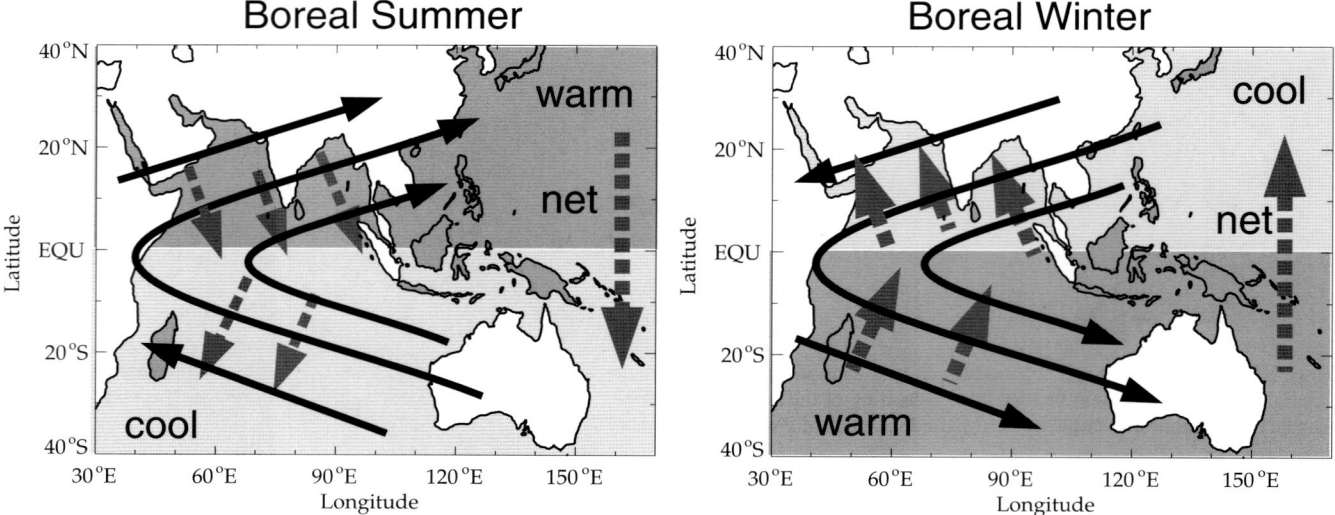

FIGURE 6 A regulatory model of the annual cycle of the Indian Ocean–monsoon system depicted for summer (June–September), and winter (December–February). Curved black arrows denote the wind forced by the large-scale differential heating denoted by "warm" and "cool". The grey arrows are the Ekman transports forced by the winds. The vertical grey arrow to the right of the panels shows the net ocean heat transport which reverses between boreal summer and winter. The net effect of the combined wind forced ocean circulation is to transport heat to the winter hemisphere thus modulating the SST differences between the hemispheres.

element of the meridional heat transport is that it is accomplished to a large degree by Ekman processes, a point that has been noted in previous work. For example, Levitus (1987) was the first to note that meridional Ekman transports are southward in the Indian Ocean on both sides of the equator and in both the boreal summer and the annual mean. Wacongne and Pacanowski (1996) describe a process where the upwelled waters within the western boundary currents are transported across the equator via the action of Ekman transports. Loschnigg and Webster (2000) showed that the seasonal reversal of the Ekman transports are central to the maintenance of the heat balance of the Indian Ocean.

Figure 6 shows a schematic of the monsoon system. The two panels represent summer and winter with relatively warm water (shaded) in the northern basin during summer and in the southern basin during winter. Surface winds similar to those shown in Fig. 3 are superimposed on the figure. In each season there is a strong flow from the winter to the summer hemisphere with a characteristic monsoon "swirl". The divergent part of the wind field (not shown: see Webster *et al.*, 1998) is also from the winter to the summer hemisphere. The grey arrows represent the ocean Ekman transports associated with the surface wind forcing. Irrespective of the season, the Ekman transports are from the summer to the winter hemisphere. The total effect of the feedback is to cool the summer hemisphere and warm the winter hemisphere, thus reducing the SST gradient between the summer and winter hemispheres. These transports are sufficiently large to be responsible for reducing the heating of the upper layers of the summer hemisphere to values less than shown in Table 1. In the manner described in Figure 6, the amplitudes of the seasonal cycle of the monsoon are modulated through the negative feedbacks between the ocean and the atmosphere.

C. INTERANNUAL VARIABILITY OF THE MONSOON

Figure 7 shows the difference of the near-surface wind fields between strong and weak monsoon years, defined in terms of the monsoon index shear index defined by Webster and Yang (1992). Using the index criteria, 1979, 1983, 1987 and 1992 were categorized as "weak" monsoon years and 1980, 1981, 1984, 1985 and 1994 as "strong" monsoon years. In agreement with earlier analyses of Webster and Yang (1992) and Webster *et al.* (1998), the difference fields show a tendency for an increase in westerlies across the Indian Ocean region during strong years or, alternatively, an increase is low-level easterlies during weak years. There are important local manifestations of these difference fields. In strong monsoon years the southwesterlies towards East Africa are enhanced while, at the same time, there is an increase in onshore flow towards

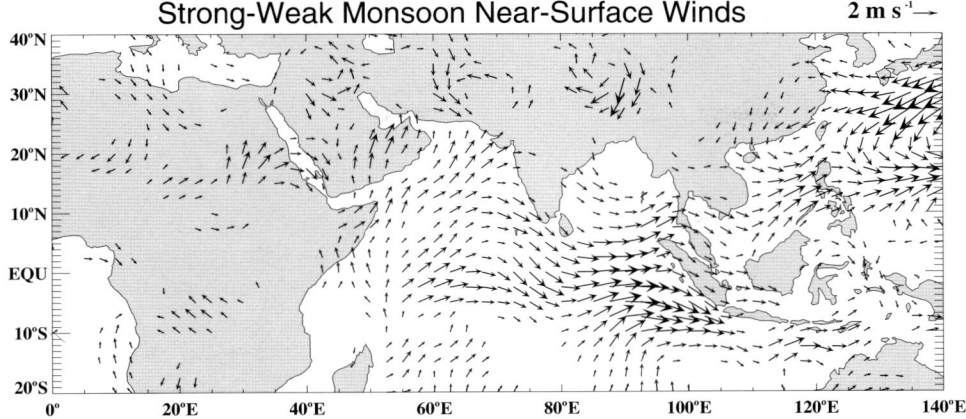

FIGURE 7 Difference in the surface winds in JJA between strong and weak monsoon years. Strong and weak monsoon years were determined using the monsoon shear index of Webster and Yang (1992) which specified 1979, 1983, 1987 and 1992 as "weak" years and 1980, 1981, 1984, 1985 and 1994 as "strong" monsoon years.

Sumatra. In weak monsoon years there is the reversal of the wind vectors. The form of the difference fields shown in Fig. 7 will turn out to be of critical importance in understanding the manner in which the monsoon is regulated. This is because the changes in winds between the monsoon extremes occur in regions where major upwelling occurs.

1. Modes of Interannual Variability in the Monsoon

Statistical analyses indicate specific bands of monsoon variability (e.g. Webster et al., 1998). Here we consider major frequency bands longer than the annual cycle. These are the biennial period (first discussed by Yasunari, 1987, 1991; Rasmusson et al., 1990; Barnett, 1991), and the multiyear variability that appears on ENSO timescales. Within the biennial period we include the newly discovered Indian Ocean dipole (Webster et al., 1999; Saji et al., 1999; Yu and Rienecker, 1999, 2000).

a. Biennial Variability

The interannual variability of monsoon rainfall over India and Indonesia–Australia shows a biennial variability during certain period of the data record (Fig. 2). It is sufficiently strong and spatially pervasive during these periods to show prominent peaks in the 2–3-year period range, constituting a biennial oscillation in the rainfall of Indonesia (Yasunari and Seki, 1988) and east Asia (Tian and Yasunari, 1992; Shen and Lau, 1995) as well as in Indian rainfall (Mooley and Parthasarathy, 1984). The phenomenon is referred to as the tropospheric biennial oscillation (TBO) in order to avoid confusion with the stratospheric quasi-biennial oscillation (QBO), appears to be a fundamental characteristic of the Asian–Australian monsoon rainfall.

The rainfall TBO, apparent in Fig. 2, appears as part of the coupled ocean–atmosphere system of the monsoon regions, increasing rainfall in one summer and decreasing it in the next. The TBO also possesses a characteristic spatial structure and seasonality (Rasmusson et al., 1990; Ropelewski et al., 1992). Meehl (1994a) stratified ocean and atmospheric data relative to strong and weak Asian monsoons. He found specific spatial patterns of the TBO with a distinct seasonal sequencing. Anomalies in convection and SST migrate from south Asia toward the southeast into the western Pacific of the Southern Hemisphere following the seasons. Lower-tropospheric wind fields associated with the TBO in the SST fields possess an out-of-phase relation between the Indian Ocean and the Pacific Ocean basins (Ropelewski et al., 1992) with an eastward phase propagation from the Indian Ocean toward the Pacific Ocean (Yasunari, 1985; Kutsuwada, 1988; Rasmusson et al., 1990; Ropelewski et al., 1992; Shen and Lau, 1995) providing possible links between monsoon variability and low-frequency processes in the Pacific Ocean (Yasunari and Seki, 1992; Clarke et al., 1998).

Explanations for the TBO fall into two main groups:

(i) The TBO results from feedbacks in the seasonal cycle of the atmosphere–ocean interaction in the warm water pool region, especially in the western Pacific Ocean. For example, Clark et al. (1998) suggest the oscillation may be produced by an air–sea interaction instability involving the mean seasonal wind cycle and evaporation. They argue that similar instabilities are not possible in the Indian Ocean and that Indian Ocean oscillations found there are the result of Pacific instabilities.

(ii) Biennial oscillations occur as a natural variability of the monsoon coupled ocean–atmosphere–land monsoon system. As distinct from the views expressed in (i), the source of the biennial oscillation is thought to reside in the Indian Ocean. Nicholls (1983) noted a seasonal change in the feedback between the wind field and surface pressure. In the monsoon westerly (wet) season the wind speed anomaly is negatively correlated to the pressure anomaly, while in the easterly (dry) season it is positively correlated. The wind speed anomaly, on the other hand, is negatively correlated to the SST change throughout the year through physical processes such as evaporation and mixing of the surface ocean layer. Nicholls suggested that a simple combination of these two feedbacks in the course of the seasonal cycle induces an anomalous biennial oscillation. Meehl (1994a,b, 1997) substantiated Nicholls' hypothesis but focused on the memory of oceanic mixed layer. That is, when large-scale convection over the warm water poor region, associated with seasonal migration of ITCZ and the monsoon, is stronger (weaker), the SST will eventually become anomalously low (high) through the coupling processes listed above. The anomalous state of the SST, thus produced, would be maintained through the following dry season and even to the next wet season. In turn, the SST anomaly produces weaker (stronger) convection. In this class of hypotheses the ocean–atmosphere interaction over the warm water pool appears to be of paramount importance.

A schematic diagram of the four phases of Meehl's biennial monsoon system (Meehl, 1994a) is shown in Fig. 8. Panel (A) depicts the winter season prior to the first monsoon season showing anomalously warm SST in the central and western Indian Ocean and cooler SSTs in the eastern Indian Ocean and the Indonesian seas. Anomalously warm SSTs in the Indian Ocean herald a stronger monsoon supposedly by a heightened

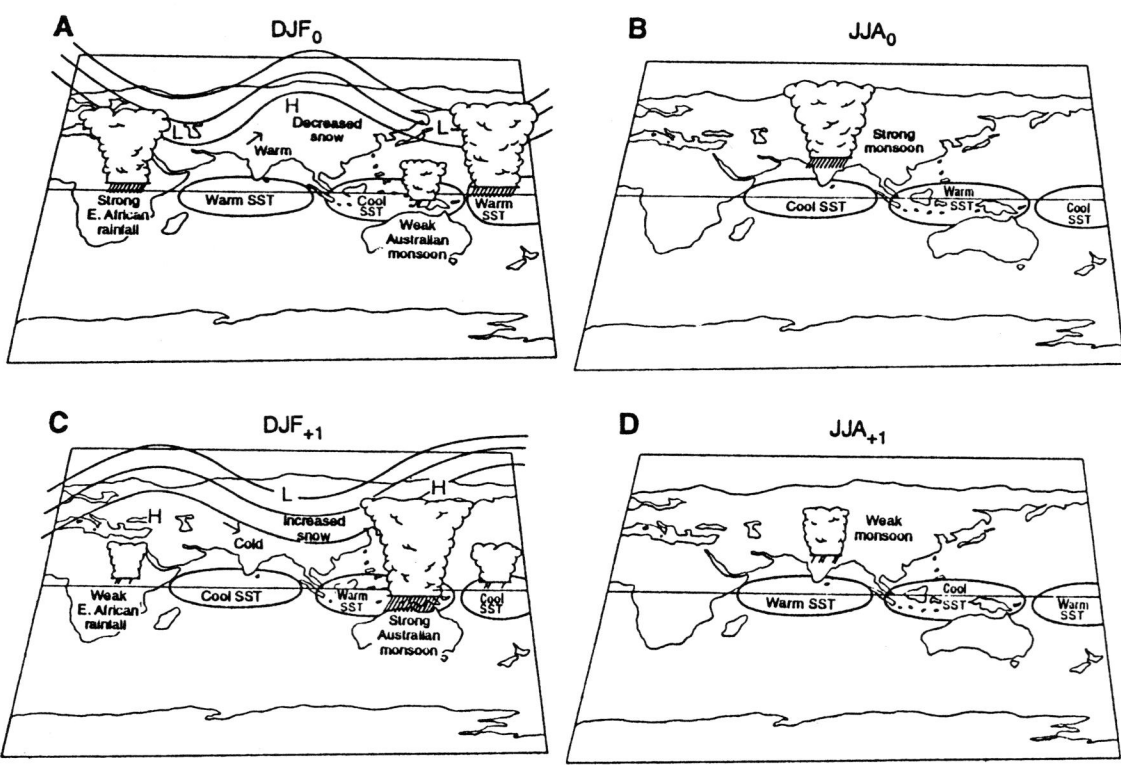

FIGURE 8 Schematic diagram of the Meehl-biennial mechanism. Diagram shows the evolution before a strong monsoon (A), through a strong monsoon season (B), to the northern winter after the strong monsoon before the weak monsoon (C), to the following weak monsoon (D) (From Meehl, 1997.)

surface hydrological cycle (panel B). This is consistent with the empirical results of Clark *et al.* (2000a). A stronger monsoon is accompanied by stronger wind mixing and evaporation, which leads to cooler SSTs in the central and eastern Indian Ocean. A reversal of the east–west SST gradient produces a stronger North Australian monsoon (panel C). In turn, the colder than normal Indian Ocean leads to a weak Indian summer monsoon (panel D). The theory also notes that a strong South Asian monsoon is preceded by a strong East African monsoon and a weak South Asian monsoon by a weak East African monsoon. Presumably, the oscillation of the East African monsoon is associated with the change of the longitudinal SST gradient.

The sequence of SST change shown in Fig. 8 follows observations quite closely including the oscillation of precipitation from year to year (Fig. 2). Whereas it is very clear that the oscillation of the SST in the Indian Ocean is indelibly tied to the variability of the monsoon rains, there are two problems with the theory. First, there is no satisfactory explanation for the change in the SST gradient along the equator. Second, it is difficult to account for the persistence of SST anomalies for the 9 months between the end of one summer season and the start of the next. Clearly, this cannot be accomplished by thermodynamical processes alone. In fact, the e-folding time of 50 m mixed layer with a 1 K anomaly at 303 K is between 40–60 days. There must be dynamical ocean processes at work in the Indian Ocean that increase the persistence of the anomalies.

b. The Indian Ocean Dipole

Between July 1997, and the early summer of 1998, the strongest seasonal SST anomalies ever recorded occurred in the Indian Ocean. During this period, the equatorial gradients of both SST and sea surface height (SSH) reversed with cooler surface temperature in the eastern basin and warmer in the west. These anomalies occurred in conjunction with strong easterly wind anomalies across the equatorial Indian Ocean. The anomalies in SST and SSH persisted for almost a year and coincided with the 1997–1998 El Niño. This event has received considerable attention (e.g. Webster *et al.*, 1999; Saji *et al.*, 1999; Yu and Rienecker, 1999, 2000). Following the summer of 1998, the pattern changed polarity, with anomalously, warm water occupying the eastern Indian Ocean and cold water in the west. In 1996, the SST distribution was very similar to that found in 1998. In fact, over the last few years there has been a sequence of positive and negative dipole events in the Indian Ocean.

Analysis of Indian Ocean SST data reveals that the 1997–98 and 1961 events were members of a longitudinal SST anomaly oscillation in the tropical Indian Ocean. Reverdin *et al.* (1986) and Kapala *et al.* (1994) refer specifically to an event of similar magnitude in 1961. Figure 9a plots a September–November (SON) dipole index defined by Clark *et al.* (2000b) which is very similar to the index defined by Saji *et al.* (1999). The index oscillates between positive (anomalously warm west, cold east) and negative (anomalously cold west, warm east) values. Figure 10 compares mean seasonal anomalous SON SST distributions for the 1997 positive dipole event with the negative dipole occurring the previous year. In both years the anomalous SST difference exceeded 2°C.

Because of the relative abundance of data and the magnitude of the event, the 1997–1998 positive event

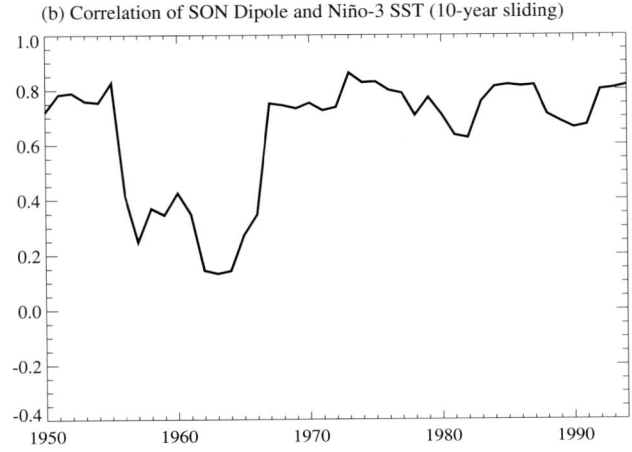

FIGURE 9 Statistics of the Indian Ocean dipole. (a) Time section of the dipole index. Index is defined as the SST difference in the areas 5°S–10°N and west of 60°E, and 10°S–5°N and east of 100°E for the October–November period. The index, defined by Clark *et al.* (2000b) is similar to that defined by Saji *et al.* (1999). (b) Ten-year running correlations between the dipole index and the October–November Niño-3 SST.

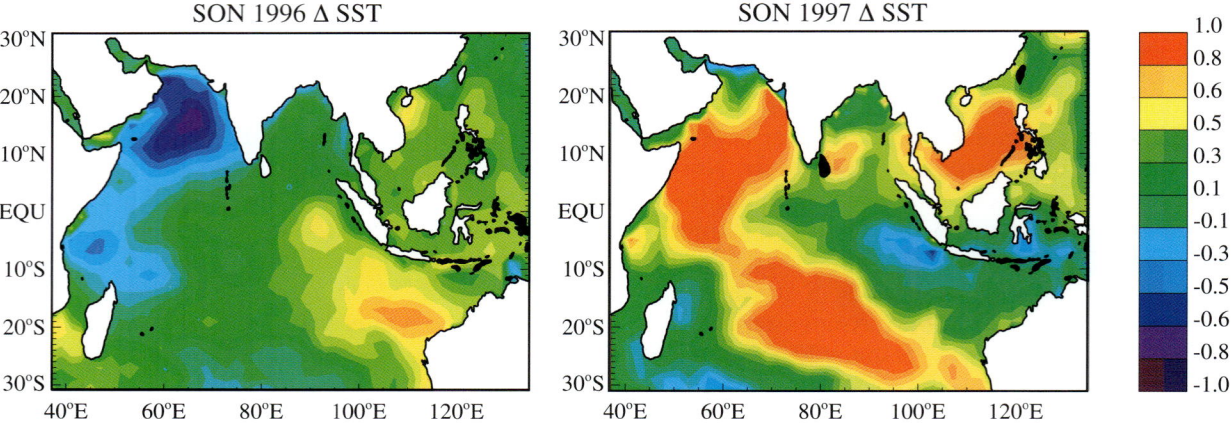

FIGURE 10 The anomalous SST fields for the boreal autumn (SON) over the Indian Ocean for a positive dipole year (1997) and a negative dipole year (1998).

was studied extensively. There are a number of points that have emerged from these studies which are pertinent to subsequent discussion:

(i) The association of the Indian Ocean dipole with the El Niño of 1997–1998, or with the ENSO phenomena in general, remains unclear. Saji *et al.* (1998) for example, states that there is little or no relationship between ENSO and the dipole. However, there are moderate correlations between the dipole index and the Niño 3 SST index which may be seen from the 10-year sliding correlation between the SON dipole and the Niño 3 SST (Fig. 9b). An overall correlation of about 0.55 exists. Based on correlations between ENSO and the dipole, Reason *et al.* (2000) argue that the dipole is simply an extension of the ENSO influence in the Indian Ocean.

(ii) It is clear that ENSO variability may be associated with some dipole events, but it may not be involved in all. Figure 9b shows significantly reduced correlation in the 1950s and 1960s. Furthermore, during 1997–1998 the climate patterns around the Indian Ocean rim were very different from those normally associated with El Niño. Although the El Niño was the strongest in the century, monsoon rains were normal in South Asia and North Australia when drought may have been expected. Instead of slightly increased rainfall in East Africa, the rainfall was the largest positive anomaly of the century, even larger than the 1961 excursion. Arguably, in 1997–1998, climate anomalies around the basin could be more associated with the anomalous conditions in the Indian Ocean than in the Pacific. Finally, large dipole events have occurred that are not matched by ENSO extrema, most notably the 1961 event documented by Reverdin *et al.* (1986) and Kapala *et al.* (1994).

(iii) Saji *et al.* (1999) note that the signature of the dipole commences in the late summer. Such a commencement occurs for both the positive and the negative phases of the dipole.

(iv) Webster *et al.* (1999) suggest that the dipole is a coupled ocean–atmosphere instability that is essentially self-maintaining. Analyses suggest that the initial cooling of the eastern Indian Ocean sets up an east-to-west SST gradient that drives near-equatorial anomalous easterlies. In turn, these winds change the SSH to tilt upwards to the west. Relaxation of SSH anomalies, in the form of westwardly propagating and downwelling ocean Rossby waves, depress the thermocline in the west and enhance the warming of the western Indian Ocean. The slow propagation of these modes (1–2 m s^{-1}), and the manner in which they maintain the warm water by deepening the thermocline, assures that the dipole is a slowly evolving phenomenon.

c. Interannual Variability and Monsoon–ENSO Relationships

Table 2 shows the long-term relationships between Indian and North Australian rainfall and ENSO variability. It is clear from the statistical analyses displayed in the table that the monsoon undergoes oscillations in conjunction with the ENSO cycle. Specifically, when the Pacific Ocean SST is anomalously warm, the Indian

TABLE 2 Relationship Between the Phases of ENSO (El Niño and La Niña) and the Variability of the Indian Summer Monsoon (June–August) and the North Australian Summer Monsoon (December–February)

Rainfall	All India summer			North Australian summer		
	Total	El Niño	La Niña	Total	El Niño	La Niña
Below Average	53	24	2	49	20	4
Above Average	71	4	19	58	5	17
Deficient	22	11	2	18	9	0
Heavy	18	0	7	17	2	5

Rainfall in these two monsoon regimes are represented by indices of total seasonal rainfall. "Deficient" and "heavy" rainfall seasons refer to variability less than, or greater than, one standard deviation, respectfully. (From Webster et al., 1998.)

rainfall is often diminished in the subsequent year. The correlation of the AIRI and the SOI over the entire period is −0.5 (Torrence and Webster, 1999). Shukla and Paolina (1983) were able to show that there was a significant relationship between drought and ENSO. In fact, all El Niño years in the Pacific Ocean were followed by drought years in the Indian region. Not all drought years were El Niño years, but out of the total of 22 El Niño years between 1870 and 1991, only two were associated with above average rainfall. La Niña events were only associated with abundant rainy seasons, and only two were associated with a monsoon with deficient rainfall. A large number of wet years were not associated with cold events, just as many drought years were not associated with warm events. Although the relationship is far from perfect, it is clear that the monsoon and ENSO are related in some fundamental manner. Similar relationships occur between the north Australian rainfall and ENSO.

The statistical relationships discussed above indicate a common co-occurrence of monsoon variability and ENSO extremes. But at what part of the annual cycle do anomalously strong and weak monsoon seasons emerge? To help answer this question mean monthly circulation fields were composited for the weak and strong monsoon years. The upper and lower tropospheric zonal wind fields in the south Asian sector for the composite annual cycle of the strong and weak monsoons are shown in Fig. 11. At the time of the summer monsoon when the anomalous monsoon is defined, both the low-level westerlies and the upper-level easterlies are considerably stronger during strong monsoon years than during weak years as depicted in Fig. 7. But what is very striking is that the anomalous signal of upper level easterlies during strong years extends back until the previous winter with 5–6 m s^{-1} less westerly during strong years. However, in the lower troposphere the difference between strong and weak years occurs only in the late spring and summer. Thus there is a suggestion that the anomalies signify

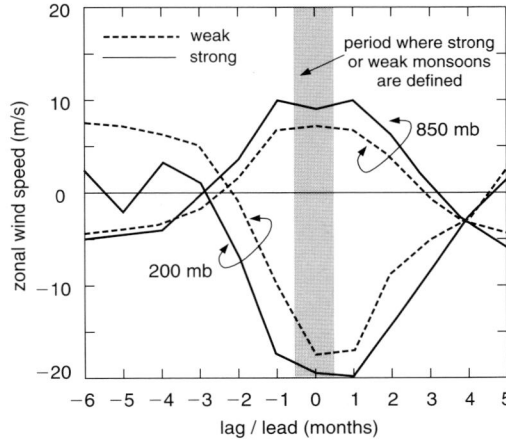

FIGURE 11 Variation of the anomalous 850 mb and 200 mb zonal wind components relative to strong and weak monsoon years defined at month zero (July). The curves indicate a different circulation structure in the Indian Ocean region prior to strong and weak summer monsoons up to two seasons ahead. (After Webster and Yang, 1992.)

external influences from a broader scale into the monsoon system. This surmise is supported by what is known about tropical convective regions. Generally, enhanced upper tropospheric winds will be accompanied by enhanced lower tropospheric flow of the opposite sign. But this is clearly not the case prior to the strong monsoon, suggesting that the modulation of the upper troposphere probably results from remote influences. This is troublesome in relation to ENSO–monsoon relationships. The ENSO cycle is phase-locked with the annual cycle and an El Niño, for example, develops in the boreal spring. Figure 11 suggests that there are precursors to anomalous monsoons that develop before the growth of an ENSO event. Whether or not this is a result of problems in the definition of an anomalous monsoon or a result from which insight may be gained regarding monsoon variability is not known.

2. Interannual Modes in Ocean Heat Transport

If the monsoon is indeed a coupled ocean–atmosphere system then there should be oceanic variability occurring on the same timescales as atmospheric variability. To test this hypothesis, Loschnigg and Webster (2000) forced the McCreary et al. (1993) dynamic ocean model with NCEP winds. Radiative forcing was obtained from Bishop and Rossow (1991). In the results discussed here, 5-day average fields were used to force the model.

Figure 12a shows time sections of the mean monthly cross-equatorial heat transports, storage change and net heat flux into the Indian Ocean north of the equator for the period 1984–1990 (Loschnigg and Webster, 2000). The period was chosen because it included the years 1987 and 1988 which corresponded to an intense El Niño and a weak monsoon followed by a La Niña and a strong monsoon, respectively. Figure 12b shows a sequence of annual averages of these fields. Averaged over the entire 7 years, the net heating of the North Indian Ocean is balanced to a large degree by the southward heat transport across the equator. Storage change averages roughly to zero through the period. However, these long-term balances do not occur during 1987 and 1988. During 1987, the net surface heat flux is about 30% higher than the 7-year average, presumably because of the weaker than average winds (less evaporation and mixing) and greater surface radiational flux because of less cloud and convection. During 1997 the rises in storage indicates an increase in the mean temperature of the upper North Indian Ocean. Also, the southward heat transport is less than average which is consistent with lighter winds and less southward Ekman transports. In 1988, a strong monsoon year, the net surface heat flux is lower than the 7-year average by 30% as a result of greater evaporation and by the reduced solar heating. Cross-equatorial heat transports are almost double the previous year, again reflecting the impact of the stronger than average winds. Storage decreases indicates a net cooling of the North Indian Ocean. The results of this experiment indicate clear differences in ocean behaviour between strong and weak monsoon years.

Figures 12c and d show details of the meridional heat transport as a function of time and latitude for 1987 and 1988. The format is the same as in Fig. 5b except 5-day average forcing is used. The general annual cycle of heat transport is apparent in both years with northward transport during the winter and early spring and southward transport during summer. Both years also have strong intraseasonal variability suggesting a response by the ocean to higher frequency forcing. However, there are also significant differences. The stronger southward transport in 1988 noted in Fig. 12a can be seen clearly in the enhanced transports during the summer. However, in 1987, the weaker southward transports are made up by a combination of stronger northward heat transport in winter and early spring and weaker southward transport during summer.

Cherikova and Webster (2001) have repeated the interannual integrations of the ocean model using the 40-year period of the NCEP reanalysis data. As independent surface radiative flux data is not available for the entire period, the NCEP reanalysis data is used for the radiative heating fields as well as for surface wind forcing. Figure 13a shows a 40-year time section of the annually averaged anomalous meridional ocean heat transports plotted between 30°S and 25°N. The long-term 40-year mean has been removed leaving anomaly fields. Three specific features can be noticed:

(i) The magnitude of the variability from year to year is the same order of magnitude as found by Loschnigg and Webster (2000). Variations from year to year are about ± 0.1 PW.
(ii) With only a few exceptions, the anomalies are coherent in latitude. That is, the anomalies during a particular year are usually either positive or negative from 30°S to 25°N.
(iii) The variability is strongly oscillatory. Figure 13b shows the spectra of the transport across the equator. Three significant bands located at 2, 2.6 and 4 years are apparent suggesting a strong biennial and ENSO periodicity. Each is significant at the 95% level.

In summary, the ocean shows variability in both its thermodynamical and dynamical structures on interannual timescales. Furthermore, the response of the ocean appears to be compensatory suggesting that there exist negative feedbacks on interannual timescales that regulate or govern the system.

3. Interannual Regulation of the Monsoon

Figure 6 suggested a coupled ocean–atmosphere regulatory system of the monsoon annual cycle that reduced the amplitudes of seasonal variability in both hemispheres. It is a simple matter to extend this system to explain the oceanic compensations acting from year to year, as noted in the last section. For example, let us assume that the North Indian Ocean SST were warmer than normal in the boreal spring. The ensuing stronger monsoon flow would produce greater fluxes of heat to the southern hemisphere in the same manner found in Figure 12b. If the North Indian Ocean were cooler then

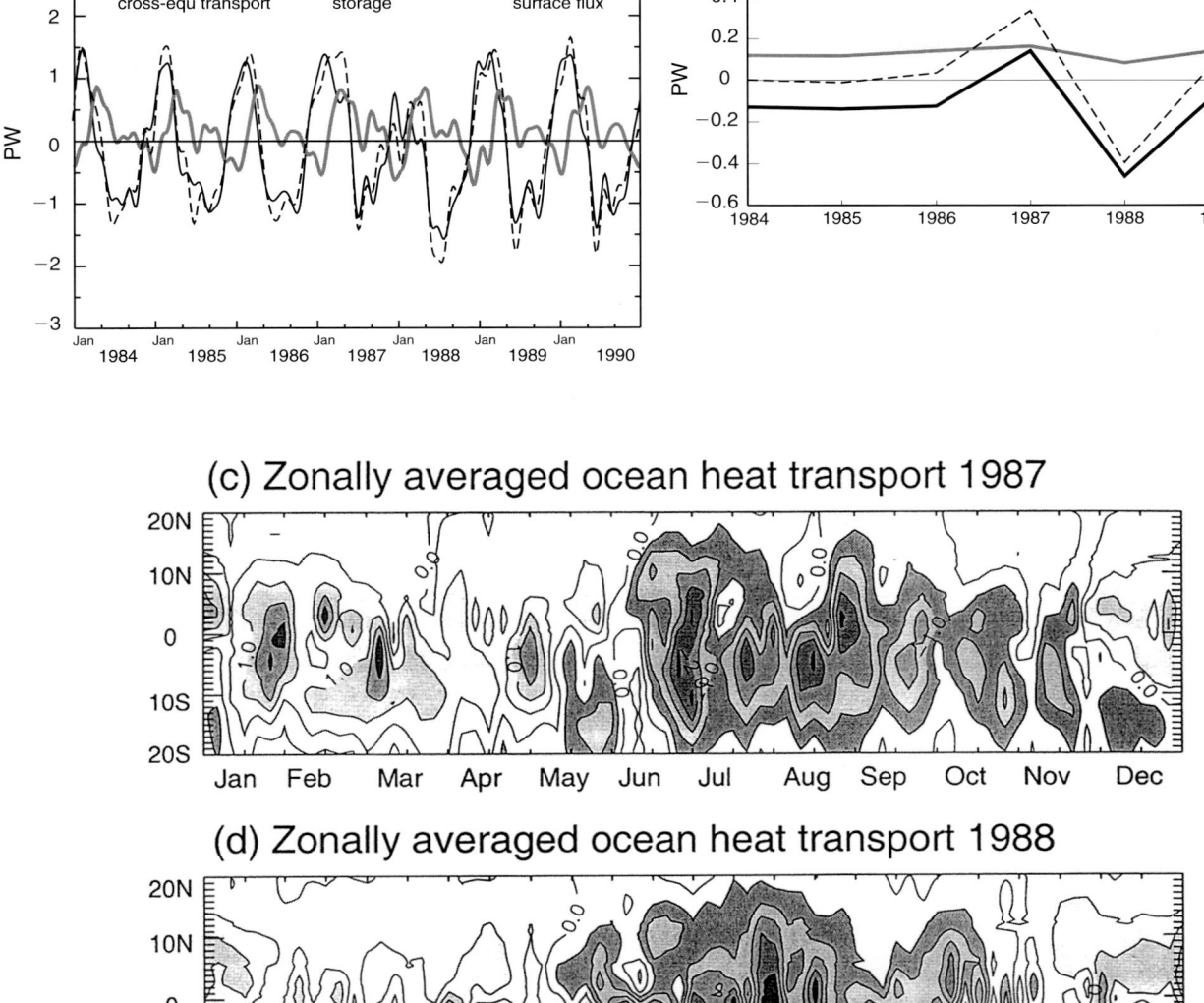

FIGURE 12 (a) Evolution of the components of the heat balance of the North Indian Ocean. Mean monthly values are plotted from January 1984 through December 1990. (b) Annual averages of the components of the heat balance of the Indian Ocean basin north of the equator for model simulations of the period 1984–1990 showing the cross-equatorial heat transport, and the heat storage and surface flux into the North Indian Ocean. (c) Time–latitude section of the total northward transport for 1987. This year was a weak monsoon year coinciding with a strong El Niño. (d) Same as (c) but for 1998 which was a strong monsoon year and also the year of a strong La Niña. Model was forced by 5-day average NCEP surface winds (Kalnay *et al.*, 1996) and ISCCP fluxes (Bishop and Rossow, 1991).

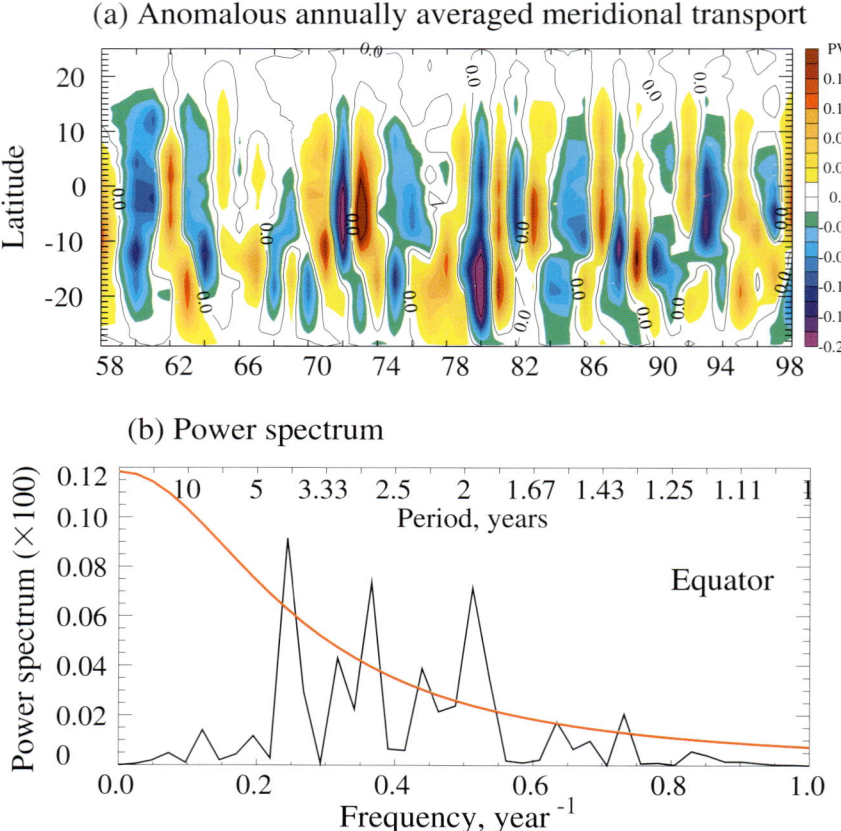

FIGURE 13 (a) Time–latitude section of the interannual anomalies of the annually averaged northward heat flux in the Indian Ocean. Field was computed by subtracting out the long-term average annual northward heat transport from the individual yearly values. Variations from year to year are about ± 0.05 to 0.1 PW. Note the latitudinal coherence of the anomalies. (b) The spectra of the anomalies of the annually averaged northward heat flux in the Indian Ocean. The 95% significance line relative to a red-noise null hypothesis is shown. Three major significant peaks emerge at 2, 2.5 and 4 years.

one would expect a reversal in compensation. In a sense, the interannual regulation described here is very similar to the Meehl (1994a) theory. Like the Meehl theory, the natural timescale of the oscillation is biennial. In addition, because ocean heat transport is an integral part of the theory, it adds a dynamic element to Meehl's theory.

D. GENERAL THEORY OF THE REGULATION OF THE COUPLED OCEAN–ATMOSPHERIC MONSOON SYSTEM

In the last section, we modified Meehl's theory by noting that the anomalous monsoon winds will induce ocean heat transports that will reverse the sign of the monsoon anomaly. However, there are still a number of issues that need to be considered. For example:

(i) Whereas there are dynamic elements added to the Meehl theory, one is still faced with the problem of maintaining an upper ocean temperature anomaly from one year to the next.

(ii) The regulation theories, either the Meehl theory or the modified Meehl theory, do not involve the Indian Ocean dipole. It could be possible, of course, that the dipole is an independent phenomenon. However, the similarity of the basic time period of the dipole to that of monsoon variability (essentially biennial) and the fact that the dipole emerges during the boreal summer monsoon suggests an interdependence. The problem, though, is how to incorporate an essentially zonal phenomenon (the dipole) with the essentially meridional phenomenon (the oceanic heat transport) into a general theory of the monsoon.

A theory which takes into account these two problems is now developed. It is shown schematically in Fig. 14. The figure displays a sequence through two monsoon seasons starting (arbitrarily) in the boreal spring. There are three columns in the figure representing the anomalous meridional oceanic heat transports (column 1), the influence of the anomalous monsoon circulation on the ocean (column 2), and the evolution of the dipole (column 3). During the two-year period, the monsoon goes through both weak and strong phases. At the same time, the dipole progresses through a positive and negative phase. We will now argue that the morphology of the dipole and the monsoon are intimately related.

(i) The left-hand column of Fig. 14 describes essentially the regulation theory discussed in Section C.3. The sequence starts with an anomalously cold North Indian Ocean in the boreal spring (March–May of the first year: MAM:1) which often precedes a weak monsoon (e.g., Sadhuram, 1997; Harzallah and Sadourny, 1997; Clark et al., 2000a) in the summer of the first year (JJA:1). A weak monsoon is associated with a reduced southward heat transport leading to the SST distribution in the first boreal fall (SON:2) shown in the second figure of the row. If the anomaly persists through to the second spring (MAM:2), it will lead to a strong monsoon in the second summer (JJA:2). Enhanced southward transports and reduced net heating of the Indian Ocean leads to an anomalously cold northern Indian Ocean.

(ii) Stronger and weaker summer monsoons also influence the ocean system in other ways. For example, the anomalous monsoon circulation (Fig. 7) will influence the upwelling patterns in two major areas: along the east Africa coast north of the equator, and along the western coast of Sumatra. The reduced southwesterly flow of Africa will decrease upwelling, while offshore flow near Sumatra will enhance upwelling. Thus, during the weak monsoon of JJA:1 the changes in the monsoon circulation will create anomalously warm water in the west and colder water in the east. On the other hand, the circulation associated with the strong monsoon in JJA:2 will produce cooler water in the eastern basin (enhanced southwesterlies) and warmer water in the east (onshore winds along the Sumatra coast). In summary, changes in the monsoon winds between strong and weak monsoons can create zonal anomalies in the SST distribution.

(iii) The east–west SST gradients caused by the anomalous monsoon intensities can lead to enhancements of the zonal SST gradients by coupled ocean–atmosphere instabilities, as described in detail in Webster et al. (1999) and summarized in Section C.1.b. Simply, the SST gradients force zonal wind anomalies which change the distribution of low-latitude sea-level height distribution. During JJA:1, the sea-level height will slope upwards to the west, while during JJA:2 it will slope upwards to the east. Relaxation of the sea-level height takes place in the form of equatorial modes. For example, during the period JJA:1–DJF:1 the relaxation will be in the form of downwelling Rossby waves. Besides having a slow westward propagation they are downwelling and deepen and warm the western Indian Ocean. In turn, the enhanced SST gradient will produce stronger easterly winds which will continue to maintain the east-to-west slope of the surface. Between JJA:2 and DJF:2 the onshore winds towards Sumatra will deepen the thermocline and enhance the zonal west-to-east SST gradient. In turn, the winds themselves will be enhanced by a wind field responding to an increasing SST gradient. The important aspect of the dipole is that it introduces slow dynamics into the system.

(iv) Careful inspection of Fig. 14 shows that the impact of the dipole is to enhance the SST distributions associated with meridional heat transports shown in the first column. For example, the dipole that develops in the period JJA:1–DJF:1 will increase the SST in the northwest equatorial Indian Ocean. This SST enhancement can be seen by following the sequence (a)–(e) in Fig. 14. On the other hand, the second dipole will cool the SST in the same location. This is the region found by Sadhuram (1997), Hazzallah and Sadourny (1997) and Clark et al. (2000a) to correlate most strongly in the winter with the following monsoon. Thus, the role of the dipole is to enhance and prolong the SST patterns necessary to regulate the intensity of the monsoon system.

E. CONCLUSIONS

In the preceding paragraphs, we have developed a theory that regulates the monsoon on both annual and interannual timescales. The study was motivated by noting that the surface heat balances in the Indian Ocean do not match the observed evolution of the SST indicating the importance of ocean heat transports. Furthermore, it appeared curious that the year-to-year variability of the South Asian monsoon is relatively small. Hence, we developed a theory of regulation of

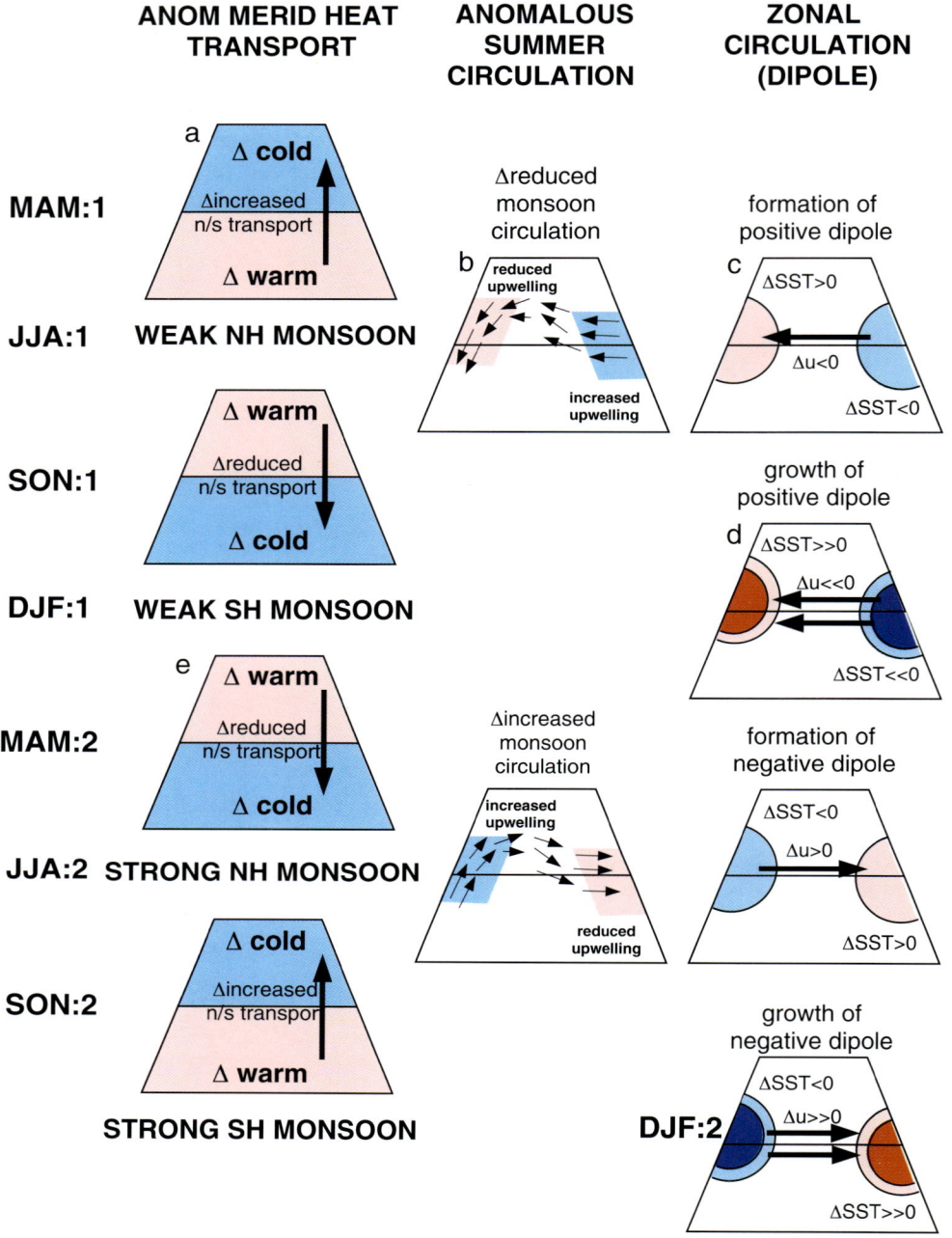

FIGURE 14 Schematic of a general theory of an ocean–atmosphere regulation system for the monsoon and the Indian Ocean. Each column indicates a set of processes. The first column shows modulation of the monsoon variability by changes in the heat transport induced by the monsoon winds. In essence this sequence represents the Meehl (1997) biennial oscillation mechanism but with ocean dynamics. The second column shows the impact of the strong and weak monsoons on the upwelling regions of the ocean basin. The wind patterns are schematic representations of those computed in Fig. 7. The third column represents the development of the Indian Ocean dipole relative to the upwelling patterns developed by the anomalous monsoon wind fields. Growth of the dipole anomaly is assumed to follow the coupled ocean–atmosphere instability described by Webster *et al.* (1999). Taken as a whole, the figure suggests that there are multiple components that regulate the monsoon with each component acting in the same sense. One important role of the dipole (either positive or negative) is to provide slow dynamics (or memory) to the SST anomalies induced by the strong or weak monsoons. For example, the sequence (a)–(e) helps to perpetuate the Northern Hemisphere anomalously warm temperatures created by the weak monsoon during the previous summer.

the monsoon that rests on negative feedbacks between the ocean and the atmosphere. We are now in a position to address some of the questions raised in the introduction.

Perhaps the most important conclusion is that it is clear that the ocean involves itself in the monsoon circulation in a dynamic manner. Although the ocean is responding to forcing from the atmosphere, the response is such that there is a strong feedback to the atmosphere. This feedback governs the amplitude and phase of the annual cycle and also modulates interannual variability.

One of the problems that emerged in earlier theories of monsoon amplitude regulation (e.g., Meehl, 1997) is that it is difficult to understand how an SST anomaly pattern produced by an anomalous monsoon can perpetuate from one year to the next. By involving ocean dynamics in the regulation process we have managed to introduce mechanisms that allow SST anomalies to persist from one year to the next. This was accomplished by noting that the Indian Ocean dipole is also parented by an anomalous monsoon through the generation of zonal temperature gradients between upwelling regions. The slow dynamics of the dipole act to enhance the zonal SST gradient initiated by the anomalous monsoon irrespective of the sign of the initial perturbation. As the dipole grows, the SST anomalies produced occur in locations that are conducive to the generation of a reverse anomaly in the monsoon. In other words, the dipole adds the slow dynamics needed in the Meehl theory.

An immediate question is whether or not the dipole is an independent entity or a function of forcing from the Pacific Ocean. First, there is irrefutable statistical evidence (Table 2) that ENSO variability in the Pacific Ocean produces a response in monsoon variability. Also, a substantial amount of the variance of the dipole can be explained in terms of the Niño-3 SST variability. However, there are periods when the dipole and ENSO are unrelated statistically when it is difficult to find an ENSO extremum to match the development of the dipole. Such a year was 1961. However, there may be a way of resolving this apparent paradox. Let us assume that the dipole is a natural mode of oscillation in the Indian Ocean in much the same way as El Niño is a natural oscillation of the Pacific. In fact, Webster et al. (1999) described a coupled ocean–atmosphere instability that supports this hypothesis. We also recall that the dipole is initiated by an anomalous monsoon. It is quite possible that a weak monsoon induced by El Niño will induce, in turn, a positive dipole which then acts to reverse the impacts of ENSO during the following year. But, according to the hypothesis, a dipole will develop relative to an anomalous monsoon no matter how the monsoon is perturbed. Perhaps in 1961, and other such years, other factors could have perturbed the monsoon.

The results presented in this study are incomplete. They have depended on a number of empirical studies and experiments with stand-alone ocean models forced with atmospheric fields. The complete associations inferred from the conclusions above are difficult to establish from empirical studies because of the very convoluted nature of the phenomenon. However, a considerable effort has been to ensure that the results are consistent with a multifaceted phenomenon. Further work will probably have to await experimentation with fully coupled ocean–atmosphere land models.

ACKNOWLEDGEMENTS

The research undertaken in this chapter was supported by the Division of Atmospheric Science of the National Science Foundation (NSF) under Grant ATM-9525847.

References

Bishop, J. K. B. and W. B. Rossow, 1991: Spatial and temporal variability of global surface solar irradiance. *J. Geophys. Res.*, **96**, 16839–16858.

Barnett, T. P., 1991: The interaction of multiple time scales in the tropical climate system. *J. Climate*, **4**, 269–285.

Clark, C. O., J. E. Cole and P. J. Webster, 2000a: SST and Indian summer rainfall: predictive relationships and their decadal variability. *J. Climate*, **13**, 2503–2519.

Clark, C. O., P. J. Webster and J. E. Cole, 2000b: Interannual and interdecadal East African rainfall variability and the Indian Ocean dipole. Submitted to *J. Climate*.

Clarke, A., J. Liu and S. Van Gorder, 1998: Dynamics of the biennial oscillation in the equatorial Indian and far western Pacific Oceans. *J. Climate*, **11**, 987–1001.

Cherikova, G. and P. J. Webster, 2001: Interannual variability in Indian Ocean heat transports. To be submitted to *J. Climate*.

Fu, R., A. D. Del Geuio, W. B. Rossow and W. T. Liu. 1992: Cirrus-cloud thermostat for tropical sea surface temperatures tested using satellite data, *Nature*, **358**, 394–397.

Gadgil, S. and S. Sajani, 1998: Monsoon precipitation in the AMIP runs. *Climate Dynam.*, **14**, 659–689.

Godfrey, J. S. *et al.*, The role of the Indian Ocean in the global climate system: Recommendations regarding the global ocean observing system. Report of the Ocean Observing System Development Panel, *Background Report 6*. Texas A&M Univ., College Station, 1995.

Godfrey, J. S., R. A. Houze, R. H. Johnson, R. Lukas, J. L. Redelsperger, A. Sumi and R. Weller, 1998: Coupled Ocean–Atmosphere Response Experiment (COARE): An interim report, *J. Geophys. Res. Oceans*, **103**, 14395–14450.

Hartmann, D. L. and M. L. Michelsen, 1993: Large-scale effects on the regulation of tropical sea surface temperature. *J. Climate*, **6**, 2049–2062.

Hastenrath, S., 1986a: On climate prediction in the tropics. *Bull. Am. Meteorol. Soc.*, **67**, 692–702.

Hastenrath, S., 1986b: Tropical climate prediction: A progress report 1985–90. *Bull. Am. Meteorol. Soc.*, **67**, 819–825.

Hastenrath, S., 1994: *Climate Dynamics of the Tropics: An Updated Edition of Climate and Circulation of the Tropics*, Kluwer Academic Publishers, Norwell, MA.

Hastenrath, S. and L. Greischar, 1993: The monsoonal heat budget of the hydrosphere–atmosphere system in the Indian Ocean sector. *J. Geophys. Res.*, **98**, 6869–6881.

Hastenrath, S. and P. Lamb, 1978: On the dynamics and climatology of surface flow over equatorial oceans. *Tellus*, **30**, 436–448.

Hsiung, J., R. E. Newell and T. Houghtby, 1989: The annual cycle of oceanic heat storage and oceanic meridional heat transport. *Q. J. R. Meteorol. Soc.*, **115**, 1–28.

Harzallah, A. and R. Sadourny, 1997: Observed lead–lag relationships between Indian summer monsoon and some meteorological variables. *Climate Dynam.*, **13**, 635–648.

Kalnay, E., M. Kanamitsu, W. Collins *et al.*, 1996: The NCEP/NCAR 40-year reanalysis project, *Bull. Amer. Meteo. Soc.*, **77**(3), 437–471.

Kantha, L. H. and C. A. Clayson, 1994: An improved mixed layer model for geophysical applications. *J. Geophys. Res.*, **99**, 225–266.

Kapala, A., K. Born and H. Flohn, 1994: Monsoon anomaly or an El Niño event at the equatorial Indian Ocean? Catastrophic rains 1961/62 in east Africa and their teleconnections. In *Proceedings of the International Conference on Monsoon Variability and Prediction*, Tech. Doc. 619, pp. 119–126, World Meteorological Organization, Switzerland.

Kutsuwade, K., 1988: Spatial characteristics of interannual variability in wind stress over the western north Pacific. *J. Climate*, **1**, 333–345.

Latif, M., D. Dommenget, M. Dima *et al.*, 1999: The role of Indian Ocean sea surface temperature in forcing east African rainfall anomalies during December–January 1997/98. *J. Climate*, **12**, 3497–3504.

Lavery, B., G. Joung and N. Nicholls, 1997: An extended high-quality historical rainfall data set for Australia. *Aust. Meteorol. Mag.*, **46**, 27–38.

Levitus, S., 1987: Meridional Ekman heat fluxes for the world ocean and individual ocean basins. *J. Phys. Oceanogr.*, **17**, 1484–1492.

Loschnigg, J. and P. J. Webster, 2000: A coupled ocean–atmosphere system of SST regulation for the Indian Ocean. *J. Climate*, in press.

McCreary, J. P., P. K. Kundu and R. L. Molinari, 1993: A numerical investigation of the dynamics, thermodynamics and mixed-layer processes in the Indian Ocean. *Progr. Oceanogr.*, **31**, 181–244.

Meehl, G. A., 1994a: Influence of the Land Surface in the Asian Summer monsoon: external conditions versus internal feedback. *J. Climate*, **7**, 1033–1049.

Meehl, G. A., 1994b: Coupled ocean–atmosphere–land processes and south Asian Monsoon variability. *Science*, **265**, 263–267.

Meehl, G. A., 1997: The south Asian monsoon and the tropospheric biennial oscillation. *J. Climate*, **10**, 1921–1943.

Mohanty, U. C., K. J. Ramesh and M. C. Pant, 1996: Certain seasonal characteristics features of oceanic heat budget components over the Indian seas in relation to the summer monsoon activity over India. *Int. J. Climatol.*, **16**, 243–264.

Mooley, D. A. and B. Parthasarathy, 1984: Fluctuations in all-India summer monsoon rainfall during 1871–1978. *Climate Change*, **6**, 287–301.

Nicholls, N., 1983: Air–sea interaction and the quasi-biennial oscillation. *Mon. Weath. Rev.*, **106**, 1505–1508.

Nicholson, S. E., 1993: An overview of African rainfall fluctuations of the last decade. *J. Climate*, **6**, 1463–1466.

Oberhuber, J. M., 1988: An atlas based on the COADS data set: The budgets of heat, buoyancy and turbulent kinetic energy at the surface of the global ocean. *Rep. 15*, Max-Planck Institut, Hamburg.

Palmer, T. N., C. Brankovic, P. Viterbo and M. J. Miller, 1992: Modeling interannual variations of summer monsoons. *J. Climate*, **5**, 399–417.

Parthasarathy, B., K. R. Kumar and D. R. Kothawale, 1992: Indian summer monsoon rainfall indices: 1871–1990. *Meteorol. Mag.*, **121**, 174–186.

Parthasarathy, B., A. A. Munot and D. R. Kothawale, 1994: All-India monthly and summer rainfall indices: 1871–1993. *Theor. Appl. Climatol.*, **49**, 219–224.

Ramanathan, V. and W. Collins, 1991: Thermodynamic regulation of ocean warming by cirrus clouds deduced from observations of the 1987 El Niño. *Nature*, **351**, 27–32.

Rasmusson, E. M. and T. H. Carpenter, 1983: The relationship between the eastern Pacific sea surface temperature and rainfall over India and Sri Lanka. *Mon. Weath. Rev.*, **111**, 354–384.

Rasmusson, E. M., X. Wang and C. F. Ropelewski, 1990: The biennial component of ENSO variability. *J. Marine Sys.*, **1**, 71–96.

Reason, C. J. C., R. J. Allan, J. A. Lindesay *et al.*, 2000: ENSO and climatic signals across the Indian Ocean Basin in the global context. Part I, interannual composite patterns. *Int. J. Climatol.*, **20**, 1285–1327.

Reverdin, G., D. Cadet and D. Gutzler, 1986: Interannual displacements of convection and surface circulation over the equatorial Indian ocean. *Q. J. R. Meteorol. Soc.*, **122**, 43–67.

Ropelewski, C. F., M. S. Halpert and X. Wang, 1992: Observed tropospherical biennial variability and its relationship to the Southern Oscillation. *J. Climate*, **5**, 594–614.

Sadhuram, Y. 1997: Predicting monsoon rainfall and pressure indices from sea surface temperature. *Curr. Sci. India,* **72**, 166–168.

Saji, N. N., B. N. Goswami, P. N. Vinayachandran and T. Yamagata, 1999: A dipole mode in the tropical Indian Ocean. *Nature*, **401**, 360–363.

Shen, S. and K. M. Lau, 1995: Biennial oscillation associated with the east Asian monsoon and tropical sea surface temperatures. *J. Meteorol. Soc. Jpn*, **73**, 105–124.

Shukla, J., 1987a: Interannual variability of monsoons. In *Monsoons* (J. S. Fein and P. L. Stephens, Eds), John Wiley and Sons, pp. 399–464.

Shukla, J., 1987b: Long-range forecasting of monsoons. In *Monsoons* (J. S. Fein and P. L. Stephens, Eds), John Wiley and Sons, pp. 523–548.

Shukla, J. and D. A Paolina, 1983: The southern oscillation and long range forecasting of the summer monsoon rainfall over India. *Mon. Weath. Rev.*, **111**, 1830–1837.

Sperber, K. R. and T. N. Palmer, 1996: Interannual tropical rainfall variability in general circulation model simulations associated with the Atmospheric Model Intercomparison Project. *J. Climate*, **9**, 2727–2750.

Stephens, G. L. and T. Slingo, 1992: An air-conditioned greenhouse. *Nature*, **358**, 369–370.

Tian, S. F. and T. Yasunari, 1992: Time and space structure of interannual variations in summer rainfall over China. *J. Meteorol. Soc. Jpn.* **70**, 585–596.

TOGA COARE International Project Office, 1994: Summary report of the TOGA COARE International Data Workshop, Toulouse, France, 2–11 August 1994, Univ. Corp. For Atmos. Res., Boulder, CO.

Torrence, T. and P. J. Webster, 1999: Interdecadal changes in the ENSO-monsoon system. *J. Climate*, **12**, 2679–2690.

Vonder Haar, T. H. and A. H. Oort, 1973: New estimate of annual poleward energy transport by the northern hemisphere ocean. *J. Phys. Oceanogr.*, **3**, 169–172.

Wacongne, S., and R. Pacanowski, 1996: Seasonal heat transport in a primitive equatuous model of the tropical Indian ocean, *J. Phys Oceanogr.* **26**(12), 2666–2699.

Wallace, J. M., 1992: Effect of deep convection on the regulation of tropical sea surface temperatures. *Nature*, **357**, 230–231.

Webster, P. J. and R. Lukas, 1992: TOGA-COARE: The Coupled Ocean–Atmosphere Response Experiment. *Bull. Am. Meteorol. Soc.*, **73**, 1377–1416.

Webster, P. J. and S. Yang, 1992: Monsoon and ENSO: Selectively Interactive Systems. *Q. J. R. Meteorol. Soc.*, **118** (July, Pt II), 877–926.

Webster, P. J., T. Palmer, M. Yanai, R. Tomas, V. Magana, J. Shukla and A. Yasunari, 1998: Monsoons: processes, predictability and the prospects for prediction. *J. Geophys. Res. (TOGA special issue)*, **103**(C7), 14451–14510.

Webster, P. J., A. Moore, J. Loschnigg and M. Leban, 1999: Coupled ocean-atmosphere dynamics in the Indian Ocean during 1997–98. *Nature*, **40**, (23 Sep.), 356–360.

Yasunari, T., 1985: Zonally propagating modes of the global east–west circulation associated with the Southern Oscillation. *J. Meteorol. Soc. Jpn.* **63**, 1013–1029.

Yasunari, T. and Y. Seki, 1992: Role of the Asian monsoon on the interannual variability of the global climate system. *J. Meteorol. Soc. Jpn*, **70**, 177–189.

Yu, L. and M. Rienecker, 1999: Mechanisms for the Indian Ocean warming during the 1997–98 El Niño. *Geophys. Res. Lett.*, **26**, 735–738.

Yu, L. S. and M. M. Rienecker, 2000: Indian Ocean warming of 1997–1998. *J. Geophys. Res. Oceans*, **105**, 16923–16939.

Laminar Flow in the Ocean Ekman Layer

John Woods

Department of Earth Science and Engineering, Imperial College, London, UK

This chapter explains why we seldom observe an Ekman spiral in the upper ocean current driven by the local wind stress. It depends on revising Ekman's assumption that the eddy viscosity K is steady and does not vary with depth. In reality, solar heating causes the eddy diffusivity to change diurnally. Two processes are involved. The first is reduction in the depth of convection. The second is the transition from turbulent to laminar flow when the Richardson number exceeds a critical value $Ri_c \approx 1$. Here we focus on the second process: solar heating introduces a buoyancy force, which quenches turbulence. This limits the classical theory for $K(Ri)$ to $Ri < 1$. Introducing that new parametrization into Ekman's theory explains the observed deviations from the Ekman spiral. This revised model has consequences for prediction in a wide range of applications, from sailing to pollution, sonar to fisheries, and climate.

A. INTRODUCTION

The Royal Meteorological Society was already 50 years old when V. W. Ekman (1905) produced his famous analytical solution for the variation of current vector with depth under a steady wind stress, the Ekman spiral. This theory has lasting heuristic value: it shows how to relate the divergence in the turbulent flux of momentum to the Coriolis acceleration in models of ocean currents. And the vertically integrated flow—called Ekman transport—lies at the heart of all theories of the wind-driven circulation. But with very few exceptions (notably under ice) attempts to observe an Ekman spiral have failed. Here, for example, is my own attempt (Fig. 1) in which the variation of current vector with depth is revealed by a set of plumes from packets of fluorescein attached at 5 m intervals along a vertical mooring line. Note that the spiral goes in the opposite direction to that predicted by Ekman's theory.

This paradox can be resolved by focusing on Ekman's assumption that the eddy viscosity is constant and uniform with depth. I shall show that solar heating quenches turbulence during the day, giving laminar flow in the Ekman layer. The evidence for that is contained in Fig. 2. The small-scale structure in the dye streak was created by Karman vortices in the wake of the sinking pellet. They have a Reynolds number of around 10, and their motion is soon killed by molecular viscosity. The photograph was taken about 2 min after the streak was formed. During that time there was horizontal displacement, with significant vertical shear, but no turbulent mixing. The flow was laminar. Why was that?

Until the 1960s, all the textbooks on oceanography stated that the flow in the sea must be universally turbulent. Defant (1960) argued that this was because the Reynolds number of ocean currents was much greater than the critical value above which molecular viscosity prevents it forming. Reynolds's famous laboratory experiment of 1883 still influenced expectations, despite the later theories of Kolmogorov (1941), who had shown that once turbulence is established, viscosity has no effect on the largest eddies. Bradshaw (1969) showed that viscosity could only quench existing turbulence if these energy-containing eddies had a Reynolds number of 10 or less. Kolmogorov's theory built on L. F. Richardson's (1920) concept of a turbulent cascade in which (at large Reynolds number) there is no flux divergence of kinetic energy through the spectrum from the large, energy-containing eddies to what we now call the Kolmogorov scale where the Reynolds number becomes one (about 1 cm in the ocean).

B. THE EFFECT OF A STABLE DENSITY GRADIENT

Richardson (1920) argued that the strength of turbulence is controlled by the energy balance of the

FIGURE 1 (a) Current structure in the Ekman layer revealed by the dyed wake of a sinking fluorescein pellet, viewed from above. The sense of rotation with depth is in the opposite direction to that of the classical Ekman spiral. The structure in the dye was created by Karman vortices shed in the wake of the pellet as it sank through water that had no ambient turbulence. (b) Looking vertically down on a dye streak left in the wake of a sinking pellet of fluorescein reveals that the azimuthal vector of the current rotates in the opposite sense to that of the Ekamn spiral.

largest eddies. They receive energy from the mean flow at a rate controlled by the square of the shear, and lose energy in two ways: (1) by the cascade to viscosity, and (2) by doing work against gravity whenever the flow is stably stratified, that is when the density increases with depth. We now give the name Richardson number (Ri) to the ratio of the Reynolds stress source and buoyant sink of turbulent kinetic energy in the biggest eddies. For a steady state, the former must be nearly an order of magnitude greater than the latter (Ellison and Turner, 1959). When that is the case, the residual 85% of the power inflow is siphoned off to viscous dissipation at small scales by the loss-free turbulent cascade. Similar ideas were in the mind of G. I. Taylor (1915). So the concept of buoyancy-controlled turbulence was firmly established well before the Society's centenary in 1950. It was well known that, as the Richardson number increases, the turbulence becomes less energetic, and the eddy diffusivity (K_b) and eddy viscosity (K_v) and their ratio (K_b/K_v) decrease. G. I. Taylor (1931) had famously established functional relationships for the variation of these eddy transport coefficients with Richardson number. That parametrization was used by Munk and Anderson (1949) in a theory for the density stratified Ekman current profile.

C. THE FATAL FLAW

However, there was a fatal flaw in that paradigm for buoyancy-controlled turbulence. The flaw was encapsulated in Taylor's parametrization of the Kattegatt data, and in the contemporary analysis by Proudman (see the account in his 1953 book). It underpinned authoritative theories linking turbulent diffusion and density stratification, notably by Monin and Obukhov (1953, 1954), and it featured in Turner's (1973) monograph *Buoyancy Effects in Fluids*. What was this fatal flaw? It was the assumption that the $K(\text{Ri})$ relationship was valid at values of $\text{Ri} \gg 1$. How was that shown to be wrong? By the classical method of fluid dynamics: by flow visualization.

D. FLOW VISUALIZATION

In the early 1960s I developed a technique of staining seawater with fluorescein dye and filming the patterns it made. We have already seen an example in Fig. 1. Another technique involved dropping a small pellet of florescien dye to create an initially vertical streak of dyed water. Bear in mind that the dye is very dilute and does not affect the density of the water significantly. The dyed water represents a material surface, which moves with the water. It overturns with billows and smaller turbulent eddies. Internal waves propagate through it, making it follow orbital motion, with diameter equal to wave height.

E. THE DISCOVERY OF LAMINAR FLOW

Close inspection showed that the dyed Karman vortex trails persist for longer than an hour (the limit set by the diver's breathing apparatus). During that time they are distorted by current shear, but not diffused by turbulence (except in rare "billow turbulence" events discussed below). If oceanic flow were universally turbulent (as required by the paradigm in the 1960s), the smallest eddies in the turbulent cascade would have a lengthscale of about 1 cm and an overturning timescale of less than 1 min (the Kolmogorov scales). Such small-scale turbulent eddying would have distorted the dye patterns in an obvious way. The fact that it was *not* detected provides unambiguous evidence that the flow is laminar: not weakly turbulent, but strictly laminar. That scientific discovery, first reported by Woods and Fosberry (1966), underpins the rest of this chapter.

F. FINE STRUCTURE

In the limited space available, I can only show a few examples of the flow patterns. In the first (Fig. 2) we see that the vertical shear is structured: with relatively uniform layers a few metres thick, separated by thin vortex sheets. We shall see shortly how the vorticity concentrated in these sheets is occasionally liberated as small patches of turbulence. But first we note that the sheet-layer structure matches that of the density profile (Fig. 2), as measured by a specially designed "fine-structure probe" (Woods, 1969c). The combination of probe and dye observations established that the thickness of the sheets is typically 10 cm. This is limited by molecular conduction.

G. WAVE-INDUCED SHEAR INSTABILITY

In Fig. 3 we see that the dyed water undulates with internal waves, which are trapped on the density sheets because their frequency lies between the Brunt–Vaisala frequencies of the sheet, on the one hand, and the adjacent layers, on the other. These trapped waves have wavelengths between 5 and 10 m. The orbital motion of the water associated with the passage of these waves modulates the shear across the sheet, with a maximum on the wave crest. Occasionally that shear is sufficient

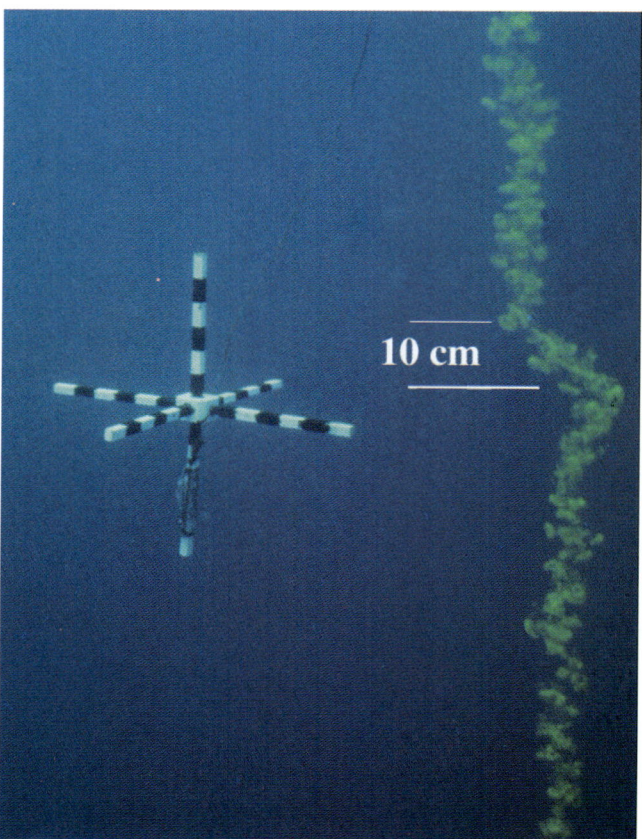

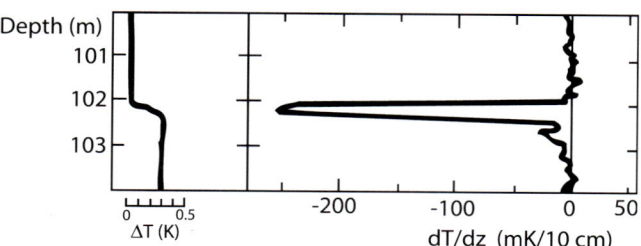

FIGURE 2 The dyed wake of a sinking pellet of fluorescein is distorted by the current fine structure to reveal a sharp shear across a 10 cm thick sheet in the density fine structure.

FIGURE 3 Internal waves trapped on fine-structure sheets of enhanced density gradient in the seasonal thermocline off Malta. These waves have a period of about 10 min, a wavelength of 5 m, and an amplitude of about 30 cm. As they go past a water particle marked by dye it undergoes a circular motion with radius equal to the wave amplitude. This orbital motion produces a shear, which has maxima at the wave crests and troughs.

to cause the local Richardson number to fall below 1/4, which is the criterion for shear instability (Phillips, 1966; Hazel, 1972). The growth rate of the instability peaks at a wavelength 7.5 times the sheet thickness; so the rapidly growing wavelets have a wavelength of less than 1 m. They grow and overturn as billows (Fig. 4), taking about 1 min. The height of these billows is twice the sheet thickness, typically 20 cm. These observations (Woods, 1968) confirmed Phillips's (1966) theoretical prediction of wave-induced shear instability.

H. BILLOW TURBULENCE

These billows become the energy-containing eddies of a local turbulent cascade. They have Reynolds numbers of about 100. The height of these eddies is consistent with Ozmidov's (1965) theory that the largest overturning scale depends on the density stratification and the overturning speed of the billow, both of which I was able to measure. Such billow turbulence events were rare in the seasonal thermocline, where I estimated

FIGURE 4 Billows generated by wave-induced shear instability, which started with a wavelet of about 75 cm wavelength. The billows are about 20 cm high, twice the thickness of the sheet, as measured by the density profile. They overturn in about 5 min.

they occupied only 5% of the volume at any instant. The rest of the flow was laminar. We shall see later that the transition to billow turbulence plays an important role in determining the eddy viscosity in the diurnal thermocline.

I. REVERSE TRANSITION

Analysis of individual billow turbulence events revealed that none persisted when the local Richardson number exceeded unity. Richardson number is scale dependent, so it must be measured on the scale of the instability, in this case the sheet thickness of 10 cm. That is why my fine-structure probe used a pair of thermometers 10 cm apart. The local Richardson number was modulated by the propagation of the internal wave trapped on the sheet, through the unstable wavelet (which has zero phase speed). The wave-induced shear is zero at the nodes. Their arrival was observed to

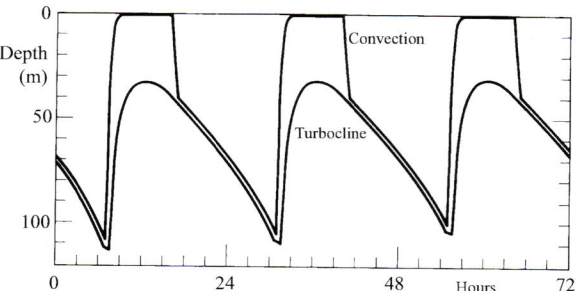

FIGURE 5 Forenoon rise in the depths of convection and of the turbocline, as computed by the model of Woods and Barkmann (1986).

quench billow turbulence, by cutting off its power supply, and occasionally to stop the growth of the unstable wavelets before they break into billows. The critical Richardson number (Ri_c 1) for reverse transition was also determined by observing the quenching of turbulence created artificially in the seasonal thermocline by emptying a bucket full of warm water (Woods 1969b).

J. REVISED PARADIGM

We conclude that the notion of K varying smoothly with Ri is valid only in the narrow range Ri < 1. And in many cases, laminar flow persists with $1 > Ri > 1/4$. Theories such as those of Monin and Obhukov, are largely irrelevant because the ocean in almost everywhere outside their range of validity. We shall now see that in the Ekman layer there is no such thing as persistent weak stability that might gently modulate the eddy transport coefficients. The flow is either vigorously turbulent or laminar.

K. ONE-DIMENSIONAL MODELLING OF THE UPPER OCEAN

Early models of ocean circulation incorporated an upper boundary layer in which the mixed layer depth was computed by convective adjustment, without taking into account turbulent entrainment. That works tolerably at high latitudes where nocturnal convection is much deeper than the Ekman depth, but it does not perform so well in the tropics where the mixed layer is strongly influenced by turbulence powered by Reynolds stresses in the Ekman layer. Munk and Anderson's (1949) model failed because it did not take account of reverse transition at Ri > 1. That fact has since been successfully introduced into a number of turbulent diffusion codes for predicting mixed layer depth. It also lies at the heart of the so-called bulk models of the mixed layer, pioneered by Kitaigorodskii (1960), which depend on diagnosing the depth at which the power input from the wind is consumed by doing work in entraining dense water from the thermocline. Kraus and Turner (1967) emphasized the subduction of water from the turbulent mixed layer into thermocline during the spring, and the density stratification that it caused. Modern mixed layer models, whether based on diffusive or bulk codes, successfully simulate the response to the weather, and annual cycle of surface fluxes.

The discovery of laminar flow did not throw a spanner in the works of mixed layer modelling, because it was already implicitly assumed that the turbulence is negligible in the stably stratified thermocline. The new paradigm was quickly assimilated without the need for further critical experiments. However, the full implications were not appreciated in the 1960s. In particular, it took many years to work out the contribution of solar heating to the flow regime.

L. DIURNAL VARIATION

The thickness of the Ekman layer, controlled by the Coriolis parameter, is normally less than 100 m. This roughly coincides with the depth above which the downward flux of solar energy exceeds 1% of that entering at the sea surface. Flow visualization studies revealed that each morning solar heating quenches turbulence (Woods, 1969a). Turbulence becomes stronger at night, powered largely by nocturnal convection. This day–night transition between laminar and turbulent flow must be taken into account in refining our theory of the Ekman layer. It occurs in two stages due to changes first in buoyant convection and then in the Richardson number.

M. BUOYANT CONVECTION

About 1 hour after sunrise the energy input from the sun first exceeds the ocean's energy loss to the atmosphere (by evaporation, conduction and net longwave radiation). The depth of convection is thereafter determined by the thermal compensation depth, defined as the depth above which solar heating equals the surface heat loss (Kraus and Rooth, 1960). The thermal compensation depth rises sharply during the forenoon to a minimum depth which depends on wind speed, but is typically less than 1 m. The turbocline at the base of the mixing layer closely follows the forenoon ascent of the thermal compensation depth (Fig. 5). As the turbocline rises, water is subducted into diurnal thermocline below. Initially this is weakly stratified by the slow rise in mixing layer temperature. Solar heating further increases the density gradient. This growing density gradient provides a buoyancy sink for the turbulence in the subducted water (Woods, 1980).

N. BILLOW TURBULENCE IN THE DIURNAL THERMOCLINE

According to classical Monin–Obhukov theory we would expect the turbulence to become weaker, and the eddy diffusivities to decrease smoothly as the density stratification rises in response to solar heating. Flow visualization revealed something much more dramatic. Immediately after subduction into the diurnal thermocline, the energy-containing eddies shrink to the size of billows we first encountered in the seasonal thermocline. This transition to tiny energy-containing eddies occurs when the density stratification is still very weak. The change is consistent with Ozmidov's criterion mentioned earlier.

The consequence for turbulent diffusion is dramatic, because the diffusivity depends on the cube of the eddy height. Although the flow in the diurnal thermocline is initially in a state of continuous turbulence, K has reduced dramatically because the biggest eddies have shrunk from tens of metres high at sunrise to less than 1 m after subduction.

As the sun rises during the forenoon, solar heating increases the density gradient in the diurnal thermocline until the Richardson number exceeds unity. The prevailing billow turbulence is then quenched, leaving laminar flow. However, the eddy viscosity had already become negligible because of the shrinking of the energy-containing eddies immediately after subduction into the diurnal thermocline.

O. CONSEQUENCES FOR THE EKMAN CURRENT PROFILE

The transition from turbulent to laminar flow each morning means that the flow in the diurnal thermocline becomes frictionless. The current at each level then starts to rotate inertially. Subduction, and therefore the transition to inertial rotation, occur later at shallower depths. This introduces a depth dependent phase lag in the rotation of the flow.

In the evening, as the sun goes down, convection deepens, and the water that has passed the day in the diurnal thermocline is re-entrained into the turbulent boundary layer. At each level the water is moving in a direction governed by the period of time during which it was rotating inertially. The duration gets longer with greater depth. So water entrained later during the night contains momentum that has experienced a greater azimuthal rotation.

During the night, vertical diffusion of momentum tends to re-establish the Ekman solution. But it has insufficient time to reach a steady-state profile before the rising sun re-stratifies the water next morning.

The result is that the current profile in the Ekman layer is in a perpetual state of transition, being driven at two timescales: the inertial period of the earth's rotation and the diurnal period of solar heating. The beating between these two periods produces a current profile of great complexity, even when the wind and cloud cover are steady (Fig. 6) (Woods and Strass, 1986). However, simulating the trajectories of drifters constrained to remain at fixed depths over many days (Fig. 7) reveals that the classical Ekman spiral is a Lagrangian phenomenon, independent of the irregular fluctuations in the Eulerian velocity profile.

P. SOLAR RADIATION

These results raise solar radiation alongside density stratification and viscosity as an agency for quenching turbulence. In view of the extreme sensitivity of turbulence to solar heating it is essential that models of the ocean boundary layer include accurate codes for computing the irradiance profile. Solar absorption by pure seawater varies strongly with wavelength, so it is necessary to describe the solar spectrum in detail; my model uses 27 wavebands from UV to IR. It is not sufficient to describe solar heating by the sum of two or three exponentials (Simpson and Dickey, 1981a,b), although that is still common practice in modelling climate.

Furthermore, the optical properties of seawater are strongly affected by the presence of particles. In the open ocean these are mainly microscopic organisms—phytoplankton rich in pigments. Predicting the impact of solar heating on turbulence requires a prediction of the equivalent chlorophyll concentration (representing all pigments), which can be done with a plankton ecosystem model that has a vertical resolution of 1 m (Woods and Barkmann, 1994). The profile of equivalent chlorophyll concentration exhibits considerable biologically generated fine structure, so it is inappropriate to simulate the absorption of sunlight in the sea by a sum of exponentials.

Assuming that an accurate profile of chlorophyll concentration is available, it is possible to compute the irradiance profile. However, this requires a sophisticated optical code. Monte Carlo codes require billions of photons per waveband to reach 50 m and become prohibitively expensive to reach 100 m. An alternative approach is to solve the radiative transfer equations (RTEs). Liu et al. (1999) have created a fast method to do that. They show that classical empirical codes cannot compute the irradiance profile with sufficient accuracy for modelling the profile of eddy diffusivity. Liu and Woods (2002) have used their RTE optical code in a plankton ecosystem model to predict the diurnally changing depth of the turbocline, the prerequisite for Ekman current simulation.

Q. APPLICATIONS

Now we come to the "so what?" question. Does it matter in any practical application of oceanography that the flow in the Ekman layer is switching back and forth between laminar and turbulent? The answer is: "Yes, it does matter in a wide variety of applications". In the little space remaining I can only very briefly refer to applications that have benefited from modelling the laminar-turbulent process.

1. Slippery Seas of Acapulco

The creation of a diurnal jet is the direct consequence of this laminar flow (Woods and Strass, 1986). In 1968, this phenomenon was exploited to predict the drift of racing yachts in the Mexico Olympics, scientific advice which contributed to three British gold medals. Woods and Houghton (1969) called the effect the Slippery Seas of Acapulco. Newsweek lauded this as "a remarkable application of science without going through the stage of technology".

2. Pollution

Lagrangian integration of the fluctuating inertial motion in the Ekman layer has been used to simulate the horizontal dispersion of a set of neutrally buoyant drifters. Statistical analysis reveals that they disperse at a rate consistent with Okubo's (1971) famous empirical graph of horizontal diffusivity with patch size. That would not be the case with Ekman's (1905) steady solution, or Munk and Anderson's (1949) $K(\mathrm{Ri})$ model.

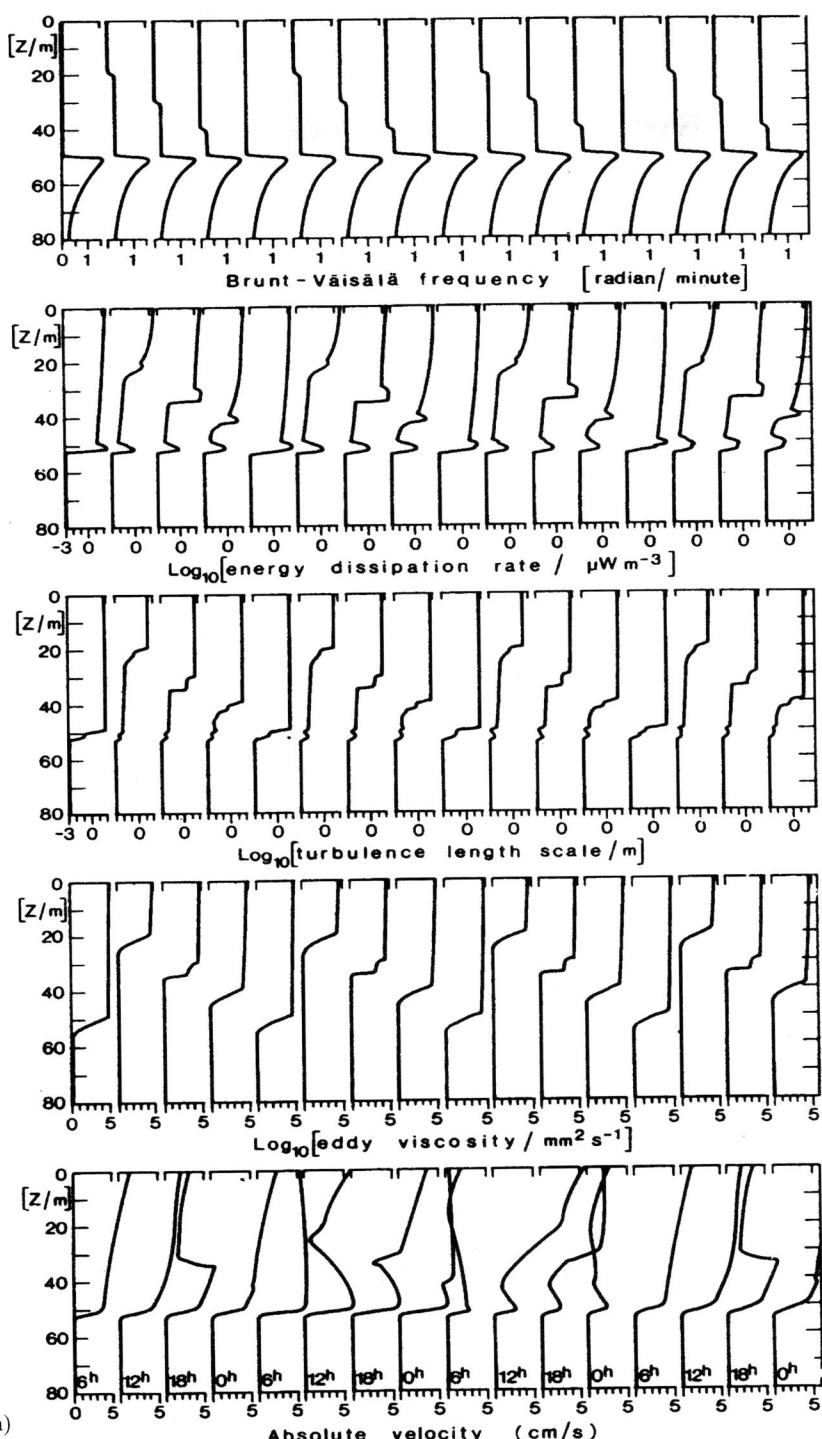

FIGURE 6 Structure of the flow in the Ekman layer, as computed by the model of Woods and Strass (1986). The 24 h cycle of subduction into the diurnal thermocline and re-entrainment into the mixing layer, beats with the inertial period of 17 h to create a complex, continually varying profile in the current vector ((b, iii), taken out of the period of the jet occurrence). This is like the pattern observed in the flow visualization experiments. (From Fig. 6C) (Reproduced from Woods and Strass, 1986.)

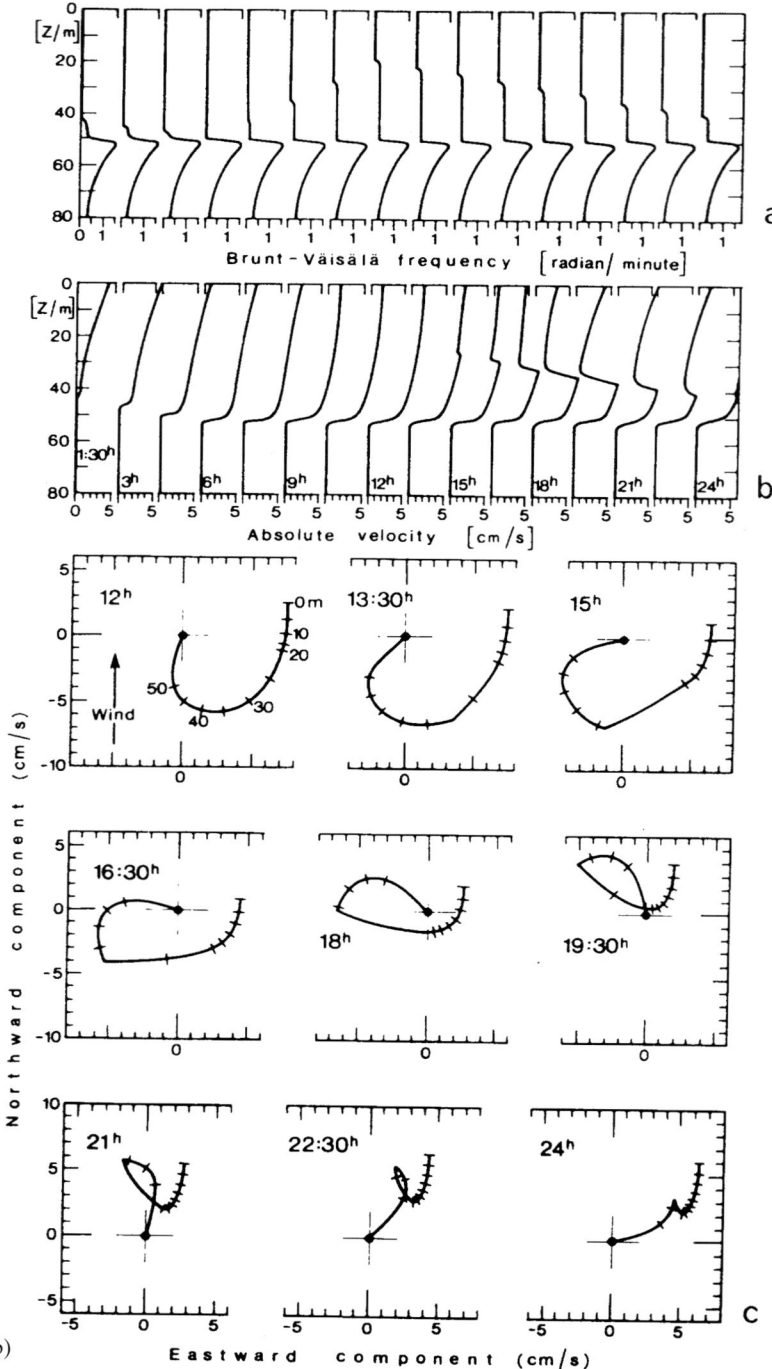

FIGURE 6 *Continued*

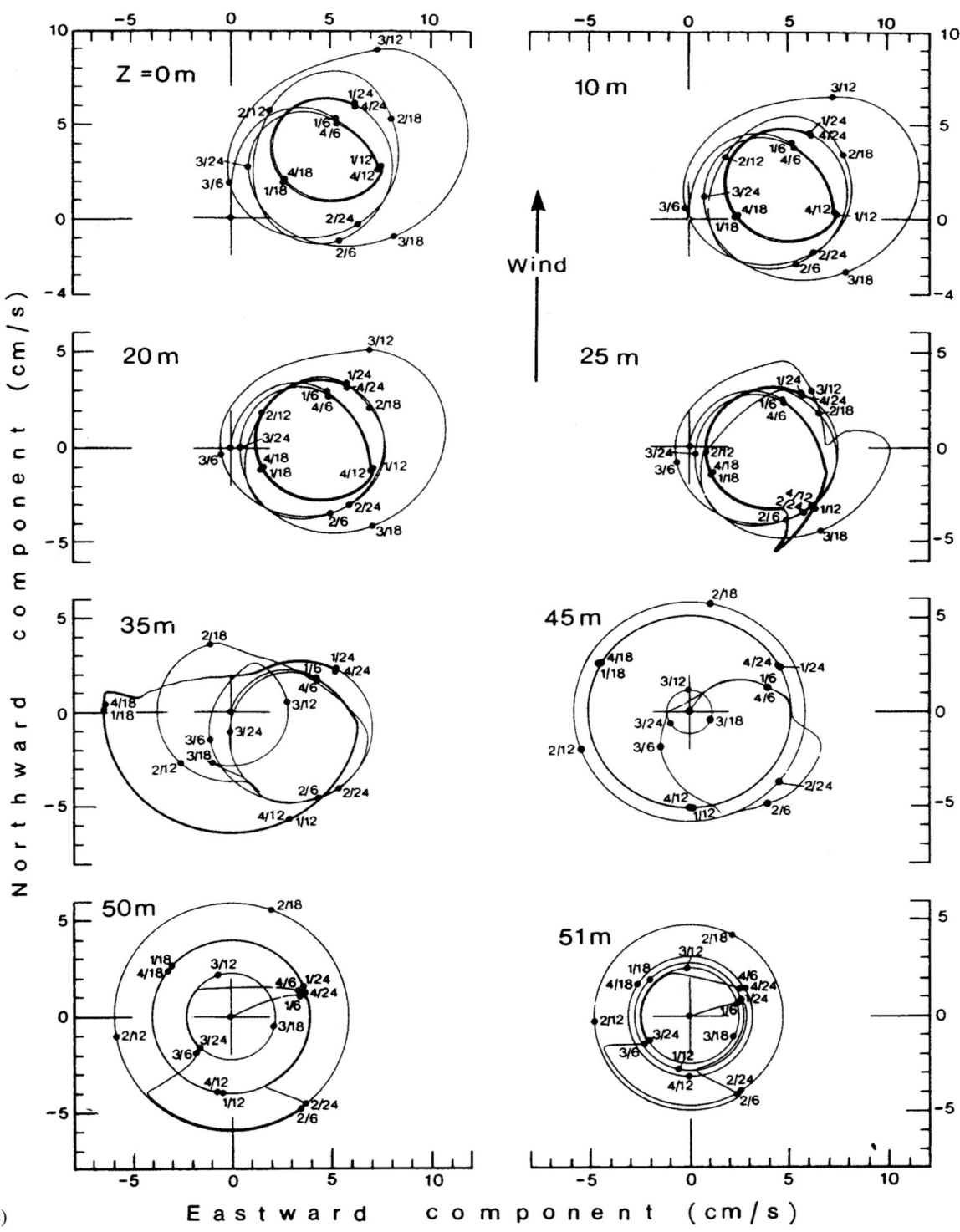

FIGURE 6 Continued

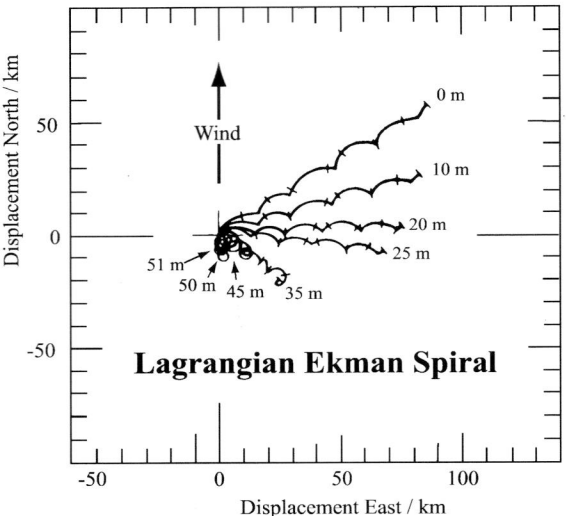

FIGURE 7 The trajectories of simulated floats constrained to remain at fixed depths in the Ekman layer, reveal that the classical Ekman spiral exists as a Lagrangian phenomenon, even when the Eulerian velocity profile exhibits irregular fluctuations. (Reproduced from Woods & Strass, 1986.)

3. Afternoon Effect in Sonar

The transition to laminar flow stops the vertical eddy diffusion of heat, allowing a large vertical gradient of temperature to develop during the afternoon in the upper ocean especially in the summer and in the tropics. This temperature gradient refracts sound waves, greatly reducing the range of hull-mounted sonars. My flow visualization research in the 1960s was initially stimulated by the Royal Navy's need to understand the cause of this "afternoon effect". Naval acousticians were already talking about "nocturnal convection" and therefore by implication daytime suppression of convection (Albers, 1965), but complete understanding had to await the identification of reverse transition and laminar flow in the diurnal thermocline, which emerged from flow visualization (Woods, 1969a).

4. Patchiness

The existence of daytime laminar flow near the sea surface led to an explanation of puzzling observation in aircraft remote sensing of sea surface temperature. Repeated flights over the same location during the hours of daylight on a calm day revealed a hundred-fold rise in the horizontal variance of temperature on scales of 1 to 100 km (P. Saunders, personal communication). It is now known that this is due to mesoscale upwelling at high Rossby number interacting with the solar heating profile (Salmon, 1985). That mechanism can only work when the laminar flow in the diurnal thermocline outcrops at the sea surface, which occurs on calm, sunny days. A similar mechanism explains the observed patchiness in ocean colour (Woods, 1988).

5. Fisheries

One of the greatest challenges in fisheries stock forecasting is how small interannual variation in weather leads to large variation in the fraction of fish larvae that survive to become adults that can reproduce in future years. The most plausible conjecture, Cushing's (1990) *match–mismatch* theory, assumes that the success of fish larvae depends on the timing of their hatching and the peak concentration of their food supply, microscopic copepods. Cushing's theory is consistent with available data, but it has proved extraordinarily difficult to construct models that can plausibly test the conjecture, and impossible to use it as the basis for fish-stock prediction models. However, recent developments in plankton ecosystem models based on the Lagrangian Ensemble method suggest that they can solve the problem. This method follows the trajectories of millions of individual plankters, as they are displaced by advection and turbulence and by their own sinking and swimming. This new class of model has already solved a number of classical problems in biological oceanography. It does so by paying very careful attention to the impact of laminar-turbulent transition on the trajectories of plankton (Woods and Onken, 1982).

6. Climate

As my final example, I shall mention the role in climate change of radiative quenching of turbulence. This is the plankton multiplier mechanism of positive biofeedback in the greenhouse mechanism (Woods and Barkmann, 1993). A change in annual surface radiation of a few watts per square metre is sufficient to reduce the depth of winter convection by some 15%. That reduces the supply of nutrients to the euphotic zone by 15%, which in turn reduces primary production by the same amount. Primary production fixes dissolved inorganic carbon, which comes from atmospheric carbon dioxide dissolved in seawater. The oceanic uptake of carbon dioxide is reduced in proportion to the reduction in primary production, or rather in the reduction in the rate at which dead plankton sink to the deep ocean floor. So an increase in IR radiation of a few watts per square metre (equivalent to the next 50 years of greenhouse gas increase in the atmosphere) reduces the oceanic uptake of carbon dioxide by some 15%. This positive biofeedback in the greenhouse mechanism also works for similar changes in solar radiation, as occur in the Milankovich cycles of the ice ages. This is the only *direct* consequence of radiation change, which can alter the concentration of carbon dioxide in the atmosphere.

Interestingly, a change of solar radiation of a few watts per square metre also occurs after major volcanic eruptions, like Mount Pinatubo in 1987. The expected changes in plankton production were observed in the Red Sea on that occasion (Genin *et al.*, 1995); a striking verification of the plankton multiplier theory.

R. DISCUSSION

The discovery of laminar flow in the upper ocean changed the way we think of buoyancy forces affecting turbulence. The existence of laminar flow in the seasonal thermocline permitted flow visualization experiments, which confirmed Philipps's (1966) theory of wave-induced shear instability. It revealed that reverse transition occurs at Richardson numbers around unity, so the parametrization of $K(Ri)$ is not valid at $Ri > 1$, as had been widely believed, following the analysis of Kattegatt data by G. I. Taylor and J. Proudman. Strangely, this discovery was easily accepted by oceanographers, because it had been implicit in successful models of mixed layer depth and thermocline stratification. One remarkable consequence was that there has been no scientific publication by other authors seeking to repeat my flow visualization experiments. The conclusions about laminar flow and reverse transition were accepted without corroboration.

However, early acceptance did not mean that the story was complete. For 20 years, oceanographers continued to express surprise that the day–night cycle of solar heating provokes a strong response in upper ocean mixing. And yet the concept of "nocturnal convection", which controls the diurnal cycle, was familiar to the ocean acoustics community since the 1950s (e.g. Albers, 1965). The sensitivity of turbulence to solar heating has become widely recognized only in the last decade. It is now appreciated that sophisticated optical modelling, based of fine resolution of the spectrum of light and on simulation of the biological processes controlling turbidity, is needed to predict the upper ocean turbulence regime.

S. CONCLUSION

Solar heating is responsible for the fact that the Ekman spiral is observed in the upper ocean only under exceptional circumstances (under ice). During the day, it greatly reduces the depth of convection, and increases the Richardson number above the critical value (Ri_c 1) for reverse transition from turbulent to laminar flow. This means that the classical parametrization of eddy viscosity as a function of Richardson number, $K(Ri)$ is only valid for $Ri < 1$. Very little of the ocean lies in the range of Richardson number, so the flow is either vigorously turbulent or laminar. The former occurs in the Ekman layer at night, the latter during the day. The vertical shear in the Ekman layer controls the power input to turbulence in the diurnal thermocline: solar heating controls the dissipation of that turbulence. Accurate simulation of the transition between turbulent and laminar flow requires a sophisticated model of solar heating, involving high resolution of the spectrum, and realistic simulation of the plankton ecosystem which controls the rapidly changing turbidity profile. Models based on this revised parametrization describe the observed complex variation of current vector with depth, which differs significantly from the Ekman spiral. Taking account of reverse transition to laminar flow in the Ekman layer has consequences in a wide variety of applications, from sailing to pollution, sonar to fisheries and climate.

References

Albers, V. A., 1965: *Underwater Acoustics Handbook*. Pennsylvania State University Press.

Bradshaw, P., 1969: Conditions for the existence of an inertial subrange in turbulent flow. *Aeronaut. Res. Coun. Rep. Mem.* No. 3603, 5.

Cushing, D. H., 1990: Plankton production and year-class strength in fish populations: an update of the match/mismatch hypothesis *Adv. Marine Biol.*, **26**, 250–293.

Defant, A., 1960: *Physical Oceanography*. Pergamon Press, Oxford.

Ekman, V. W., 1905: On the influence of the Earth's rotation on ocean currents. *Ark. Math. Astron. Fys.*, **2**(11), 1–53.

Ellison, T. H. and J. S. Turner, 1959: Turbulent entrainment in stratified flows. *J. Fluid Mech.*, **6**, 423–448.

Genin, A., B. Lazar *et al.* 1995: Vertical mixing and coral death in the Red Sea following the eruption of Mount Pinatubo. *Nature*, **377**, 507–514.

Hazel, P., 1972: Numerical studies of the stability of inviscid stratified shear flows. *J. Fluid Mech.*, **51**, 39–61.

Kitaigorodskii, S. A., 1960: On the computation of the thickness of the wind-mixed layer in the ocean. *Izv. Akad. Nauk. S.S.S.R. Geophys. Ser.*, **3**, 425–431.

Kolmogorov, A. N., 1941: Local structure of turbulence in an incompressible fluid at very high Reynolds numbers. *Dokl. Akad. Nauk. S.S.S.R.*, **30**(4), 299–303.

Kraus, E. B. and C. Rooth, 1960: Temperature and steady state vertical flux in the ocean surface layer. *Tellus*, **13**, 231–239.

Kraus, E. B. and J. S. Turner, 1967: A one-dimensional model of the seasonal thermocline: II. The general theory and its consequences. *Tellus*, **19**, 98–106.

Liu, C-C. and J. D. Woods, 2002: Sensitivity of a virtual plankton ecosystem to parametrization of solar radiation. *J. Plankt. Res.* (submitted).

Liu, C.-C., J. D. Woods and C. D. Mobley, 1999: Optical model for use in oceanic ecosystem models. *Appl. Optics*, **38**(21), 4475–4485.

Monin, A. S. and A. M. Obukhov, 1953: Dimensionless characteristics of turbulence in the atmospheric boundary layer. *Dokl. Akad. Nauk. S.S.S.R.*, **93**(2), 223–226.

Monin, A. S. and A. M. Obhukov, 1954: Basic laws of turbulent mixing in the ground layer of the atmosphere. *Acad. Sci. U.S.S.R. Leningrad Geophys. Inst.*, **24**, 163–187.

Munk, W. H. and E. R. Anderson, 1949: Notes on a theory of the thermocline. *J. Marine Res.*, **7**, 276–295.

Okubo, A., 1971: Oceanic diffusion diagrams. *Deep-Sea Res.*, **18**, 789–802.

Ozmidov, R. V., 1965: On the turbulent exchange in a stably-stratified ocean. *Atmos. Ocean. Phys.*, **1**(8), 493–497.

Phillips, O. M., 1966: *Dynamics of the Upper Ocean*, Cambridge University Press.

Proudman, J., 1953: *Dynamical Oceanography*. Methuen, London.

Reynolds, O., 1883: An experimental determination of the circumstances which determine whether the motion of water shall be direct or sinuous, and the law of resistance in parallel channels. *Phil. Trans. R. Soc. Lond.*, **A174**, 935–982.

Richardson, L. F., 1920: The supply of energy from and to atmospheric eddies. *Proc. R. Soc. Lond.*, **A97**, 354–373.

Salmon, J., 1985: Measurements of diurnal variation in the spectrum of sea surface temperature in Calvi Bay. Unpublished Ph.D. thesis, Southampton University.

Simpson, J. J. and T. D. Dickey, 1981a: The relationship between downward irradiance and upper ocean structure. *J. Phys. Oceanogr.*, **11**, 309–323.

Simpson, J. J. and T. D. Dickey, 1981b: Alternative parametrizations of downward irradiances and their dynamical significance. *J. Phys. Oceanogr.*, **11**, 876–882.

Taylor, G. I., 1915: Eddy motion in the atmosphere. *Phil. Trans. R. Soc. Lond.*, **A84**, 371–337.

Taylor, G. I., 1931: Internal waves and turbulence in a fluid of variable density. *Rapp. Cons. Perm. Int. Explor. Mer*, **76**, 35–42.

Turner, J. S., 1973: *Buoyancy Effects in Fluids*. Cambridge University Press.

Woods, J. D., 1968: Wave-induced shear instability in the summer thermocline. *J. Fluid Mech.*, **32**, 791.

Woods, J. D., 1969a: Diurnal behaviour of the summer thermocline off Malta. *Deut. Hydrograph. Zeit.*, **21**, 106.

Woods, J. D., 1969b: On designing a probe to measure ocean microstructure. *Underw. J.*, **1**, 6.

Woods, J. D., 1969c: On Richardson's number as a criterion for transition from laminar to turbulent to laminar flow. *Radio Sci.*, **4**, 1289.

Woods, J. D., 1980: Diurnal and seasonal variation of convection in the wind-mixed layer of the ocean. *Q. J. R. Meteorol. Soc.*, **106**, 379–394.

Woods, J. D., 1988: Mesoscale upwelling and primary production. In (B. J. Rothschild, Ed.), *Towards a Theory of Biological–Physical Interactions in the World Ocean*. Kluwer, Dordrecht, pp. 7–38.

Woods, J. D. and W. Barkmann, 1986: The influence of solar heating on the upper ocean. I. The mixed layer. *Q. J. R. Meteorol. Soc.*, **112**, 1–27.

Woods, J. D. and W. Barkmann, 1993: The plankton multiplier—positive feedback in the greenhouse. *J. Plankt. Res.*, **15**(9), 1053–1074.

Woods, J. D and W. Barkmann, 1994: Simulating plankton ecosystems by the Lagrangian Ensemble method. *Phil. Trans. R. Soc.*, **B343**, 27–31.

Woods, J. D. and G. G. Fosberry, 1966: Observations of the behaviour of the thermocline and transient stratifications in the sea made visible by dye markers. *Underw. Assoc. Rep.*, **1**, 31.

Woods, J. D. and D. Houghton, 1969: The slippery seas of Acapulco. *New Scient.*, 134–136.

Woods, J. D. and R. Onken, 1982: Diurnal variation and primary production in the ocean—preliminary results of a Lagrangian ensemble model. *J. Plankt. Res.*, **4**, 735–756.

Woods, J. D. and V. Strass, 1986: The influence of solar heating on the upper ocean II the wind driven current. *Q. J. R. Meteorol. Soc.*, **112**, 28–43.

Ocean Observations and the Climate Forecast Problem

Carl Wunsch

Program in Atmospheres, Oceans and Climate, Department of Earth, Atmospheric and Planetary Sciences, Massachusetts Institute of Technology, Cambridge MA, USA

The widely disseminated and accepted view of the ocean as a nearly-steady, nearly-laminar system is primarily a consequence of the great difficulty of observing it, and of the intense computational cost of modelling it. Uncritical use of the steady/laminar framework has led to a gross distortion of the science, particularly the study of climate change, partly manifested by the belief that a comparatively small number of simple observations suffices to describe the system, and by the inference that oceanic behavior under changed external forcing can be deduced by pure thought without integration of the equations of motion. All of the evidence of the last 25 years shows that the behavior is much more interesting and complex than this distorted view would imply.

Real progress will involve confronting the actual system, not the fictitious one.

A. INTRODUCTION

The problems of observing the ocean have come in recent years to loom large, often because of the importance of the ocean in climate and climate change. Many of the problems are technical ones, but a number of them might be regarded as being more a matter of culture, or of misapprehension, than of science. Whatever the cause, there are consequences for international scientific planning directed at understanding the climate system. Many in the highly sophisticated meteorological community continue to have an antiquated and misleading perception of the ocean circulation, a perception that has led to some surprising assertions about the ocean.

The need for observations of the atmosphere so as to understand and forecast it is a truism of meteorology that few would be inclined to question. One might think that a similar expression concerning the need for oceanic observations would be equally widespread. Experience suggests otherwise. I offer the following quotations, all said in my presence at international meetings:*

1. "The ocean has no physics, and so there is no need for observations. Oceanographers should be forced to stay home and run models" (statement in a climate meeting by a prominent meteorologist—and repeated 13 years later).
2. "The ocean's role in climate change is very simple to determine. One just needs a few high latitude XBT measurements to determine the state of the upper branch of the 'conveyor belt' and everything else can be calculated" (assertion by a different prominent meteorologist).
3. "Observations of the deep ocean aren't necessary. The flow is what Henry Stommel told us it was, in 1958" (assertion by yet another prominent meteorologist).

Such views are hardly universally held. But given that three experienced and influential atmospheric scientists could make these statements in planning meetings is interesting, and worthy of exploration. This exploration becomes a vehicle for a brief survey of where modern physical oceanography has taken the subject, an evolution that clearly has not yet become widely appreciated. That there can be such a wide gap between perception and reality is a consequence largely of oceanographic history, and the junior partnership role physical oceanography has had until relatively recently, relative to meteorology.

B. SOME HISTORY: OBSERVATIONAL AND THEORETICAL

The opacity of the ocean to electromagnetic radiation has meant that for most of the history of the

---

* Names withheld to protect the innocent (note that the word "innocent" has two distinct meanings). At the RMS Millennial Meeting, prizes were offered to anyone able to guess one or more of the authors of these statements. There were no takers.

subject, all observations of the ocean were made by sending a ship to some specific location, and physically lowering an instrument to the depth where a measurement was desired. In the era before modern (solid-state) electronics, that is before about 1965, the only measurements that could be made this way were based almost entirely upon mechanical systems. Obtaining useful measurements of a fluid system at high pressures (10 m of water is approximately one atmosphere, and the ocean reaches to depths exceeding 6000 m in many places) without the use of electromagnetic sensors or information delivery devices is a considerable intellectual challenge! A number of extremely clever individuals solved this problem for temperature, salinity, and a few other chemical properties (e.g., crude, by modern standards, oxygen concentration measurements were possible in the 1920s). By the end of the nineteenth century, electrical (rather than electronic) signals from current meters could be sent up cables from ships. Some of the intriguing devices that were developed over the years are discussed e.g., by von Arx (1962).

Consider that oceanographic ships are very expensive devices; today (year 2000), a deep-sea-capable vessel costs about US\$25,000/day to operate (apart from the science costs). Similar great expense was incurred, in suitably deflated monetary units, throughout the history of the subject. (Its beginnings are often traced to the despatch of the Challenger Expedition by the Admiralty in 1873. Even then, it required a military agency to meet the bills.) Such ships remain quite slow, with a modern oceanographic vessel steaming at about 12 knots. At that rate, and stopping periodically to lower devices into the sea, it takes approximately 1 month to work one's way across the North Atlantic Ocean, and about 2 months to do the same in the Pacific Ocean (few oceanographic ships today have the range to do scientific work while crossing the Pacific Ocean). Consequently, no one could afford to have a ship linger very long at any one place, with the limits typically being a few hours to a day or two. Thus *time series* of oceanic variables were almost nonexistent.

To obtain some understanding of the large-scale oceanic fluid structure, oceanographers resorted to combining observations of the ocean from time intervals spanning many decades. Such a strategy can have various outcomes. At one extreme, it fails altogether, with data from one expedition being clearly incompatible with that from another. Indeed, velocity measurements from ships were impossible to use: the flow field was clearly extremely variable in both space and time, changing completely over a few hours and a few kilometres. The other extreme outcome is that the observations are very similar, permitting contouring of large-scale property fields over the entire ocean (an example is shown in Fig. 1) or even the globe. Fortunately for oceanographers, and in great contrast to the velocity field, scalar properties such as temperature, salinity, oxygen (and subsequently a whole suite of scalar properties such as silica, phosphate, and even quasi-geostrophic potential vorticity) proved to be apparently stable, decade after decade, and so, interpretable.

The picture which thus emerged early on in the subject was that of an ocean in an essentially laminar state, unchanging in time, and leading to the "ocean circulation" pictures now available in countless textbooks, both meteorological and oceanographic. In particular, the large-scale thermal and salinity fields permitted gradient calculations to be made over large distances, and the geostrophic (thermal) wind computed. Because of the perceived need to assume a so-called level of no motion to set the absolute value of the thermal wind with depth, in many such flow schemes, this level could be adjusted to produce flow fields supposedly consistent with the large-scale property gradients (see e.g., Reid, 1986) which were then presumed to be representative of the direction and magnitude of the flows.

Somewhat surprisingly, there was essentially no theory of the ocean circulation until the late 1940s (the nearly singular exception being Ekman, 1905). But beginning with the now-famous papers of Sverdrup (1947), Stommel (1948) and Munk (1950), a theory of the wind-driven circulation developed over the next decades (the almost complete failure until the 1980s of the theoretical community to consider the role of buoyancy forcing of the ocean is another remarkable phenomenon). The first models were wholly analytical, steady and laminar. The earliest computers (whose use began in the 1950s) were so small and slow, that the numerical models were also steady and laminar. By the late 1960s, theoretical and observational oceanographers had produced a quite pleasing combination of apparently consistent theory and observation in which the ocean circulation was large-scale, only slowly changing if at all, and apparently describable at least in gross outline by a slow, sticky, quasi-geostrophic theory.

C. THE PERCEPTION

Let us return now and examine the three above quotations in the light of this history, as it will be seen that they make a certain amount of sense in that context. Quotation (1) employs the word "physics" in its strange meteorological usage meaning that the ocean has only "dynamics" and apparently none of the "right-hand

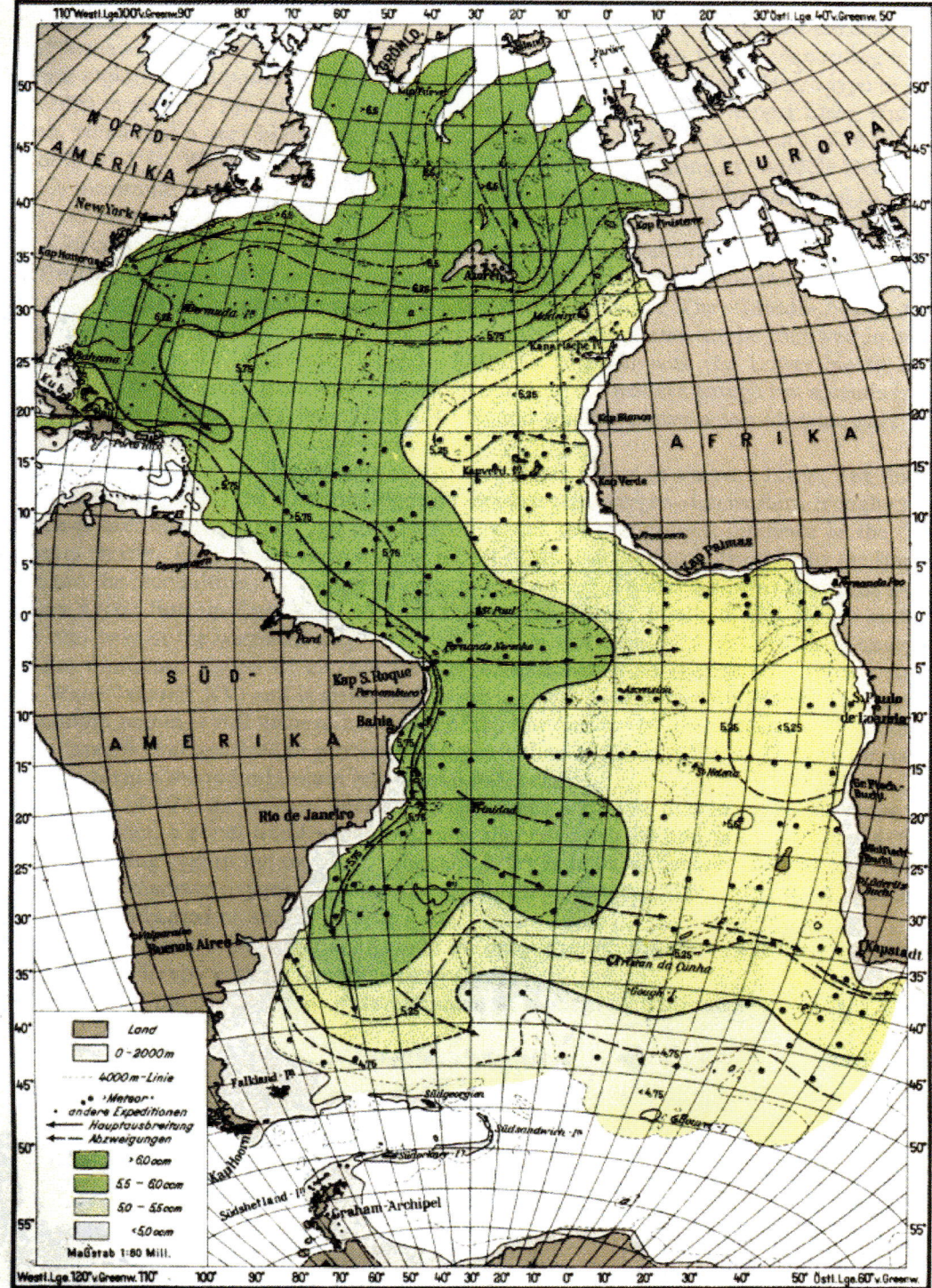

FIGURE 1 A famous picture of the oxygen concentration in the Atlantic Ocean showing the great "tongue" of excess concentration at approximately 2500 m depth (Wüst, 1935). The presence of this, and other such large-scale features, led to the interpretation of the flow as being large-scale (on the scale of the feature itself), slow, and steady.

side" of the meteorological equations. The latter consists of radiative processes, cloud physics, etc. But this assertion, even if one accepts the premise, makes absolutely no sense. The history of fluid dynamics is one of constant interplay of theory and observation (experiment). Pure thought, unattenuated by tests against observation, seems often to lead one seriously astray. Casual perusal of the *Journal of Fluid Dynamics* and similar publications shows how tightly bound the experimental side is with the theoretical, even in the absence of the meteorologist's "physics". Of course, this ignorant comment also fails to recognize how much in ocean general circulation models represents necessary parameterizations of unresolved processes (mixing of several kinds, topography, eddy fluxes, boundary layers, etc.). The mind boggles at the speaker's thought process.

Quotation (2) has perhaps a bit more justification. W. Broecker (1991) created the notion of a "global conveyor belt" which represents the extreme end-view of the ocean as simple, laminar, and steady. Indeed it is a geologist's view of a simple slab-like flow field which is easy to understand and describe. (I decline to reproduce this picture, which most readers will have seen anyway, for fear of reinforcing a misconception that will likely require a generation to pass before it is expunged from the collective subconscious.) Many physical oceanographers, who should know better, have adopted this terminology and picture for their own ends* and thus the wider community has assumed that it must be an accurate depiction of the ocean circulation (in that wider community I include not only meteorologists, but also biological and chemical oceanographers, who may perhaps be forgiven). What is right about the global conveyor belt and what is wrong? Without going into detail, I would make three lists (Box 1).

Below, I will justify some of the statements about the conceptual difficulties, but full discussion of the lists in Box 1 would require more space than is available. For the moment, I will say only that in general, the

---

BOX 1
THE "GLOBAL CONVEYOR"

Accurate Elements
There *is* a net meridional overturning mass and property transport
(although the two are not the same).
Water *does* sink at high latitudes in the North Atlantic Ocean.
The resulting bottom water *does* return eventually to the seasurface in a closed system.

Difficulties. Missing
Water also sinks at high southern latitudes, in amounts roughly equal to that in the North Atlantic.
Water sinks to intermediate depths in many places including the North and South Pacific Oceans.
Much of the water which sinks at high latitudes in the North Atlantic is recycled back to the surface in the Southern Ocean before proceeding.
The bulk of the surface flow of water in the ocean is in the Antarctic Circumpolar Current (about 130×10^9 kg s^{-1}), the largest current on earth.
Much of the surface water entering the South Atlantic comes from the Circumpolar Current.

The North Pacific Ocean is a region of apparently minimal upwelling mass transport.
The fluid passing westward through the Indonesian passages appears to recirculate, primarily, around Australia.
The time average circulation is in large part dominated by narrow, strong boundary currents (Gulf Stream, Kuroshio, Agulhas) and in the southern ocean, the heat flux is dominated by eddy processes.
The kinetic energy of the ocean circulation is dominated (two or more orders of magnitude) by the time-varying component (Fig. 2).

Difficulties. Conceptual
The mean circulation is an integral over many small scale, extremely energetic, space and time-varying components.
A change in the high-latitude sinking rate need not translate into a simple shift in a global laminar flux rate.
Stoppage of convection in the North Atlantic does not necessarily mean that it cannot shift to some other ocean.

---

* The conveyor belt picture *is* a wonderful cocktail party metaphor for nonscientists.

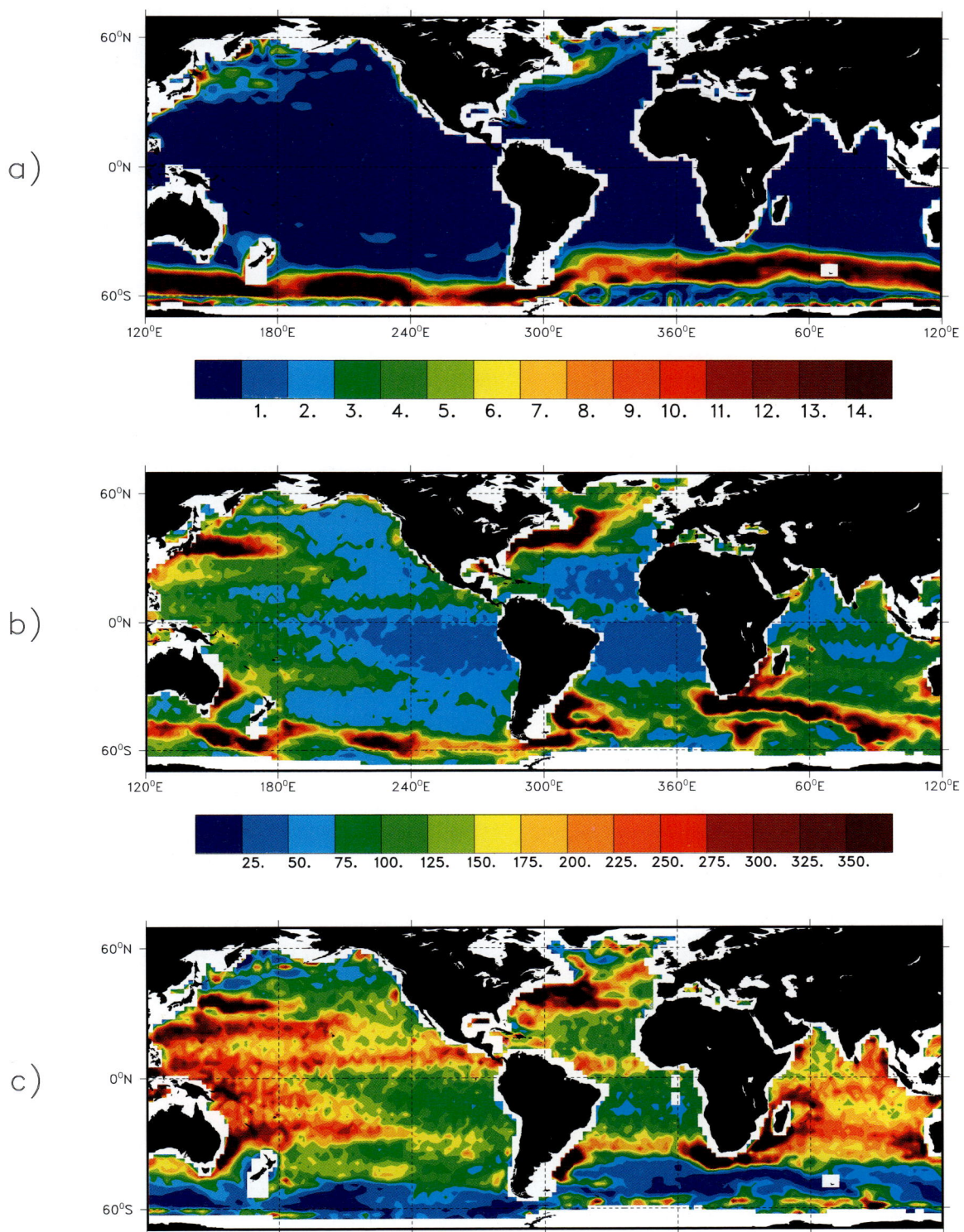

FIGURE 2 (a) The estimated kinetic energy of the seven-year time average ocean circulation as computed geostrophically from the absolute seasurface elevation of the TOPEX/POSEIDON altimetric satellite, and multiplied by f^2 to eliminate the equatorial singularity. (b) Same as the top panel except for the variability about the time mean. (c) The ratio of the kinetic energy of the variability to that of the mean, showing that almost everywhere, the variability dominates the mean. There are bias errors present, particularly in the calculation of the mean flow, arising from spatial smoothing and geoid errors; nonetheless, the figure is at least semi-quantitatively accurate.

FIGURE 3 The compelling abyssal circulation scheme created by Stommel and Arons in the 1950s (Stommel, 1958). Unfortunately, there is no observational support for the interior flow depicted. What evidence does exist suggests that the interior flows are primarily zonal, with meridional transports taking place in association with intense boundary mixing.

reduction of the complex, heterogeneous, turbulent, time-varying general circulation of a rotating, stratified fluid in a complicated geometry to something diagnosable by pure thought is a remarkable achievement in fluid dynamics—if it is true that it has been done.

Quotation (3) is based upon a famous picture shown in Fig. 3. The abyssal circulation scheme created by Stommel and Arons in the 1950s, another quite compelling, basically simple idea (Stommel, 1948) in what was an intellectual tour-de-force: he recognized that the geostrophic vorticity equation required fluid in the oceanic interior emanating from high-latitude sinking regions, to flow poleward, and thus toward the fluid source. Everything else followed from the requirement of mass and vorticity balance overall, achieved through the presence of boundary currents on the western edges of the ocean. What is usually called the "Stommel–Arons" theory is a beautiful and satisfying construct, and it is commonly asserted (as speaker (3) did) to *be* the deep ocean circulation. Observational evidence for the existence of deep western boundary currents is quite compelling. Unfortunately for this beautiful theory, there is not a shred of observational evidence for the interior circulation scheme that results, and the continued *assumption* that it describes the abyssal ocean circulation has become a major obstacle to progress.

D. THE REALITY

Beginning in the early 1970s, modern electronics had evolved to the point that vacuum tubes were no longer necessary, and pressure cases, and tape-recording technology had improved to the point that one could (nearly) routinely obtain *time series* of oceanographic data. Over the past 30 years, it has become possible to obtain multiyear observations of the velocity field from both moored and freely drifting instruments at great depth. A very large literature has developed on this subject (e.g., Robinson, 1983). What was reasonably evident by about 1975 from records such as the one shown in Fig. 4 that the ocean is intrinsically turbulent. A classical example is the 11-year-long current meter record described by Müller and Siedler (1992) which failed to produce a statistically significant non-zero time

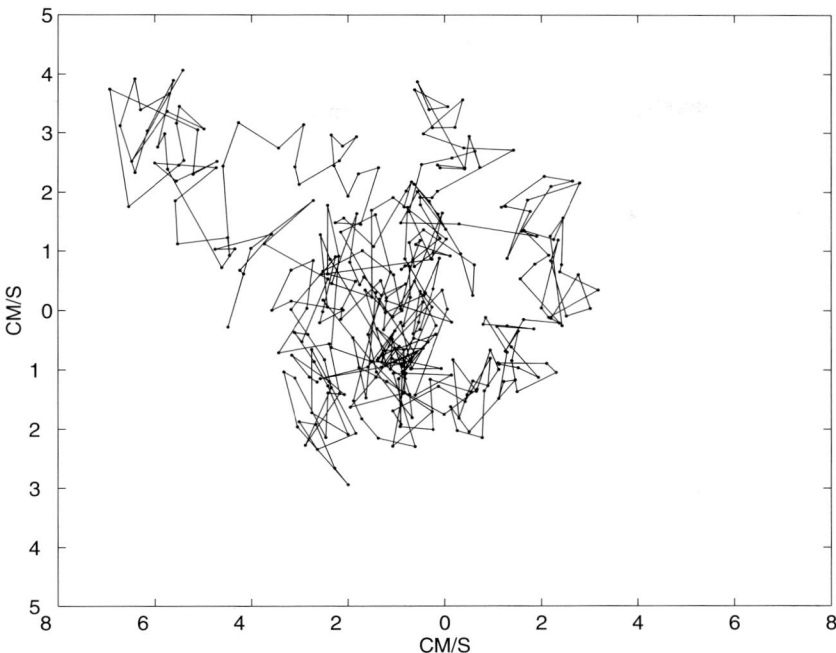

FIGURE 4 Current meter hodograph from 24 h average velocities over 1 year. The record was obtained at 1528 m depth at a position 27.3°N, 40.8°W in the North Atlantic as part of the Polymode program. The result is typical of open ocean records (although this one is quiet). This particular record does produce a mean displacement toward the northwest. If the high frequency components (periods shorter than 1 day) are included (not shown), the figure is an extremely dense cloud.

average of the highly variable velocity measured at depth in the eastern North Atlantic. (Early physical oceanographers (e.g., Maury, 1855; Helland-Hansen and Nansen, 1920) had clearly been aware of the presence of time-varying elements in the ocean circulation; little could be done to understand them, and ultimately the collective subconscious tended to treat them simply as a "noise".)

With the acquisition over a generation of a slowly growing data base, it has become clear that few if any elements of the ocean circulation are truly steady. The literature of the past 25 years is filled with studies expressing astonishment that new observations suggested that almost everything, when examined in detail, appeared to change and that the circulation of the ocean is not actually steady. One example (Fig. 5) will have to suffice. It shows the movement of neutrally buoyant floats (at a depth of 2500 m) in the Brazil Basin, exactly the region where the great oxygen tongue (Fig. 1) has been used to infer a laminar drift toward the south along the western boundary.

Perhaps the single most concrete evidence of the degree of temporal variability in the ocean has come from satellite altimetry (e.g., Fu *et al.*, 1994; Wunsch and Stammer, 1998). In particular, the TOPEX/POSEIDON spacecraft, which was flown by the US and French space agencies in the teeth of fierce arguments from much of the oceanographic community that it would be useless, has demonstrated a truly remarkable degree of variability in the oceans and has been assimilated along with other data into a global general circulation model (Stammer *et al.*, 2000) using the method of Lagrange multipliers (adjoint method). This variability is best appreciated from animations (available from the author), but one static figure will have to do here: Fig. 6 shows the assimilated model surface elevation anomaly at 2-day intervals during January 1994. Each centimetre of elevation difference corresponds to approximately 7×10^9 kg s^{-1} barotropic water transport in 4000 m of water in midlatitudes. These rapid changes are indeed largely barotropic. That the elevation changes significantly over a few days is meant to be apparent (the animations are much more striking).

There is a large-scale oceanic circulation, which appears to be stable over decades, but expected to be slowly changing everywhere in ways we do not understand because we do not have adequate measurements of it. To what extent does this slowly changing part of the circulation dominate the movement of properties in

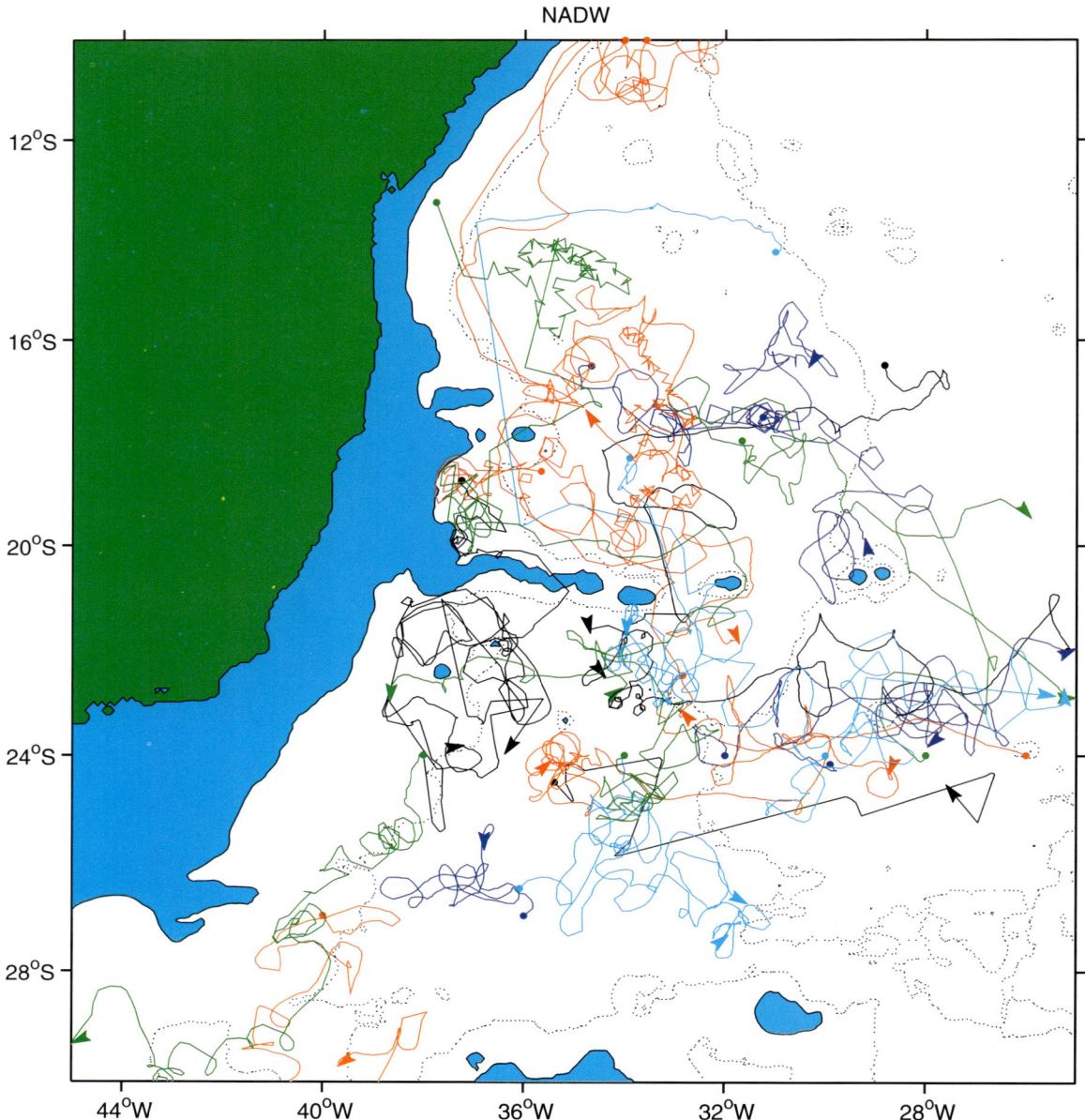

FIGURE 5 Movement over 800 days of neutrally buoyant floats in the region of the great oxygen tongue of Fig. 1—the depth of the North Atlantic Deep Water. This water mass has always been supposed to be moving slowly, more or less uniformly, southward, but the floats show that the movement is actually more east–west and with an extremely complicated structure (Hogg and Owens, 1999; N. Hogg, personal communication, 2000).

the ocean of importance to climate? This question can only be answered by a quantitative knowledge of the property transports of both the time-varying and quasi-time independent components.

With the benefits of hindsight, one perceives that the large-scale property structures in the ocean (the great "tongues") need not imply any large-scale time-mean flow. Quite turbulent flows are capable of producing large-scale structures in scalar properties, representing integrals over very long times and distances of the fluctuating movement of the circulation. One's eye is captured by the largest scale structures (it is apparently an evolutionary advantage that humans developed the capacity to see subtle large-scale patterns superimposed upon very complex backgrounds). But in most cases, it is the spatial gradients of these structures, not the struc-

D. THE REALITY

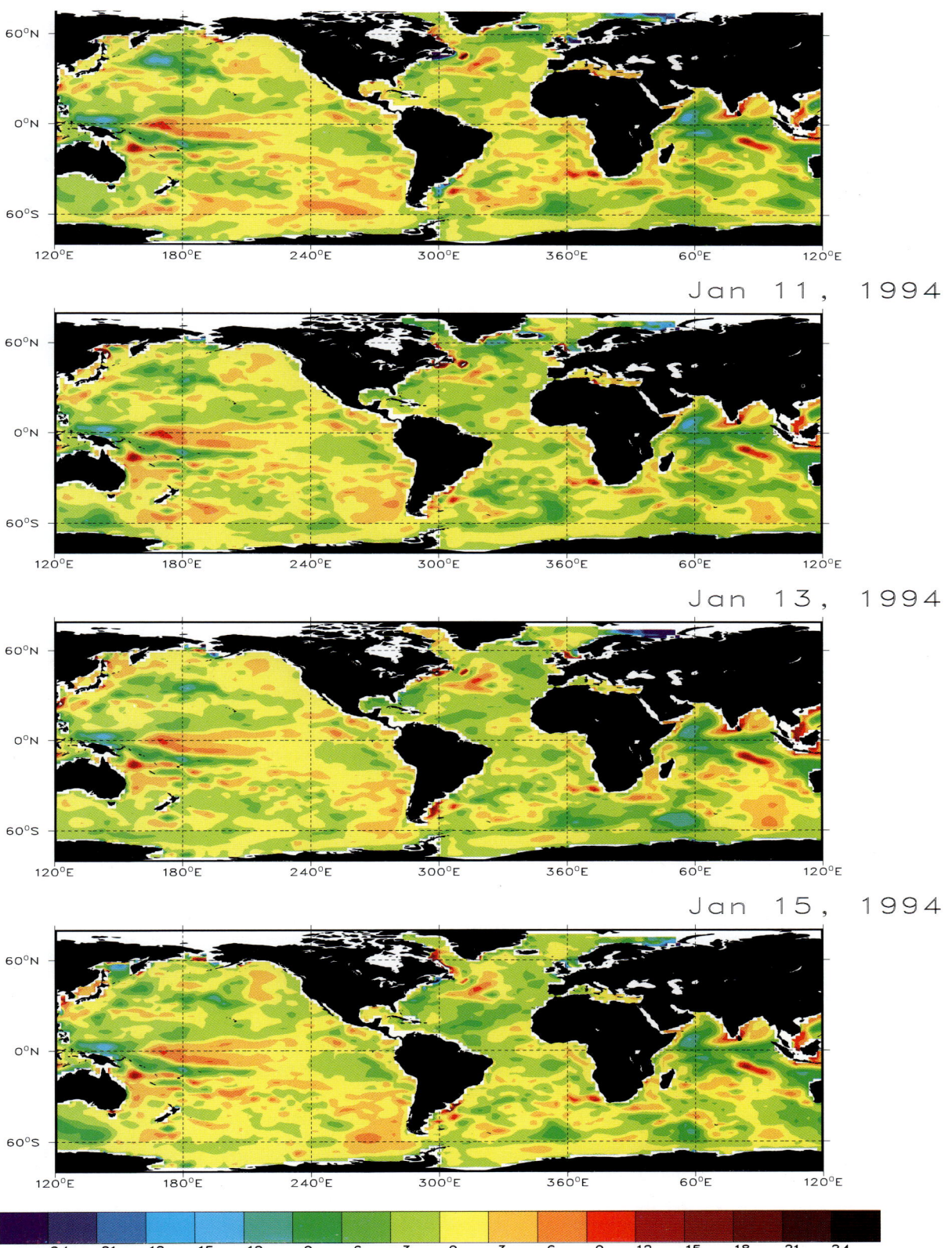

FIGURE 6 Sea surface elevation anomaly (relative to a six-year mean) from a fully assimilated ocean general circulation model. Even at intervals as short as 2 days, there are significant, and dynamically important shifts in the flow field (see Stammer *et al.*, 2001).

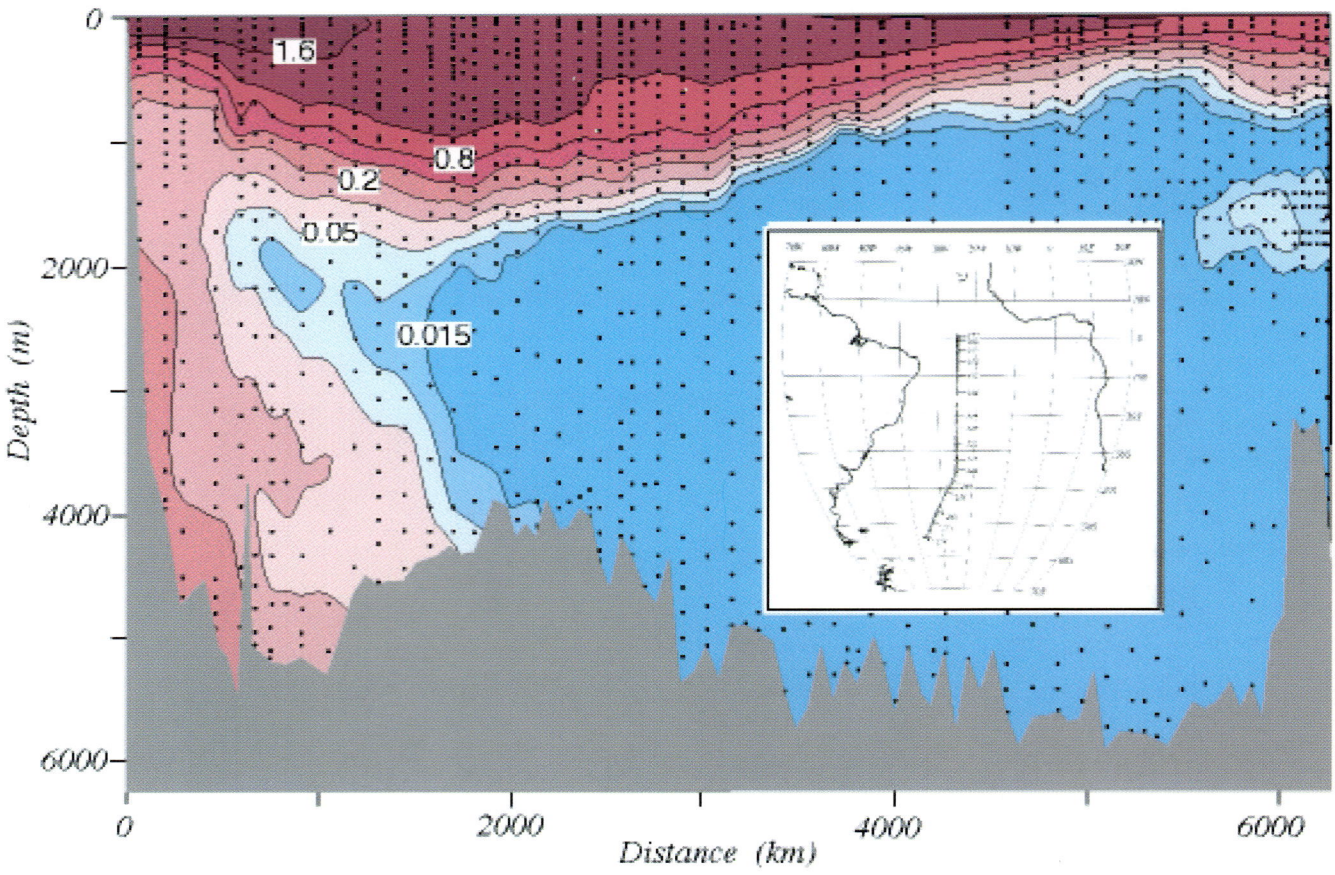

FIGURE 7 Freon, F-11, as observed during 1989 by Weiss *et al.* (1993). Within 30 years of the steep increase in atmospheric concentration, the tracer is found in the abyssal ocean far from its source. Disruption or enhancement of such connections could obviously have serious consequences for climate.

tures themselves, which are of dynamical and kinematical significance, and which are very difficult to estimate by eye. No one, to my knowledge, has attempted to produce global contours of the lateral *gradients* of the large-scale fields.

The history of oceanography is littered with appealing, simplifying, ideas, that had ultimately, to be painfully dislodged. The problem is further compounded by the fact that models have become so sophisticated and interesting, it is tempting to assume they must be skilful. This is a very dangerous belief!

E. CONCLUSIONS BASED UPON MISAPPREHENSIONS

The distorted view of oceanic behaviour based upon the various historical pictures has led to some ultimately crippling misunderstandings about how the ocean operates. For example, it is still common to hear it said that oceanic timescales are so long that it cannot change for hundreds, if not thousands, of years. This statement is commonly coupled with the assertion that one can ignore the ocean abyss for understanding climate change in time periods shorter than hundreds to thousands of years. Yet the fallacy of this view has been in evidence for at least 25 years. As a single example consider Fig. 7 (from Weiss *et al.*, 1993) showing Freon, F-11, as observed along a meridional section in the South Atlantic Ocean. Within 10–20 years of the onset of this transient tracer in the atmosphere, it is found on the floor of the ocean, hundreds of kilometres from the injection point. One might like to assume that these surface/abyssal connections (they occur all over the world ocean) are for some reason, unchangeable. But why should they be unchangeable? If one disrupted this vertical exchange, the near-surface properties could become entirely different very quickly. (The reader is reminded too (see, e.g., Boyle 1990) that the palaeoceanographic evidence is for massive changes in the ocean circulation on timescales hardly exceeding 10 years, a timescale consistent with what is known about the vertical connections such as seen in Fig. 7)

Another, very serious, climate issue arises from the distorted picture of the ocean. If the ocean is a simple laminar system, then very coarse general circulation models can be used to calculate how it will change under changing external conditions, e.g. doubled CO_2. It is not uncommon to see published calculations of future climate states obtained using ocean models with a spatial resolution as coarse as 4° laterally. Although the writers of such papers would undoubtedly deny that they are producing "forecasts", the reader is usually given little or no guidance as to the actual expected skill of such models. Is it plausible that a 4° or even 1° ocean model can be integrated with skill for 1000 years? If there is real skill, then the modelling community has solved one of the most difficult of all problems in turbulence: that of a rotating, stratified fluid in a complex geometry. What is the evidence for its truth?

Some simple "back-of-the-envelope" calculations show the scope of the problem. Consider one example. Much of "climate" is governed by the movement through the ocean of fluid properties (temperature, salt, carbon, etc.). Suppose one's model has a 1 mm s^{-1} *systematic* error in the computed velocity (Lagrangian) of a fluid particle. Then at the end of 100 years, one has a 3000 km position error for that particle. In terms of where enthalpy, carbon, etc. are located in the ocean, and where and how they may re-enter the atmosphere are concerned, errors of this magnitude can completely reverse the sign of the atmosphere–ocean exchange. Do ocean models have errors of this size? I have no idea, as it seems not to have been worthy of study, because "everyone knows" the ocean is laminar and simple. F. Bryan (personal communication, 2000) has shown that the so-called POP-model (Smith *et al.*, 2000) undergoes a sign reversal in the air–sea heat flux in the crucial area of the Grand Banks when the model resolution is shifted from 0.2° laterally to 0.1°. In general, ocean models are not numerically converged, and questions about the meaning of nonnumerically converged models are typically swept aside on the basis that the circulations of the coarse resolution models "look" reasonable.

There are many other examples. I have written elsewhere (Wunsch, 2001) about the conflicting "paradigms" of the ocean circulation and the consequent difficulties in designing proper programs to understand the system.

F. WHERE DO WE GO FROM HERE?

To some degree, everything said above can be reduced to the statement that "absence of evidence was taken as evidence of absence". The great difficulty of observing the ocean meant that when a phenomenon had not been observed, it was assumed to be not present. The more one is able to observe the ocean, the more the complexity and subtlety that appears. To the extent that estimates of the climate state omit physical (and chemical and biological) processes which have either never been observed, or not understood, any forecast, or even statement that "we understand it" will remain vulnerable to suggestions that the omitted processes are (1) present, and (2) represent serious systematic errors which fatally compromise any possibility of a skilful climate forecast.

At its worst, the assumption that the system is much simpler than it actually is, leads to the corruption of an entire literature. Readers of palaeoclimate papers in particular, will notice that extraordinarily complicated and far-reaching changes in the climate system are often reduced to simple assertions about how the "global conveyor" changed. In effect, these authors believe that the equations of motion governing the general circulation of the ocean can be solved and integrated for thousands or millions of years in one's head. One might be suspicious of such claims about the atmospheric circulation. I am unaware of any concrete evidence that modelling the ocean is any simpler than modelling the atmosphere. Indeed, one might readily infer that it is much more difficult: the deformation radius in the ocean is an order of magnitude smaller than in the atmosphere, producing small-scale permanent features capable of transporting properties over large distances (e.g., the boundary jets; see Fig. 8). The ocean has a "memory" in some elements of at least hundreds of years, meaning that long-ago interactions with the atmosphere are still in principle manifest in the ocean we see today.

Oceanography is a branch of fluid dynamics on a grand scale. I believe that the lessons of fluid dynamics tell us that one should remain quite sceptical of any theory or model which has not been carefully compared to observation. There are too many examples, even within oceanography itself, where some important phenomenon, which in principle could have been predicted by theory or model, was not thought of until observations showed its importance (Wunsch, 2001). Examples are temperature and velocity microstructure, the intricate current regime near the equator, the dominance of high-latitude barotropic fluctuations, and the recent realization that the ocean probably mixes

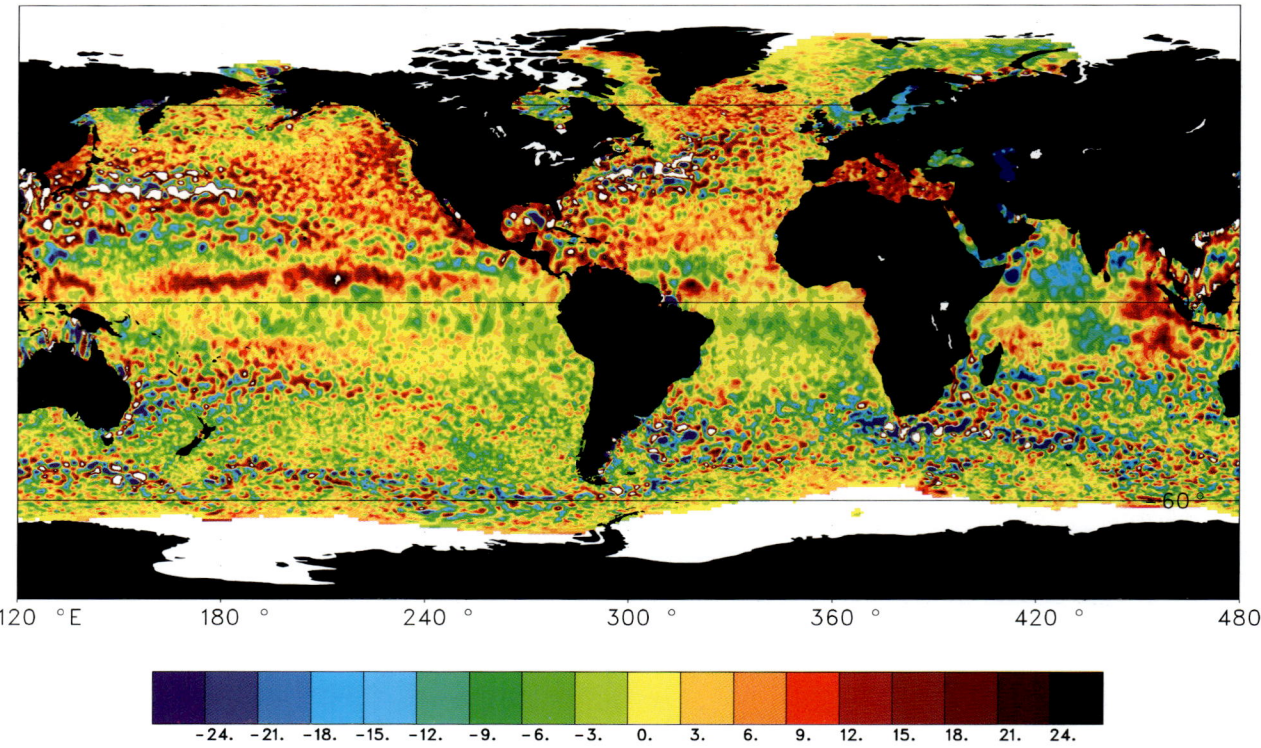

FIGURE 8 Surface elevation anomaly (cm) for the ocean during one 10-day period obtained by combining data from two altimeteric satellite systems (TOPEX/-POSEIDON and ERS-2) and using an objective analysis scheme with a 10-day *e*-folding decorrelation time. The extraordinary fine detail is none the less still a somewhat smoothed version of the true field, owing to the incomplete coverage and blurring by the 10-day interval.

primarily at its boundaries—in flagrant conflict with almost all GCMs.

The moral of the tale is that we need, and will surely continue to need for many years, an adequate set of observations of the ocean. Advancing technology is making the ocean ever-more accessible, although much cleverness will still be required. To a very large degree, however, the problems are less technical today than they are sociological, as alluded to in the Introduction. Oceanographers, in bleak contrast to meteorologists, do not have a customer for sustained observations. Meteorologists are both blessed and cursed by the need to produce weather forecasts. They long-ago convinced governments that for weather forecasting to be feasible, one must have adequate observations and consequently, vast sums are expended each year around the world to sustain *in situ* and satellite measurements of the atmosphere. Nowhere in the world do oceanographers have a governmental commitment to sustained large-scale observation systems. Indeed, the great bulk of large-scale measurements are still made today by individual scientists. Although the satellites have made a large difference to oceanography in the past decade, none of the oceanographic satellites of the most urgent importance (altimeters, scatterometers) is yet regarded as operational. They are thus vulnerable to shifts in government priorities and fashions.

Lest my opening quotations be perceived as unfairly pointing a finger at meteorologists, let me end with another quotation:

> "Programs like WOCE [the World Ocean Circulation Experiment] represent a Fall from Grace. They make oceanography too much like meteorology. Real oceanographers make their own observations."

This comment (in a widely disseminated email message) was made by an oceanographer, albeit a highly theoretical one. If adequate observations are regarded as incurring Original Sin, evidently the cultural shifts necessary to address the ocean fluid problem will not come easily. In the absence of operational government agencies, the oceanographic problem can only be solved with the direct and immediate assistance of the corresponding meteorological-cum-climate agencies. The first order of business is evidently, to be clear that everyone understands the problem, and to recognize the great influence past assumptions exercise over future necessity.

ACKNOWLEDGEMENTS

Supported in part by the NOPP-ECCO Consortium, ONR Contract N00014–99–1–1050. D. Stammer made some helpful comments on the manuscript.

References

von Arx, W. S., 1962: *An Introduction to Physical Oceanography.* Addison-Wesley, Reading, MA.

Boyle, E. A., 1990: Quaternary deepwater paleoceanography. *Science*, **249**, 863–870.

Broecker, W. S., 1991: The great ocean conveyor. *Oceanography*, **4**, 79–89.

Ekman, V. W. 1905: On the influence of the earth's rotation on ocean-currents. *Ark. Mat. Astron. Fys.*, **2**(11).

Fu, L.-L., E. J. Christensen, C. A. Yamarone, M. Lefebvre, Y. Ménard, M. Dorrer and P. Escudier, 1994: TOPEX/POSEIDON mission overview. *J. Geophys. Res.*, **99**, 24369–24382.

Helland-Hansen, B. And F. Nansen, 1920: Temperature variations in the North Atlantic Ocean and in the atmosphere. *Smithsonian Misc. Coll.*, **70**(4).

Hogg, N. G. And W. B. Owens, 1999: Direct measurement of the deep circulation within the Brazil Basin. *Deep-Sea Res.*, **46**(2), *G. Siedler Volume*, 335–353.

Maury, M. F., 1855. *The Physical Geography of the Sea and Its Meteorology.* Harper, New York.

Müller, T. J. and G. Siedler, 1992: Multi-year current time series in the eastern North Atlantic Ocean. *J. Marine Res.*, **50**, 63–98.

Munk, W., 1950: On the wind-driven ocean circulation. *J. Meteorol.*, **7**, 79–93.

Reid, J. L., 1986. On the total geostrophic circulation of the South Pacific Ocean: flow patterns, tracers and transports. *Progr. Oceanogr.*, **16**, 1–61.

Robinson, A. R. (Ed.), 1983: *Eddies in Marine Science.* Springer-Verlag, Berlin.

Smith, R. D., M. E. Maltrud, F. O. Bryan and M. W. Hecht, 2000. Numerial simulation of the North Atlantic Ocean at 1/10°. *J. Phys. Ocean.*, **30**, 1532–1561.

Stammer, D., C. Wunsch, R. Giering, C. Eckert, P. Heimbach, J. Marotzke, A. Adcroft, C. N. Hill and J. Marshall, 2001: The global ocean state during 1992–1997, estimated from ocean observations and a general circulation model. Part I. Methodology and estimated state. Submitted for publication.

Stommel, H., 1948: The westward intensification of wind-driven ocean currents. *Trans. Am. Geophys. Un.*, **29**, 202–206.

Sverdrup, H. U., 1947. Wind-driven currents in a baroclinic ocean; with application to the equatorial currents on the eastern Pacific. *Proc. Nat. Acad. Sci. U.S.A.*, **33**, 318–326.

Weiss, R. F., M. J. Warner, P. K. Salameli, F. A. Van Voy and K. G. Harrison, 1993: South Atlantic Ventilation Experiment: SIO Chloroflurocarbon Measurements, *SIO Reference 93–49*.

Wunsch, C., 2001: Global problems and global observations. In *Ocean Circulation and Climate. Observing and Modelling the Global Ocean.* G. Siedler, J. Church and J. Gould, Eds., Academic, San Diego, 47–80.

Wunsch, C. And D. Stammer, 1998: Satellite altimetry, the marine geoid and the oceanic general circulation. *Ann. Rev. Earth Plan. Sci.*, **26**, 219–254.

Wüst G., 1935: *Schichtung und Zirkulation des Atlantischen Ozeans. Die Stratosphare. Wissenschaftliche Ergebnisse der Deutschen Atlantischen Expedition auf dem Forschungs-und Vermessungsschiff "Meteor" 1925–1927,* 6: 1st Par, 2. (Reprinted as The *Stratosphere of the Atlantic Ocean* (W. J. Emery, Ed.), 1978, Amerind, New Delhi.)

Part 4. The Biogeochemical System

Biogeochemical Connections Between the Atmosphere and the Ocean
P. S. Liss

Modelling Vegetation and the Carbon Cycle as Interactive Elements of the Climate system
P. M. Cox, R. A. Betts, C. D. Jones, S. A. Spall and I. J. Totterdell

Biogeochemical Connections Between the Atmosphere and the Ocean

Peter S. Liss

School of Environmental Sciences, University of East Anglia, Norwich, UK

In comparison with other planets in the solar system, the chemical composition of the earth's atmosphere clearly shows the influence of biological processes. The level of carbon dioxide is much lower and of oxygen much higher due to the photosynthesis of land and marine plants, and many reduced gases exist at concentrations far in excess of what purely inorganic processes would allow. In this chapter the emphasis is on the role of biological activity in the oceans in affecting the chemical (and physical) properties of the atmosphere. Air–sea exchanges of ozone, carbon dioxide, dimethyl sulphide, dimethyl selenide and ammonia are described with regard to their importance for processes in the atmosphere, together with the role of atmospheric dust inputs on ocean biology with its potential for concomitant change in trace gas fluxes to the atmosphere.

A. INTRODUCTION

Just over 30 years ago mankind had the opportunity to view planet earth from the moon. It can be argued that this represented a major turning point in how we have regarded our home planet ever since. Seen from the moon, the earth looks very isolated in space. From that viewpoint its unity is obvious, which reinforced the idea that our planet must be seen and studied as a whole, rather than split into its component reservoirs, as is so often done in academic research. The obvious dominance of the oceans in terms of coverage relative to land led to the idea that the earth should really have been called "ocean". Finally, the blue oceans, green/brown land and white clouds all looked very different to the colouring of the other planets we can see using telescopes from earth. It cannot be said that any of this understanding was new, since intellectually all of it had been known for many years. What was new was the perspective that viewing our planet from an outside observation post provided. Thus the isolation, beauty and very special properties of earth were made visually apparent.

The extraordinary properties made manifest in the blues, greens, browns and whites of the earth as seen from space are, of course, due in large measure to the presence of living organisms on the planet. In this chapter some of the ways in which the functioning of the earth's (particularly marine) biota control important properties of the atmosphere, oceans and land will be discussed. For the atmosphere, comparison of its chemical composition with that of other planets in the solar system strongly shows the influence of biological processes. Many reduced gases exist, at concentrations far in excess of what purely inorganic processes would allow. In the oceans the biota play a strong role in determining the amounts of gases important in controlling the earth's radiation balance, such as carbon dioxide and dimethyl sulphide (DMS), as well as the chemistry of the atmosphere and geochemical cycling of several key elements including nitrogen, selenium and the halogens. It has recently been shown how delicately poised the oceanic biota are in terms of availability of the element iron, and hence in their ability to control levels of several gases of importance in the atmosphere.

This chapter begins with a brief overview of the effect of the biosphere on the overall composition of the earth's atmosphere. There follows a section on the role of the marine biota in producing or consuming various trace gases which are important for the chemistry (and physics) of the atmosphere. After this there is a complementary section on the role that inputs from the atmosphere can play in affecting the level of biological activity in the oceans, which activity can in turn affect the gaseous emissions discussed in the previous section. This cyclic interaction between the atmospheric and oceanic reservoirs leads to a short concluding section on the global nature of biogeochemical inter-reservoir transfers; a concept encouraged by the holistic view of the earth obtained by observing it from the moon.

B. THE CHEMICAL COMPOSITION OF THE EARTH'S ATMOSPHERE

The gaseous composition of the earth's atmosphere clearly shows the influence of biological processes on land and in the sea. This is well illustrated in Fig. 1, which shows a comparison of the present composition of the atmosphere (i.e. with life present), with that predicted if no life existed on the earth.

The prediction of the lifeless composition is made by assuming the earth and its atmosphere to be at thermodynamic equilibrium, with the presence of life disturbing that equilibrium. It should be noted that in Fig. 1 the percentage contribution of each gas to the total composition is shown as a histogram on a logarithmic scale for both the life present and the life absent cases. A clear example of the effect of life on atmospheric composition is the 3–4 orders elevation of O_2 and concomitant decrease in CO_2 arising from the uptake of the latter and release of the former as a result of photosynthesis by land plants and marine phytoplankton. Furthermore, several trace gases, including the important greenhouse gases CH_4, N_2O, occur 2–3 orders of magnitude higher in concentration in our present atmosphere than would be predicted in the absence of life. Other gases including NH_3, HCl and DMS (not shown) are present in low but measurable amounts in the current atmosphere, whereas they would be at significantly lower levels or totally absent in the air of a lifeless earth. Some important gases, e.g. water vapour and ozone, are omitted from the figure because, although the biosphere and its products are important for their cycling, it is difficult to estimate concentration levels for them in the pre-life atmosphere.

C. AIR–SEA EXCHANGE OF GASES OF IMPORTANCE IN THE ATMOSPHERE

Many gases cross the atmosphere–ocean interface, and in Table 1 a summary is given of the gases involved, as well as an indication of their importance in the atmosphere, both troposphere and stratosphere. For most of the gases, the net flux direction is from sea to air, i.e. the oceans are a source of the gases to the atmosphere. However, there are some exceptions, two of which are now discussed.

1. Ozone

Ozone (O_3) is one exception, since for this gas the oceans are an absolute sink (i.e., there is no re-emission, due to the high reactivity of ozone with iodide ions and organic material at the sea surface, leading to its destruction in the water (Garland *et al.*, 1980). Although the rate of uptake of O_3 by the oceans is small compared with deposition to land surfaces, their large area means that they may still be a significant sink in the global budget of the gas (Ganzeveld and Lelieveld, 1995). Another possible exception is ammonia where the net flux direction is somewhat uncertain and is probably highly variable, a situation which will be discussed further later.

2. Manmade Carbon Dioxide

Another exception is manmade CO_2 where the oceans take up about 40% of the total amount of the gas emitted to the atmosphere from the burning of fossil fuels (IPCC, 1996). This contrasts with the situation for natural CO_2 where, although in some locations and for some seasons the net fluxes are both in and out of the oceans, averaged for the whole of the oceans and over a yearly cycle there is near balance between the gross up and down flows. The observed seasonal and spatial variability is due to both temperature affecting the solubility of CO_2 in seawater, and biological uptake and release of the gas during photosynthesis and respiration, respectively. However, with the increase in atmospheric CO_2 resulting from anthropogenic activities the natural balanced two-way fluxes have a net downwards component superimposed on them.

The importance of biological activity is not only for

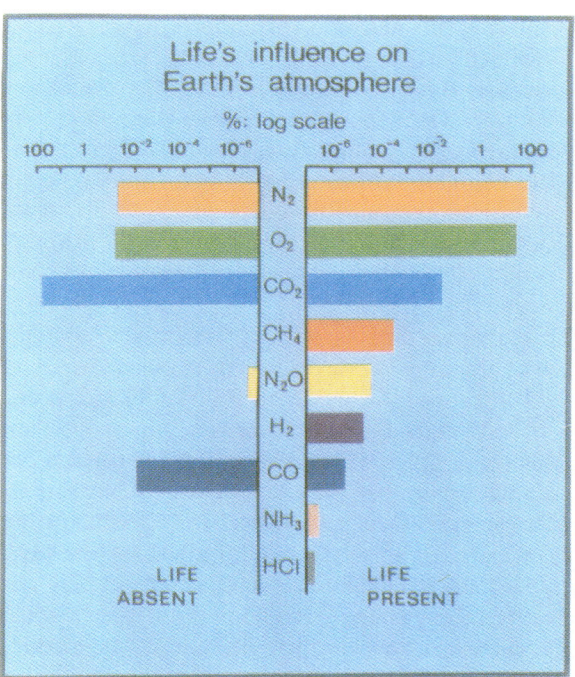

FIGURE 1 Life's influence on the composition of the earth's atmosphere. Percentage composition on a logarithmic scale with and without life present on the planet (IGBP, 1992, after Margulis and Lovelock, 1974).

TABLE 1 Air–Sea Exchange of Gases of Importance in the Atmosphere

Gas	Net flux direction	Importance in atmosphere	Troposphere	Stratosphere
Dimethyl sulphide (DMS)	↑	Acidity	✓	
		Cloud condensation nuclei	✓	
		Budgeting S cycle	✓	
Carbonyl sulphide (COS)	↑	Particles	✓	✓
Carbon disulphide (CS_2)	↑	Particles	✓	✓
Ammonia (NH_3) and methylamines	↑ (↓?)	Alkalinity	✓	
		Aerosol composition	✓	
Organohalogens (natural)	↑	Oxidation capacity	✓	✓
		Budgeting I (and Br?) cycle(s)	✓	
Carbon monoxide	↑	Oxidation capacity	✓	
Non-methane hydrocarbons	↑	Oxidation capacity	✓	
Nitric oxide (NO)	↑	Oxidation capacity	✓	
Ozone (O_3)	↓	Oxidation capacity	✓	✓
		Radiation balance		
Carbon dioxide (CO_2) (manmade)	↓	Radiation balance	✓	?
Nitrous oxide (N_2O)	↑	Radiation balance	✓	?
Methane (CH_4)	↑	Radiation balance	✓	?
Dimethyl selenide (DMSe)	↑	Budgeting Se cycle	✓	
Elemental mercury (Hg)	↑	Budgeting Hg cycle	✓	

the small-scale and short-term changes in CO_2 uptake indicated above. Changes in ocean productivity also have the potential to strongly influence levels of CO_2 in the atmosphere on much longer time-scales. This is illustrated in Fig. 2. The figure shows the results of a study by Watson and Liss (1998) in which a simple model (Sarmiento and Toggweiler, 1984) was used to examine how far changes in marine biological activity can affect atmospheric CO_2 partial pressures. In each run the model was initialized at pre-industrial steady state (approximately 260 ppm). At $t = 100$ years a perturbation was applied. For the "Strangelove ocean" all marine biological activity was stopped instantaneously. "Cold surface water productivity = 0" and "cold surface water productivity = max" show the possible extremes to which atmospheric CO_2 can be driven by modulating the efficiency of the cold-water marine biota. After 1000 years the difference between the atmospheric CO_2 level produced by the lifeless and maximum productivity oceans is about 250 ppm (180–430 ppm). Although this is clearly an extreme range, it does illustrate that modulation of biological activity in the oceans has the potential to have a very substantial effect on atmospheric CO_2 concentrations.

Turning now to gases in Table 1 whose direction of net flux is from the oceans to the atmosphere, several examples are considered.

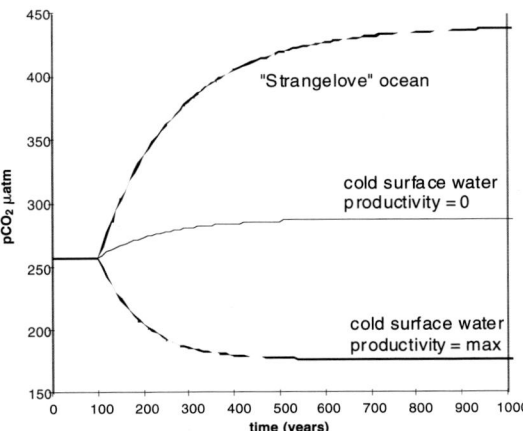

FIGURE 2 Effects of marine biota on atmospheric CO_2 from a simple model (Sarmiento and Toggweiler, 1984); the various scenarios are explained in the text (Watson and Liss, 1998).

3. Dimethyl Sulphide

Dimethyl sulphide (DMS) is a gas produced in ocean surface waters by the activities of phytoplankton. The DMS itself is formed by the enzymatic cleavage of a precursor molecule (dimethyl sulphoniopropionate, DMSP), which the organisms make as an osmolyte or as a cryoprotectant. The commonest way for the DMSP to DMS transition to occur is on death of the plankton by zooplankton grazing or attack by viruses. Once released to the seawater there is a range of bacterial and photochemical transformations which the DMS can undergo (see Liss et al., 1997, for a review of production and destruction processes of DMS in seawater). However, there is almost always a measurable residual concentration of DMS in the water and it is this which drives the net flux of the gas to the atmosphere. Because of its rapid dispersion and reaction in the atmosphere, air concentrations are generally only a few percent of the equivalent water values and are often ignored in air–sea flux calculations with little loss of accuracy.

Once in the atmosphere, DMS is subject to oxidation by free radicals such as hydroxyl and nitrate to form a variety of products (Ravishankara et al., 1997), the most important of which appear to be methane sulphonic acid (MSA) and sulphur dioxide, which can be further oxidized to sulphate either in water droplets or in aerosols, including sea-salt particles (see e.g., Sievering et al., 1992). All of these products are acidic and give the natural acidity to aerosol particles, rain and other forms of precipitation in the marine atmosphere (Charlson and Rodhe, 1982). Of course, over industrialized/urbanized parts of the globe, particularly in the Northern Hemisphere, human activities have strongly perturbed the natural sulphur cycle, with concomitant increase in both sulphate aerosol numbers and decrease in pH of precipitation and particulates. In areas remote from heavily populated land, particularly in the Southern Hemisphere, the natural cycle is still dominant. Chin and Jacob (1996) have calculated that in terms of the total atmospheric burden of sulphur the percentage contributions of the three main sources are currently: biogenic (mainly from oxidation of DMS) 42%, anthropogenic 37% and volcanic 18%.

The transfer of sulphur in the form of DMS from the sea to the marine atmosphere, with some of the oxidation products ultimately deposited on land, is a vital component of the budget of the element in the earth system, since material balance cannot be achieved without it (Brimblecombe et al., 1989).

A further important property of DMS and specifically submicron particles formed via its oxidation is in affecting climate. Such particles can interact both directly with sunlight (Shaw, 1987) and indirectly by acting as nuclei on which cloud droplets form (termed cloud condensation nuclei, CCN). The number density of CCN is a major determinant in cloud formation and thus of the albedo of the planet (Charlson et al., 1987). Over land there is generally an abundance of CCN resulting from soil dust blown into the air and from pollution sources. However, over the oceans, far from land (i.e. the majority of the globe), the main source of CCN appears to be from oxidation of marine-produced DMS. This is supported by various lines of evidence, including the chemical composition of the CCN being largely sulphate (neutralized to some extent by uptake of ammonia gas from the atmosphere, see later), and the coherence in the seasonality of both CCN number

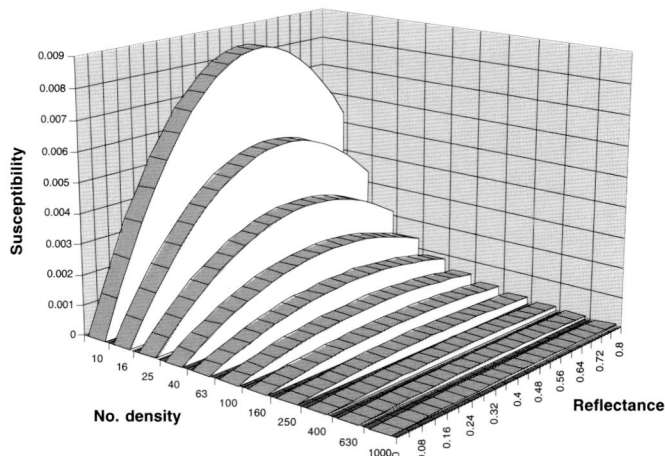

FIGURE 3 Susceptibility of cloud reflectance to changes in CCN number density for different total cloud reflectances. Susceptibility is the increase in reflectance per additional droplet cm^{-3}. (Redrawn in Watson and Liss (1998) from the equations given by Twomey (1991).)

density and cloudiness (as measured by optical depth), which in turn correspond to the times when marine plankton are actively producing DMS in seawater (Boers et al., 1994; Liss and Galloway, 1993).

Twomey (1991) has attempted to quantify the magnitude of the effect of CCN number density on cloud albedo (and hence reflectance), as illustrated in Fig. 3. This shows that the CCN–albedo relationship is highly nonlinear. Thus, the effect of change in CCN number on the incremental increase in albedo for each particle added (called susceptibility in Fig. 3) is much more pronounced at low compared with high particle numbers. For example, once the air contains several hundred particles or droplets per cubic centimetre, addition of further particles has little effect on cloud reflectance. Over and close to land, CCN numbers generally exceed this level. Thus, we must look to oceanic areas, and particularly the Southern Hemisphere, with its large ratio of sea to non-ice-covered land, for the greatest potential climatic impact of changing CCN numbers, whether the particles arise, for example, from alteration in oceanic DMS emissions or from inputs of SO_2 from (increased) fossil fuel burning. To graphically illustrate the point. Twomey cites a calculation which indicates that equal quantities of sulphur entering the atmosphere in the two hemispheres will have 25 times the effect on cloud albedo in the Southern compared with the Northern Hemisphere.

4. Dimethyl Selenide

Selenium is the element immediately below sulphur in Group VI of the Periodic Table. For this reason similarities in behaviour between the two elements might be expected, in this case for their methylated forms DMS and DMSe. For almost a quarter of a century it has been known that, as in the case for sulphur already discussed, there is a discrepancy in the global budget of selenium which can only be rectified by a flux of some, probably volatile, form of the element from the ocean to the atmosphere (Mosher and Duce, 1987). However, until recently, difficulties with the sensitivity of the analytical techniques available for measuring the very low levels at which selenium exists in the environment have meant that it has not been possible to confirm this and establish the size and form of the sea to air flux by direct measurement. Notwithstanding these difficulties, a recent measurement campaign initiated by the author and Drs David Amouroux and Olivier Donard of the University of Pau has thrown significant light on this topic (Amouroux et al., 2001). Volatile forms of selenium were measured at picomolar concentration levels in water samples collected on a research cruise in the north Atlantic in the summer of 1998. The results show that the main volatile forms of selenium are DMSe and the mixed selenium–sulphur compound DMSeS. In addition, simultaneous measurements of DMS and biological parameters clearly show a correlation between the concentrations of DMSe and DMS and strongly indicate a biogenic source for the volatile selenium gases measured. Calculation of the size of the sea-to-air flux of volatile selenium and extrapolation to the oceans as a whole indicates that it is likely to be sufficient to balance the global budget for the element. This implies that the sea is a source of selenium to the land, a result with important implications for human health, as recently reviewed by Rayman (2000).

5. Ammonia

Ammonia (NH_3) is another example of a gas found in the atmosphere which would not occur there but for biological activity. In Fig. 1 the lifeless earth's atmosphere shows essentially no NH_3, whereas with life present there is a small but significant amount. The ammonia comes from a variety of sources on land as well as from the oceans; here we will consider only the latter source because it can be treated as a companion air–sea flux to that of DMS discussed previously. As in the case of sulphur, the nitrogen cycle has also been substantially amended through a variety of human activities.

In order to examine the role and behaviour of NH_3 in its natural setting, we now have to go to parts of the globe furthest from man's activities. Since such areas are also far from land, this enables the marine part of the cycle to be examined without major influence from terrestrial source regions (this is possible because the atmospheric lifetime of NH_3 after emission from land (or sea) is only a matter of days). Figure 4 shows measurements made by Savoie et al. (1993) on atmospheric aerosol particles collected over a 5-year period at a coastal sampling site in Antarctica. A clear seasonal cycle is evident for ammonium (the form of ammonia in the aerosol), with highest values in the spring and summer seasons and minimum levels in autumn and winter. Also remarkable in Fig. 4 is the similarity of the ammonium to the MSA and non-sea-salt sulphate (both oxidation products of DMS) seasonal cycles. These similarities strongly suggest a marine biological origin for the NH_3, as well as for the DMS products. Support for this idea comes from a more detailed analysis of the data in Fig. 4, from which it can be calculated that the molar ratio of ammonium to non-sea-salt sulphate in the aerosols is close to 1:1, implying a chemical composition of ammonium bisulphate (NH_4HSO_4). The final piece of the jigsaw comes from measurements made in the North and South Pacific Ocean by Quinn et al. (1990), from which they were able to calculate the fluxes of both NH_3 and DMS being emitted from the ocean to the atmosphere. Although both fluxes show quite a lot of spatial variability, the mean values are again reasonable close to a 1:1 ratio, supporting the idea that the aerosol composition in terms of both nitrogen and sulphur (the main constituents of the total mass) can be explained by marine biogenic emissions.

D. IMPACT OF ATMOSPHERIC DUST ON OCEAN BIOGEOCHEMISTRY

As discussed in the previous section, trace gases produced or consumed by biological activity in the oceans clearly affect the chemical and physical properties of the atmosphere. However, the atmosphere also provides input of nongaseous material to the oceans in the form of land-derived mineral dust and particulates formed during burning and the use of fertilizers in agriculture. These solids can have a potentially large impact on marine productivity, particularly in certain parts of the oceans.

As with land plants, marine plankton need nutrient elements (e.g. nitrogen, phosphorus, iron, etc.) for optimal growth. It is generally agreed that of these nutrients it is deficits of nitrogen and iron which present the greatest potential for limitation and for them atmospheric inputs are also likely to be important. In the case of nitrogen there is growing concern about the impact of the increasing input of human-derived nitrogen compounds to regions of the open oceans where nitrogen is the nutrient-limiting biological activity, such

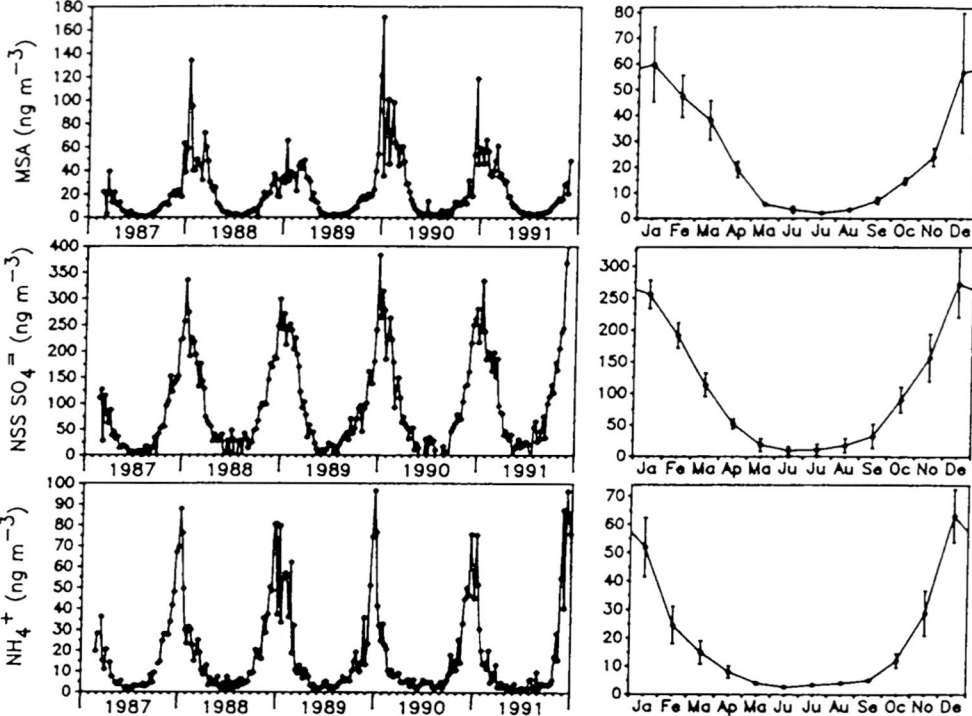

FIGURE 4 Graphs on the left show the actual time series of weekly-average concentrations of methane sulphonic acid (MSA), non-sea-salt sulphate (NSS SO_4^{2-}) and ammonium NH_4^+) at Mawson, Antarctica. Graphs on the right illustrate the composted seasonal cycles, showing the monthly means and their 95% confidence intervals (from Savoie *et al.*, 1993).

as the subtropical gyres of the North and South Pacific. While estimates suggest that, at present, atmospheric nitrogen accounts for only a few percent of the annual new nitrogen delivered to surface waters in these regions, the atmospheric input to the ocean is highly episodic, often coming in large pulses extending over only a few days when it can play a much more important role. Further, such inputs are increasing as the developing nations of the world increase their production and use of fixed nitrogen. For example, a modelling study by Galloway *et al.* (1994) developed maps of the "recent" (1980) and expected (2020) annual deposition of oxidized forms of nitrogen, which includes nitrate, nitrite and nitric acid derived from combustion processes, from the atmosphere to land and ocean surfaces. In Fig. 5 the projected ratios of the estimated deposition of oxidized nitrogen in 2020 to the values in 1980 are shown. From 1.5 to 3 times, and in some limited areas up to 4 times, the recent rate of deposition will occur by 2020 over large areas of coastal and open oceans, and these estimates do not include possible changes (almost certainly increases) in reduced and organic nitrogen inputs. This leads to the possibility of regional biogeochemical impacts in both coastal and open ocean areas.

Turning now to the case of iron, it has been known for almost a hundred years that marine biological activity, as judged by underutilization of "traditional" plant nutrients in the water such as phosphate and nitrate, appears limited in areas of the ocean far from land. Examples of these high-nutrient low-chlorophyll (HNLC) areas are the southern oceans, and the equatorial and sub-Arctic Pacific. It has been speculated for a long time that the nutrient limiting biological activity in these areas is iron. The fact that the underutilization of nitrate and phosphate occurs most clearly in waters remote from land is in accord with the idea that an important source of the iron is from deposition of soil dust carried from the continents via the atmosphere. Figure 6 shows the distribution of the global flux of iron to the oceans. The distribution is clearly very nonlinear (the scale on the figure is logarithmic), with highest deposition close to deserted continental areas and the remote HNLC regions listed above being amongst the regions receiving least.

However, because of the difficulties of measuring iron at the very low concentrations at which it occurs in seawater, it has until recently been difficult to test whether it is indeed an important limiting nutrient for plankton growth. The first experiments were done onboard ship using flasks of seawater containing plankton and the

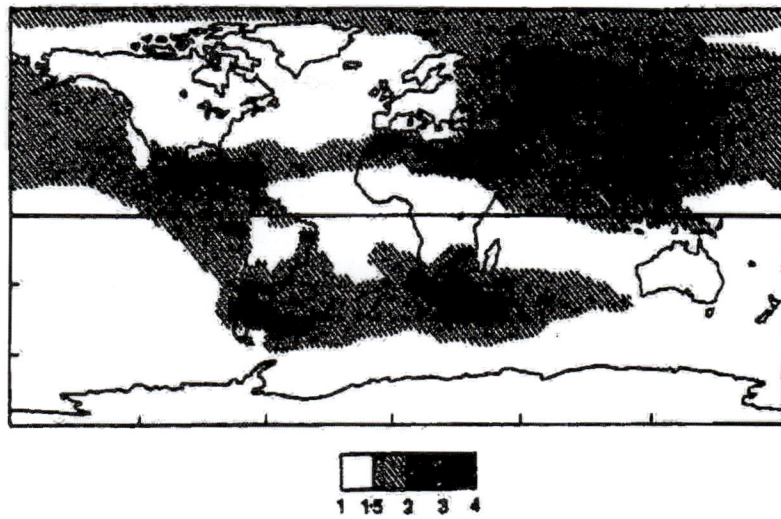

FIGURE 5 Ratio of the estimated deposition of oxidized forms of nitrogen to ocean and land surfaces in 2020 relative to 1980 (adapted from Galloway *et al.*, 1994).

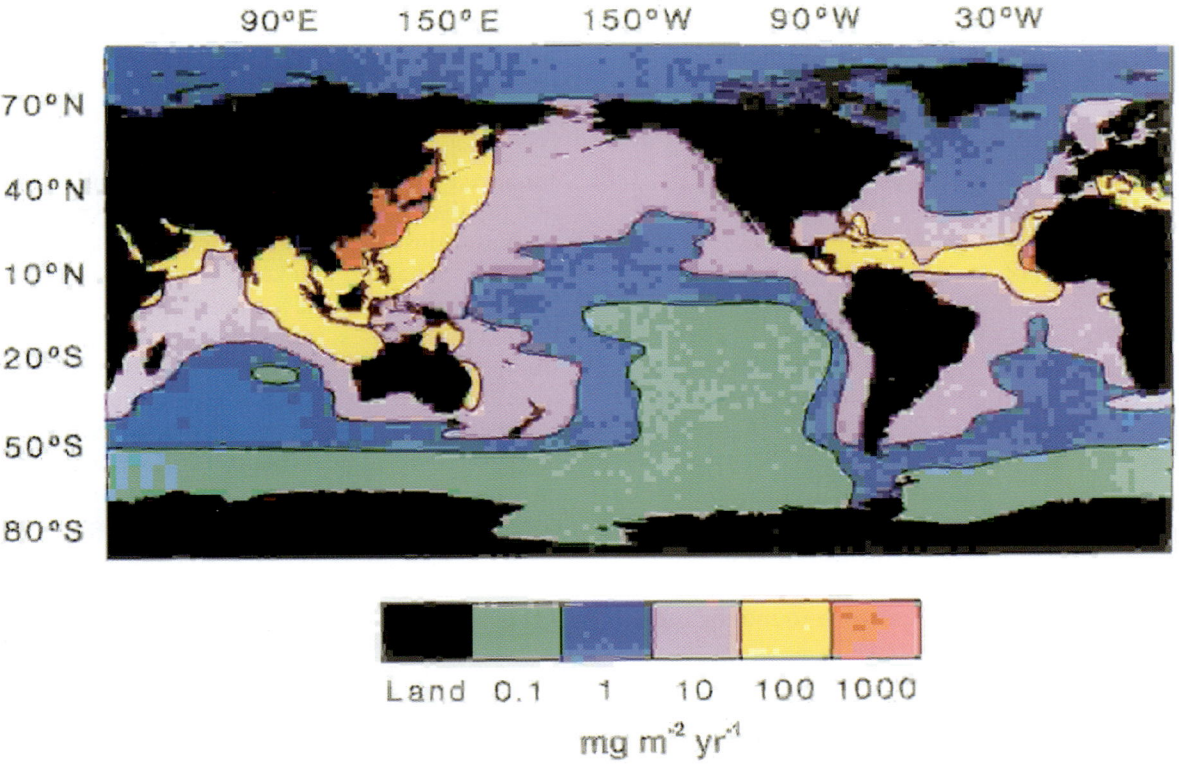

FIGURE 6 Regional atmospheric deposition flux of total iron to the ocean calculated from the atmospheric deposition of mineral dust given in Duce *et al.* (1991). The calculation assumes that 3.5% of the mineral dust is iron (the average concentration reported by Taylor and McLennon, 1985). (From Donaghay *et al.*, 1991.)

results seemed to confirm that adding iron did indeed enhance plant growth (e.g., Martin and Fitzwater, 1988). However, questions remained about how well the flasks mirrored conditions in the oceans and there were also problems of contamination during collection of the seawater (Fitzwater *et al.*, 1996). These difficulties led to the idea of conducting such experiments by enriching a small patch of ocean water with iron *in situ* (Watson *et al.*, 1991). Several such experiments have been carried out to date, two in the equatorial Pacific (Martin *et al.*, 1994; Coale *et al.*, 1996) and most recently in the southern ocean (Boyd *et al.*, 2000). In all cases there was clear evidence in support of the iron limitation hypothesis. From the viewpoint of ocean biology affecting the properties of the atmosphere, these *in situ* studies showed that the enhanced biological activity arising from iron amendment leads to enhanced drawdown of carbon dioxide (Cooper *et al.*, 1996; Watson *et al.*, 2000) and increased production of DMS (Turner *et al.*, 1996).

E. GLOBAL PERSPECTIVES ON BIOGEOCHEMICAL FLUXES ACROSS THE AIR–SEA INTERFACE

We conclude this chapter with two examples which attempt to place the air–sea exchange of DMS and iron, discussed earlier, into a broader whole earth context.

1. DMS and the CLAW Hypothesis

The case that in the unpolluted atmosphere DMS is the major source of acidity and aerosol particles (including CCN) was discussed earlier. However, some major workers in this field (Charlson, Lovelock, Andreae and Warren, 1987; the acronym CLAW derives from the initial letters of their surnames) have gone further and proposed that there is a feedback mechanism through which production of DMS by plankton can act as a climate regulating mechanism. This is illustrated diagramatically in Fig. 7. On the left-hand side of the figure the production of DMS in the sea, its emission to the atmosphere and conversion to a variety of products is shown. Following round the cycle, this leads to the role of sulphate CCN in cloud formation with its obvious link to climate. The novel proposal made by Charlson *et al.* (1987) is shown on the right-hand side of the diagram. In this it is suggested that a change in atmospheric temperature, arising from, for example, altered solar input or concentration of greenhouse gases, would lead to a change in DMS production in the sea, the sign of which acts to back-off the initial atmospheric perturbation. Thus, an increase in air temperature would give rise to the plankton making more

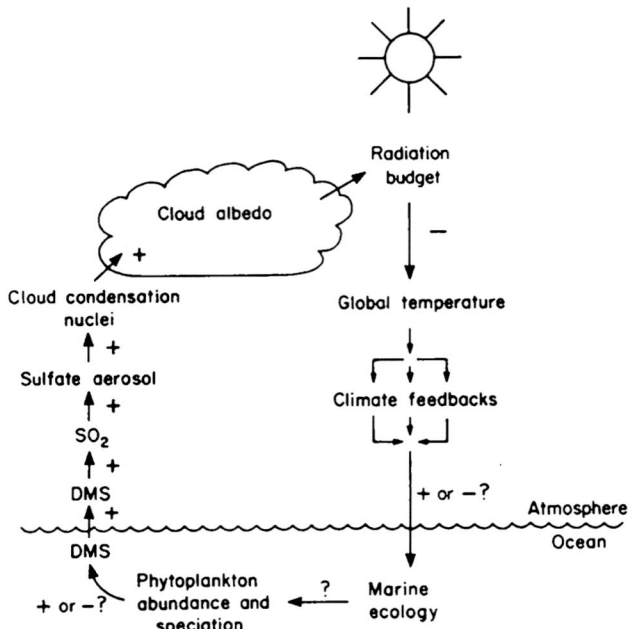

FIGURE 7 Proposed feedback cycle between climate and marine DMS production. Signs indicate whether an increase in the value of the preceding parameter in the cycle is expected to lead to an increase (+) or a decrease (−) in the value of the subsequent parameter (from Andreae, 1990).

DMS, which would lead to more CCN and hence enhanced cloudiness, which in turn would have a negative feedback on the initial temperature change. When proposed this was a very original and provocative concept, which has led to a large volume of research aimed at testing it. Although the many parts of the complex chain of processes from DMS to climate have been studied individually, it is still not clear whether the overall CLAW hypothesis is an important one, or even whether the climate–DMS feedback has the necessary negative sign. Some evidence against comes from the past record of atmospheric sulphur retained in an ice core from Antarctica and shown in Fig. 8. What this seems to indicate is that during the last ice age, atmospheric levels of both non-sea-salt sulphate and MSA were higher than now, presumably due to greater oceanic biological activity then, which is contrary to what would be predicted by the CLAW hypothesis. However, other ice core results seem to tell a somewhat different story, for example Hansson and Saltzman (1993) present data from a Greenland core which show an apparent opposite trend for MSA but a similar one for non-sea-salt sulphate. Although the ice core records potentially allow a more global view of the validity of the CLAW to be taken, at present there is insufficient data to draw definite conclusions.

2. Iron

Although the results discussed earlier clearly indicate that iron can play an important role in controlling biological activity in the oceans, extrapolation of the findings to larger areas let alone globally (typical *in situ* iron-enriched patch sizes were approximately 100 km^2) and to longer timescales (enrichment experiments lasted for less than 20 days) is clearly a major problem. However, as with the CLAW hypothesis discussed above, data from ice cores and other proxy records of past environmental conditions can prove useful in this context. Figure 8 shows a compilation of ice core and deep-sea sediment records spanning from the last ice age to the present. Figure 8a shows $\delta^{18}O$ as a marker of temperature as well as iron and MSA (a proxy for DMS, since we know of no other way in which it can be formed in the environment); both show higher levels in the last ice age. Figure 8b shows atmospheric CO_2 as recorded in air bubbles trapped in the ice, as well as several markers of ocean productivity found in deep-sea sediments (alkenones, dinosterol, TOC) all of which indicate greater production than at present. The data are concordant with an hypothesis that says that iron deposition from the atmosphere was greater during the last glaciation and that this led to higher productivity in the oceans, which resulted in lower atmospheric CO_2 and higher DMS (due to greater drawdown of CO_2 and enhanced production of DMS). Although such correlations are fascinating and certainly necessary in order to establish the role of iron in global biogeochemistry, the case is far from proved since we lack confirmatory data from a range of locations and, as usual, correlation cannot of itself prove causation.

References

Amouroux, D., P. S. Liss, E. Tessier, M. Hamren-Larsson and O. F. X. Donard, 2001: Role of the oceans in the global selenium cycle. *Earth Planet. Sci. Lett.*, in press.

Andreae, M. O., 1990: Ocean-atmosphere interactions in the global biogeochemical sulfur cycle. *Marine Chemistry*, **30**, 1–29.

Boers, R., G. P. Ayers and J. L. Gras, 1994: Coherence between seasonal variation in satellite-derived cloud optical depth and boundary-layer CCN concentrations at a midlatitude Southern Hemisphere station. *Tellus B*, **46**, 123–131.

Boyd, P. W., A. J. Watson, C. S. Law and 32 coauthors, 2000: A mesoscale phytoplankton bloom in the polar Southern Ocean stimulated by iron fertilization. *Nature*, **407**, 695–702.

Brimblecombe, P., C. Hammer, H. Rodhe, A. Ryaboshapko and C. F. Boutron, 1989: Human influence on the sulphur cycle. In *Evolution of the Global Biogeochemical Sulphur Cycle* (P. Brimblecombe and A. Yu Lein, Eds), John Wiley, pp. 77–121.

Charlson, R. J. and H. Rodhe, 1982: Factors controlling the acidity of natural rainwater. *Nature*, **295**, 683–685.

Charlson, R. J., J. E. Lovelock, M. O. Andreae and S. G. Warren, 1987: Oceanic phytoplankton, atmospheric sulphur, cloud albedo and climate. *Nature*, **326**, 655–661.

Chin, M. and D. J. Jacob, 1996: Anthropogenic and natural contributions to tropospheric sulfate: a global model analysis. *J. Geophys. Res.*, **101**, 18691–18699.

Coale, K. H. *et al.* 1996: A massive phytoplankton bloom induced by an ecosystem-scale iron fertilization experiment in the equatorial Pacific Ocean. *Nature*, **383**, 495–501.

Cooper, D. J., A. J. Watson and P. D. Nightingale, 1996: Large decrease in ocean-surface CO_2 fugacity in response to *in situ* iron fertilization. *Nature*, **383**, 511–513.

Donaghay, P. L., P. S. Liss, R. A. Duce, D. R. Kester, A. K. Hanson, T. Villareal, N. W. Tindale and D. J. Gifford, 1991: The role of episodic atmospheric nutrient inputs in the chemical and biological dynamics of oceanic ecosystems. *Oceanography*, **4**, 62–70.

Duce, R. A., P. S. Liss, J. T. Merrill and 19 coauthors, 1991: The atmospheric input of trace species to the world ocean. *Glob. Biogeochem. Cycles*, **5**, 193–259.

Fitzwater, S. E., K H. Coale, R. M. Gordon, K. S. Johnson and M. E. Ondrusek, 1996: Iron deficiency and phytoplankton growth in the equatorial Pacific. *Deep-Sea Res.* Pt II, **43**, 995–1015.

Galloway, J. N., H. Levy II, and P. S. Kasibhatla, 1994: Consequences of population growth and development on deposition of oxidized nitrogen. *Ambio*, **23**, 120–123.

Ganzeveld, L. and J. Lelieveld, 1995: Dry deposition parameterization in a chemistry general circulation model and its influence on the distribution of reactive trace gases. *J. Geophys. Res.*, **100**, 20999–21012.

Garland, J. A., A. W. Elzerman and S. A. Penkett, 1980: The mech-

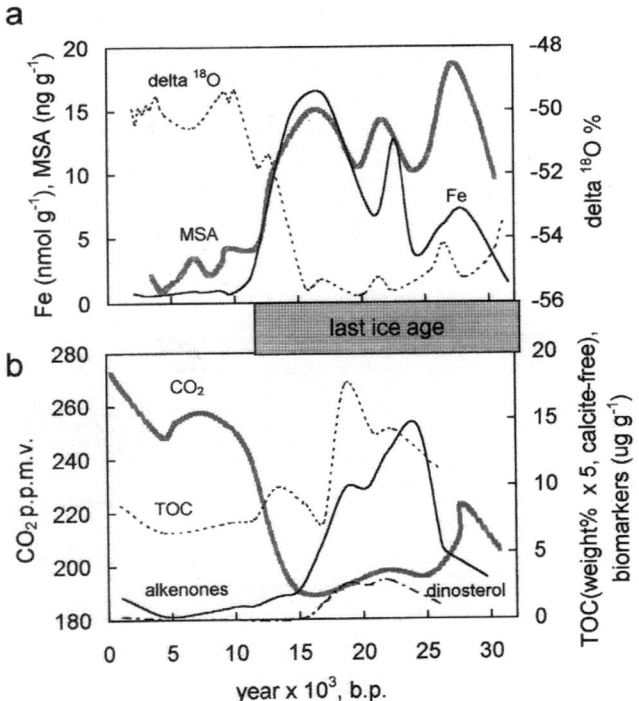

FIGURE 8 Ice core and sediment data for the Holocene and end of the last ice age. (a) Methane sulphonic acid (MSA) and $\delta^{18}O$ in an ice core from Dome C, east Antarctica; estimated Fe concentration in an ice core from Vostok, Antarctica. (b) CO_2 concentration in an ice core from Vostok, Antarctica; total organic carbon (TOC), alkenones and dinosterol in a sediment core from the eastern tropical Pacific (from Turner *et al.*, 1996).

anism for dry deposition of ozone to seawater surfaces. *J. Geophys. Res.*, **85**, 7488–7492.

Hansson, M. E. and E. S. Saltzman, 1993: The first Greenland ice core record of methanesulfonate and sulfate over a full glacial cycle. *Geophys. Res. Lett.*, **20**, 1163–1166.

IGBP, 1992: *Global Change: Reducing Uncertainties*. International Geosphere-Biosphere Programme.

IPCC, 1996: *Climate Change 1995*. The Science of Climate Change, Contribution of Working Group I to the Second Assessment Report of the Intergovernmental Panel on Climate Change (J. T. Houghton, L. G. Meiro Filho, B. A. Callander, N. Harris, A. Kattenberg and K. Maskell Eds).

Liss, P. S. and J. N. Galloway, 1993: Air–sea exchange of sulphur and nitrogen and their interaction in the marine atmosphere. In *Interactions of C, N, P and S Biogeochemical Cycles and Global Change* (R. Wollast, F. T. Mackenzie and L. Chou, Eds), Springer, pp. 259–281.

Liss, P. S., A. D. Hatton, G. Malin, P. D. Nightingale and S. M. Turner, 1997: Marine sulphur emissions. *Phil. Trans. R. Soc. Lond. B*, **352**, 159–169.

Margulis, L. and J. E. Lovelock, 1974: Biological modulation of the Earth's atmosphere. *Icarus*, **21**, 471–489.

Martin, J. H. and S. E. Fitzwater, 1988: Iron deficiency limits phytoplankton growth in the north-east Pacific subarctic. *Nature*, **331**, 341–343.

Martin, J. H., K. H. Coale, K. S. Johnson and 41 coauthors, 1994: Testing the iron hypothesis in ecosystems of the equatorial Pacific Ocean. *Nature*, **371**, 123–129.

Mosher, B. W. and R. A. Duce, 1987: A global atmospheric selenium budget. *J. Geophys. Res.*, **92**, 13289–13298.

Quinn, P. K., T. S. Bates, J. E. Johnson, D. S. Covert and R. J. Charlson, 1990. Interactions between the sulfur and reduced nitrogen cycles over the Central Pacific Ocean. *J. Geophys. Res.*, **95**, 16405–16416.

Ravishankara, A. R., Y. Rudich, R. Talukdar and S. B. Barone, 1997: Oxidation of atmospheric reduced sulphur compounds: perspective from laboratory studies. *Phil. Trans. R. Soc. Lond. B*, **352**, 171–182.

Rayman, M. P. 2000: The importance of selenium to human health. *Lancet*, **356**, 233–241.

Sarmiento, J. L. and J. R. Toggweiler, 1984: A new model for the role of the oceans in determining atmospheric CO_2. *Nature*, **308**, 621–624.

Savoie, D. L., J. M. Prospero, R. J. Larsen, F. Huang, M. A. Izaguirre, T. Huang, T. H. Snowdon, L. Custals and C. G. Sanderson, 1993: Nitrogen and sulfur species in Antarctic aerosols at Mawson, Palmer Station and Marsh (King George Island). *J. Atmos. Chem.*, **17**, 95–122.

Shaw, G. E., 1987. Aerosols as climate regulators—a climate biosphere linkage. *Atmos. Environ.*, **21**, 985–986.

Sievering, H., J. Boatman, E. Gorman, Y. Kim, L. Anderson, G. Ennis, M. Luria and S. Pandis, 1992: Removal of sulphur from the marine boundary layer by ozone oxidation in sea-salt aerosols. *Nature*, **360**, 571–573.

Taylor, S. R. and S. M. McLennon, 1985: *The Continental Crust: Its Composition and Evolution*. Blackwell Scientific, Oxford.

Turner, S. M., P. D. Nightingale, L. J. Spokes, M. I. Liddicoat and P. S. Liss, 1996. Increased dimethyl sulphide concentrations in sea water from *in situ* iron enrichment. *Nature*, **383**, 513–516.

Twomey, S., 1991. Aerosols, clouds and radiation. *Atmospheric Environment*, **25A**, 2435–2442.

Watson, A. J. and P. S Liss, 1998: Marine biological controls on climate via the carbon and sulphur geochemical cycles. *Phil. Trans. R. Soc. Lond. B.*, **353**, 41–51.

Watson, A. J., P. S. Liss and R. A Duce, 1991: Design of a small scale iron enrichment experiment. *Limnol. Oceanogr.*, **36**, 1960–1965.

Watson, A. J., D. C. E. Bakker, A. J. Ridgwell, P. W. Boyd and C. S. Law, 2000: Effect of iron supply on Southern Ocean CO_2 uptake and implications for glacial atmospheric CO_2. *Nature*, **407**, 730–733.

Modelling Vegetation and the Carbon Cycle as Interactive Elements of the Climate System

Peter M. Cox,[1] Richard A. Betts,[1] Chris D. Jones,[1] Steven A. Spall[1] and Ian J. Totterdell[2]

[1]Met Office, Hadley Centre for Climate Predictions and Research, Bracknell, UK.
[2]Southampton Oceanography Centre, Southampton, UK

The climate system and the global carbon cycle are tightly coupled. Atmospheric carbon in the form of the radiatively active gases, carbon dioxide and methane, plays a significant role in the natural greenhouse effect. The continued increase in the atmospheric concentrations of these gases, due to human emissions, is predicted to lead to significant climatic change over the next 100 years. The best estimates suggest that more than half of the current anthropogenic emissions of carbon dioxide are being absorbed by the ocean and by land ecosystems (Schimel et al., 1995). In both cases the processes involved are sensitive to the climatic conditions. Temperature affects the solubility of carbon dioxide in sea water and the rate of terrestrial and oceanic biological processes. In addition, vegetation is known to respond directly to increased atmospheric CO_2 through increased photosynthesis and reduced transpiration (Sellers et al., 1996a; Field et al., 1995), and may also change its structure and distribution in response to any associated climate change (Betts et al., 1997). Thus there is great potential for the biosphere to produce a feedback on the climatic change due to given human emissions.

Despite this, simulations carried out with General Circulation Models (GCMs) have generally neglected the coupling between the climate and the biosphere. Indeed, vegetation distributions have been static and atmospheric concentrations of CO_2 have been prescribed based on results from simple carbon cycle models, which neglect the effects of climate change (Enting et al., 1994). This chapter describes the inclusion of vegetation and the carbon cycle as interactive elements in a GCM. The coupled climate–carbon cycle model is able to reproduce key aspects of the observations, including the global distribution of vegetation types, seasonal and zonal variations in ocean primary production, and the interannual variability in atmospheric CO_2. A transient simulation carried out with this model suggests that previously-neglected climate–carbon cycle feedbacks could significantly accelerate atmospheric CO_2 rise and climate change over the twenty-first century.

A. INTRODUCTION

The biosphere influences weather and climate over a extraordinary range of timescales (see Table 1). Stomatal pores on plant leaves respond to the environment with a characteristic timescale of minutes, modifying evapotranspiration and producing detectable impacts on the diurnal evolution of the boundary layer and near-surface variables. At the other extreme, deep-rooted plants and soil microbes together amplify the concentration of carbon dioxide in the soil, possibly accelerating the weathering of silicate rocks which is believed to control the concentration of atmospheric CO_2 on geological timescales (Lovelock and Kump, 1994; Berner, 1997). In between are processes operating over decades to centuries which have great relevance to the issue of human-induced climate change.

Key amongst these is the uptake of CO_2 by the oceans and by land ecosystems, which are together responsible for absorbing more than half of the human CO_2 emissions (Schimel et al., 1996). Marine biota modify oceanic uptake by producing carbonate shells which sink to depth, producing a biological carbon pump. The photosynthetic rates of land plants are enhanced under high CO_2, which acts to increase carbon storage in vegetation and soils.

However, the atmosphere–ocean and atmosphere–land exchanges of carbon are also known to be sensitive

TABLE 1 Biospheric Feedbacks on Weather and Climate

Process	Feedback mechanism	Timescale
Stomatal response	Surface energy partitioning	Minutes
Leaf seasonality	Surface characteristics	Months–seasons
DMS from plankton	Cloud albedo	Months–years
Vegetation distribution	Surface characteristics	1–10^3 years
Land carbon storage	Atmospheric CO_2	1–10^3 years
Ocean carbon storage	Atmospheric CO_2	10–10^4 years
Enhanced rock weathering	Atmospheric CO_2	$>10^5$ years

to climate. Warming reduces the solubility of carbon dioxide in sea water, and increases soil and plant respiration. As a result, climate warming alone may result in a reduction of carbon storage. The competition between the direct effects of increasing CO_2 (which tends to increase carbon storage in the ocean and on land), and the effects of any associated climate warming (which may reduce carbon storage, especially on land), currently seems to favour CO_2 uptake, but how might this change in the future?

To address this question we have included dynamic vegetation and an interactive carbon cycle in the Hadley Centre GCM. The structure of the coupled climate–carbon cycle model is described in Section B, and its pre-industrial simulation is analysed in Section C. Results from the first transient climate change simulation are presented in Section D. Finally, the associated uncertainties and implications are discussed in Section E.

B. MODEL DESCRIPTION

The model used here (HadCM3LC) is a coupled ocean–atmosphere GCM with an interactive carbon cycle and dynamic vegetation. Figure 1 is a schematic showing the coupling of these components, and the new feedbacks which have been introduced.

1. Ocean–Atmosphere GCM (HadCM3L)

The ocean–atmosphere GCM is based on the third generation Hadley Centre coupled model, HadCM3 (Gordon et al., 2000), which is one of the first 3-D OAGCMs to be used predictively without flux adjustments. This advance was achieved partly through improved treatments of atmospheric processes and sub-grid ocean transports (Gent and McWilliams, 1990), but was primarily a result of the use of an enhanced ocean horizontal resolution of 1.25° latitude by 1.25° longitude. The additional computational expense of including an interactive carbon cycle made it necessary to reduce the ocean resolution to that used in the second generation Hadley Centre model (Johns et al., 1997). Thus, the model used here has a horizontal resolution for both atmosphere and ocean of 2.5° latitude by 3.75° longitude, with 19 vertical levels in the atmosphere and 20 in the ocean. The atmospheric physics and dynamics packages are identical to those used in HadCM3 (Pope et al., 2001), including a new radiation scheme (Edwards and Slingo, 1996), a parametrization of momentum by convective processes (Kershaw and Gregory, 1997), and an improved land surface scheme that simulates the effects of soil water phase change and CO_2-induced stomatal closure (Cox et al., 1999). The reduced ocean resolution required some modification to the diffusion coefficients (Gent and McWilliams, 1990) and to the parametrization of Mediterranean outflow, but otherwise the ocean physics is as described by Gordon et al. (2000).

HadCM3L was tested initially without flux adjustments, but the resulting errors in the climate simulation were found to have unacceptably large impacts on offline simulations of the land and ocean carbon cycles. It was therefore decided to use flux adjustments for this first series of coupled climate–carbon cycle experiments.

2. The Hadley Centre Ocean Carbon Cycle Model (HadOCC)

The ocean is a significant sink for anthropogenic carbon dioxide, with the suggestion that a third of anthropogenic emissions of CO_2 are taken up by the ocean (Siegenthaler and Sarmiento, 1993). Surface waters exchange CO_2 with the atmosphere, but it is the transfer of CO_2 to depth that determines the capacity of the ocean for long-term uptake. There are several mechanisms for this transfer of CO_2 to deeper waters.

Phytoplankton, the plants of the ocean ecosystem, take up CO_2 during their growth, converting the carbon to organic forms. The sinking of a fraction of this organic carbon is a mechanism for the export of carbon to depth, where it is remineralized back to CO_2. This

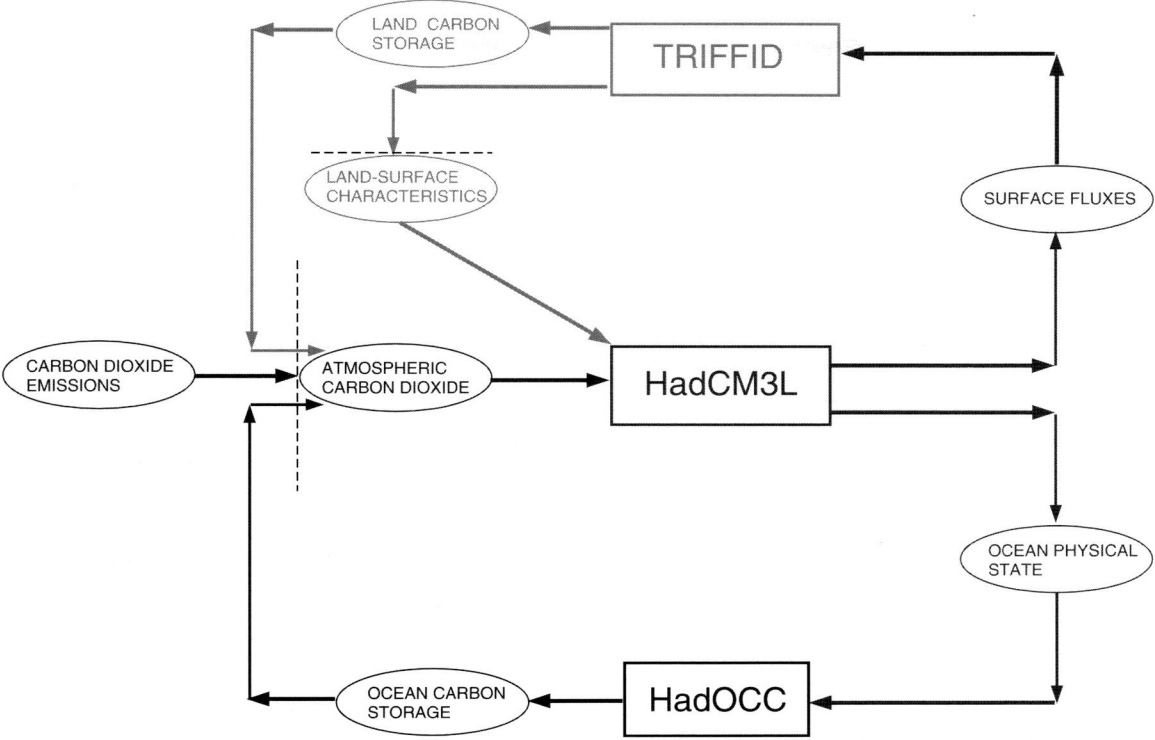

FIGURE 1 Schematic showing the structure of the coupled climate–carbon cycle model, which consists of the HadCM3L ocean–atmosphere GCM, the Hadley Centre Ocean Carbon Cycle Model (HadOCC) and the TRIFFID dynamic global vegetation model. The coupled model enables two feedback loops to be closed; one associated with climate-driven changes in atmospheric CO_2, and the other with biophysical feedbacks resulting from changes in the structure and distribution of vegetation. The dotted black lines denote where the loops are typically broken in standard GCM experiments.

process is known as the biological pump. Reproducing primary production (growth of phytoplankton) and the export of organic carbon is important if a model is to represent the fate of CO_2 in the ocean.

The Met Office ocean model has already been used in the estimate of oceanic uptake of anthropogenic CO_2 (Taylor, 1995). This model included only inorganic cycling of carbon, in which a solubility pump acts to increase the deep ocean carbon concentration. Carbon dioxide is absorbed in cold, dense polar waters which sink to the deep ocean. Equatorial upwelling allows deep waters to warm and release carbon dioxide back to the atmosphere. The simulations were found to be useful in studying carbon uptake, but it was noted that the neglect of ocean biology led to some unrealistic results such as the incorrect phase of seasonal variations in partial pressure at high latitudes. Ocean biology was added to the carbon model in order to improve the simulation of carbon uptake and to increase the range of possible simulated responses to scenarios of climate change (Palmer and Totterdell, 2001).

The resulting model, HadOCC (Hadley Centre Ocean Carbon Cycle model), simulates the movements of carbon within the ocean system, including exchanges of carbon dioxide gas with the atmosphere, the circulation of dissolved inorganic carbon (known as DIC or tCO_2) within the ocean, and the cycling of carbon by the marine biota. The principle components of the model are handled as tracers within the physical ocean model. They are: (nitrogenous) nutrient, phytoplankton, zooplankton, detritus, tCO_2 and alkalinity.

The air-to-sea flux of carbon dioxide is calculated using standard parametrizations:

$$F_{AS} = K\,(c_a - c_o) \qquad (1)$$

where c_a and c_o are respectively the partial pressures of CO_2 in the atmosphere and ocean at a given location. K parametrizes the effect of the wind speed on the gas transfer velocity, using the formulation of Wanninkhof (1992). Winds are obtained from the atmospheric model via the coupler. The partial pressure of CO_2 in the surface waters is determined by solving equations representing the sea water acid–base system. The expressions for the dissociation constants of carbonic acid, hydrogen carbonate, boric acid and water and for the solubility of CO_2 in sea water are taken from DOE (1994). Using the salinity-dependent boron concentration of Peng (1987), the acid–base system is solved using the method of Bacastow and Keeling (1981) to yield the concentration of carbonic acid and hence the partial pressure of CO_2. The temperature and salinity values used in these calculations are the local values from the ocean model.

The biological model is an explicit ecosystem model consisting of four components: nutrient (assumed to be nitrate), phytoplankton, zooplankton and (sinking) detritus. The complexity of the model was restricted to just four components in order for it to be economical enough for use in long integrations. This means that the behaviours of many different species and size-fractions are aggregated into a single component for each of phytoplankton and zooplankton. The model calculates the flow of nitrogen between the four components of the ecosystem at each grid box, and also computes the associated transfers of carbon and alkalinity. The carbon flows have no direct effect on the behaviour of the ecosystem as growth of phytoplankton is not limited by availability of carbon.

The phytoplankton population changes as a result of the balance between growth, which is controlled by light level and the local concentration of nutrient, and mortality, which is mostly as a result of grazing by zooplankton. Detritus, which is formed by zooplankton excretion and by phyto- and zooplankton mortality, sinks at a fixed rate and slowly remineralizes to reform nutrient and dissolved inorganic carbon. Thus both nutrient and carbon are absorbed by phytoplankton near the ocean surface, pass up the food chain to zooplankton, and are eventually remineralized from detritus in the deeper ocean.

3. The Dynamic Global Vegetation Model (TRIFFID)

Over the last decade a number of groups have developed equilibrium biogeography models which successfully predict the global distribution of vegetation based on climate (Prentice *et al.*, 1992; Woodward *et al.*, 1995). Such models have been coupled "asynchronously" to GCMs in order to quantify climate–vegetation feedbacks. This involves an iterative procedure in which the GCM calculates the climate implied by a given land-cover, and the vegetation model calculates the land-cover implied by a given climate. The process is repeated until a mutual climate–vegetation equilibrium is reached (Claussen, 1996; Betts *et al.*, 1997). Such techniques have yielded interesting results, but are not appropriate for simulating transient climate change, for which the terrestrial biosphere may be far from an equilibrium state.

In order to fully understand the role of climate–vegetation feedbacks on these timescales we need to treat the land-cover as an interactive element, by incorporating dynamic global vegetation models (DGVMs) within climate models. The earliest DGVMs were based on bottom-up "gap" forest models, which explicitly model the growth, death and competition of individual plants (Friend *et al.*, 1993; Post and Pastor, 1996). Such models can produce very detailed predictions of vegetation responses to climate, but they are computationally expensive for large-scale applications. Also, GCM climates are not likely to be sensitive to the details of the species or age composition of the land-cover. For this study it is more appropriate to adopt a "top-down" DGVM approach, in which the relevant land-surface characteristics, such as vegetated fraction and leaf area index, are modelled directly (Foley *et al.*, 1996). A model of this type, called TRIFFID (Top-down Representation of Interactive Foliage and Flora Including Dynamics), has been developed at the Hadley Centre for use in these coupled climate–carbon cycle simulations.

TRIFFID defines the state of the terrestrial biosphere in terms of the soil carbon, and the structure and coverage of five plant functional types (Broadleaf tree, Needleleaf tree, C_3 grass, C_4 grass and shrub) within each model gridbox. The areal coverage, leaf area index and canopy height of each type are updated based on a carbon balance approach, in which vegetation change is driven by net carbon fluxes calculated within the MOSES 2 land surface scheme. MOSES 2 is a tiled version of the land surface scheme described by Cox *et al.* (1999), in which a separate surface flux and temperature is calculated for each of the land-cover types present in a GCM gridbox. In its standard configuration, MOSES 2 recognizes the five TRIFFID vegetation types plus four nonvegetation land-cover types (bare soil, inland water, urban areas and land ice). Carbon fluxes for each of the vegetation types are derived using the coupled photosynthesis–stomatal conductance model developed by Cox *et al.* (1998), which utilizes existing models of leaf-level photosynthesis in C_3 and C_4 plants (Collatz *et al.*, 1991, 1992). Plant respiration is broken down into a growth component, which is proportional to the photosynthetic rate, and a maintenance component which is assumed to increase exponentially with temperature ($q_{10} = 2$). The resulting rates of photosynthesis and plant respiration are dependent on both climate and atmospheric CO_2 concentration. Therefore, with this carbon-balance approach, the response of vegetation to climate occurs via climate-induced changes in the vegetation to atmosphere fluxes of carbon.

Figure 2 is a schematic showing how the MOSES 2 land-surface scheme is coupled to TRIFFID for each vegetation type. The land–atmosphere fluxes (above the dotted line) are calculated within MOSES 2 on every 30 min GCM timestep and time-averaged before being passed to TRIFFID (usually every 10 days). TRIFFID (below the dotted line of Fig. 2) allocates the average net primary productivity over this coupling period into the growth of the existing vegetation (leaf, root and wood biomass), and to the expansion of the vegetated area in each gridbox. Leaf phenology (bud burst and leaf drop) is updated on an intermediate timescale of 1 day, using accumulated temperature-dependent leaf turnover rates. After each call to TRIFFID the land surface parameters required by MOSES 2 (e.g. albedo, roughness length) are updated based on the new vegetation state, so that changes in the biophysical properties of the land surface, as well as changes in terrestrial carbon, feedback onto the atmosphere (see Fig. 1). The land surface parameters are calculated as a function of the type, height and leaf area index of the vegetation.

The natural land-cover evolves dynamically based on competition between the types, which is modelled using a Lotka–Volterra approach and a (tree–shrub–grass) dominance hierarchy. We also prescribe some agricultural regions, in which grasslands are assumed to be dominant. Carbon lost from the vegetation as a result of local litterfall or large-scale disturbance is transferred into a soil carbon pool, where it is broken down by microorganisms which return CO_2 to the atmosphere. The soil respiration rate is assumed to double for every 10 K warming (Raich and Schlesinger, 1992), and is also dependent on the soil moisture content in the manner described by McGuire *et al.* (1992).

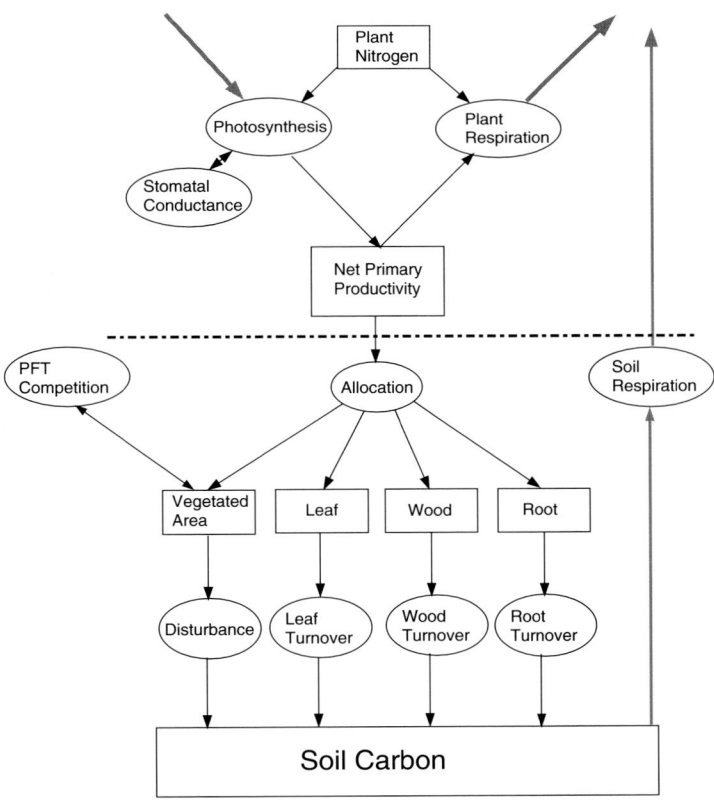

FIGURE 2 Schematic showing TRIFFID carbon flows for each vegetation type. Processes above the dotted line are fluxes calculated in the MOSES 2 land surface scheme every atmospheric model timestep ($\approx$ 30 min). In dynamic mode, TRIFFID updates the vegetation and soil carbon every 10 days using time averages of these fluxes.

C. PRE-INDUSTRIAL STATE

1. Spin-Up Methodology

The present-day natural carbon sink of about 4 GtC yr^{-1} is a small fraction of the gross carbon exchanges between the Earth's surface and the atmosphere. It is therefore vitally important to reach a good approximation to a pre-industrial equilibrium, since even a small model drift could easily swamp this signal in a climate change experiment. With this in mind, a spin-up methodology was designed to obtain a carbon cycle equilibrium to within ±10% of the current natural carbon sink (i.e. ±0.4 GtC yr^{-1}). The methodology was required to produce a consistent climate–carbon cycle equilibrium state, without the computational expense of running the entire coupled model for thousands of simulated years.

Table 2 shows the multistage process used. The first two stages are primarily concerned with producing an equilibrium distribution of dissolved carbon in the ocean model (Jones and Palmer, 1998). In the first stage the ocean–atmosphere GCM was integrated forward for approximately 60 years, using a prescribed "pre-industrial" CO_2 concentration of 290 ppmv. The initial ocean physical state was based on temperature and salinity data from the Levitus climatology (Levitus and Boyer, 1994; Levitus et al., 1994). During this stage the modelled sea surface temperature and salinity were relaxed to the climatology (i.e. "Haney forced"), with a relaxation timescale of about two weeks. This stage also included the TRIFFID dynamic global vegetation model used in "equilibrium" mode. Vegetation and soil variables were updated iteratively every five model years, using an implicit scheme with a long internal timestep ($\approx$ 1000 years). This approach is equivalent to a Newton–Raphson algorithm for approaching equilibrium, and offline tests have shown that it is very effective in producing equilibrium states for the slowest variables (e.g. soil carbon and forest cover).

The second stage was a long run of approximately 2000 years in ocean-only mode using the distorted physics technique (Bryan, 1984; Wood, 1998), which allows a lengthened timestep of 24 h to be used. This enables a long period to be simulated quickly and cheaply, allowing the deep ocean structure to come close to equilibrium. Climatological temperature and salinity data were again used to Haney force the ocean surface with the same coefficients as before. Other fluxes were derived from monthly mean values calculated from the last 10 years of the coupled phase, with daily variability imposed from the last year of the run. The daily variability was found to be crucial in allowing realistic mixed layer depths to be maintained in the ocean only phase (Jones and Palmer, 1998). The forcing fluxes used were; penetrative solar heat flux, nonpenetrative heat flux, wind mixing energy, wind stress (N-S and E-W components) and net freshwater flux (precipitation minus evaporation). The driving fluxes were modified to include the effects of sea-ice in the coupled phase, allowing the ocean-only phase to be carried out without an interactive sea-ice model.

In the third stage the ocean was recoupled to the atmosphere model, using the ocean state from the end of the ocean-only phase, but still Haney-forcing the ocean surface state back to the climatology. TRIFFID was switched to its default dynamic mode, in which the vegetation cover is updated every 10 days. A 90-year run was carried out to allow the coupled model to adjust to the introduction of short timescale vegetation variability, and to define heat and freshwater "flux adjustments" for the subsequent stages.

Stage 4 saw the introduction of internally generated SST variability for the first time, as the Haney forcing terms were switched off and replaced by these fixed adjustments. This was the point when the impact of El Niño events on the carbon balance became obvious (Jones et al., 2001). A 60-year simulation was completed to give the system ample time to adjust.

Finally, the atmospheric CO_2 was treated as an interactive variable in stage 5. All previous stages had used a prescribed CO_2 concentration. A short fully interac-

TABLE 2 The Stages Used to Spin-Up the Coupled Model to Its Pre-industrial Equilibrium

Physical model		Biology		Atmos. CO_2	Model years
Mode	Ocean forcing	HadOCC	TRIFFID		
Coupled	Haney forced	No	Equil.	Fixed	~60
Ocean-only	Haney forced	Yes	No	Fixed	~2000
Coupled	Haney forced	Yes	Dynamic	Fixed	~90
Coupled	Flux-adjusted	Yes	Dynamic	Fixed	~60
Coupled	Flux-adjusted	Yes	Dynamic	Interactive	~20

tive climate–carbon cycle run (of about 20 years) was carried out to ensure that there were no significant drifts in the atmospheric CO_2 concentration. Results from the subsequent pre-industrial simulation of 100 years are discussed in the next section.

2. The Mean Pre-industrial State

a. The Ocean Carbon Cycle

HadOCC explicitly models the ocean ecosystem in order to represent the biological pump. The two main processes of this pump are primary production and the sinking flux of organic carbon. The HadOCC model has been shown to compare well to observations for both these processes (Palmer and Totterdell, 2001). The global annual total production in the model is about 53 GtC yr^{-1}, while recent satellite-derived estimates are 39.4 (Antoine et al., 1996), 48.5 and 49.4 (Longhurst et al., 1990) GtC yr^{-1}. Considering that these estimates refer to net production (including respiration losses) and that the model's figure is for gross production, the model is well within the range of observational estimates.

Figure 3 shows the zonal total of primary production for each month averaged over the 100 years of the control simulation. Also shown for comparison is the zonal total of primary production from the climatology of Antoine et al. (1996). The seasonal cycle in primary production is captured by the ocean ecosystem model. The magnitude in the model is higher than suggested by Antoine et al. (1996) but that study has the lowest total production of the three climatologies mentioned above. In the equatorial Pacific the model overestimates primary production. This is a result of excessive upwelling of nutrients at the equator, but also the model does not represent more complex limitations on production. These include, for example, iron limitation and ammonia inhibition, which have been suggested as important processes in high-nutrient low-chlorophyll (HNLC) regions such as the equatorial Pacific (Loukas et al., 1997).

b. The Land Carbon Cycle

The performance of the terrestrial component is best assessed by comparing its simulation of the global distribution of vegetation types to that observed. Figure 4 shows the fractional coverage of the five TRIFFID plant functional types and bare soil from the end of the pre-industrial control. The locations of the major forests and deserts are generally well captured. However, the boreal forests tend to be a little too sparse, especially in Siberia where the atmospheric model has a cold bias in winter. The absence of a cold-deciduous needleleaf plant functional type in TRIFFID may also be a contributory factor here. If anything, the simulated tropical forests are too extensive, perhaps because of the neglect of various disturbance factors (e.g. fire and anthropogenic deforestation). This TRIFFID simulation includes some prescribed agricultural regions, in which grasslands are assumed to dominate (elsewhere forests dominate where they can survive). This ensures that the heavily cultivated mid-latitudes are simulated as grasslands in agreement with the observations. The competition between C_3 and C_4 grasses is simulated in the model, producing a dominance of C_4 in the semi-arid tropics and a dominance of C_3 elsewhere. However, this balance may be expected to shift as atmospheric CO_2 increases.

Figure 4g compares this simulation to the IGBP land-cover dataset derived from remote sensing. The fractional agreement in each gridbox, f_{agree}, is calculated as the maximum overlap between the vegetation fractions from the model (v^{mod}) and the observations (v^{obs}):

$$f_{agree} = \sum_{i=1}^{5} \min \{v_i^{mod}, v_i^{obs}\} \qquad (2)$$

where i labels the vegetation type. By this measure, TRIFFID reproduces 63% of the IGBP land-cover in this coupled simulation. The level of agreement is lower in the regions of north-east Eurasia which are dominated by cold deciduous spruce forests and in the savanna regions of Africa, but higher in the midlatitudes and the tropical forests.

The simulated global totals for vegetation carbon (493 GtC) and soil carbon (1180 GtC) are consistent with other studies (Schimel et al., 1995; Zinke et al., 1986; Cramer et al., 2001). The integrated land net primary productivity (NPP) associated with this vegetation distribution is about 60 Gt C yr^{-1}, which is also within the range of other estimates (Warnant et al., 1994; Cramer et al., 2001) and almost identical to the value suggested by Melilto et al., 1996 for the present-day carbon cycle. Since the value simulated here corresponds to a "pre-industrial" CO_2 concentration of 290 ppmv, we expect slightly higher NPP to be simulated for the present-day.

c. Interannual Variability

Figure 5 shows the annual mean CO_2 flux to the atmosphere over 100 years of the pre-industrial control simulation (bottom panel), along with the contributions to this from the ocean and the terrestrial biosphere. The 20-year running mean of the total flux lies within the quality bounds of ± 0.4 GtC yr^{-1} which we required for the equilibrium. It is also very clear that the modelled carbon cycle displays significant interannual variability about this equilibrium state. Variability in the total

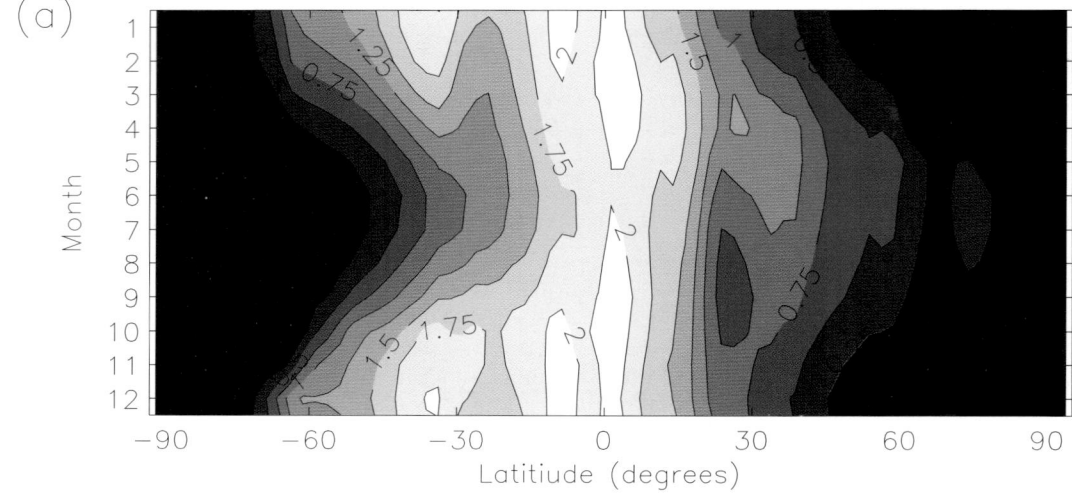

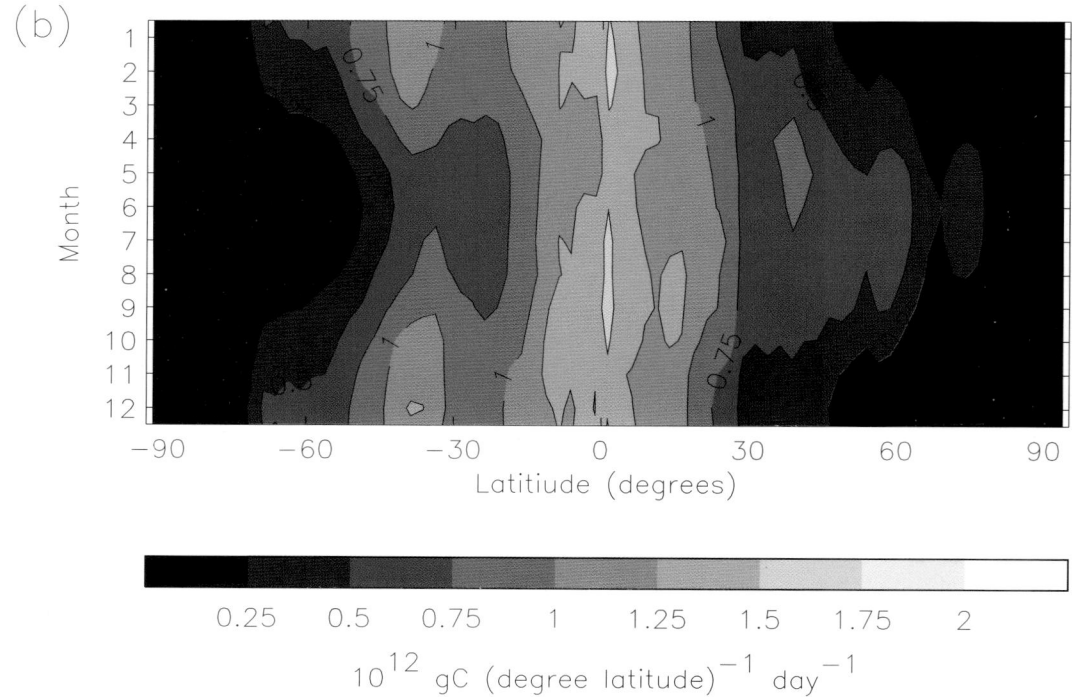

FIGURE 3 Ocean primary production as a function of latitude and month of the year. (a) Mean over 100 years of the control period. (b) The Antoine et al. (1996) satellite-derived climatology.

flux, and thus the atmospheric CO_2 concentration, is dominated by the contribution from the terrestrial biosphere. The net fluxes from the ocean and the land tend to have opposite signs, but the magnitude of the variability in the land fluxes is far larger.

This interannual variability in the modelled carbon cycle is found to be correlated with the El Niño Southern Oscillation (ENSO). This is illustrated by considering the Niño3 index. The Niño3 index is a measure of the ENSO cycle and is the sea surface temperature (SST) anomaly over the region 150°W-90°W, 5°S-5°N in the equatorial Pacific. Under El Niño con-

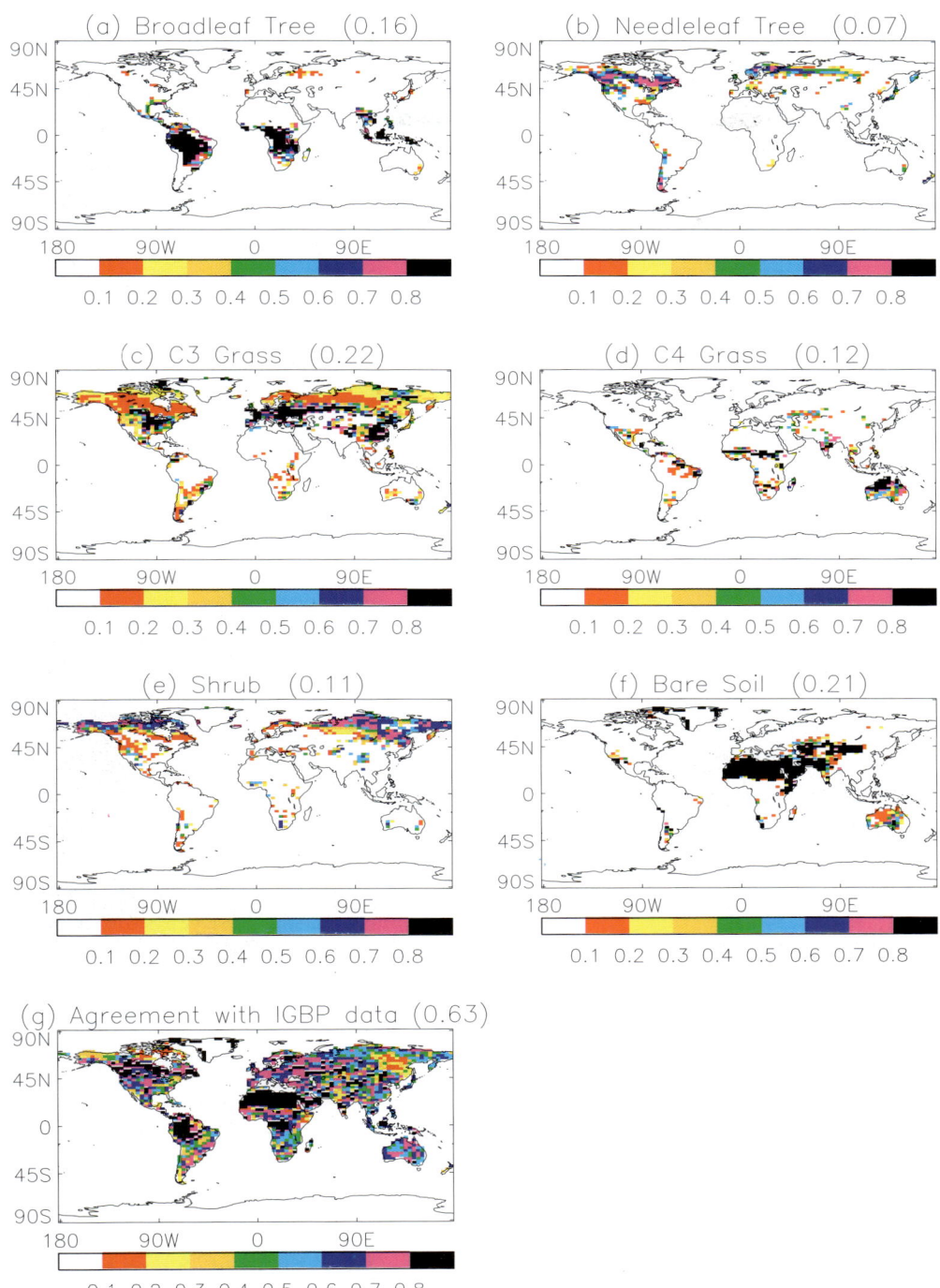

FIGURE 4 Annual mean fractional coverage of vegetation types ((a)–(e)) and bare soil (f) from 100 years of the pre-industrial control simulation. Plot (g) compares this simulation with the fractions of broadleaf tree, needleleaf tree, grass ($C_3 + C_4$) and shrub from the IGBP land-cover dataset. The numbers in parentheses are means over the entire land area.

ditions, there is reduced ocean upwelling, less cooler water being brought to the surface, a positive SST anomaly and positive Niño3 index. Conversely, La Niña conditions are characterized by greater upwelling, a negative SST anomaly and negative Niño3 index. When the Niño3 index is positive the model simulates an increase in atmospheric CO_2 resulting from the combined effect of the terrestrial biosphere acting as a large

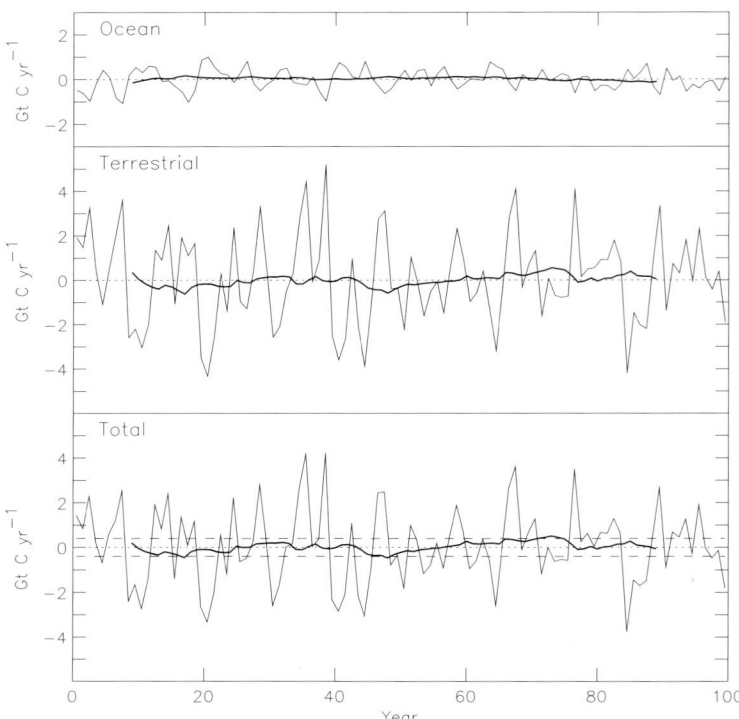

FIGURE 5 The flux of CO_2 to the atmosphere as a function of year during the control period, together with contributions to this from the ocean and the terrestrial biosphere. The thin lines show the annual mean data and the thick lines are the 20-year running means of the annual data. In the plot for total flux, the dashed lines indicate the quality criterion required for the pre-industrial equilibrium of ± 0.4 GtC yr^{-1}.

source (especially in Amazonia Tian et al., 1998), which is only partially offset by a reduced outgassing from the tropical Pacific. The opposite is true during the La Niña phase.

How realistic is this ENSO-driven variability in CO_2? In order to answer this question Jones et al. (2001) analysed the long-term records of atmospheric CO_2 concentration taken at Mauna Loa, Hawaii since 1958 (Keeling, 1989). In these records, superimposed on the seasonal cycle and the long-term upward trend is interannual variability which cannot be readily explained by changes in fossil fuel burning. The correlation between these interannual changes in atmospheric CO_2 and the ENSO cycle was first reported in the 1970s (Bacastow, 1976; Bacastow et al., 1980). Plate 31 plots the anomaly in the growth rate of atmospheric CO_2 against the Niño3 index, from the pre-industrial control simulation and the Mauna Loa observations. In both cases the atmospheric CO_2 anomaly is positive during El Niño (positive Niño3) and decreases during La Niña (negative Niño3). The model appears to slightly overestimate the interannual variability in CO_2 because the amplitude of its internally generated ENSO is too large.

The gradients of the blue dashed and red dashed lines represent the sensitivity of the carbon cycle to ENSO, as given by the model and the observations respectively. Here, we have excluded observations which immediately follow major volcanic events (black stars), since during these years the carbon cycle will also have been perturbed by the induced tropospheric cooling (Jones and Cox, 2001). These lines are almost parallel, suggesting that the modelled carbon cycle has a realistic sensitivity to ENSO. This is most important, since a realistic response to the internally generated climate anomalies is required before we can have any confidence in using the model to predict the response of the carbon cycle to anthropogenic forcing.

D. A FIRST TRANSIENT CLIMATE–CARBON CYCLE SIMULATION

A transient coupled climate–carbon cycle simulation was carried out for 1860–2100, starting from the pre-industrial control state, and using CO_2 emissions as given by the IS92a scenario (Houghton et al., 1992). These emissions included a contribution from net land-use change (e.g. deforestation, forest regrowth), since TRIFFID does not yet model the effect of these interactively. The concentrations of the other major green-

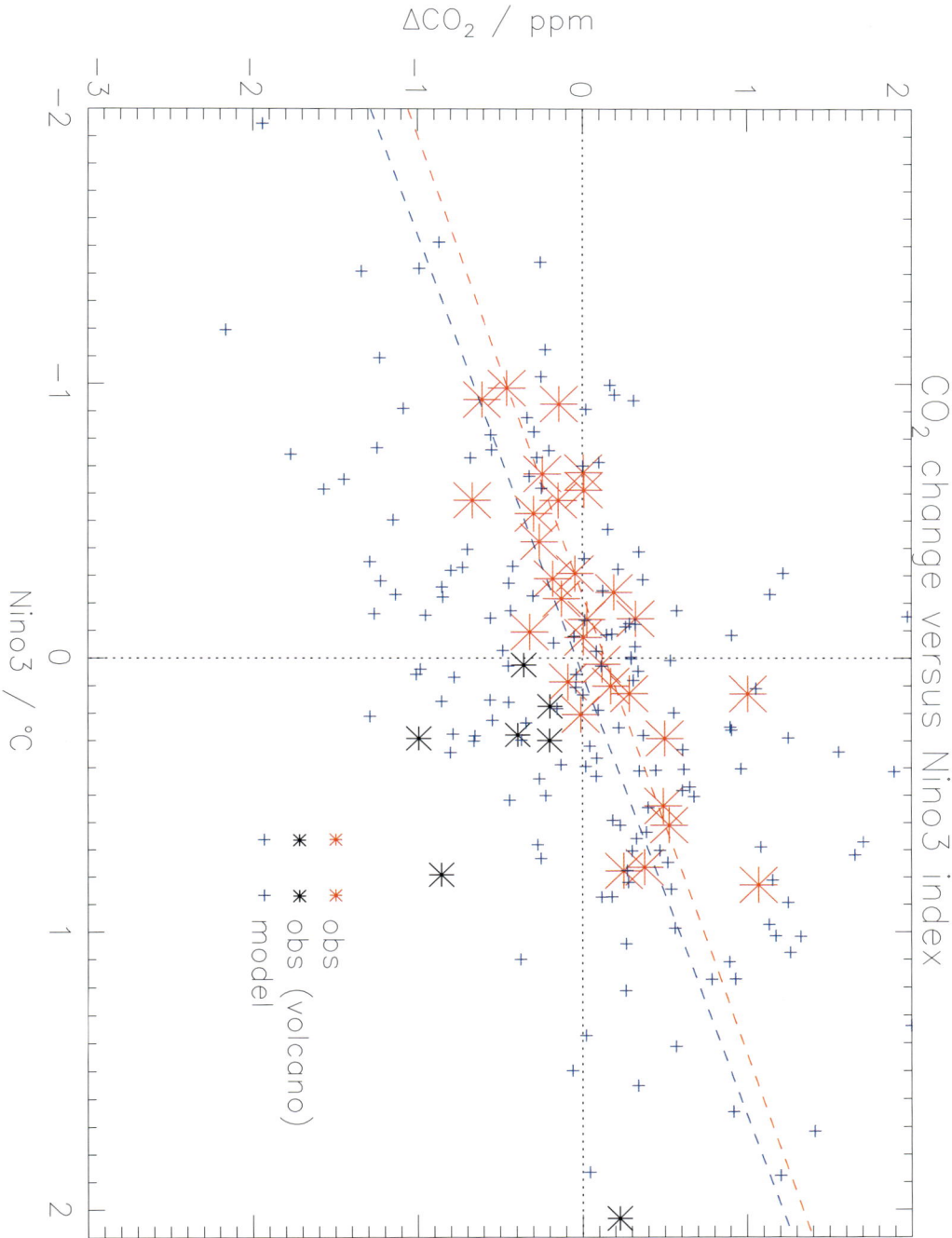

FIGURE 6 Anomaly in the growth rate of atmospheric CO_2 versus Niño3 index (i.e. the annual mean sea-surface temperature anomaly in the tropical Pacific, 150°W–90°W, 5°S–5°N), from the pre-industrial control simulation (blue crosses) and the Mauna Loa observations (red stars). The gradients of the red dashed and blue dashed lines represent the sensitivity of the carbon cycle to ENSO, as given by the observations and the model respectively. We have excluded observations which immediately follow major volcanic events (black stars), since during these years the carbon cycle may have been significantly perturbed by the induced tropospheric cooling.

house gases were also prescribed from IS92a, but the radiative effects of sulphate aerosols were omitted in this first simulation.

Figure 7 shows the changing carbon budget in the fully coupled run (Cox et al., 2000). The continuous black line represents the simulated change in CO_2 since 1860 (in GtC), and the coloured lines show the contribution to this change arising from emissions (red), and

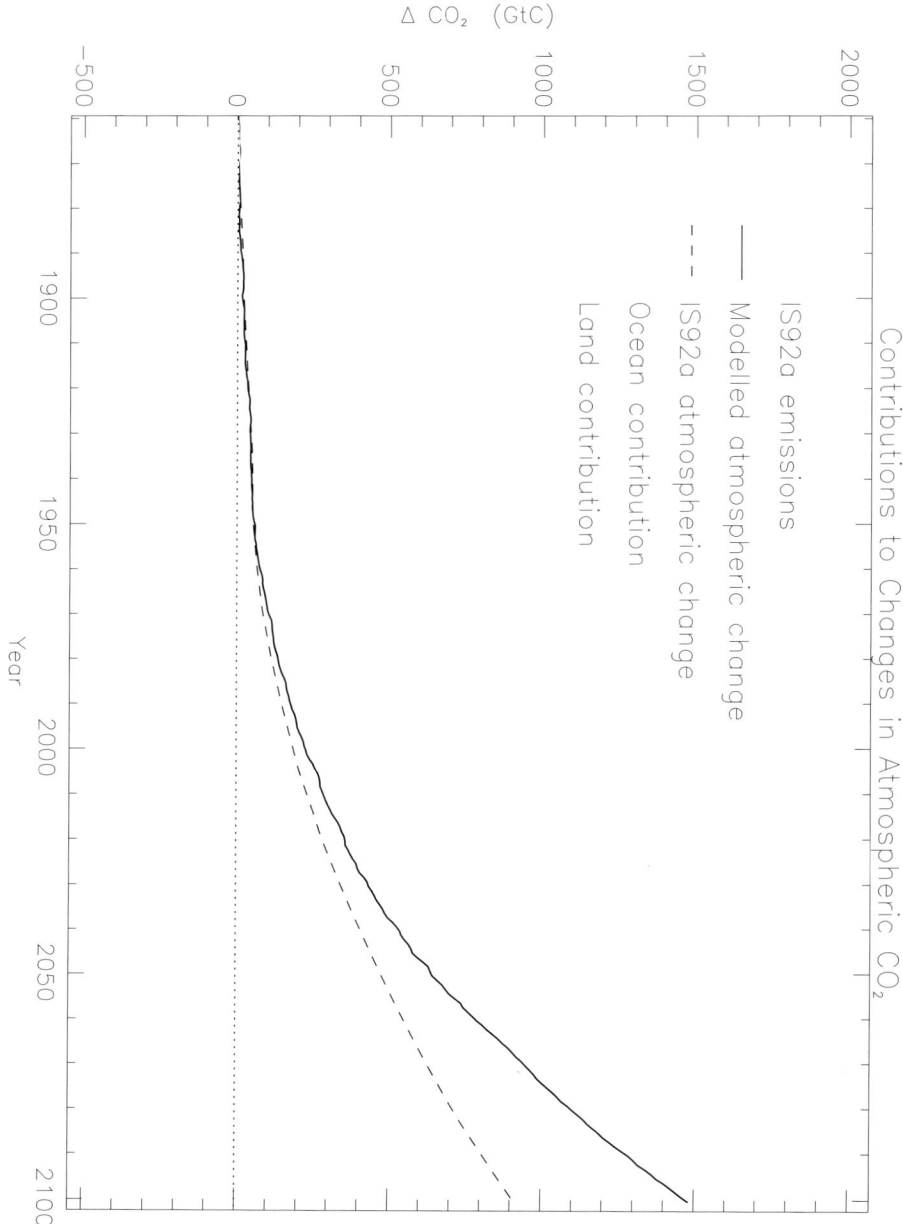

FIGURE 7 Budgets of carbon during the coupled climate–carbon cycle simulation. The continuous black line shows the simulated change in atmospheric CO_2 (in GtC). The red, green and blue lines show the integrated impact of the emissions, and of land and ocean fluxes respectively, with negative values implying net uptake of CO_2. For comparison the standard IS92a atmospheric CO_2 scenario is shown by the black dashed line. Note that the terrestrial biosphere takes up CO_2 at a reducing rate from about 2010 onwards, becoming a net source at around 2050.

changes in the carbon stored on the land (green) and in the oceans (blue). Net uptake by the land or the ocean since 1860 is represented by negative values, since this acts to reduce the change in atmospheric carbon. The gradients of these lines are equivalent to the annual mean fluxes in GtC yr^{-1}.

The dotted black line shows the change in atmospheric carbon given by the standard IS92a CO_2 *concentration* scenario, which has been often used to drive GCM climate predictions. This scenario was calculated with offline carbon cycle models which neglected the effects of climate change (Enting *et al.*, 1994). A simulation with our coupled climate–carbon cycle model produced similar increases in CO_2, when the carbon cycle components were artificially isolated from the climate change. The difference between the continuous black line and the dashed black line therefore represents an estimate of the effect of climate–carbon cycle feedbacks on atmospheric CO_2.

1. 1860–2000

From 1860 to 2000, the simulated stores of carbon in the ocean and on land increase by about 100 GtC and 75 GtC respectively (Figure 7). The land uptake is largely due to CO_2-fertilization of photosynthesis, which is partially counteracted by increasing respiration rates in the warming climate (Cramer et al., 2001). The ocean uptake is also driven by the increasing atmospheric CO_2, which results in a difference in the partial pressure of CO_2 across the ocean surface (equation 1). During this period, both components of the modelled carbon cycle are therefore producing a negative feedback on the CO_2 increase due to anthropogenic emissions.

However, the modelled atmospheric CO_2 is 15 to 20 ppmv too high by the present day, which corresponds to a timing error of about 10 years. Possible reasons for this include an overestimate of the prescribed net land-use emissions and the absence of other important climate forcing factors. The modelled global mean temperature increase from 1860 to 2000 is about 1.4 K (Fig. 10b), which is higher than observed (Nicholls et al., 1996). This may have led to an excessive enhancement of respiration fluxes over this period. The overestimate of the historical warming is believed to be due to the neglect of the cooling effects of anthropogenic aerosols (Mitchell et al., 1995).

We have recently completed a further coupled climate–carbon cycle experiment which includes the effects of aerosols from volcanoes, as well as those from human activity, along with the additional radiative forcing factor from solar variability. This experiment produces a much better fit to the observed increase in global mean temperature, halving the overestimate of current CO_2 to 5–10 ppmv. The remaining "error" is almost certainly within the bounds of uncertainties associated with the prescribed net land-use emissions.

Nevertheless, the coupled model does a good job of simulating the recent carbon balance. For the 20 years centered on 1985, the mean land and ocean uptake are 1.5 and 1.6 GtC yr^{-1} respectively (cf. best estimates for the 1980s of 1.8 ± 1.8 and 2.0 ± 0.8 GtC yr^{-1} (Schimel et al., 1995).

2. 2000–2100

The simulated atmospheric CO_2 diverges much more rapidly from the standard IS92a concentration scenario in the future. The ocean takes up about 400 GtC over this period, but at a rate which is assymptoting towards 5 GtC yr^{-1} by 2100. This reduced efficiency of oceanic uptake is partly a consequence of the nonlinear dependence of the partial pressure of dissolved CO_2 (pCO_2) on total ocean carbon concentration (TCO_2), but may also have contributions from climate change (Sarmiento et al., 1998). Although the thermohaline circulation weakens by about 25% from 2000 to 2100, this is much less of a reduction than seen in some previous simulations (Sarmiento and Le Quere, 1996), and the corresponding impact on ocean carbon uptake is less significant. In this experiment, increased thermal stratification due to warming of the sea surface suppresses upwelling, which reduces nutrient availability and lowers primary production by about 5%. However, ocean-only tests suggest a small climate-change impact on oceanic carbon uptake, since this reduction in the biological pump is compensated by reduced upwelling of deepwaters which are rich in TCO_2.

The most marked impacts of climate–carbon cycle feedbacks are seen in the terrestrial biosphere, which switches from being a weak sink to become a strong source of carbon from about 2050. (Note how the gradient of the green line in Fig. 7 becomes positive from this time.) There are two main reasons for this. Firstly, the GCM predicts an appreciable drying and warming in the Amazon basin, which results in "dieback" of the tropical rainforest and a release of carbon to the atmosphere. Similar CO_2-induced climate changes have been predicated by other GCMs for this region (Gedney et al., 2000), but the magnitude of these seems largest in the latest Hadley Centre models. This may be partly because the Hadley Centre models include CO_2-induced stomatal closure, which seems particularly to reduce evaporation and rainfall in Amazonia (Cox et al., 1999). A similar terrestrial response was also produced by a very different ecosystem model when driven offline by anomalies from HadCM3 (White et al., 1999). We might expect an even more dramatic change in HadCM3LC, since for the first time the climate-induced dieback of the forest is able to produce further reductions in precipitation and increases in surface temperature, via biophysical feedbacks.

The second reason for the sink-to-source transition in the terrestrial biosphere is a widespread loss of soil carbon. Broadly speaking, a rise in CO_2 alone tends to increase the rate of photosynthesis and thus terrestrial carbon storage, providing other resources are not limiting (Melillo et al., 1995; Cao and Woodward, 1998). However, plant maintenance and soil respiration rates both increase with temperature. As a consequence, climate warming (the indirect effect of a CO_2 increase) tends to reduce terrestrial carbon storage (Cramer et al., 2001), especially in the warmer regions where an increase in temperature is not beneficial for photosynthesis. At low CO_2 concentrations the direct effect of CO_2 dominates, and both vegetation and soil carbon increase with atmospheric CO_2. However, as CO_2 rises further, terrestrial carbon begins to decrease, since the direct effect of CO_2 on photosynthesis saturates but the specific soil respiration rate continues to increase with temperature. The transition between these two regimes occurs abruptly at around 2050 in this experiment.

Figure 8 shows how zonal mean soil and vegetation carbon change through the simulation. The model suggests that carbon accumulation in natural ecosystems to date has occurred in the soils and vegetation of the northern midlatitudes, and in the vegetation of the tropics. In the simulation, carbon continues to be accumulated in the Northern Hemisphere vegetation, but at a reducing rate from about 2070 onwards. The loss of tropical biomass beginning in about 2030 is very clearly shown in Fig. 8a. The rapid reduction in soil carbon seems to start around 2050 in the tropics, and around 2070 in the midlatitudes. Note, however, that soil carbon continues to increase in the far north until the end of the run.

Figure 9 shows maps of the changes in vegetation and soil carbon from 1860 to 2100. The increase of biomass in the Northern Hemisphere can be seen to arise from a thickening of the boreal forests. The massive reduction in Amazonian vegetation is also obvious. Losses of soil carbon, relative to 1860, are more widespread with reductions in most areas south of 60°N. South-east Asia, and the tundra-covered areas of Siberia and Alaska, are amongst the few regions to have more soil carbon in 2100 than in 1860.

Overall, the land carbon storage decreases by about 170 GtC from 2000 to 2100, accelerating the rate of CO_2 increase in the next century. By 2100 the modelled CO_2 concentration is about 980 ppmv in the coupled experiment, which is more than 250 ppmv higher than the standard IS92a scenario (Fig. 10a). This difference is equivalent to about 500 GtC (rather than 170 GtC), because the standard scenario implicitly assumes a continuing terrestrial carbon sink, accumulating to about 300–400 GtC by about 2100.

As a result of these climate–carbon cycle feedbacks, global-mean and land-mean temperatures increase from 1860 to 2100 by about 5.5 K and 8 K respectively, rather than about 4 K and 5.5 K, under the standard concentration scenario (Fig. 10b).

E. DISCUSSION

These numerical experiments demonstrate the potential importance of climate–carbon cycle feedbacks, but the magnitude of these in the real Earth system is still highly uncertain. The strongest feedbacks, and therefore the greatest uncertainties, seem to be associated with the terrestrial biosphere. The cause of the present-day land carbon sink is still in doubt, with CO_2-fertilization, nitrogen deposition and forest regrowth all implicated in certain regions. The location of this sink is even more debatable, perhaps because this is subject to great interannual variability. Whilst increases in atmospheric CO_2 are expected to enhance photosynthesis (and reduce transpiration), the associated climate warming is likely to increase plant and soil respiration. Thus there is a competition between the direct effect of CO_2, which tends to increase terrestrial carbon storage, and the indirect effect, which may reduce carbon storage.

The outcome of this competition has been seen in a range of DGVMs (Cramer et al., 2001), each of which simulate reduced land carbon under climate change alone and increased carbon storage with CO_2 increases only. In most DGVMs, the combined effect of the CO_2 and associated climate change results in a reducing sink towards the end of the twenty-first century, as CO_2-induced fertilization begins to saturate but soil respiration continues to increase with temperature. The manner in which soil and plant respiration respond in the long-term to temperature is a key uncertainty in the projections of CO_2 in the twenty-first century (Giardina and Ryan, 2000).

1. Sink-to-Source Transitions in the Terrestrial Carbon Cycle

In this subsection we introduce a simple terrestrial carbon balance model to demonstrate how the conversion of a land CO_2 sink to a source is dependent on the responses of photosynthesis and respiration to CO_2 increases and climate warming. We consider the total carbon stored in vegetation and soil, C_T, which is increased by photosynthesis, Π, and reduced by the total ecosystem respiration, R:

$$\frac{dC_T}{dt} = \Pi - R \qquad (3)$$

where Π is sometimes called Gross Primary Productivity (GPP), and R represents the sum of the respiration fluxes from the vegetation and the soil. In common with many others (McGuire et al., 1992; Collatz et al., 1991, 1992; Sellers et al., 1996b; Cox et al., 1998b), we assume that GPP depends directly on the atmospheric CO_2 concentration, C_a, and the surface temperature, T (in °C):

$$\Pi = \Pi_{max} \left\{ \frac{C_a}{C_a + C_{0.5}} \right\} f(T) \qquad (4)$$

E. DISCUSSION

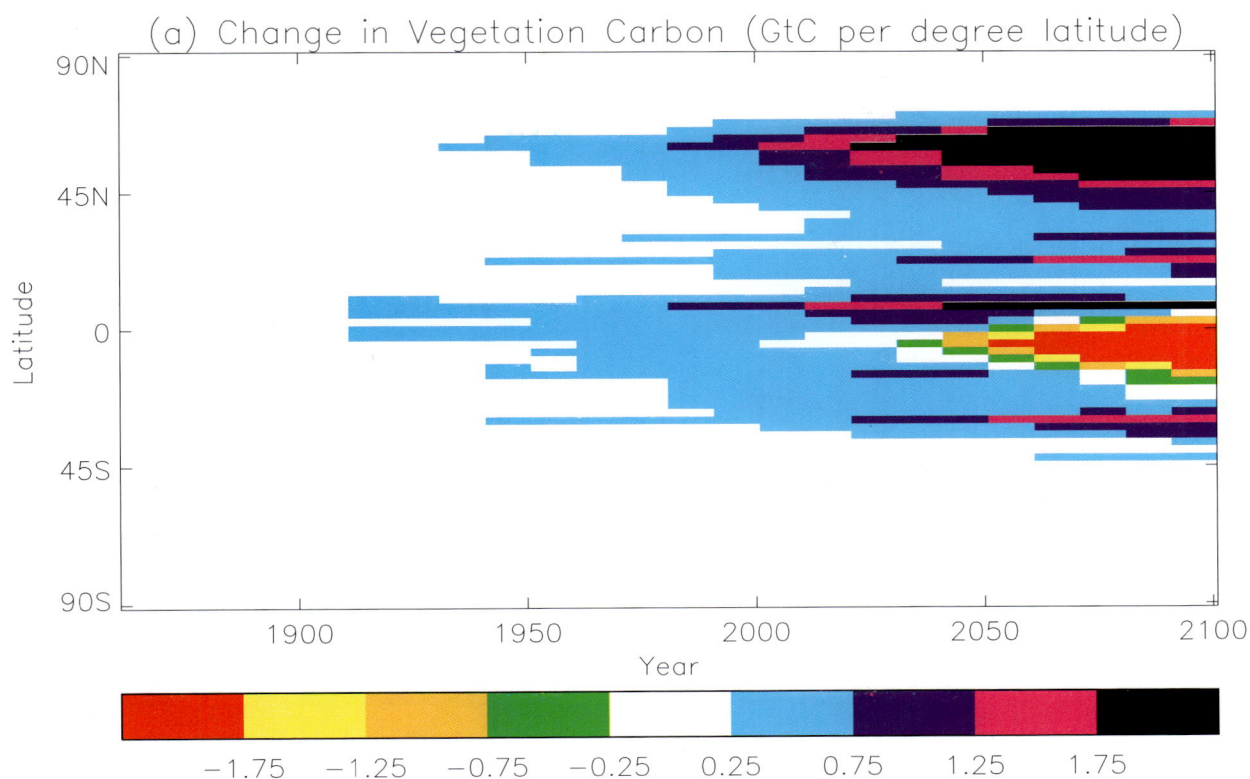

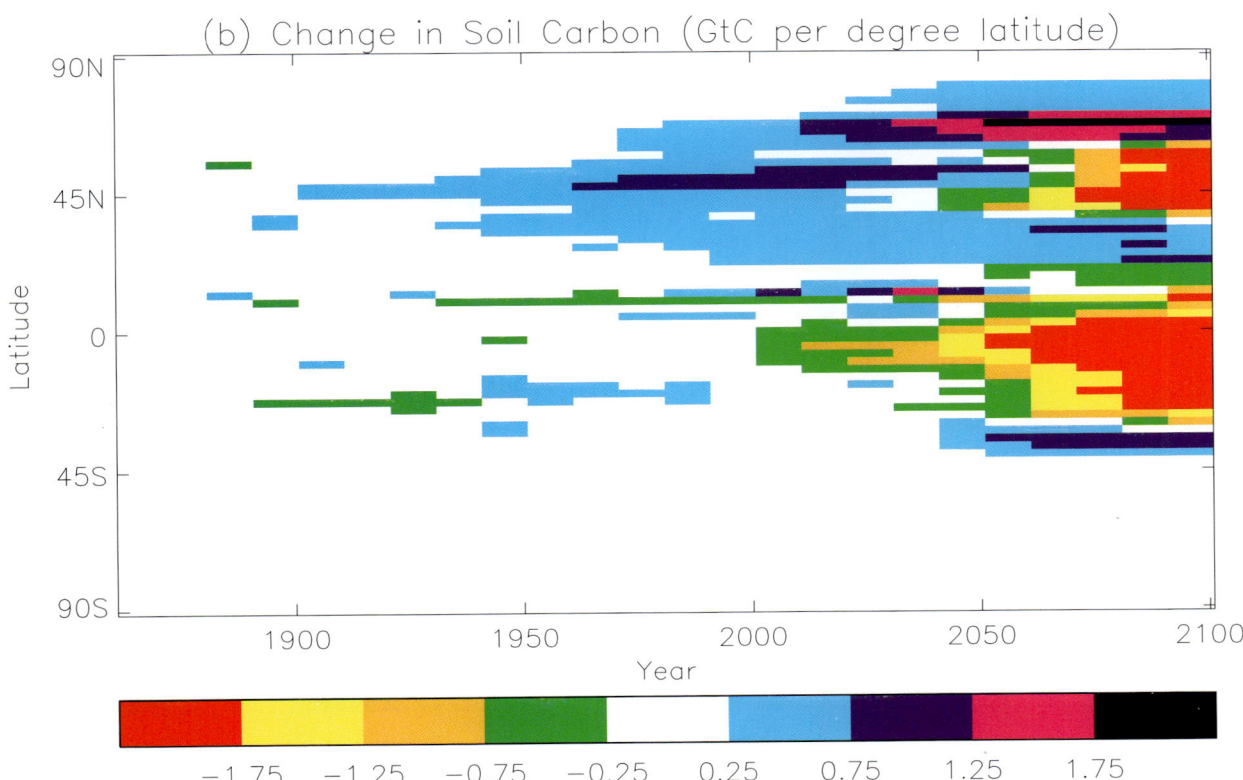

FIGURE 8 Latitude–time plot of modelled changes in vegetation carbon (a) and soil carbon (b) relative to 1860.

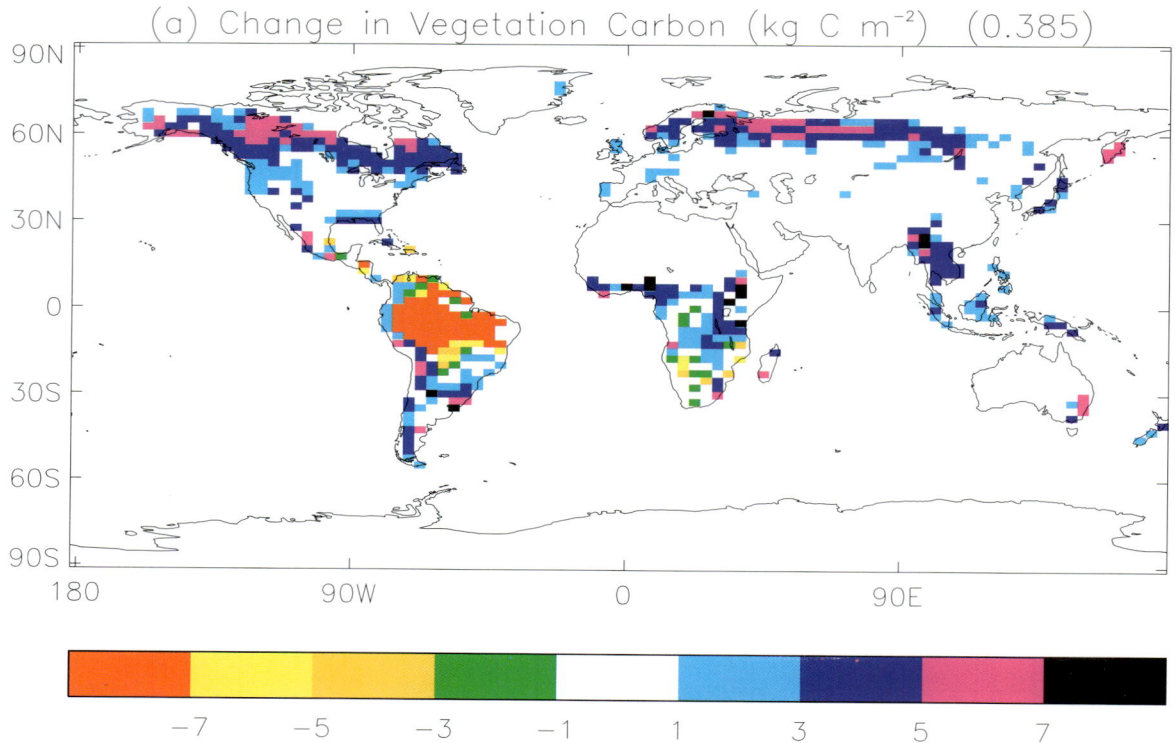
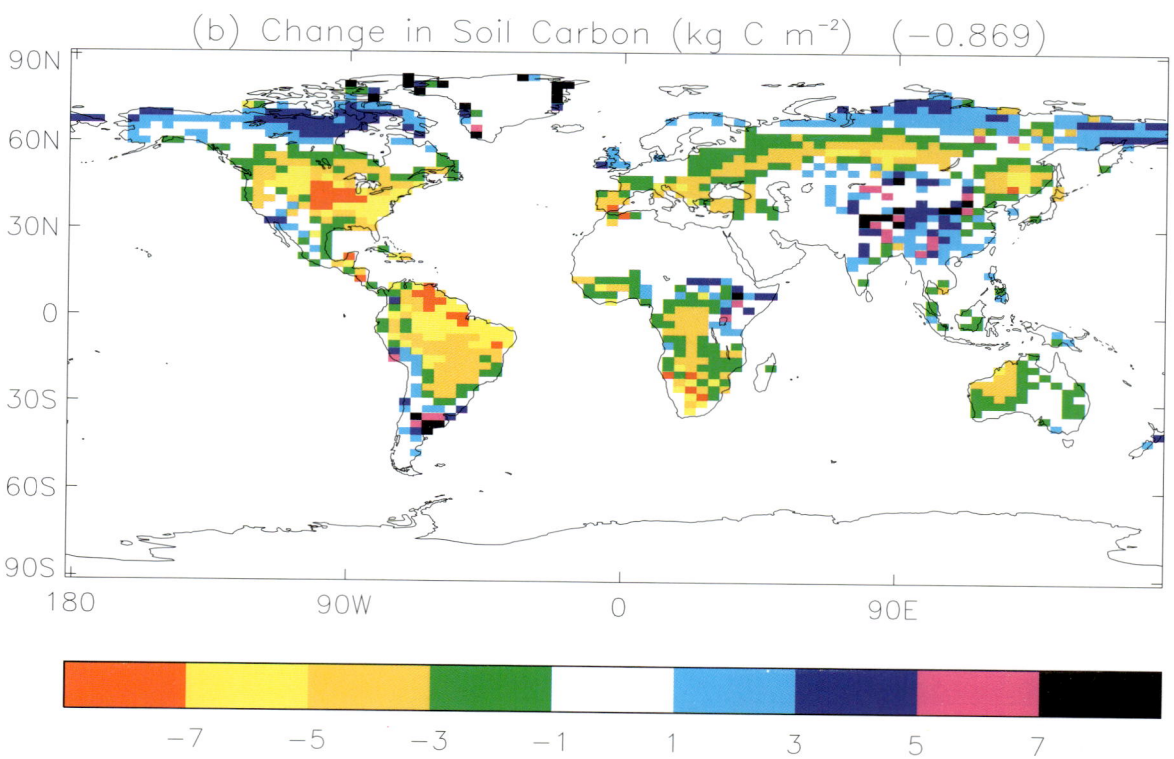

FIGURE 9 Modelled changes in vegetation carbon (a) and soil carbon (b) throughout the transient simulation (calculated as the difference between the mean for the 2090s minus the mean for the 1860s).

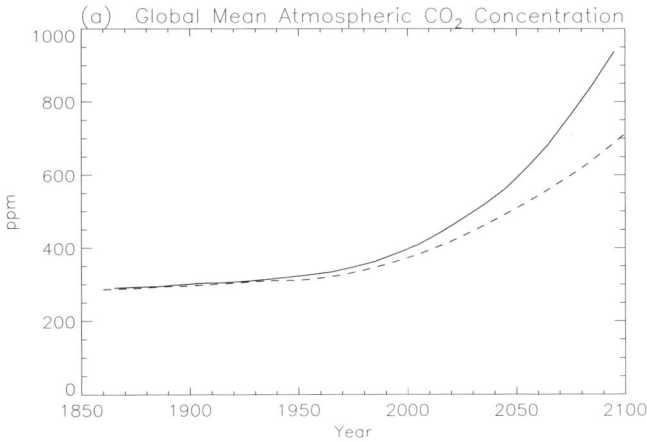

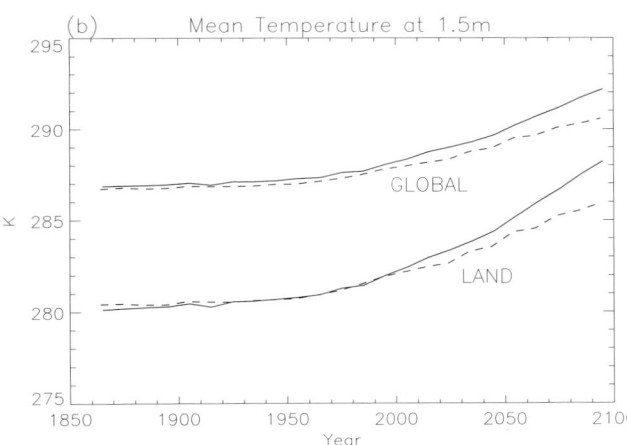

FIGURE 10 Comparison between the transient simulation with interactive CO_2 and dynamic vegetation (continuous lines), and the standard HadCM3 run with prescribed CO_2 and fixed vegetation (dashed lines). (a) Global mean CO_2 concentration and (b) global mean and land mean temperature, versus year.

where Π_{max} is the value which GPP assymptotes towards as $C_a \to \infty$, $C_{0.5}$ is the "half-saturation" constant (i.e. the value of C_a for which Π is half this maximum value), and $f(T)$ is an arbitrary function of temperature. We also assume that the total ecosystem respiration, R, is proportional to the total terrestrial carbon, C_T. The specific respiration rate (i.e. the respiration per unit carbon) follows a "Q10" dependence, which means that it increases by a factor of q_{10} for a warming of T by 10°C. Thus the ecosystem respiration rate is given by

$$R = r \, C_T \, q_{10}^{(T-10)/10} \qquad (5)$$

where r is the specific respiration rate at $T = 10°C$. It is more usual to assume separate values of r and q_{10} for different carbon pools (e.g. soil/vegetation, leaf/root/wood), but our simpler assumption will still offer good guidance as long as the relative sizes of these pools do not alter significantly under climate change. Near surface temperatures are expected to increase approximately logarithmically with the atmospheric CO_2 concentration, C_a (Huntingford and Cox, 2000):

$$\Delta T = \frac{\Delta T_{2 \times CO_2}}{\log 2} \log \left\{ \frac{C_a}{C_a(0)} \right\} \qquad (6)$$

where ΔT is the surface warming, $\Delta T_{2 \times CO_2}$ is the climate sensitivity to doubling atmospheric CO_2, and $C_a(0)$ is the initial CO_2 concentration. We can use this to eliminate CO_2-induced temperature changes from equation (5):

$$R = r_0 \, C_T \left\{ \frac{C_a}{C_a(0)} \right\}^\alpha \qquad (7)$$

where $r_0 \, C_T$ is the initial ecosystem respiration (i.e. at $C_a = C_a(0)$) and the exponent α is given by

$$\alpha = \frac{\Delta T_{2 \times CO_2}}{10} \frac{\log q_{10}}{\log 2} \qquad (8)$$

We can now use equations (3), (4) and (7) to solve for the equilibrium value of terrestrial carbon C_T^{eq}:

$$C_T^{eq} = \Pi_{max} \left\{ \frac{C_a}{C_a + C_{0.5}} \right\} \left\{ \frac{C_a(0)}{C_a} \right\}^\alpha \frac{f(T)}{r_0} \qquad (9)$$

The land will tend to amplify CO_2-induced climate change if C_T^{eq} decreases with increasing atmospheric CO_2 (i.e. $dC_T^{eq} / dC_a < 0$). Differentiating equation with respect to C_a yields

$$\frac{dC_T^{eq}}{dC_a} = C_T^{eq} \left[\frac{(1 - \alpha_*)}{C_a} - \frac{1}{C_a + C_{0.5}} \right] \qquad (10)$$

where

$$\alpha_* = \frac{\Delta T_{2 \times CO_2}}{\log 2} \left\{ \frac{\log q_{10}}{10} - \frac{1}{f} \frac{df}{dT} \right\} \qquad (11)$$

The condition for the land to become a source of carbon under increasing CO_2 is therefore

$$C_a > \frac{1 - \alpha_*}{\alpha_*} C_{0.5} \qquad (12)$$

This means that there will always be a critical CO_2 concentration beyond which the land becomes a source, as long as:

(i) CO_2 fertilization of photosynthesis saturates at high CO_2, i.e. $C_{0.5}$ is finite.
(ii) $\alpha_* > 0$, which requires:
  (a) climate warms with increasing CO_2, i.e. $\Delta T_{2 \times CO_2} > 0$
  (b) respiration increase more rapidly with temperature than GPP, i.e.

$$\frac{\log q_{10}}{10} > \frac{1}{f} \frac{df}{dT} \qquad (13)$$

Conditions (i) and (ii)(a) are satisfied in the vast

majority of climate and terrestrial ecosystem models. Detailed models of leaf photosynthesis indicate that $C_{0.5}$ will vary with temperature from about 300 ppmv at low temperatures, up to about 700 ppmv at high temperatures (Collatz et al., 1991). Although there are differences in the magnitude and patterns of predicted climate change, all GCMs produce a warming when CO_2 concentration is doubled. The global mean climate sensitivity produced by these models ranges from 1.5 K to 4.5 K (Houghton et al., 1996), but mean warming over land is likely to be a more appropriate measure of the climate change experienced by the terrestrial biosphere. We estimate a larger range of $2K < \Delta T_{2 \times CO_2} < 7$ K, because the land tends to warm more rapidly than the ocean (Huntingford and Cox, 2000).

There is considerable disagreement over the likely long-term sensitivity of respiration fluxes to temperature, with some suggesting that temperature-sensitive "labile" carbon pools will soon become exhausted once the ecosystem enters a negative carbon balance (Giardina and Ryan, 2000). However, condition (ii)(b) is satisfied by the vast majority of existing land carbon cycle models, and seems to be implied (at least on the 1–5 year timescale) by climate-driven interannual variability in the measured atmospheric CO_2 concentration (Jones and Cox, 2001; Jones et al., 2001).

Most would therefore agree that the terrestrial carbon sink has a finite lifetime, but the length of this lifetime is highly uncertain. We can see why this is from our simple model (equation 12). The critical CO_2 concentration is very sensitive to α_* which is itself dependent on the climate sensitivity, and the difference between the temperature dependences of respiration and GPP (equation 11).

We expect the temperature sensitivity of GPP to vary regionally, since generally a warming is beneficial for photosynthesis in mid and high latitudes (i.e. $df/dT > 0$), but not in the tropics where the existing temperatures are near optimal for vegetation (i.e. $df/dT \leq 0$). As a result, we might expect global mean GPP to be only weakly dependent on temperature ($df/dT \approx 0$). We can therefore derive a range for α_*, based on plausible values of climate sensitivity over land (2 K $< \Delta T_{2 \times CO_2} < 7$ K) and respiration sensitivity ($1.5 < q_{10} < 2.5$). This range of $0.1 < \alpha_* < 0.9$ translates into a critical CO_2 concentration which is somewhere between 0.1 and 9 times the half-saturation constant (equation 12).

Therefore on the basis of this simple analysis the range of possible critical CO_2 values spans almost two orders of magnitude. Evidently, the time at which the sink-to-source transition will occur is extremely sensitive to these uncertain parameters. This may explain why many of the existing terrestrial models do not reach this critical point before 2100 (Cramer et al., 2001).

Fortunately we can reduce the uncertainty range further. Critical CO_2 values which are lower than the current atmospheric concentration are not consistent with the observations, since the "natural" land ecosystems appear to be a net carbon sink rather than a source at this time (Schimel et al., 1995). For a typical half-saturation constant of $C_{0.5} = 500$ ppmv this implies that all combinations of q_{10} and $\Delta T_{2 \times CO_2}$ which yield values of $\alpha_* < 0.6$ are unrealistic. Also, sensitivity tests with our coupled model indicate that $q_{10} = 2$ provides an almost optimal fit to the observed variability in atmospheric CO_2 due to ENSO (Jones et al., 2001) and volcanic eruptions (Jones and Cox, 2001), suggesting that the probability distribution for possible q_{10} values is peaked quite sharply about this value. Similarly the complete range of climate sensitivity values are not all equally probable, since the most advanced GCMs tend to produce values clustered around the centre of the range. It is therefore meaningful to produce a central estimate for the critical CO_2 value. Using $q_{10} = 2$, $C_{0.5} = 500$ ppmv, and $\Delta T_{2 \times CO_2} = 4.8$ K (which is consistent with the warming over land in our coupled model,) yields a critical CO_2 value of about 550 ppmv, which is remarkably close to the sink-to-source transition seen in our experiment (see Figs 7 and 10a).

We draw two main conclusions from this section. The recognized uncertainties in climate and respiration sensitivity imply a very large range in the critical CO_2 concentration beyond which the land will act as a net carbon source. However, the central estimates for these parameters suggest a significant probability of this critical point being passed by 2100 in the real Earth system, under a "business as usual" emissions scenario, in agreement with the results from our coupled climate–carbon cycle model.

F. CONCLUSIONS

The ocean and the land ecosystems are currently absorbing about half the human emissions of CO_2, but many of the uptake processes are known to be sensitive to climate. GCM climate change predictions typically exclude interactions between climate and the natural biosphere, since they use fixed vegetation distributions and CO_2 concentrations which are calculated offline neglecting climate change. We have developed a 3-D coupled climate–carbon cycle model (HadCM3LC), by coupling models of the ocean carbon cycle (HadOCC) and the terrestrial carbon cycle (TRIFFID) to a version of the Hadley Centre GCM. The methodology developed to spin-up the model has been shown to yield a pre-industrial equilibrium to good accuracy. The climate–carbon cycle model is able to reproduce key aspects of the observations, including the global distribution of vegetation types, seasonal and zonal variations in ocean primary production, and the interannual variability in atmospheric CO_2.

A transient simulation has been carried out with this model for 1860–2100 using the IS92a ("business as usual") CO_2 emissions scenario. This experiment suggests that carbon cycle feedbacks could significantly accelerate atmospheric CO_2 rise and climate change over the next 100 years. The modelled terrestrial biosphere switches from being an overall sink for CO_2 to become a strong source from about 2050, as soil carbon starts to decline sharply and local climate change leads to significant loss of the Amazon rainforest. By 2100 the modelled CO_2 is approximately 250 ppmv higher than that usually assumed in GCM climate-change experiments. The corresponding global-mean warming for 1860–2100 is about 5.5 K, as compared to 4 K without carbon cycle feedbacks.

The quantitative aspects of this experiment must be treated with caution, owing to uncertainties in the components of the coupled model. In particular, climate models differ in their responses to greenhouse gases, and terrestrial carbon models differ in their responses to climate change (VEMAP Members, 1995; Cramer et al., 2001). However, we have presented a simple analysis to demonstrate that a sink-to-source transition of the terrestrial biosphere is assured beyond some critical atmospheric CO_2 concentration, provided that a few simple conditions apply. Qualitatively then, the eventual saturation of the land carbon sink and its conversion to a carbon source, is supported by our existing understanding of terrestrial ecosystem processes.

Unfortunately, the precise point at which the land biosphere will start to provide a positive feedback cannot yet be predicted with certainty. This depends on a number of poorly understood processes, such as the long-term response of photosynthesis and soil respiration to increased temperatures (Giardina and Ryan, 2000), and the possible acclimation of photosynthesis to high CO_2. Our results suggest that accurate prediction of climate change over the twenty-first century, will be as dependent on advances in the understanding and modelling of these physiological and ecological processes, as it is on the modelling of the physical processes currently represented in GCMs.

ACKNOWLEDGEMENTS

This work was supported by the UK Department of the Environment, Transport and the Regions, under contract PECD 7/12/37.

References

Antoine, D., J.-M. Andre and A. Morel, 1996: Oceanic primary production 2. Estimation at global scale from satellite (Coastal Zone Color Scanner) chlorophyll. *Glob. Biogeochem. Cycles*, **10**, 57–69.

Bacastow, R., 1976: Modulation of atmospheric carbon dioxide by the Southern Oscillation. *Nature*, **261**, 116–118.

Bacastow, R. and C. Keeling, 1981: Atmospheric CO_2 and the southern oscillation: effects associated with recent El Niño events. *WMO/ICSU/UNEP Scientific Conference on Analysis and Interpretation of Atmospheric CO_2 Data, World Climate Programme*, Vol. WCP 14, WMO, World Climate Programme Office, Geneva.

Bacastow, R., J. Adams, C. Keeling, D. Moss, T. Whorf and C. Wong, 1980: Atmospheric carbon dioxide, the southern oscillation, and the weak 1975 El Niño. *Science*, **210**, 66–68.

Berner, R., 1997: The rise of plants and their effect on weathering and atmospheric CO_2. *Nature*, **276**, 544–546.

Betts, R. A., P. M. Cox, S. E. Lee and F. I. Woodward, 1997: Contrasting physiological and structural vegetation feedbacks in climate change simulations. *Nature*, **387**, 796–799.

Bryan, K., 1984: Accelerating the convergence to equilibrium of ocean–climate models. *J. Phys. Oceanogr.*, **14**, 666–673.

Cao, M. and F. I. Woodward, 1998: Dynamic responses of terrestrial ecosystem carbon cycling to global climate change. *Nature*, **393**, 249–252.

Claussen, M., 1996: Variability of global biome patterns as a function of initial and boundary conditions in a climate model. *Climate Dynam.*, **12**, 371–379.

Collatz, G. J., J. T. Ball, C. Grivet and J. A. Berry, 1991: Physiological and environmental regulation of stomatal conductance, photosynthesis and transpiration: a model that includes a laminar boundary layer. *Agric. Forest Meteorol.*, **54**, 107–136.

Collatz, G. J., M. Ribas-Carbo and J. A. Berry, 1992: A coupled photosynthesis-stomatal conductance model for leaves of C_4 plants. *Aust. J. Plant Physiol.*, **19**, 519–538.

Cox, P. M., C. Huntingford and R. J. Harding, 1998: A canopy conductance and photosynthesis model for use in a GCM land surface scheme. *J. Hydrol.*, **212–213**, 79–94.

Cox, P. M., R. A. Betts, C. B. Bunton, R. L. H. Essery, P. R. Rowntree and J. Smith, 1999: The impact of new land surface physics on the GCM simulation of climate and climate sensitivity. *Climate Dynam.*, **15**, 183–203.

Cramer, W., A. Bondeau, F. Woodward, I. Prentice, R. Betts, V. Brovkin, P. Cox, V. Fisher, J. Foley, A. Friend, C. Kucharik, M. Lomas, N. Ramankutty, S. Stitch, B. Smith, A. White and C. Young-Molling, 2001: Global response of terrestrial ecosystem structure and function to CO_2 and climate change: results from six dynamic global vegetation models. *Glob. Change Biol.*, in press.

DOE, 1994: *Handbook of Methods for the Analysis of the Various Parameters of the Carbon Dioxide System in Sea Water; Version 2.* US DOE.

Edwards, J. M. and A. Slingo, 1996: Studies with a flexible new radiation code. I: Choosing a configuration for a large-scale model. *Q. J. R. Meteorol. Soc.*, **122**, 689–720.

Enting, I., T. Wigley and M. Heimann, 1994: Future emissions and concentrations of carbon dioxide; key ocean/atmosphere/land analyses. Division of Atmospheric Research Technical Paper 31, CSIRO.

Field, C., R. Jackson and H. Mooney, 1995: Stomatal responses to increased CO_2: implications from the plant to the global scale. *Plant Cell Environ.*, **18**, 1214–1225.

Foley, J. A., I. C. Prentice, N. Ramankutty, S. Levis, D. Pollard, S. Stitch, and A. Haxeltine, 1996: An integrated biosphere model of land surface processes, terrestrial carbon balance and vegetation dynamics. *Glob. Biogeochem. Cycles*, **10**(4), 603–628.

Friend, A. D., H. H. Shugart and S. W. Running, 1993: A physiology-based model of forest dynamics. *Ecology*, **74**, 797.

Gedney, N., P. Cox, H. Douville, J. Polcher and P. Valdes, 2000: Characterizing GCM land-surface schemes to understand their responses to climate change. *J. Climate*, **13**, 3066–3079.

Gent, P. R. and J. C. McWilliams, 1990: Isopycnal mixing in ocean circulation models. *J. Phys. Oceanogr.*, **20**, 150–155.

Giardina, C. and M. Ryan, 2000: Evidence that decomposition rates of organic carbon in mineral soil do not vary with temperature. *Nature*, **404**, 858–861.

Gordon, C., C. Cooper, C. A. Senior, H. Banks, J. M. Gregory, T. C. Johns, J. F. B. Mitchell and R. A. Wood, 2000: The simulation of SST, sea ice extents and ocean heat transports in a version of the Hadley Centre coupled model without flux adjustments. *Climate Dynam.*, **16**, 147–168.

Houghton, J. T., B. A. Callander and S. K. Varney, 1992: *Climate Change 1992. The Supplementary Report to the IPCC Scientific Assessment.* Cambridge University Press.

Houghton, J. T., L. G. Meira Filho, B. A. Callander, N. Harris, A. Kattenberg and K. Maskell, 1996: *Climate Change 1995—The Science of Climate Change.* Cambridge University Press.

Huntingford, C. and P. Cox, 2000: An analogue model to derive additional climate change scenarios from existing GCM simulations. *Clim. Dynam.*, **16**, 575–586.

Johns, T. C., R. E. Carnell, J. F. Crossley, J. M. Gregory, J. F. B. Mitchell, C. A. Senior, S. F. B. Tett and R. A. Wood, 1997: The second Hadley Centre coupled ocean-atmosphere GCM: Model description, spinup and validation. *Climate Dynam.*, **13**, 103–134.

Jones, C. and P. Cox, 2001: Modelling the volcanic signal in the atmospheric CO_2 record. *Glob. Biogeochem. Cycles*, accepted.

Jones, C. D. and J. R. Palmer, 1998: Spin-up methods for HadCM3L. Climate Research Technical Note 84, Hadley Centre for Climate Prediction and Research.

Jones, C., M. Collins, P. Cox and S. Spall, 2001: The carbon cycle response to ENSO: A coupled climate—Carbon cycle model study. *J. Climate*, accepted.

Keeling, C., R. Bacastow, A. Carter, S. Piper, T. Whorf, M. Heimann, W. Mook and H. Roeloffzen, 1989: A three dimensional model of atmospheric CO_2 transport based on observed winds: 1, Analysis of observational data. In *Aspects of Climate Variability in the Pacific and the Western Americas*, (D. Peterson, Ed.), no. 55 in Geophys. Monogr. Ser., AGU, pp. 165–236.

Kershaw, R. and D. Gregory, 1997: Parametrization of momentum transport by convection. Part I. Theory and cloud modelling results. *Q. J. R. Meteorol. Soc.*, **123**(541), 1133–1151.

Levitus, S. and T. P. Boyer, 1994: *World Ocean Atlas 1994*, Vol. 4: *Temperature*. NOAA/NESDIS E/OC21. U.S. Department of Commerce.

Levitus, S., R. Burgett and T. P. Boyer, 1994: *World Ocean Atlas 1994*, Vol. 3: *Salinity*. NOAA/NESDIS E/OC21. U.S. Department of Commerce.

Longhurst, A., S. Sathyendranath, T. Platt and C. Caverhill, 1990: An estimate of global primary production in the ocean from satellite radiometer data. *J. Plankt. Res.*, **17**, 1245–1271.

Loukas, H., B. Frost, D. E. Harrison and J. W. Murray, 1997: An ecosystem model with iron limitation of primary production in the equatorial Pacific at 140 W. *Deep-Sea Res. II*, **44**, 2221–2249.

Lovelock, J. and L. Kump, 1994: The failure of climate regulation in a geophysiological model. *Nature*, **369**, 732–734.

McGuire, A., J. Melillo, L. Joyce, D. Kicklighter, A. Grace, B. Moore III and C. Vorosmarty, 1992: Interactions between carbon and nitrogen dynamics in estimating net primary productivity for potential vegetation in North America. *Glob. Biogeochem. Cycles*, **6**, 101–124.

Melillo, J. M., I. C. Prentice, G. D. Farquhar, E. D. Schulze and O. E. Sala, 1995: Terrestrial biotic responses to environmental change and feedbacks to climate. In *Climate Change 1995* (J. T. Houghton, L. G. M. Filho, B. A. Callander, N. Harris, A. Kattenberg and K. Maskell, Eds.), Cambridge University Press.

Mitchell, J. F. B., T. C. Johns, J. M. Gregory and S. F. B. Tett, 1995: Climate response to increasing levels of greenhouse gases and sulphate aerosols. *Nature*, **376**, 501–504.

Nicholls, N., G. Gruza, J. Jouzel, T. Karl, L. Ogallo and D. Parker, 1996: *Climate Change 1995. The Science of Climate Change*, Ch. 3, Observed climate variability and change. Cambridge University Press.

Palmer, J. R. and I. J. Totterdell, 2001: Production and export in a global ocean ecosystem model. *Deep-Sea Res.*, **48**, 1169–1198.

Peng, T.-H., 1987: Seasonal variability of carbon dioxide, nutrients and oxygen in the northern North Atlantic surface water: observations and a model. *Tellus*, **39B**, 439–458.

Pope, V. D., M. L. Gallani, P. R. Rowntree and R. A. Stratton, 2001: The impact of new physical parametrizations in the Hadley Centre Climate Model—HadAM3. *Climate Dynam.*, **16**, 123–146.

Post, W. M. and J. Pastor, 1996: LINKAGES—an individual-based forest ecosystem model. *Climat. Change*, **34**, 253–261.

Prentice, I., W. Cramer, S. P. Harrison, R. Leemans, R. A. Monserud and A. M. Solomon, 1992: A global biome model based on plant physiology and dominance, soil properties and climate. *J. Biogeogr.*, **19**, 117–134.

Raich, J. and W. Schlesinger, 1992: The global carbon dioxide flux in soil respiration and its relationship to vegetation and climate. *Tellus*, **44B**, 81–99.

Sarmiento, J. and C. L. Quere, 1996: Oceanic carbon dioxide uptake in a model of century-scale global warming. *Nature*, **274**(5291), 1346–1350.

Sarmiento, J., T. Hughes, R. Stouffer and S. Manabe, 1998: Simulated response of the ocean carbon cycle to anthropogenic climate warming. *Nature*, **393**(6682), 245–249.

Schimel, D., D. Alves, I. Enting, M. Heimann, F. Joos, D. Raynaud, T. Wigley, M. Prather, R. Derwent, D. Enhalt, P. Fraser, E. Sanhueza, X. Zhou, P. Jonas, R. Charlson, H. Rodhe, S. Sadasivan, K. P. Shine, Y. Fouquart, V. Ramaswamy, S. Solomon, J. Srinivasan, D. Albritton, R. Derwent, I. Isaksen, M. Lal and D. Wuebbles, 1995: Radiative forcing of climate change. *Climate Change 1995. The Science of Climate Change* (J. T. Houghton, L. G. M. Filho, B. A. Callander, N. Harris, A. Kattenberg and K. Maskell, Eds), Cambridge University Press, pp. 65–131.

Sellers, P. J., L. Bounoua, G. J. Collatz, D. A. Randall, D. A. Dazlich, S. O. Los, J. A. Berry, I. Fung, C. J. Tucker, C. B. Field and T. G. Jensen, 1996a: Comparison of radiative and physiological effects of doubled atmospheric CO_2 on climate. *Science*, **271**, 1402–1406.

Sellers, P., D. Randall, C. Collatz, J. Berry, C. Field, D. Dazlich, C. Zhang and G. Collelo, 1996b: A revised land surface parameterisation (SiB2) for atmospheric GCMs. Part I: Model formulation. *J. Climate*, **9**, 676–705.

Siegenthaler, U. and J. L. Sarmiento, 1993: Atmospheric carbon dioxide and the ocean. *Nature*, **365**, 119–125.

Taylor, N., 1995: Seasonal uptake of anthropogenic CO_2 in an ocean circulation model. *Tellus*, **47B**, 145–169.

Tian, H., J. Melillo, D. Kicklighter, A. McGuire, J. Helfrich III, B. Moore III and C. Vorosmarty, 1998: Effects of interannual climate variability on carbon storage in Amazonian ecosystems. *Nature*, **396**, 664–667.

VEMAP Members, 1995: Vegetation/ecosystem modelling and analysis project: comparing biogeography and biogeochemistry models in a continental-scale study of terrestrial responses to climate change and CO_2 doubling. *Glob. Biogeochem. Cycles*, **9**, 407–437.

Wanninkhof, R., 1992: Relationship between wind speed and gas exchange over the ocean. *J. Geophys. Res.*, **97**, 7373–7382.

Warnant, P., L. Francois, D. Strivay and J.-C. Gerard, 1994: CARAIB: A global model of terrestrial biological productivity. *Glob. Biogeochem. Cycles*, **8**(3), 255–270.

White, A., M. Cannell and A. Friend, 1999: Climate change impacts on ecosystems and on the terrestrial carbon sink: a new assessment. *Glob. Environ. Change*, **9**, S21–S30.

Wood, R. A., 1998: Timestep sensitivity and accelerated spinup of an ocean GCM with a complex mixing scheme. *J. Atmos. Ocean. Technol.*, **15**, 482–495.

Zinke, P. J., A. G. Stangenberger, W. M. Post, W. R. Emanuel and J. S. Olson, 1986: Worldwide organic soil carbon and nitrogen data. NDP-018, Carbon Dioxide Information Center, Oak Ridge National Laboratory, Oak Ridge, Tennessee, USA.

Part 5. Middle Atmosphere and Solar Physics: Other Planets and Epochs

Some Fundamental Aspects of Atmospheric Dynamics, with a Solar Spinoff
M. E. McIntyre

Atmospheric Dynamics of the Outer Planets
A. P. Ingersoll

Palaeoclimate Studies at the Millennium: the Role of the Coupled System
P. J. Valdes

Some Fundamental Aspects of Atmospheric Dynamics, with a Solar Spinoff

Michael E. McIntyre

Centre for Atmospheric Science at the Department of Applied Mathematics and Theoretical Physics, Cambridge, UK
http://www.atm.damtp.cam.ac.uk/people/mem/

Global-scale atmospheric circulations illustrate with outstanding clarity one of the grand themes in physics: the organization of chaotic fluctuations in complex dynamical systems, with nontrivial, persistent mean effects. This is a theme recognized as important not just for classical textbook cases like molecular gas kinetics but also for an increasingly wide range of dynamical systems with large phase spaces, not in simple statistical–mechanical equilibrium or near-equilibrium. In the case of the atmosphere, the organization of fluctuations is principally due to three wave-propagation mechanisms or quasi-elasticities: gravity/buoyancy, inertia/Coriolis, and Rossby/vortical. These enter the fluctuation dynamics along with various forms of highly inhomogeneous turbulence, giving rise to a "wave–turbulence jigsaw puzzle" in which the spatial inhomogeneity—characteristic of "wave-breaking" understood in a suitably generalized sense—exhibits phase coherence and is an essential, leading-order feature as illustrated, also, by the visible surf zones near ocean beaches. Such inhomogeneity is outside the scope of classical turbulence theory, which assumes spatial statistical homogeneity or small departures therefrom. The wave mechanisms induce systematic correlations between fluctuating fields, giving rise to mean fluxes or transports of momentum quite different from those found in gas kinetics or in classical turbulence theory. Wave-induced momentum transport is a long-range process, effective over distances far exceeding the fluctuating material displacements or "mixing lengths" characteristic of the fluid motion itself. Such momentum transport, moreover, often has an "anti-frictional" character, driving the system away from solid rotation, not toward it, as underlined most plainly by the existence of the stratospheric quasi-biennial oscillation (QBO) and its laboratory couunterparts.

Wave-induced momentum transport drives the Coriolis-mediated "gyroscopic pumping" of meridional circulations against radiative relaxation, as illustrated by the Murgatroyd–Singleton mesospheric circulation and the Brewer–Dobson (misnamed Hadley) stratospheric circulation. The Brewer–Dobson can be contrasted with the convectively driven tropospheric Hadley circulation and with the oceans' thermohaline circulation. The ocean is an opposite extreme case in the sense that radiative relaxation has no counterpart. If the stratosphere is compared to a tape recorder whose motor is the gyroscopic pump and whose recording head is the tropical tropopause with its seasonal water-vapour signal, then the distribution of radiative-equilibrium temperatures across the tropics may be compared to the guidance wheels influencing the upward path of the tape. In other words, solar heating does not *drive* the tropical stratospheric upwelling, but does *influence* the way in which the upwelling mass flux demanded by the wave-driven pumping is distributed across the tropics. Because the tropical mean circulation problem is nonlinear, one cannot think of the tropical part of the circulation as a linear superposition of thermally and mechanically driven "contributions".

Insights into the dynamics of atmospheric circulations have recently led to a breakthrough in an astrophysical problem, that of understanding the differential rotation, meridional circulation, and helium distribution within the sun, with strong implications for helioseismic inversion and the so-called lithium and beryllium problems. This is an example of meteorological understanding informing solar and stellar physics.

A. INTRODUCTION

I want to revisit the fluid dynamics of global-scale atmospheric circulations in a way that emphasizes physical fundamentals, and to show how this has led to new insights into the workings of the sun's interior and other stellar interiors. In all these problems we are dealing with chaotic fluid motion at vast Reynolds numbers, involving various kinds of turbulence. But a leitmotif running through the whole story is the inapplicability of the classical turbulence paradigm, not only quantitatively but also qualitatively. Classical turbulence theory was historically important and is still implicitly relied on both in the meteorological literature and in the astrophysical literature, whenever the concept of "eddy viscosity" is used in one or another of its versions. The chaotic fluctuations that are most important in shaping global-scale circulations are organized, by contrast, in a very nonclassical way, outside the scope of classical turbulence theory but now recognized as essential even to a qualitative understanding of those circulations.

Our understanding is most secure in the case of the middle atmosphere, the region consisting of the stratosphere below 50 km and mesosphere above. One reason is that we now have a wealth of Lagrangian information from long-lived chemical tracers, thanks to the effort to understand ozone-layer chemistry. The middle atmosphere is more accessible to remote sensing than the troposphere, because of its chemical diluteness and lack of moist convection. Ozone, water vapour, and methane mixing ratios are measured in parts per 10^6, and most other trace chemicals in parts per 10^9 or less. Clouds and aerosols are relatively speaking thin or nonexistent, though crucial to the chemistry when present (Albritton et al., 1998). These circumstances, taken advantage of by an array of clever and sophisticated observing methods, have made the middle atmosphere into a wonderful outdoor laboratory for the study of basic processes in stratified, rotating fluid motion, such as must also exist beneath the convection zones of sun-like stars.

In the fluid dynamics of the middle atmosphere, the Coriolis and buoyancy effects associated with rotation and stable stratification are of course immensely strong, in relation to mean circulations and their chemical consequences. Typical Coriolis timescales, of the order of hours over a wide range of latitudes, and buoyancy timescales, of the order of minutes, are several orders of magnitude less than the timescales for mean circulations and chemical transports. Coarse-grain gradient Richardson numbers are large. Thus mean circulations feel an enormous stiffness from the Coriolis and buoyancy effects. This implies stiffness of the fluid-dynamical equations, in the numerical-analytical sense—the same stiffness that famously spoiled L. F. Richardson's pioneering attempt at numerical weather prediction. It also reminds us of the possible importance of wave propagation mechanisms—the reason why the classical turbulence paradigm fails.

B. THERMAL AND CHEMICAL EVIDENCE

Before saying more about the circulation dynamics, let us recall two well-known facts that constrain any theory. One is that, at solstice, the summer pole is the sunniest place on Earth. Diurnally averaged insolation—the solar energy arriving per unit time per unit horizontal area—is a maximum there, because of the substantial tilt of the earth's axis, 23.5° or 0.41 radian. The other fact is that, at altitudes between about 80 km and 90 km, the summer pole has been observed again and again to be the coldest place on earth (excluding laboratories) despite being the sunniest.

This is the famous cold summer polar mesopause. Even in a heavily averaged climatological picture (Fleming et al., 1990), temperatures dip below about 150 K—colder even than the high Antarctic plateau in midwinter, which seldom if ever gets below 170 K. And there are large fluctuations about climatological values: rocket-launched falling sphere measurements have recorded, on at least one occasion (Lübken, 1999), temperatures as low as 105 K. Partly because of these extremes of cold, the summer polar mesopause—more precisely, the region a few kilometres below it—is the region of formation of, by far, the world's highest clouds, the so-called "polar mesospheric", "noctilucent", or "night shining" clouds (Gadsden and Schröder, 1989; Thomas et al., 1989; Schröder, 1999a). They occur at altitudes usually within the range 80–90 km and have often been detected by space-based instruments, including visible-wavelength imagers on geosynchronous meteorological satellites (Gadsden, 2000). The clouds are also seen, less often, by ground-based observers between latitudes ~50° to 60°, around midnight on some clear summer nights. They have been noticed ever since the "Krakatau twilights" of 1883, according to Schröder (1999). I have seen them myself, looking north around 2 a.m. from a low hill near Cambridge, which at 52.2°N is near the equatorward limit of observability.

Now noctilucent clouds are believed on good evidence, including *in situ* rocket measurements, to be made of ice crystals (G. Witt and E. Kopp, personal communication). Their formation therefore depends not only on the low temperatures but also, crucially, on a supply of water vapour from below (Thomas et al., 1989). Water vapour is photochemically destroyed on timescales of days by the very hard solar ultraviolet that can penetrate to the upper mesosphere (Brasseur and Solomon, 1984, Fig. 4.40). The necessary supply of water vapour depends on the existence of a systematic rising motion over the summer polar cap, as suggested schematically by the heavy dashed curve in Fig. 1. Air and water vapour are thus brought up from the less fiercely irradiated, and slightly less dry, lower layers. Water vapour mixing ratios near 50 km altitude, for instance, are maintained relatively stably at just over 6 ppmv (parts per million by volume), partly by direct supply of water and partly by the oxidation of methane, both coming from the troposphere far below, via the freeze-drying tropical tropopause.

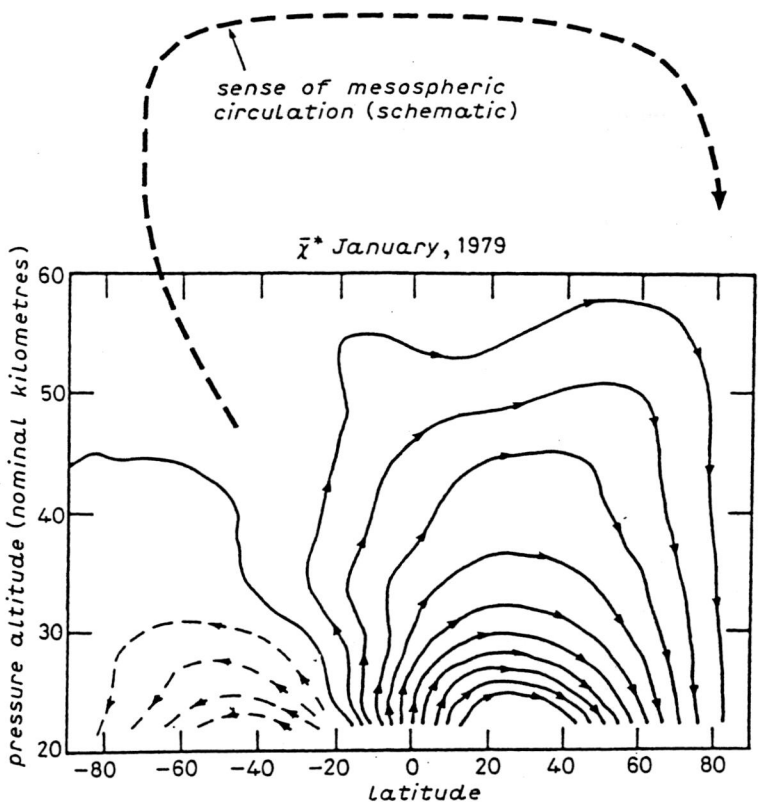

FIGURE 1 Mass transport streamlines of the global-scale mean circulation for January 1979, solid curves, from Solomon et al. (1986). This was estimated from LIMS satellite radiometer data in conjunction with detailed calculations of radiative heating and cooling rates. The picture gives the typical latitude and height dependence of the zonal and time averaged mass circulation of the stratosphere, defined in a quasi-Lagrangian sense (Section H) giving a simplified, but roughly correct, indication of the vertical advective transport of chemical constituents (Section H). The heavy dashed streamline indicates the qualitative sense of the mesospheric or Murgatroyd-Singleton branch of the circulation, deduced from other observational and theoretical evidence (Andrews et al., 1987) The lower pair of circulation cells is often referred to as the Brewer-Dobson circulation and the mesospheric branch as the Murgatroyd-Singleton circulation. The altitude scale is nominal log-pressure altitude, based on a nominal pressure scale height of 7 km. The right-hand cell of the circulation is a little stronger than usual, in this picture; the wintertime stratosphere was dynamically very active in January 1979. The northward mean velocities at top right (within the frame, not counting the heavy dashed streamline) are of the order of 2 or $3\,\mathrm{m\,s^{-1}}$. On the heavy dashed streamline they would be more like $10\,\mathrm{m\,s^{-1}}$.

The rising motion in the summer polar mesosphere is part of a self-consistent picture that checks out very well in other ways. For instance the rising motion is an essential part of *why* summer polar mesopause temperatures are so low. There is a refrigerating action due to the expansion of the air as it rises through the stable stratification. The consequent adiabatic cooling opposes the net radiative heating, which has positive contributions both from the sun and from infrared radiation emitted by warmer layers near the stratopause below (the infrared photons having long free paths at these altitudes, many of them escaping directly to space, but a few being reabsorbed on their way up). The result is that temperatures around 80–90 km are pulled down, by the refrigerating action, by as much as ~100 K below radiative equilibrium, though quantitative estimates are difficult because of uncertainties about the detailed photochemistry, and about departures from local thermodynamic equilibrium (e.g. Andrews et al., 1987; Ward and Fomichev, 1996).

Temperatures in certain other parts of the middle atmosphere also depart significantly from radiative equilibrium, for instance over the winter polar cap. There, air is compressed, and its temperature pushed above radiative equilibrium, by the systematic descending motion seen on the right of Fig. 1. This contrasts with the summer stratopause, the temperature maximum at pressure altitudes near 1 hPa, or about 50 km or 7 pressure scale heights, which is relatively close to radiative equilibrium, probably within 10 K or so, the circulation being relatively weak. The high temperatures at the summer stratopause are well explained by the absorption, by ozone, of solar ultraviolet at wavelengths ~200–300 nm, which can penetrate down to 50 km or lower (Brasseur and Solomon, 1984, Fig. 4.40; Andrews *et al.*, 1987). Implicit in all the foregoing is, of course, the notion of a radiative equilibrium temperature T_{rad} toward which global-scale temperatures, in the absence of any circulation, would tend to relax. This notion is well justified by detailed studies of radiative transfer (e.g. Andrews *et al.*, 1987; Fels, 1985), relaxation times being of the order of days to weeks.

A circulation like that in Fig. 1 and its seasonal variants is consistent not only with observed temperatures and the observed behaviour water vapour and methane but also with that of a number of other chemical constituents (nearly all of which, incidentally—having molecules larger than diatomic—are greenhouse gases, with quasi-bending modes of vibration at infrared frequencies).

For instance, man-made chlorofluorocarbons (CFCs) are chemically inert under everyday conditions. Together with their low water-solubility and small rate of uptake by the oceans, this means that, as has been checked and rechecked by countless observations (Albritton *et al.*, 1998), they are mixed fairly uniformly throughout the troposphere. The troposphere can therefore be thought of as a big tank or reservoir whose contents are recirculated through the middle atmosphere at rates of the order suggested by Fig. 1, or a fraction less. This accounts rather well for the observed destruction rates of CFCs, which correspond to e-folding atmospheric lifetimes of order a century—actually somewhat shorter, by a factor 2 or so, for CFC11 ($CFCl_3$) (Albritton *et al.*, 1998), and somewhat longer for CFC12 (CF_2Cl_2), the former being a bit less stable because of its extra chlorine atom. It is mainly at altitudes $\gtrsim 25$ km that the CFCs are destroyed, by ultraviolet photolysis; and so rates of CFC destruction are closely tied to rates of circulation. In terms of mass flow that reaches altitudes $\gtrsim 25$ km, these rates amount to a few percent of the tropospheric mass per year. This is consistent with circulation times of the order of several years (Hall *et al.*, 1999) and CFC e-folding lifetimes of the order of a century because of the falloff of mass density with altitude, roughly in proportion to pressure, by a factor e^{-1} for each pressure scale height, i.e. for each 7 km or so. Thus mass densities at, e.g., 30 km, are about 1% of those at sea level.

It is easy to check (Holton *et al.*, 1995) that the orders of magnitude in this picture require upward velocities at, say, 20 km in the tropical lower stratosphere (bottom centre of Fig. 1) of the order of 0.2 mm s^{-1} or 6 km per year, about one pressure scale height per year. This sort of velocity is far too small to observe directly—or so it was thought until a few years ago. Figure 2, from Mote *et al.* (1998; see also Mote *et al.*, 1996)), comes close to qualifying as a direct observation, and gives us a powerful independent consistency check on the whole picture.

Figure 2 was derived from satellite data obtained quite recently, long after results like that of Fig. 1 had been obtained, indirectly, with the help of an earlier

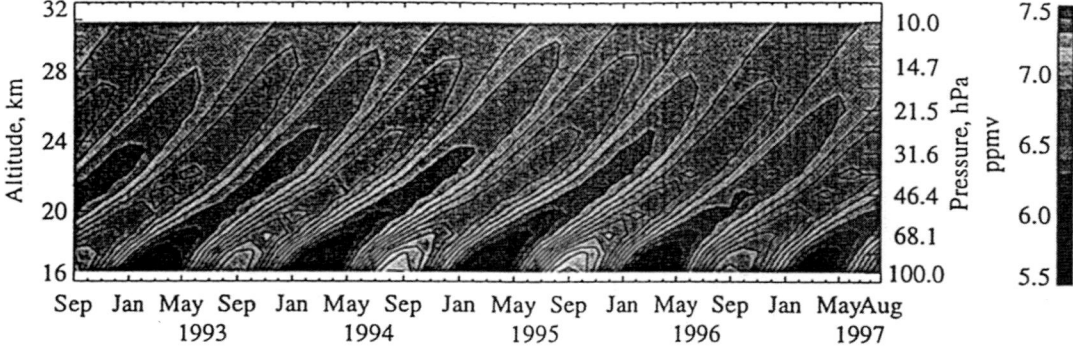

FIGURE 2 The tropical stratosphere as a "tape recorder" (from Mote *et al.*, 1998; *q.v.* for a careful discussion of the data analysis and interpretation). The time–altitude plot is from HALOE, a solar occultation limb sounder on the Upper Atmosphere Research Satellite (UARS), showing the signal from the annual cycle in water vapour being advected upward by the large-scale mean velocity $\bar{w}^*$. Near 20 km, $\bar{w}^* \approx 0.2$ mm s^{-1}. Soundings between 14°N and 14°S were used. The quantity plotted is "total hydrogen", water vapour plus twice methane, in parts per million by volume. Because of methane oxidation chemistry, this is to excellent approximation a passive tracer in the altitude range shown. The raw data have been fitted to a suitable set of extended empirical orthogonal functions (EEOFs) to extract the annual cycle in an objective way; the plot shows the contributions from the dominant two EEOFs, accounting for 68% of the variance. But the tape-recorder signal is still easily visible without any such processing (see Plate 1 of Mote *et al.*, 1998)—and in data from other instruments, notably the passive Microwave Limb Sounder on UARS (Plate 1a of Mote *et al.*, 1996).

generation of satellites. It shows the water vapour signature of the annual cycle being carried up in the rising flow just like a signal recorded on a moving magnetic tape. At altitudes around 20 km, the velocity apparent in the picture is very close to $0.2\,\mathrm{mm\,s^{-1}}$. Mote *et al.* (1998) follow a careful model-fitting procedure to verify that upward advection at this speed is indeed the main effect near 20 km, as distinct from, e.g., the vertical eddy diffusion due to small-scale turbulence or infrared random walking (Sparling *et al.*, 1997).

But what, then, is driving the global-scale circulation illustrated in Fig. 1? As with ordinary mechanical refrigerators or compressors, something has to pump the air around. It will not, by itself, move vertically against the strong stable stratification. To make noctilucent clouds you have to do work against buoyancy forces and and pull air upward. To warm the winter pole you have to push air downward. So what is doing the pulling and pushing?

The answer, as hinted earlier, is *wave-induced momentum transport*—more precisely, the irreversible transport of momentum and angular momentum by wave motions generated in one place (mostly in the more massive troposphere below) and dissipated in another (mostly higher up), the dissipation mechanisms including various kinds of *wave breaking*, in a sense to be explained in Section I, p. 294. This, in combination with Coriolis effects, acts as a global-scale mechanical pump, as we shall be reminded in Section H. One implication, confirmed by an analysis of MSU (Microwave Sounding Unit) channel 4 satellite data (Yulaeva *et al.*, 1994), is that the seasonal variation of the total upwelling mass flux in the tropical stratosphere is tied to seasonal variations in extratropical rather than in tropical conditions.

The way the momentum transport works has features that may seem surprising, and historically did seem surprising, especially from a statistical–mechanical or turbulence–theoretical viewpoint. Such surprise was expressed, for instance, in the closing pages of E. N. Lorenz's (1967) landmark review on the general circulation. For if we regard the wave fields as examples of chaotic fluctuations, what a plasma physicist might call "weak turbulence"—and the real wave fields are often, indeed, more or less chaotic—then we find on analysing their properties that they have a distinct tendency to behave "anti-frictionally" in the sense of driving the mean state of the atmosphere not toward, but away from, solid rotation.

This is the opposite to what is predicted by classical turbulence theory, which says that chaotic fluctuations about the mean should give rise to a viscosity, in much the same way as in gas kinetics. Hence the notion of "eddy viscosity". Of course there was never the slightest reason why fluctuations should necessarily behave in this classical way. Chaotic though they may be, many of the fluctuating motions strongly feel one or more of the wave propagation mechanisms (Rossby, gravity, inertia–gravity) that are neglected in classical turbulence theory but are characteristic of stratified rotating fluid systems. By its nature, any wave propagation mechanism tends to promote systematic correlations between the different fluctuating fields, hence systematic mean stresses. In other words the fluctuations, however chaotic or otherwise they may be, tend to be organized dynamically in such a way that momentum is systematically transported through the fluid—over distances limited only by the distances over which waves can propagate.

This is the fundamental point. Wave-induced momentum transport is, by its nature, a long-range process, acting over distances that can be vastly greater than the "mixing lengths", or length scales for typical material displacements, that characterize local turbulent eddy motion. By contrast, the notion of "eddy viscosity" makes sense only when the fluctuation-induced momentum transport is at an opposite extreme—a short-range process dominated by mixing length scales that are smaller than length scales characterizing the mean state, as with mean free paths in gas kinetics.

C. WAVE-INDUCED MOMENTUM TRANSPORT: SOME SIMPLE EXAMPLES

Wave-induced momentum transport has been well known for two centuries or more as a generic phenomenon, characteristic of many types of waves in a vacuum or in fluid media. Photons in a vacuum provide the simplest example, with wave amplitude a proportional to a typical electric or magnetic field strength: we can think of photons as being just like bullets having a certain momentum and carrying that momentum from one place to another, at rates $O(a^2)$. Waves in fluid media are often described in the same way, but it is important to understand that the picture is not literally true any more. The analogy with photons in a vacuum is only an analogy, indeed only a partial analogy (McIntyre, 1981, 1993; Bühler and McIntyre, 2001). This is because the presence of the material medium not only vitiates Lorenz invariance of the wave problem—there is now a preferred frame of reference—but also significantly complicates wave-mean momentum budgets correct to $O(a^2)$. A material medium can support $O(a^2)$ mean stresses in ways that a vacuum (in classical physics) cannot.

Careful analyses of problems of this kind correct to

$O(a^2)$, with attention to boundary conditions as well as to the nonlinearities in the fluid-dynamical equations, show that the notion of a wave packet possessing a definite momentum no longer makes sense except in some very restricted circumstances. It makes no sense to talk for instance about the waves in material media depositing "their momentum", or exchanging "it" with the mean flow. But the notion of a wave-induced momentum *flux* (a) does generally make sense, (b) is more often than not correctly quantified by the photon analogy, though not always, and (c) is in any case the relevant thing for our purpose. It is in momentum fluxes, i.e. transport rates, that we are interested. To make the literature on atmospheric waves make sense, one sometimes has to insert the word "flux" or "transport" after the word "momentum".*

The underpinning mathematical theory is conceptually and technically tricky, especially in its most general, three-dimensional form. It is still under development today; among recent contributions an especially important one is that of Bühler (2000). However, the theory for the simplest and, for present purposes, most important case, that of waves on a zonally symmetric mean flow, is relatively straightforward and is available in detail in the standard literature (e.g. Andrews *et al.*, 1987). Some key points will be recalled briefly in Section F onwards.

---

*These points, including point (b), were in essence made long ago by Léon Brillouin. Following him, I have discussed them more fully in earlier publications. See McIntyre (1981, 1993), and their bibliographies, tracing the problem back to the "Abraham–Minkowski controversy" about light waves in a refractive medium, and to a rare mistake by Lord Rayleigh. There is a perpetual source of confusion insofar as two distinct conservable quantities enter the theory, *momentum* on the one hand and *quasimomentum* or *pseudomomentum*, often misleadingly called "wave momentum", on the other. They are the same in a vacuum but differ when a material medium is present. Pseudomomentum is an $O(a^2)$ wave property, calculable from linearized wave theory alone. Momentum is calculable only if the state of the material medium is known correct to $O(a^2)$. Both conservable quantities are associated with a translational symmetry, via Noether's theorem, hence look superficially similar within the mathematical formalisms of Hamiltonian and Lagrangian dynamics, a simple example being Hamilton's equations for ray tracing. But the two translational symmetries are physically different when a material medium is present. One of them is translational invariance of the *physics*, giving rise to conservation of momentum. The other is translational invariance, i.e., homogeneity, of the *medium*, giving rise to conservation of pseudomomentum. When ray tracing is a valid approximation, one can usually take the wave-induced momentum flux to be *equal* to the pseudomomentum flux (which in turn can be equated to the outer product of the group velocity and the pseudomomentum density) without incurring serious errors in calculating wave-induced forces. This is point (b) above, sometimes called the "pseudomomentum rule". It is a consequence of Kelvin's circulation theorem (Andrews and McIntyre, 1978; McIntyre, 2000a; Bühler, 2000). For the most recent update see Bühler and McIntyre (2001).

A simple experiment can be used to shortcut the theoretical issues and illustrate the physical reality, and robustness, of wave-induced momentum transport. Figure 3a shows a version convenient for lecture demonstrations, in which a small cylinder is oscillated vertically and radiates capillary–gravity waves anisotropically to either side. If chalk dust or other floating powder is sprinkled on the surface, where the fluctuations are concentrated, one sees that the water flows away from the wavemaker in a persistent and conspicuous mean motion that sweeps the surface dust along with it. The experiment can be done on an overhead projector, and always works robustly. I have demonstrated it countless times when lecturing on this topic.

The observed mean flow depends on strong dissipation of the waves. This results in an irreversible flux or transport of momentum from the wavemaker to the fluid on either side of it, where the waves are dissipating. The wave dissipation is enhanced by the dirty water surface, which gives rise to a strong viscous boundary layer just under the surface, which may be intermittently turbulent. That is, the waves can be *breaking* even though no air entrainment takes place. The fact that irreversible momentum transport is involved can be seen by carefully stopping the wavemaker, and observing that the mean flow persists much longer than does the outgoing wave field. There is a significant contribution to the mean flow that is not a mere Stokes drift or other reversible mean-flow contribution dependent on the continued presence of the outgoing waves (and on the choice of averaging operator to define "mean flow"). Such reversible contributions return to zero as soon as the waves propagate out of the region of interest, or become directionally randomized or finally dissipated.

The fact that the flow is due to wave-induced momentum transport and not to the oscillatory boundary layer on the wavemaker itself is also clear in the experiment, if one uses vertical oscillations of the wavemaker. The Rayleigh–Schlichting boundary-layer streaming generated at the surface of the wavemaker is then directed toward the wavemaker, i.e., in a sense opposite to that observed. So the observed mean flow is due, rather, to the wave-induced momentum transport. Moreover, with a larger tank, one can do versions of the experiment in which the waves are generated by a larger, concave wavemaker that focuses them at a more distant spot, as suggested in Fig. 3b. This larger-scale demonstration is set up every year at the Cambridge Summer School in Geophysical and Environmental Fluid Dynamics. Sure enough, the momentum is transported, then, over the greater distance. The strongest mean flow is seen where the waves are focused.

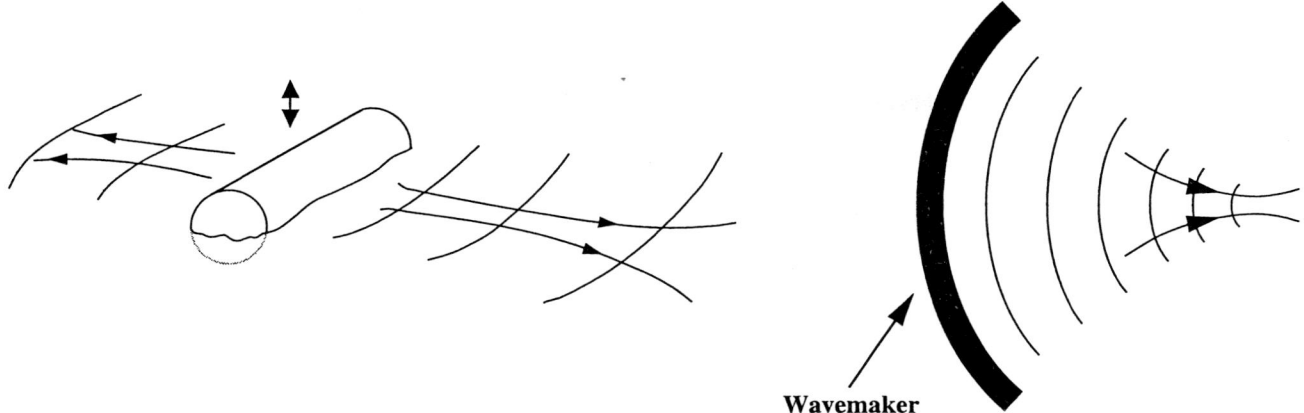

FIGURE 3 Simple experiments with water waves, illustrating the wave-induced momentum transport (here manifested by the appearance of a strong mean flow directly driven by it) that results from the generation of waves in one place and their dissipation in another (Section C). The mean flow can be made visible by sprinkling a little powder such as chalk dust on the surface of the water. In experiment (a), a cylinder about 10 cm long and 4 cm in diameter is oscillated vertically. Experiment (b) uses a curved wavemaker about 60 cm in radius and arc length. Good results are easily obtained with capillary–gravity waves having frequencies ~5 Hz. (From McIntyre and Norton, 1990.)

A spectacular case of wave-induced momentum transport over still greater distances was demonstrated by Walter Munk and colleagues (Snodgrass *et al.*, 1966) in a famous observational study of surface gravity waves propagating from the Southern Ocean across the entire Pacific, all the way to beaches in Alaska, where the waves, breaking in the near-shore surf zone, generate longshore mean currents in fundamentally the same way as in Fig. 3. In other words, mean forces exerted on the Southern Ocean, by winds in the latitudes of the "roaring forties" and "fifties", were shown to drive mean currents off the coast of Alaska.* Even more than Fig. 3b, this underlines the point that wave-induced momentum transport is a long-range process.

D. THE QUASI-BIENNIAL OSCILLATION (QBO)

Fundamentally similar effects occur with, for example, internal gravity waves in a stably stratified fluid. Figure 4 shows schematically the apparatus for the celebrated Plumb–McEwan experiment (Plumb and McEwan, 1978), in which the effects manifest themselves in an especially interesting form, which I think must surprise anyone encountering them for the first time. A flexible membrane is oscillated in a standing wave, at the bottom of an annular container filled with a stably stratified salt solution. The imposed conditions are close to mirror-symmetric. There is no externally-imposed difference between the clockwise and anti-clockwise directions around the annulus; in this experiment, Coriolis effects are negligible. When the bottom boundary is oscillated, with a period exceeding $2\pi/N$, the buoyancy or Brunt–Väisälä period of the stable stratification, internal gravity waves are generated. They can be regarded as a superposition of two progressive modes, clockwise and anticlockwise. They propagate upward and dissipate mainly by viscosity, salt diffusivity being negligible for this purpose. In a larger-scale apparatus they would dissipate by wave breaking, but to my knowledge such an experiment has not been done.

Naive symmetry arguments suggest that nothing interesting will happen. But, when the experiment is run at sufficient wave amplitude, the mirror symmetry is broken spontaneously and a conspicuous mean flow around the annulus is generated. This goes robustly into a quasi-periodic sequence of states, in which the mean flow keeps reversing, and only the long-time-average state is approximately mirror-symmetric. Sufficient amplitude means sufficient to overcome viscous drag on the mean flow. The experiment has been repeated at the University of Kyoto, and the results are available on the web along with advice on experimental technique; see caption to Fig. 4.

The mechanism operating in the experiment is simple and well understood (Plumb, 1977). It involves irreversible wave-induced transport of angular momentum, dependent on the viscous dissipation of the waves, and on wave refraction and Doppler shifting by the

---

* For a careful discussion of why the photon analogy is still only an analogy, see especially McIntyre (1993).

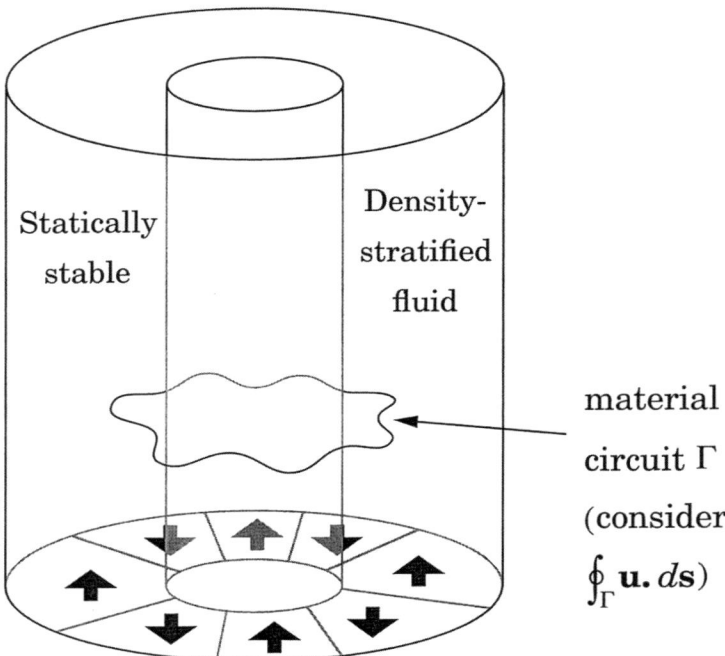

FIGURE 4 Schematic of the Plumb–McEwan experiment (Plumb and McEwan, 1978), showing how a reversing mean flow, first clockwise then anticlockwise around the annulus, can be generated by nothing more than standing oscillations of a flexible membrane at the bottom of an annular container filled with a stably stratified salt solution. The standing oscillations generate internal gravity waves at periods around $20\,\text{s} > 2\pi/N$, the buoyancy or Brunt–Väisälä period of the stable stratification. The annular gap width in the original apparatus was 12 cm, which is smaller than sketched here, relative to the outer radius, 30 cm. Also, the original apparatus had twice the number of flexible segments, giving 8 full wavelengths around the annulus. When gap widths are made significantly less than 12 cm, with water as the working fluid, the experiment fails because of viscous drag on the mean flow. For a useful set of notes on how the experiment was successfully repeated at Kyoto University, see the "Inside Stories" and "Tech Tips" at http://www.gfd-dennou.org/library/gfd_exp/exp_e/ under the heading "QBO", where a movie of the experiment running (.avi file format) can also be seen.

resulting shear flow, operating in a feedback loop. The Doppler shifting leads to enhanced dissipation, reduced vertical group velocity, and reduced vertical penetration—or *filtering*, as it is often called—of the clockwise and anticlockwise progressive wave components alternately. This gives rise to a characteristic space–time signature in which features such as zero crossings in the mean velocity profile propagate inexorably downward. The behaviour is clearly anti-frictional: if one calculates an eddy viscosity as momentum flux divided by mean shear, then one gets nonsensical negative and infinite values. The fluctuations drive the system away from, not toward, solid rotation.

It is now recognized as no accident that a similar space–time signature is observed in the real atmosphere, in the form of the famous quasi-biennial oscillation (QBO) of the zonal winds in the equatorial lower to middle stratosphere. The overwhelming weight of evidence, from observation and modelling, says that this flow in the real atmosphere is a wave-driven mean flow, like the laboratory experiment, even though there are still, even today, large uncertainties as to exactly which wave types are significant (e.g. Baldwin et al., 2001), and large uncertainties as to the nature and strength of their sources and as to whether some of them propagate horizontally as well as vertically (Andrews and McIntyre, 1976; Dunkerton, 1983).

By a strange coincidence, the whole oscillation, with features descending at variable rates of the order of one or two scale heights per year, takes place in air *rising* at nearly the same rate, as illustrated in Fig. 2. Wave-induced angular momentum fluxes have to be almost twice what would otherwise be required (Gray and Pyle, 1989; Dunkerton, 1991). This point seems to have been missed in early work on the QBO, which also tended to consider too small a set of wave types. The important wave types include, but are not restricted to, the equatorial Kelvin wave and other vertically propagating waves trapped in an equatorial waveguide (Matsuno, 1966; Gill, 1982) by the "potential well" in the Coriolis parameter $f = 2|\Omega|\sin(\text{latitude})$, twice the vertical component of the earth's angular velocity Ω, which is picked out by the atmosphere's stable stratification $N^2 \gg 4|\Omega|^2$.

E. THE MICHELSON–MORLEY PRINCIPLE

So how, then, despite the uncertainties about wave types and wave sources, can we be so confident that the real QBO is a wave-driven mean flow? In my view an important part of the answer goes back to the seminal paper of Wallace and Holton (1968). It points up the importance of negative results in science—one might call it the Michelson–Morley principle—a principle neglected today, but neglected at our peril.

I have discussed this more fully elsewhere (McIntyre, 1994, 2000a); but the essence of the matter is that strenuous efforts to model QBO dynamics without wave-induced momentum transport—by scientists who well knew what they were doing and were well aware of observational constraints—failed in a significant way. That failure greatly adds to our confidence that wave-induced momentum transport is essential. Such transport had not yet been thought of at the time; and Wallace and Holton were led to a surprising, and no doubt disconcerting, conclusion—that it was impossible to explain the QBO without postulating a mysterious zonal force field that descended as time progressed and displayed an anti-frictional character. Despite being unable to guess what physical processes might give rise to such a force field, Wallace and Holton rather courageously published their negative result. It was only then recognized, increasingly clearly, through the work of Lindzen and others (Lindzen and Holton, 1968; Wallace and Kousky, 1968; Holton and Lindzen, 1972; Plumb, 1977), that there is just one kind of physical process in the atmosphere, wave-induced momentum transport, that can and should produce the otherwise mysterious force through filtering by wave refraction and Doppler shifting. With a suitable spread of phase speeds (Saravanan, 1990), one easily gets anti-frictional behaviour and a space–time signature of just the kind observed.

The essential behaviour does not, incidentally, depend on the existence of "critical levels" (an extreme case of wave filtering), nor on influence from above, by the semi-annual oscillation for instance, as was once thought (Plumb, 1977). These points are underlined by the results of the laboratory experiment. In the experiment, there are no critical levels, nor any phenomenon resembling the semi-annual oscillation. The experiment also illustrates anti-frictional behaviour in its purest form: the system is driven away from solid rotation by literally nothing other than the fluctuations generated by the oscillations of the lower boundary.

F. THE NONACCELERATION CONSTRAINT

Theoretical modelling of the feedback loop or wave–mean interaction giving rise to the anti-frictional behaviour is usually based on zonal averaging. The most convenient formulation uses the so-called TEM (transformed Eulerian-mean) equations, in which the rate of wave-induced momentum transport is quantified by an accurate standard diagnostic, the Eliassen–Palm flux (Andrews et al., 1987; Andrews and McIntyre, 1976). According to the TEM equations, one may picture the dynamics in terms of a zonally symmetric "mean fluid motion" in which notional rings of fluid are pushed eastward or westward by the wave-induced force, $\bar{F}$ say. The force is given quantitatively by the divergence of the Eliassen–Palm flux, which vanishes under so-called "nonacceleration conditions".

The most important ways for such conditions to be violated—this is sometimes called "breaking the nonacceleration constraint" (McIntyre, 1980; McIntyre and Palmer, 1985)—are for the waves to be breaking or otherwise dissipating. The nonacceleration constraint can be regarded as coming from Kelvin's circulation theorem, which is applicable to closed material contours or circuits undulated by the wave motion, such as the contour Γ sketched in Fig. 4. For instance the mean-flow changes in the Plumb–McEwan experiment are associated with changes in the Kelvin circulation, $\oint_\Gamma \mathbf{u} \cdot d\mathbf{s}$, caused by the viscous forces associated with wave dissipation. The Kelvin circulation would otherwise be invariant. Here $\mathbf{u}$ is the three-dimensional velocity field and $d\mathbf{s}$ the element of arc length. The role of material contours like Γ is also, incidentally, the reason why Lagrangian means arise naturally in the theory of wave–mean interaction (Andrews and McIntyre, 1978; Bühler and McIntyre, 1998; McIntyre, 2000a; Bühler, 2000).

G. EXTRATROPICAL LATITUDES

With wave-induced angular momentum transport so conspicuously important near the equator, as illustrated by the real QBO, it is perhaps not surprising that such transport should turn out to have a role in higher latitudes as well—though just how central a role was not recognized until recently, outside small communities of specialists who gradually developed the needed fluid-dynamical understanding. With hindsight, those developments can be traced back to (a) the work of Kelvin, Hough, and Rossby on what are now called Rossby waves or vorticity waves, which turn out to be the principal wave type involved; (b) the discovery by Rossby (1936, 1940) and Ertel (1942) of the concept of potential vorticity (PV), measuring the Kelvin circulation $\oint_{\Gamma'} \mathbf{u} \cdot d\mathbf{s}$, for small circuits Γ' lying on an isentropic surface; (c) the recognition by Charney (1948) and Kleinschmidt (1950) of what is now called PV invertibility, allowing Rossby-wave dynamics to be understood intuitively, beyond idealized cases (Hoskins *et al.*, 1985); (d) the recognition of large-scale anti-frictional behaviour, principally by Jeffreys in the 1920s and by Starr and co-workers in the 1950s; and (e) the attempts by Taylor, Rossby, and others (echoed for instance in Eady's 1950 RMS Centenary paper) to develop a "vorticity-transfer" view of turbulence stemming from Taylor's classic paper on eddy motion in the atmosphere (Taylor, 1915). The TEM theory, providing important conceptual simplifications within a coherent mathematical framework, began to take shape in the work of Eliassen and Palm (1961), Charney and Drazin (1961) and Dickinson (1969). Dickinson's seminal and perceptive work gave strong hints, moreover, that it would be fruitful to consider the waves and turbulence together—as complementary, interacting pieces of a single jigsaw puzzle. The first fully consistent theoretical model to capture all the essential aspects of such a jigsaw was the "nonlinear Rossby wave critical-layer theory" of Stewartson, Warn and Warn (Killworth and McIntyre, 1985), hereafter "SWW theory", which used matched asymptotic expansions to represent the dynamics of the interacting pieces for certain idealized cases of breaking Rossby waves. We return briefly to the SWW theory in Section J.

H. GYROSCOPIC PUMPING

Away from the equator the response to the zonal force $\bar{F}$ depends on Coriolis effects, which, as noted earlier, are immensely strong for present purposes. The equator is exceptional because everything is dominated there by the still stronger stable stratification, as reflected in the sine-latitude dependence of the Coriolis parameter f and in the large values of Prandtl's ratio

$$N/|2\Omega| \approx 10^2 \qquad (1)$$

with N the buoyancy frequency as before, making the horizontal component of Ω unimportant. Haynes (1998) has recently clarified what "away from the equator" means for present purposes, through scale-analytical estimates and numerical experiments with given zonally symmetric zonal forces $\bar{F}$. When typical parameters are substituted into his estimates, including height scales, say $\gtrsim 10$ km, it is found that "away from the equator" means further than about $15°$ latitude. This number is insensitive to the precise choice of parameters because of a quarter-power dependence on time and height scales and on radiative relaxation times. An interesting corollary of Haynes' results is that there is no need for the QBO to depend exclusively on forces $\bar{F}$ confined to the tropics.

Away from the equator, the most important part of the response to a given $\bar{F}$ is what is now called the "wave driving", or "gyroscopic pumping", of the mean meridional circulation. If a ring of fluid feels a persistent westward push, for instance, as suggested in Fig. 5, then Coriolis effects tend to deflect it persistently poleward—a mechanical pumping action that depends on the earth's rotation. This is the essence of what is happening at most altitudes and latitudes in Fig. 1. At top left, above the frame of Fig. 1, near the summer mesopause, the situation is similar apart from sign: because of gravity-wave filtering, the wave-driven push is eastward and the air is pumped equatorward. This in turn pulls air upward in the polar cap, producing the strong refrigeration already noted.

We can learn more about the fluid dynamics involved through thought experiments and numerical experiments with given forces $\bar{F}$, applied to a zonally symmetric model atmosphere away from the equator. Because of the tendency toward anti-frictional behaviour of the real wave-induced forces, this is arguably a more useful experiment than the kind in which $\bar{F}$ is taken to be a friction force. The mean circulation that arises in response to the given $\bar{F}$ can be identified with the "residual mean" meridional and vertical velocities $(\bar{v}^*, \bar{w}^*)$ that appear in the TEM equations.

The vertical velocity $\bar{w}^*$ is not, incidentally, the same as a Lagrangian mean. But it has in common with Lagrangian means—in the restricted circumstances in which the latter are well defined (Sparling *et al.*, 1997; Thuburn and McIntyre, 1997)—the property of pushing or pulling temperatures above or below radiative equilibrium temperatures T_{rad}, as assumed in Section B. This property can be seen from the TEM equations, in which eddy heat fluxes can be neglected under typical parameter conditions (Andrews *et al.*, 1987). It is this same property that was used in deriving Fig. 1 from observed temperatures, with the help of a sophisticated radiation scheme. The $\bar{w}^*$ field and the associated meridional velocity $\bar{v}^*$ required to produce the observed departures of zonal-mean temperatures from T_{rad} are what were actually calculated, with appropriate corrections to ensure mass conservation (Shine, 1989), then converted into the corresponding stream function. For

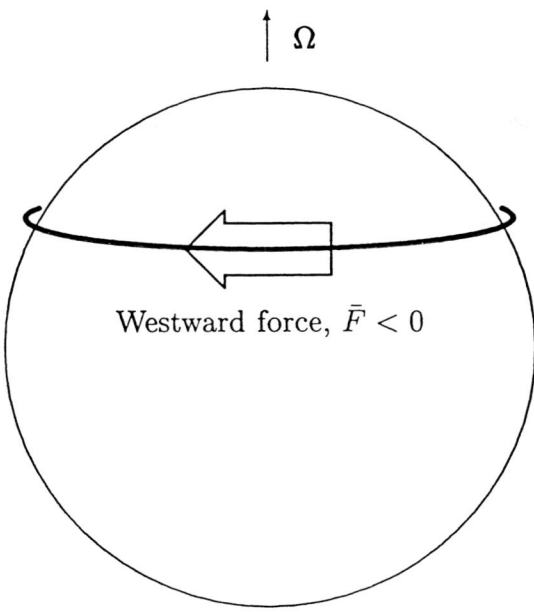

FIGURE 5 The gyroscopic pumping mechanism, in the case $\bar{F} < 0$ (appropriate to Rossby waves everywhere and to internal gravity waves in the winter mesosphere). When a ring of air located away from the equator is pushed westwards, Coriolis effects try to deflect it poleward. This is the mechanical pumping action that drives the mean circulation against radiative relaxation. The picture reminds us that the old idea of a sun-driven "Hadley" circulation in the middle atmosphere cannot, for instance, explain why the Brewer–Dobson cells look nothing like the Murgatroyd-Singleton cell at solstice.

related reasons, $\bar{w}^*$ is also a better measure of the mean vertical advection of chemical tracers than is the Eulerian mean, $\bar{w}$.

What happens when one starts the gyroscopic pump? More precisely, what do the equations say will happen in a thought experiment in which one starts exerting a given, zonally symmetrical, steady westward force, $\bar{F} < 0$, in some region occupying finite ranges of altitude and latitude away from the equator? To a first approximation the answer is that one sees a time-dependent adjustment in which the motion gradually settles toward steady poleward flow $\bar{v}^*$ through the forcing region, whose Coriolis force $f\bar{v}^*$ comes into balance with the force $\bar{F}$ as the steady state is approached. Away from the equator, this behaviour is robust.

In the final steady state, the mass streamlines close entirely downward (Haynes *et al.*, 1991, 1996). This last aspect of the response pattern is sometimes called the "downward control" property. The word "control" does not mean anything absolute, of course, but simply refers to what happens in this particular thought experiment, or numerical experiment, in which $\bar{F}$ is given and in which attention is focused on the circulation pattern $(\bar{v}^*, \bar{w}^*)$ that arises in response to the given $\bar{F}$.

The steady-state response can be pictured as resembling the lower part of Fig. 1, in a typical experiment, if we choose an $\bar{F}$ distribution with suitable negative values. If, for instance, we choose $\bar{F}$ to be negative within, but zero beneath, the frame of the figure then, during the adjustment toward the steady state, the lower part of the streamline pattern gradually burrows downward until it reaches the earth's surface, where it closes off in a frictional boundary layer. This downward burrowing is an aspect of the downward control property that will be important for our discussion of the solar interior.

The downward burrowing is fastest, and the steady state is approached fastest, for broad, deep, global-scale $\bar{F}$ distributions covering a large range of latitudes, having low-order zonally symmetrical Hough function structure. The approach to steadiness is then usually, for practical purposes, somewhat faster than seasonal timescales in the middle and upper stratosphere at least. Subglobal scales, on the other hand, represented by higher-order Hough functions, take much longer to adjust. Haynes *et al.* give a careful analysis and discussion of the finer points of the adjustment process and of its timescales and their relation to latitudinal structure and to f^{-1}, N^{-1}, and radiative relaxation times. There is some further discussion of the time-dependent behaviour in Holton *et al.* (1995) and in Haynes (1998).

In the most accurate form of the theory, the steady-state balance $\bar{F} = -f\bar{v}^*$ is replaced by $\bar{F} = -f_{\bar{M}}\bar{v}^*$ where $f_{\bar{M}}$ is the latitudinal gradient of $\bar{M}$, the zonal-mean specific angular momentum, divided by the perpendicular distance to the earth's axis. One can then have steady-state "sideways control" (Dunkerton, 1991), to the extent that the isopleths of $\bar{M}$ slope sideways as must happen most significantly in parts of the tropics, where f is relatively small.

The downward control property is related to the finiteness of the pressure and density scale heights, i.e., to the exponential falloff of pressure and density with altitude. Again glossing over some technicalities (Haynes et al., 1991), we have a situation in which the finite mass of atmosphere above the forcing region is incapable of absorbing infinite amounts of angular momentum. In the parameter regimes of interest the system solves this problem, so to speak, by approaching a steady state in which all the angular momentum being supplied by $\bar{F}$ is sent downward into the solid earth, whose moment of inertia is regarded as infinite for this purpose. Of course the ability to approach a steady state at all depends on the relaxational character of infrared radiation.

If one gets too close to the equator, then not only can the $\bar{M}$ isopleths slope sideways, but their pattern can be substantially reshaped as part of the response to $\bar{F}$. Advection of the relative contribution to $\bar{M}$, as distinct from the Coriolis contribution, makes the problem nonlinear. A possible consequence is that the response to the given $\bar{F}$ may no longer be robust, or even unique, though mass conservation still demands that air must be pulled up from somewhere not too far from the equator. This is at the research frontier of atmospheric dynamics and is not yet well understood, though progress is being made (Haynes, 1998; Scott and Haynes, 1998; Sankey, 1998; Plumb and Eluszkiewicz, 1999). There are simplified (shallow-water) models that do exhibit nonuniqueness in the form of hysteretical behaviour—multiple equilibria, regime flipping (Sankey, 1998)—in the upward path taken by the tropical or subtropical branch of the circulation that arises in response to a given $\bar{F}$, though we do not yet have a clear view of the significance for stratified flow.

The path taken is also influenced, within the tropics, by the distribution of radiative equilibrium temperatures T_{rad} and therefore of solar heating across the tropics. In terms of the tape-recorder analogy, changing the solar declination is a bit like moving the guidance wheels of the tape recorder, again affecting the path taken by the rising air within the tropics but not, to a first approximation, the total mass flux demanded by the gyroscopic pumping that is extracting the air from the tropics. Such a picture is consistent with the observational evidence from MSU channel 4, mentioned in Section B, on the seasonal variation in the upwelling mass flux (Yulaeva et al., 1994).

In the analogy we may think of the gyroscopic pumping action as the tape recorder's "motor". Of course from an old-time audio enthusiast's viewpoint the tape recorder is a poor quality one, with no "capstan" and with much "wow and flutter" in the form of seasonal, intraseasonal, and interannual variability. And the nonlinearity of the tropical upwelling problem, and its possible hysteretical behaviour, mean that the guidance wheels are sloppy and allow lots of "play" in the path taken.

I. ROSSBY AND GRAVITY WAVE BREAKING

Which wave types are most important for the global-scale circulation in the real extratropical middle atmosphere? Here, as hinted earlier, we think we know more than in the case of the tropics and the QBO. Below 50 km or so in the winter hemisphere, and below 25 km or so in the summer hemisphere, it seems clear from observational and modelling studies that the circulation is driven mainly by large-scale Rossby waves propagating or diffracting upwards.

Because of their strong chirality or handedness, tied to the sense of the earth's rotation, Rossby waves have ratchet-like properties that tend to produce persistently westward, or retrograde, mean forces whenever and wherever the waves are dissipated. This explains the persistently poleward mean motion in the Brewer–Dobson cells illustrated in Fig. 1. The two principal ways in which Rossby waves are dissipated are by infrared radiative damping and by wave breaking. Usually, in the real atmosphere, there is an intimate combination of both. For Rossby waves, as for all the other wave types of interest, "breaking" is most aptly defined as the rapid and irreversible deformation of those material contours that would otherwise undulate reversibly in the manner described by linearized wave theory (McIntyre and Palmer, 1984, 1985), as with the contour Γ sketched in Fig. 4. "Rapid" means on timescales comparable to an intrinsic wave period or less.

This definition of wave breaking is very widely applicable. It applies to the surface capillary–gravity waves of Fig. 3 and to breakers on Alaska beaches, as well as to Rossby waves and to the internal gravity waves of Fig. 4. It applies regardless of whether wave breaking is mediated by dynamical instabilities, as is usual with internal gravity waves, or not, as with plunging ocean-beach breakers. In ocean-beach breakers there is no identifiable instability process. What is common to all cases is that the irreversible deformation of material contours like Γ is another way to "break the nonacceleration constraint". When Rossby waves break, the resulting motion can often be described as layerwise two-dimensional turbulence, constrained, like the waves themselves, by the strong stratification. The typical consequence is again to drive the system away from, not toward, solid rotation as we shall see shortly.

Observational and model studies quantitatively diagnosing the Eliassen–Palm flux show that the forces due to breaking Rossby waves are indeed sufficient to drive the poleward flows in Fig. 1, in the Brewer–Dobson cells (Rosenlof and Holton, 1993; Holton et al., 1995), though internal gravity waves may make a contribution that could also be significant. Gravity waves may be especially significant in the summertime upper stratosphere where Rossby-wave activity is weak and $\bar{v}^*$ is small (Rosenlof, 1996; Alexander and Rosenlof, 1996), and perhaps also in the subtropical stratosphere where a relatively small $\bar{F}$ can produce significant gyroscopic pumping, because of the relatively small values of $\bar{M}$ gradients. Above 50 km or so, internal gravity waves take over as the most important wave type (e.g. Lindzen, 1981; Holton, 1982; Hamilton, 1997). Unlike Rossby waves, internal gravity waves can of course produce mean forces that are either prograde or retrograde, depending on wave filtering, as in the Plumb–McEwan experiment. Observations of internal gravity waves in the mesosphere, mostly by radar and lidar techniques, supplemented by rocket soundings, suggest that in order-of-magnitude terms, at least, such waves are well able to drive a Murgatroyd–Singleton circulation qualitatively like the heavy dashed curve in Fig. 1 and strong enough to make noctilucent clouds. The waves are observed to dissipate mainly by breaking due to unstable convective overturning, of heavy air over light, as clearly seen in the formation of adiabatic layers (e.g. Fritts et al., 1988) and in the mixing of certain chemical tracers such as atomic oxygen (C. R. Philbrick, personal communication).

J. THE JIGSAW PUZZLE: BAROTROPIC MODELS

Rossby-wave dynamics, including propagation, breaking, and the associated momentum and angular-momentum transport—the whole wave–turbulence jigsaw puzzle—can be illustrated by the simple textbook case of barotropic nondivergent vortex dynamics in Rossby's "beta plane" or "approximately flat earth" model. In place of the PV we have the absolute vorticity $Q(x,y,t)$, say. Invertibility holds exactly: a stream function $\psi(x,y,t)$ describing the incompressible velocity field can be derived by from the Q field by inverting a Poisson equation, given suitable boundary conditions such as evanescence at infinity:

$$\psi = \nabla^{-2}(Q - f) \qquad (2)$$

Here $f(y)$ is again the Coriolis parameter, and

$$\mathbf{u} = (u,v), \quad u = -\partial\psi/\partial y, \quad v = \partial\psi/\partial x \qquad (3)$$

where (x,y) are eastward and northward Cartesian coordinates and (u,v) the corresponding components of the velocity vector $\mathbf{u}(x,y,t)$. There is just one evolution equation

$$DQ/Dt = 0 \qquad (4)$$

where D/Dt is the two-dimensional material derivative, defined by

$$D/Dt = \partial/\partial t + \mathbf{u}\cdot\nabla = \partial/\partial t + u\partial/\partial x + v\partial/\partial y \qquad (5)$$

If desired, one can model dissipation by adding terms to the right of equation (4). Note incidentally that invertibility, as always, entails nonlocality, i.e. instantaneous action-at-a-distance—perfectly accurate in this case because all fast waves, such as acoustic waves and gravity waves, have been suppressed by giving them infinite propagation speeds. In this implicit sense, the system is infinitely stiff: there is no delay in the pushing of one fluid element by another, all the way to infinity. Note also the single time derivative, in the single evolution equation (4), pointing to the chirality and ratchet-like properties of Rossby waves.

We may regard the system (2)–(5) as describing motion on an isentropic or constant-θ surface in a strongly stratified atmosphere, in the limit of infinite buoyancy frequency N, where θ is potential temperature. Figure 6 shows the characteristic pattern of a Rossby-wave disturbance, in which constant-Q contours, $\Gamma_{Q\theta}$, say, lying on the isentropic surface—which according to (4) are also material contours—undulate in the $\pm y$-direction, corresponding to north–south on the real earth and implying a pattern of Q anomalies from which inversion, (2), (3), yields a velocity field qualitatively as shown by the solid arrows (Hoskins et al., 1985). If one now makes a mental movie of the resulting motion, one sees that the undulations will propagate westward because the velocity field is a quarter wavelength out of phase with the displacement field. This is the Rossby-wave propagation mechanism, sometimes called "quasi-elasticity" to emphasize that it is a restoring effect, tending to return parcels to their equilibrium latitudes, but perhaps better called "semi-elasticity" because of the chirality.

The standard Rossby wave theory—linearize equation (4) about rest, take $\beta = df/dy =$ constant, look for

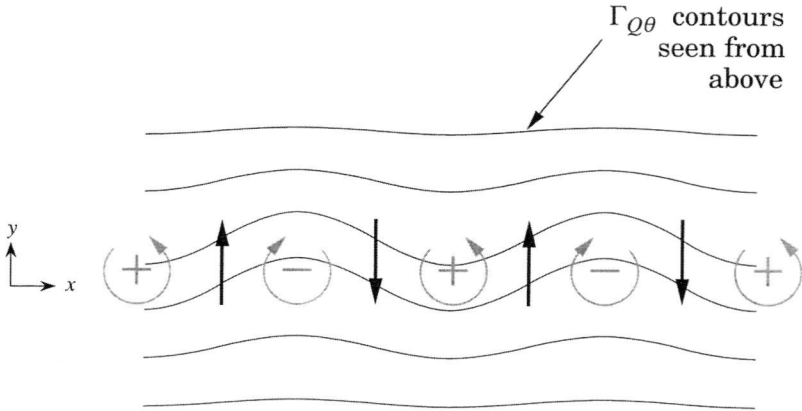

FIGURE 6 Plan view of the chiral or one-way restoring mechanism or "quasi-elasticity" to which Rossby waves owe their existence (diagram taken from Hoskins et al., 1985). The mechanism has a central role in nearly all large-scale dynamical processes in the atmosphere and oceans, including barotropic and baroclinic instabilities, vortex coherence (e.g. blocking), and vortex interactions.

solutions $\propto \exp(ikx + ily - i\omega t)$ so that $\nabla^{-2} = -(k^2 + l^2)^{-1}$, and deduce $\omega = -\beta k(k^2 + l^2)^{-1}$—presumes that the sideways displacements and slopes of the undulating contours $\Gamma_{Q\theta}$ are small and that the undulations are reversible. Rossby-wave breaking is the opposite extreme situation, in which this condition of reversible undulation is violated, and the contours $\Gamma_{Q\theta}$ deform irreversibly; the relevant arrow-of-time "paradox" is discussed in McIntyre and Palmer (1984).

Now if the Rossby wave undulations take place in a mean or background shear flow $\bar{\mathbf{u}} = (\bar{u}(y), 0)$, and have a monochromatic zonal (x) phase speed that coincides with the background flow velocity at some value of y—by convention called a critical line—then there is always some surrounding region in which Rossby-wave breaking, in the foregoing sense, is important. This region is sometimes called a "Rossby-wave surf zone". Such zones can be broad or narrow according as the wave amplitude is large or small. In the narrow case, the SWW theory and its further development in Killworth and McIntyre (1985) and in Haynes (1989) has given us a comprehensive understanding of how the wave field and surf zone interact, when modelled using equations (2)–(5). The results illustrate in detail how the nonacceleration constraint can be broken even in a dissipationless fluid system such as (2)–(5).

The idealized problems considered in the SWW theory and its further development are problems in which a small-amplitude Rossby wave, excited by an undulating side boundary, encounters a critical line in a basic flow initially having constant shear. A narrow surf zone forms in which the contours $\Gamma_{Q\theta}$ deform irreversibly in a recirculating flow, a so-called Kelvin cat's-eye pattern, straddling the critical line. As the contours wrap around, the Q distribution changes and, partly through the action-at-a-distance implied by the inversion, i.e. by the ∇^{-2} operator in equation (2), induces a change in the flow outside the surf zone. This affects phase gradients with respect to y in the outer flow, hence trough tilts and Reynolds stresses $\overline{u'v'}$, in such a way that the surf zone appears as an absorber to the Rossby waves outside it during the early stages of wave breaking, but later becomes a reflector because of the finite scope for rearrangement of the Q distribution within a surf zone of finite width. The way in which the surf zone interacts with its surroundings is precisely expressed by the use of matched asympotic expansions. Full details of the analysis are too lengthy to reproduce here but can be found in the review material of Killworth and McIntyre (1985), which also establishes a general theorem implying that the absorption–reflection behaviour is generic.

The essence of what happens is, however, very simple, and is summarized in Fig. 7. The sketch on the left, Fig. 7a, shows the background $\bar{Q}$ profile, $\bar{Q} = \text{const.} + \beta y$, superposed on the result of rearranging it in such as way as to homogenize it over some finite y-interval. This mimics in simplified form what happens in the surf zone. Figure 7(b) is the zonally (x-)averaged $\bar{Q}$ profile from an actual solution, in a case where the flow in the surf zone is chaotic because of the onset of secondary instabilities (Haynes, 1989, and personal communication). To get the corresponding zonal mean momentum change, the inversion implied by (2)–(3) is trivial: one merely has to integrate with respect to y the mean change $\delta \bar{Q}$ in $\bar{Q}$:

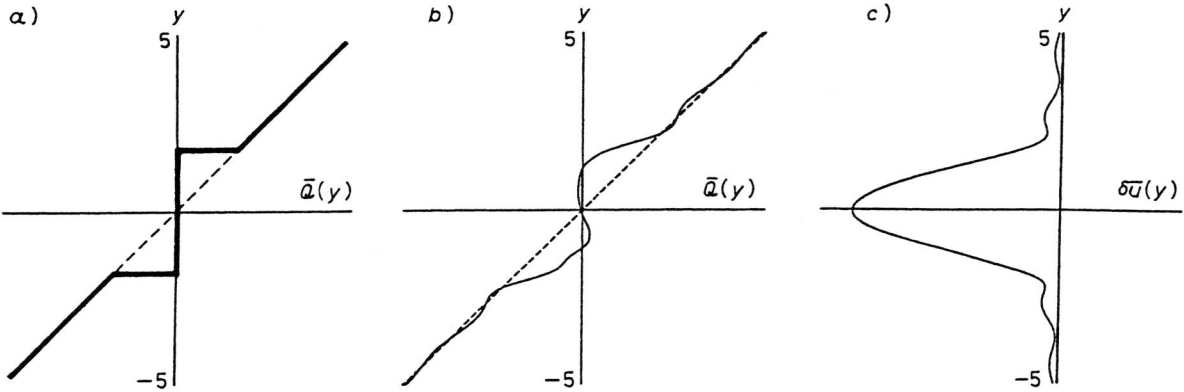

FIGURE 7 The relation between mean wave-induced force and Q-rearrangement by a breaking Rossby wave, in the simplest relevant model system, the dynamical system (2)–(5). (Courtesy P. H. Haynes; for mathematical details see Killworth and McIntyre (1985) and Haynes (1989).) Plot (a) shows idealized $\bar{Q}$ distributions before and after mixing Q in some y-interval or latitude band; (b) shows the x-averaged $\bar{Q}$ distribution in an actual model simulation using equations (2)–(5); (c) shows the resulting mean momentum deficit, equation (6), whose profile would take a simple parabolic shape in the idealized case corresponding to (a). Notice that the momentum has been transported against its own mean gradient, because the wave source is at positive y and the background shear is positive.

$$\delta\bar{u}(y) = \int_y^\infty \delta\bar{Q}(\tilde{y})d\tilde{y} \qquad (6)$$

The result is shown in Fig. 7c. The momentum change corresponding to the simplified, left-hand $\bar{Q}(y)$ profile in Fig. 7a is a parabolic shape, not shown, qualitatively similar to the right-hand graph. The parabolic shape results from integrating the linear $\delta\bar{Q}$ profile. More generally, it is plain from (6) that any Q rearrangement that creates a surf-zone-like feature with weakened mean Q gradient will give rise to a momentum deficit. This is another clearcut example of anti-frictional behaviour, because $\delta\bar{u}(y)$ has to be added to a background flow $\bar{u}(y) \propto y$ with positive shear $\partial\bar{u}(y)/\partial y > 0$ (not shown in the figure), and the wave source is located at positive y, outside the domain of the figure. Thus the momentum has been transported against its own mean gradient.

The mean momentum tendency is given by
$$\frac{\partial \bar{u}}{\partial t} = -\frac{\partial}{\partial y}(\overline{u'v'}) = \overline{v'Q'} \qquad (7)$$
where the overbars denote the Eulerian zonal mean and primes denote fluctuations about that mean. This makes explicit the relation between the mean momentum tendency, the wave-induced Reynolds stress convergence, and the eddy flux of Q measuring the rate of rearrangement of the $\bar{Q}$ distribution in the surf zone. The second equality, a corollary of (3) and the relation $Q' = \partial^2\psi'/\partial x^2 + \partial^2\psi'/\partial y^2$ from (2), was given in Taylor's (1915) classic paper and is often called the "Taylor identity".

The essential point, then, is that whenever the Q distribution is rearranged by Rossby-wave breaking, a momentum deficit appears—signalling the presence of a retrograde force due to irreversible wave-induced momentum transport. This illustrates not only anti-frictional behaviour but also the ratchet-like character of the whole process, ultimately a consequence of the single time derivative in (4). The momentum change where Rossby waves dissipate by breaking is always retrograde, i.e., against the background rotation, westward in the case of the earth. If there is some forcing effect that tends to restore the background Q gradient (requiring a nonzero term on the right of (4)), then the Q rearrangement and the irreversible momentum transport can persist.

A feature that is robust, and appears to carry over to larger-amplitude cases, is the strong spatial inhomogeneity of the "jigsaw puzzle". The flow is more "wavelike" in some places (the outer flow) and more "turbulent" in others (the surf zone), where "turbulent" means that the material contours deform irreversibly. Each region strongly affects the others dynamically. More realistic cases with larger wave amplitude typically show the same generic features. Figure 8 shows some material tracer fields for such a case, mimicking conditions in the real lower to middle stratosphere by going to a more realistic spherical geometry and to divergent (shallow-water) barotropic flow (see caption and Norton, 1994; McIntyre and Norton, 2000). The inhomogeneity is very striking indeed. The midlatitude surf zone is strongly turbulent, while its borders are relatively wave-like. This is especially so at the inner

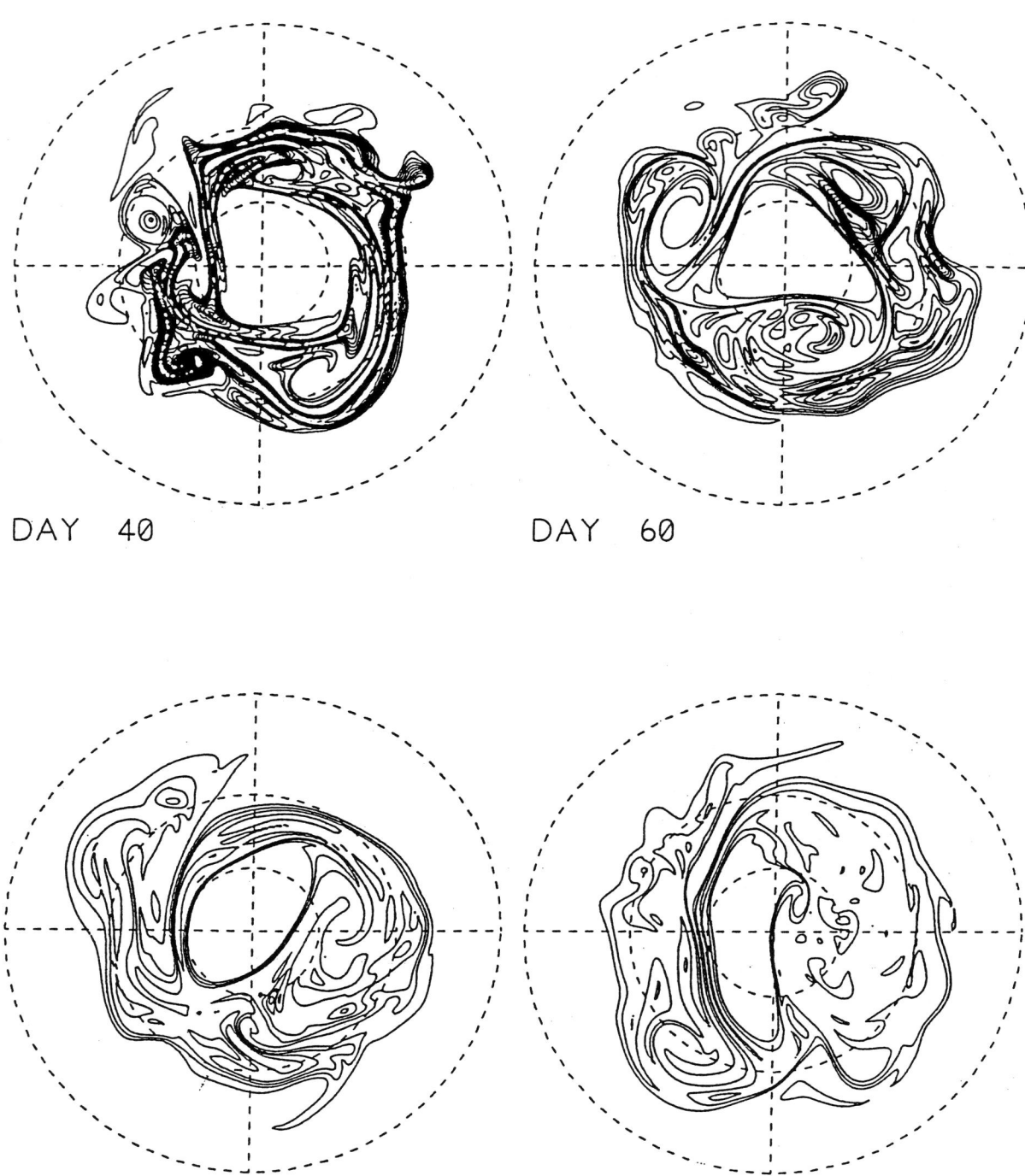

FIGURE 8 Model flow on the sphere, closely resembling flow in the real wintertime stratosphere at altitudes around 25 or 30 km. The map projection is conformal (polar stereographic), with the equator and the 30°N and 60°N latitude circles shown dashed. The flow is visualized by passive tracer released as a compact blob into the midlatitude stratospheric surf zone, clearly showing the fast two-dimensional turbulent mixing in that region, despite which the stratospheric polar vortex remains almost completely isolated from its surroundings, and likewise, to a lesser extent, the tropics (helping to explain what is seen in Fig. 2). The isolation of the (core of the) polar vortex recalls classic smoke rings and is of great importance to stratospheric polar chemistry, including the Antarctic ozone hole and its (so far less severe) Arctic counterpart. The isolation is due to the combined effects of the Rossby-wave restoring mechanism and the strong shear just outside the edge. (Courtesy of Dr W. A. Norton, from whom an animated video of the model run is available (Dept of Atmospheric, Oceanic, and Planetary Physics, Clarendon Laboratory, Parks Road, Oxford, OX1 3PU, UK).) Details of the model and the model run are given in Norton (1994). A "shallow-water" model is used, with mean depth 4 km, giving behaviour qualitatively close to that of (2) except that the inversion operator is replaced by one with shorter range (Rossby length ≈2000 km in middle latitudes); for justification of this last assertion, see McIntyre and Norton (2000).

border of the surf zone, marking the edge of the polar vortex—whose resilence makes it appear to behave, in animated versions of Fig. 8, almost like an elastic band. This is a clear example of what is now called an "eddy-transport barrier". The barrier effect results from the combined effects of the Rossby-wave restoring mechanism and strong shear just outside the polar vortex (Juckes and McIntyre, 1987). All this has implications not only for chemistry, in a well-known way, but also, for instance, for the dynamics of the much-discussed Holton–Tan effect, as will be noted shortly.

K. GENERALIZATION TO REALISTIC STRATIFIED FLOW

The whole picture just summarized generalizes to layerwise two-dimensional motion in a finitely stratified atmosphere provided that we reinterpret Q as the Rossby–Ertel PV

$$Q = \rho^{-1}(2\Omega + \nabla \times \mathbf{u}) \cdot \nabla \theta \quad (8)$$

where ρ is mass density, and replace ∇^{-2} in equation (2) by a more complicated, three-dimensional inversion operator (quasi-geostrophic or higher-order (Hoskins et al., 1985; Davis, 1992)). Qualitatively, everything is much the same as before: equation (4) is unchanged, and its single time derivative implies the same chirality as before; Rossby quasi-elasticity still works as shown in Fig. 6; and the rearrangement of PV on isentropic surfaces is still robustly associated with an angular momentum deficit (Robinson, 1988), and with ratchet-like, irreversible angular momentum transport in the sense required to drive the Brewer–Dobson cells. The Taylor identity still holds except that the Reynolds stress divergence is replaced, to a first approximation, by a form of the Eliassen–Palm flux divergence.

The wave–turbulence inhomogeneity illustrated in Figs 7 and 8 is also found to be typical of the real, stratified atmosphere. One reason for the generic character of the inhomogeneity is its tendency to be self-reinforcing. Where Q contours are crowded together, typically at the edges of surf zones—idealized as the "corners" in the $\bar{Q}$ profile in Fig. 7a—one has a strengthening of the Rossby quasi-elasticity. Even more importantly, there is a weakening within surf zones. Thus mixing becomes easier in surf zones once they begin to be established, and *vice versa*. Other things being equal, then, the inhomogeneity tends to perpetuate itself. The structure can of course be changed by sufficiently drastic changes in circumstance, such as the introduction of strong Rossby-wave motion with a different phase speed from the original one, tending to carve out a surf zone near a different y value in the background shear (Bowman, 1996). Likewise, the QBO-related changes in tropical winds can change relative or intrinsic phase speeds and hence modulate the tropical and subtropical structure, a fact that, together with the absorption–reflection behaviour of surf zones, must change the dynamics of the winter-stratospheric "Rossby-wave cavity" (Matsuno, 1970). Such a mechanism of modulation by the QBO—a highly nonlinear mechanism—probably underlies the celebrated Holton–Tan effect (Baldwin et al., 2001).

Another reason to expect the wave–turbulence inhomogeneity is, as I have argued more carefully elsewhere (McIntyre, 1994, 2000a), the fundamental integral constraint on PV rearrangement on a (topologically spherical) isentropic surface S enclosing the earth. Using the definition (8) together with Stokes' theorem on the closed isentropic surface S, we have

$$\iint_S bQ\,dA = 0 \quad (9)$$

where dA is the area element and b is defined such that $bdAd\theta$ is the mass element. That is, $b = \rho/|\nabla\theta|$. Because b is positive definite, (9) tells us at once that the only way to mix Q to homogeneity on the surface S is to make Q zero everywhere—a fantastically improbable state on a planet as rapidly rotating as the earth. Since real Rossby waves do break, and do mix Q, they must do so imperfectly and produce spatial inhomogeneity and anti-frictional tendencies.

To be sure, one can imagine a thought experiment in which the air on and near the isentropic surface S begins by rotating solidly with the earth and then has its angular velocity uniformly reduced by some incident Rossby-wave field, in such a way as to give a uniformly reduced pole-to-pole latitudinal profile of $\bar{Q}$. But the tailoring of a Rossby-wave field to do this would be a more delicate affair than standing a pencil on its tip; and the natural occurrence of such a wave field would be another fantastically improbable thing. To summarize, then, these fluid systems naturally exhibit a strong wave–turbulence inhomogeneity, and with it a marked tendency for the associated wave-induced momentum transport to behave anti-frictionally, in the sense of driving the system away from, not toward, solid rotation.

L. THE SUN'S RADIATIVE INTERIOR

1. Inevitability of a Magnetic Field

When put together with new helioseismic data, our understanding of stratospheric circulations has recently enabled us to learn something new and significant about the sun's deep interior. Here I mean the radiative interior, at radii $< 5 \times 10^5$ km, as distinct from the overlying convection zone, whose upper surface is the visible surface at about 7×10^5 km. We can now answer in the affirmative, with high confidence, an age-old question: does a dynamically significant magnetic field pervade the interior? The mere existence of such a field has immediate and far-reaching consequences, as will be explained, for our ability to sharpen theoretical hypotheses about other aspects of the sun's interior and to make inferences from helioseismic inversion.

Of course the possibility that there *could* be such an interior field has long been recognized, if only because magnetic diffusion times ($\sim 10^{10}$ y) are comparable to, or somewhat longer than, the age of the sun (Cowling, 1945), the latter currently being estimated as close to 4×10^9 y. Thus the interior needs no dynamo action: it could contain a magnetic field left over from the sun's formative years. Indeed, when the sun began to form as a protostar, the interstellar magnetic field must have been greatly concentrated as material collapsed inward (e.g., Gough, 1990, and references therein), dragging the field with it and spinning up to high rotation rates. On the other hand, it is also possible, and has often been hypothesized, that any such interior field was expelled or annihilated long ago—perhaps by some kind of turbulent motion within the early sun—leaving only the rapidly oscillating field associated with the convection zone and its 22-year cycle, for which dynamo action is a plausible explanation. Be that as it may, models of solar spindown—the sun's evolution from its early, rapidly rotating state, exporting angular momentum through the solar wind—have tended to assume that the interior was free of dynamically significant interior magnetic fields for most of the relevant timespan of 4×10^9 y (see, for instance, Spiegel and Zahn, 1992, and references therein).

The argument now to be sketched (for more detail, see McIntyre, 1994 and Gough and McIntyre, 1998) depends crucially on recent helioseismic results concerning the sun's differential rotation (e.g. Thompson *et al.*, 1996; Kosovichev *et al.*, 1997). These strongly indicate, for one thing, that classical laminar spindown models will not fit observations. The convection zone is in quite strong differential rotation, as suggested schematically in Fig. 9 and its caption. But the radiative interior, by contrast, turns out to be approximately in solid body rotation except possibly near the poles. (The helioseismic technique, which depends on observing the rotational splitting of acoustic vibrational modes of the sun, is insensitive to details near the rotation axis.) Now this near-solid interior rotation, of which the first indications began to emerge about a decade ago, is incompatible not only with classical laminar spindown but also with practically any other purely fluid-dynamical picture. How can anyone claim such a thing? The basis is our observational and theoretical knowledge of terrestrial middle-atmospheric dynamics.

The sun's interior is strongly stratified, and, for present purposes, is a fluid-dynamical system very like the terrestrial stratosphere, with short buoyancy and Coriolis timescales. Even the ratio of those timescales is similar: throughout most of the interior, Prandtl's ratio $N/2|\Omega| \approx 10^2$. So, in a laminar spindown model for instance, the dynamics is qualitatively like that in a zonally symmetrical terrestrial model stratosphere. The main difference is one of detail: because photon mean free paths are small, radiative heat transfer is diffusive, and also rather weak for fluid-dynamical purposes (diffusivities $\sim 10^3$ m^{2}s^{-1}).

Now if laminar spindown were the only thing happening, it would drive the interior away from solid rotation toward some differentially rotating state that matches the differential rotation of the convection zone. If, on the other hand, layerwise two-dimensional turbulence were present in the interior, perhaps through the breaking of Rossby waves emitted from the convection zone, then the effect would again be to drive the interior away from solid rotation in, probably, some other way (McIntyre, 1994), subject to the same kinds of dynamical constraints as those with which we are familiar in the terrestrial stratosphere, not least the integral constraint (9). Broadband gravity waves emitted from the convection zone would tend to do the same thing, drive the interior away from solid rotation, in yet another way, through anti-frictional behaviour as in the terrestrial QBO. In summary, one can get all kinds of forms of departures from solid rotation, but one cannot get solid rotation out of the fluid dynamics except by some fantastically improbable accident. Recall again, for instance, the spontaneous symmetry-breaking in the Plumb–McEwan experiment.

I am claiming, therefore, that there appears to be no purely fluid-dynamical process that can make the interior rotate nearly solidly as observed. This is the basis for concluding that there must be an interior poloidal magnetic field. Such a field, a few lines of which are sketched in the right half of Fig. 9, can do the job very easily, through Alfvénic torques. Moreover, it is the only known way to do the job—to stop the interior differential rotation that fluid-dynamical processes would otherwise bring about.

Recent refinements to the helioseismic results, from the MDI (Michelson Doppler Interferometer) instrument on the SOHO spacecraft (Solar and Heliospheric Observatory) have strongly confirmed the near solid interior rotation and have furthermore begun to re-

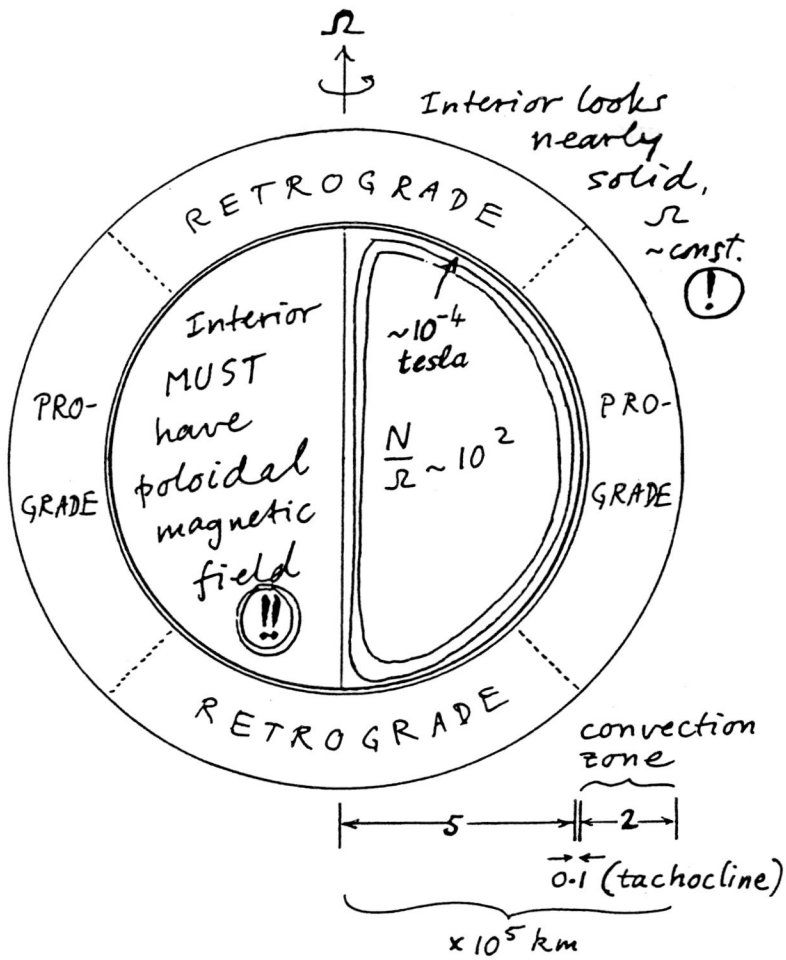

FIGURE 9 Schematic of proposed new model for the sun's interior (Gough and McIntyre, 1998). The strong differential rotation of the convection zone, now well known from helioseismic inversion (Thompson et al., 1996; Kosovichev et al.,1997), is roughly depth-independent and is indicated schematically by the words "PROGRADE" and "RETROGRADE". Angular velocities range over about ±10% of an intermediate value, $|\Omega| \approx 2.7 \times 10^{-6}$ s^{-1}, close to the typical interior value. Interior rotation rates are much more nearly uniform. The poloidal magnetic field occupying the radiative interior is a necessary dynamical ingredient, crucial to explaining the observed depth of the tachocline, the thin shear layer separating the base of the convection zone from the interior, whose depth Δ is roughly of the order 0.1×10^5 km; cf. the sun's radius, 7×10^5 km. The estimate of magnetic field strength $|B| \approx 10^{-4}$ T (tesla) is an extremely rough, preliminary estimate, refinement of which will depend on intricate nonlinear modelling not yet done, and on the fullest possible use of helioseismic data now being accumulated. Because of the expected approximate scaling (Gough and McIntyre, 1998), which turns out to involve a ninth power law, $|B| \propto \Delta^{-9}$, it will be especially crucial to work toward refined estimates of Δ. The latest attempt at a quantitative estimate (Elliott and Gough, 1999) tentatively replaces 0.1×10^5 km by 0.13×10^5 km. The next refinement will be to make Δ a function of latitude.

solve the detailed vertical structure at the top of the interior, just under the base of the convection zone. There is strong vertical shear in a layer of small but now just-resolvable thickness, ≲2% of the solar radius; but the magnitudes involved (Richardson numbers $N^2/|\text{vertical shear}|^2 \gg 1$) imply overwhelmingly strong control by the stable stratification, leading to layerwise two-dimensional motion. Older ideas that the region might be shear-unstable, and three-dimensionally turbulent, appear to be definitely ruled out. Consequently, the shear layer, which solar physicists call the "tachocline", must have a relatively quiescent dynamical character like the rest of the interior, i.e., at worst layerwise two-dimensionally turbulent like the wintertime terrestrial stratosphere, or possibly as quiescent, even, as the summertime upper stratosphere.

On this basis, and from the helioseismic results, one can estimate the circulation of the tachocline in much the same way as the circulation shown in Fig. 1. It turns out that the circulation or ventilation time is about 10^6

times longer: not several years but more like several million years (Gough and McIntyre, 1998). But this is a mere instant in comparison with 4×10^9 y, the age of the sun and the timescale of its spindown. Rates of change due to spindown can therefore be neglected in theories of tachocline structure and dynamics: the dynamics can be treated as quasi-steady. Furthermore, wave-induced forces, which are hard to estimate but are generally thought to be significant, if at all, on longer timescales ~10^8 y to 10^9 y (see Garcia López and Spruit, 1991, and references therein) can likewise be neglected, almost certainly, in the tachocline dynamics with its 10^6 y circulation timescale.

The only remotely plausible candidate for driving a circulation that is so fast, relative to the other timescales involved, is gyroscopic pumping by Reynolds stresses within the dynamically vigorous convection zone above. As in the terrestrial stratosphere, this must produce a downward-burrowing circulation. The burrowing has to be stopped, however, not by a frictional boundary layer at a solid surface but by a magnetic diffusion layer capping the interior magnetic field. If the burrowing is not stopped somehow, then we are back to a classical laminar spindown scenario, which would produce a tachocline much deeper than observed (Haynes *et al.*, 1991; Spiegel and Zahn, 1992; Gough and McIntyre, 1998).

2. The Tachopause and the Lithium and Beryllium Problems

An idealized boundary-layer theory for the magnetic diffusion layer (Gough and McIntyre, 1998) confirms that the magnetic field is, indeed, capable of stopping the burrowing, and furthermore that the boundary layer is thin, more than an order of magnitude thinner than the tachocline itself. This is turn implies that the tachocline must end in a relatively sharp "tachopause". Such information can be fed back into helioseismic inversion models. In particular, the fast tachocline circulation implies that there has to be a near-discontinuity in helium abundance, hence sound speed, at the tachopause; such a structure has recently been shown to fit the helioseismic data very well and to lead to a refined estimate of the tachocline thickness (Elliott and Gough, 1999); see caption to Fig. 9.

The boundary layer theory also gives us a first, rather crude, estimate, *ca.* 10^{-4} T, of typical magnetic field strengths just below the tachopause. Stronger fields would imply smaller tachocline depths Δ, and vice versa. However, this estimate is likely to need revision, partly because it turns out that Δ is only weakly dependent on field strength (caption to Fig. 9), and partly because a quantitative estimate will require solution of a rather complicated, and strongly nonlinear, magnetohydrodynamic problem for the upwelling branch of the tachocline circulation. This is work in progress. In the upwelling branch, which has to be in middle latitudes to be consistent with the observed pattern of differential rotation, we expect the interior field to be skimmed up into the convection zone, where it will be subject to reconnection. There will probably be significant zonal Lorenz forces in this upwelling region.

The implications are, however, already far-reaching in at least three senses. First, if the interior is constrained by the simplest possible magnetic field, well diffused from its primordial state and aligned with the rotation axis, hence in the form of a poloidal field with field lines configured as in Fig. 9, as if wrapped around a torus, or doughnut, then what is known as *Ferraro's law of isorotation* applies. This says that the angular velocity on the surface of each torus containing closed field lines must be constant, which is equivalent to saying that no Alfvénic torsional oscillations are excited. (There is an argument saying that such torsional oscillations will be damped rather quickly relative to the sun's age, essentially because of the near-degeneracy of the problem, with singular eigenfunctions in the form of a delta function confined to the surface of one torus (Hollweg, 1978; McIntyre, 1994.)) Because some of the toroidal surfaces penetrate deep into the interior, this offers for the first time a chance of obtaining information about differential rotation near the sun's core—information that is unobtainable directly from helioseismic inversion.

The second implication is a better understanding of spindown itself. The spindown problem under the constraint of Ferraro's law (with or without wave-induced forces) becomes very different from the classic spindown problem, again bringing new understanding to what we observe of the present differential rotation.

The third is a new possibility for solving the notorious lithium-burning problem (e.g. Garcia López and Spruit, 1991; S. Vauclair, personel communication). To explain observed lithium abundances at the surfaces of various populations of stars, one needs constituent-transporting mixing mechanisms or circulations that reach below the tachocline. The strong gyroscopic pumping by the convection zone might have such a role. On the simplest picture sketched in Fig. 9, there are just two places where the circulation might be able to burrow to sufficient depth for the purpose: they are the two poles, where the magnetic field must vanish, if confined by the convection zone. The downward-burrowing circulation might therefore be able to "dig" a kind of "polar pit" in the magnetic field, at each pole (Gough and McIntyre, 1998), penetrating much more deeply than elsewhere. It can probably afford to take its time over this, say $\lesssim 10^9$ y. This could still be enough to burn the lithium. Even if the interior magnetic field were more complicated, the hairy-sphere theorem tells us that weak points vulnerable to "pit digging" must exist. Detailed predictions will again depend on numerical solution of a highly nonlinear magnetohydrodynamic problem, yet to be attempted. There is a similar problem with beryllium, sharpening the observational constraints on the numerical modelling in the same way as with multiple chemical species in the earth's middle atmosphere.

I shall end on a much more speculative note. It is conceivable that, despite their generally high damping rates, interior Alfvénic torsional oscillations might nevertheless be significant especially if the field somehow configures itself to minimize the damping. Could there be, one wonders, a QBO-like torsional oscillation in the sun's interior excited by broadband gravity waves from the convection zone, despite the likely weakness of such waves? This would be a bit like the balance wheel of a watch. My present feeling is that it is less likely than when I argued for it in 1994 (McIntyre, 1994)—but then again, people who study isotopes in palaeoclimatic records tell us that there is evidence for solar oscillations on many timescales, centuries to millennia, far slower than the sunspot cycle. Perhaps all these fluctuations on various timescales originate, without exception, in the convection zone. That is entirely possible, in such a strongly nonlinear dynamical subsystem. But then again, perhaps some of these time scales come from the radiative interior. They are compatible with the range of possible interior field strengths.

ACKNOWLEDGEMENTS

This chapter draws on material from invited lectures given previously to the 12th Yukawa Memorial Symposium held in Nishinomiya, Japan (McIntyre, 1998), to the European Space Agency (McIntyre, 1999), and to the 1999 IUGG Symposium on solar variability (McIntyre, 2000b). The Yukawa Symposium was especially important in reminding me of problems in other branches of physics, engineering, and biology, such as molecular motors, that depend in interesting ways on the dynamical organization of fluctuations. I thank Joan Alexander, David Andrews, Dave Broutman, Oliver Bühler, Keith Browning, David Dritschel, Patrick Espy, Michael Gadsden, Mark Hallworth, Björn Haßler, Peter Haynes, Tony Hollingsworth, Jim Holton, Brian Kerridge, Ernst Kopp, Adrian Lee, Philip Mote, Warwick Norton, Tim Palmer, Russ Philbrick, Alan Plumb, John Pyle, Clive Rodgers, Karen Rosenlof, David Sankey, Richard Scott, Adrian Simmons, Keith Shine, Claude Souprayen, Chris Warner, Darryn Waugh, Georg Witt, Michio Yamada and Shigeo Yoden for valuable help, stimulation, comments and correspondence. My research has received generous support from the UK Natural Environment Research Council through the UK Universities' Global Atmospheric Modelling Programme, from the Isaac Newton Institute for Mathematical Sciences, and from a SERC/EPSRC Senior Research Fellowship.

References

Albritton, D. L. et al., 1998: *Scientific Assessment of Ozone Depletion 1998* (WMO Global Ozone Research and Monitoring Project Report No. 44). Geneva, World Meteorol. Org. Available from WMO, GO$_3$OS, P.O. Box 2300, 1211–Geneva–2, Switzerland.

Alexander, M. J. and J. R. Holton, 1997: A model study of zonal forcing in the equatorial stratosphere by convectively induced gravity waves. *J. Atmos. Sci.*, **54**, 408–419.

Alexander, M. J. and K. H. Rosenlof, 1996: Nonstationary gravity wave forcing of the stratospheric zonal mean wind. *J. Geophys. Res.*, **101**, 23465–23474.

Andrews, D. G. and M. E. McIntyre, 1976: Planetary waves in horizontal and vertical shear: the generalized Eliassen–Palm relation and the mean zonal acceleration. *J. Atmos. Sci.*, **33**, 2031–2048.

Andrews, D. G. and M. E. McIntyre, 1978: An exact theory of nonlinear waves on a Lagrangian-mean flow. *J. Fluid Mech.*, **89**, 609–646.

Andrews, D. G., J. R. Holton and C. B. Leovy, 1987: *Middle Atmosphere Dynamics*. Academic Press, London, New York.

Baldwin, M. P. et al., 2001: The quasi-biennial oscillation. *Revs. Geophys.*, **39**, 179–229.

Bowman, K. P., 1996: Rossby wave phase speeds and mixing barriers in the stratosphere. Part I: Observations. *J. Atmos. Sci.*, **53**, 905–916.

Brasseur, S. and S. Solomon, 1984: *Aeronomy of the Middle Atmosphere*. Reidel, Dordrecht.

Bretherton, F. P., 1969: Waves and turbulence in stably stratified fluids. *Rad. Sci.*, **4**, 1279–1287.

Bühler, O., 2000: On the vorticity transport due to dissipating or breaking waves in shallow-water flow. *J. Fluid Mech.*, **407**, 235–263.

Bühler, O. and M. E. McIntyre, 1998: On non-dissipative wave-mean interactions in the atmosphere or oceans. *J. Fluid Mech.*, **354**, 301–343.

Bühler, O. and M. E. McIntyre, 2001: The recoil from acoustic or gravity waves refracted by a vortex: a new wave–mean interaction effect. In preparation.

Charney, J. G., 1948: On the scale of atmospheric motions. *Geofysiske Publ.*, **17**(2), 3–17.

Charney, J. and P. G. Drazin, 1961: Propagation of planetary-scale disturbances from the lower into the upper atmosphere. *J. Geophys. Res.*, **66**, 83–109.

Cowling, T. G., 1945: On the Sun's general magnetic field. *Mon. Not. R. Astronom. Soc.*, **105**, 166–174.

Davis, C. A., 1992: Piecewise potential vorticity inversion. *J. Atmos. Sci.*, **49**, 1397–1411.

Dickinson, R. E., 1969: Theory of planetary wave-zonal flow interaction. *J. Atmos. Sci.*, **26**, 73–81.

Dunkerton, T. J., 1983: Laterally-propagating Rossby waves in the easterly acceleration phase of the quasi-biennial oscillation. *Atmos.–Ocean*, **21**, 55–68.

Dunkerton, T. J., 1991: Nonlinear propagation of zonal winds in an atmosphere with Newtonian cooling and equatorial wavedriving. *J. Atmos. Sci.*, **48**, 236–263.

Dunkerton, T. J., 1997: The role of gravity waves in the quasi-biennial oscillation. *J. Geophys. Res.*, **102**, 26053–26076.

Dunkerton, T. J. et al., 1981: Some Eulerian and Lagrangian diagnostics for a model stratospheric warming. *J. Atmos. Sci.*, **38**, 819–843.

Eady, E. T., 1950: The cause of the general circulation of the atmosphere. In Centenary Proceedings Royal Meteorological Society (P. A. Sheppard, J. K. Bannon, C. R. Burgess, G. Manley, G. D. Robinson and R. C. Sutcliffe, Eds), Reading, UK, Royal Meteorological Society, 156–172.

Eliassen, A. and E. Palm, 1961: On the transfer of energy in stationary mountain waves. *Geofysiske Publ.*, **22**(3), 1–23.

Elliott, J. R. and D. O. Gough, 1999: Calibration of the thickness of the solar tachocline. *Astrophys. J.*, **516**, 475–481.

Ertel, H., 1942: Ein Neuer hydrodynamischer Wirbelsatz. *Met. Z.*, **59**, 271–281. English translation in W. Schröder (Ed.) 1991: *Geophysical Hydrodynamics and Ertel's Potential Vorticity: Selected Papers of Hans Ertel*, Bremen-Rönnebeck, Interdivisional Commission of History of IAGA, Newsletter No. **12**.

Fels, S. B., 1985: Radiative-dynamical interactions in the middle atmosphere. In *Advances in Geophysics: Issues in Atmospheric and Oceanic modeling* (**28A**), S. Manabe, Ed., Academic Press, Orlando, FL, pp. 277–300.

Fleming, E. L. et al., 1990: Zonal mean temperature, pressure, zonal wind and geopotential height as functions of latitude. COSPAR International Reference Atmosphere 1986. Part II: Middle atmosphere models. *Adv. Space Res.*, **10**(12), 11–62.

Fritts, D. C. et al., 1988: Evidence of gravity wave saturation and local turbulence production in the summer meosphere and lower thermosphere during the STATE experiment. *J. Geophys. Res.*, **93**, 7015–7025.

Gadsden, M., 2000: Polar mesospheric clouds seen from geostationary orbit. *J. Atmos. Solar-Terrest. Phys.*, **62**, 31–36.

Gadsden, M. and W. Schröder, 1989: *Noctilucent Clouds*. Springer, Heidelberg.

Garcia, R. R., 1989: Dynamics, radiation, and photochemistry in the mesosphere: implications for the formation of noctilucent clouds. *J. Geophys. Res.*, **94**, 14605–14615 (Special Issue for the International Workshop on Noctilucent Clouds).

García López, R. J. and H. C. Spruit, 1991: Li depletion in F stars by internal gravity waves. *Astrophys. J.*, **377**, 268–277.

Gill, A. E., 1982: *Atmosphere-Ocean Dynamics*. Academic Press, London and New York.

Gough, D. O., 1990: On possible origins of relatively short-term variations in the solar structure. *Phil. Trans. R. Soc. Lond.*, **A330**, 627–640.

Gough, D. O. and M. E. McIntyre, 1998: Inevitability of a magnetic field in the Sun's radiative interior. *Nature*, **394**, 755–757.

Gray, L. J. and Pyle, J. A., 1989: A two-dimensional model of the quasi-biennial oscillation of ozone. *J. Atmos. Sci.*, **46**, 203–220.

Hall, T. M. et al., 1999: Evaluation of transport in stratospheric models. *J. Geophys. Res.*, **104**, 18815–18839.

Hamilton, K., (Ed.), 1997: *Gravity Wave Processes: Their Parameterization in Global Climate Models*. Springer-Verlag, Heidelberg.

Haynes, P. H., 1989: The effect of barotropic instability on the nonlinear evolution of a Rossby wave critical layer. *J. Fluid Mech.*, **207**, 231–266.

Haynes, P. H., 1998: The latitudinal structure of the quasi-biennial oscillation. *Q. J. R. Meteorol. Soc.*, **124**, 2645–2670.

Haynes, P. H. et al., 1991: On the "downward control" of extratropical diabatic circulations by eddy-induced mean zonal forces. *J. Atmos. Sci.*, **48**, 651–678; see also **53**, 2105–2107.

Haynes, P. H. et al., 1996: Reply to comments by J. Egger on 'On the "downward control" of extratropical diabatic circulations by eddy-induced mean zonal forces'. *J. Atmos. Sci.*, **53**, 2105–2107.

Hines, C. O., 1972: Gravity waves in the atmosphere. *Nature*, **239**, 73–78.

Hollweg, J. V., 1978: Some physical processes in the solar wind. *Rev. Geophys. Space Phys.*, **16**, 689–720.

Holton, J. R., 1982: The role of gravity wave induced drag and diffusion in the momentum budget of the mesosphere. *J. Atmos. Sci.*, **39**, 791–799.

Holton, J. R. and R. S. Lindzen, 1972: An updated theory of the quasi-biennial cycle of the tropical stratosphere. *J. Atmos. Sci.*, **29**, 1076–1080.

Holton, J. R. et al., 1995: Stratosphere–troposphere exchange. *Rev. Geophys.*, **33**, 403–439.

Hoskins, B. J., M. E. McIntyre and A. W. Robertson, 1985: On the use and significance of isentropic potential-vorticity maps. *Q. J. R. Meteorol. Soc.*, **111**, 877–946. Also *Corrigendum*, etc., **113** (1987), 402–404.

Houghton, J. T., 1978: The stratosphere and the mesosphere. *Q. J. R. Meteorol. Soc.*, **104**, 1–29.

Juckes, M. N. and M. E. McIntyre, 1987: A high resolution, one-layer model of breaking planetary waves in the stratosphere. *Nature*, **328**, 590–596.

Killworth, P. D. and M. E. McIntyre, 1985: Do Rossby-wave critical layers absorb, reflect or over-reflect? *J. Fluid Mech.*, **161**, 449–492.

Kleinschmidt, E., 1950: Über Aufbau und Entstehung von Zyklonen (1,2 Teil). *Met. Runds.*, **3**, 1–6, 54–61.

Kosovichev, A. G. et al., 1997: Structure and rotation of the solar interior: initial results from the MDI medium-l program. *Sol. Phys.*, **170**, 43–61.

Lindzen, R. S., 1981: Turbulence and stress owing to gravity wave and tidal breakdown. *J. Geophys. Res.*, **86**, 9707–9714.

Lindzen, R. S. and J. R. Holton, 1968: A theory of the quasi-biennial oscillation. *J. Atmos. Sci.*, **25**, 1095–1107.

Lorenz, E. N., 1967: *The Nature and Theory of the General Circulation of the Atmosphere.* World Meteorol. Org., Geneva.

Lübken, F.-J., 1999: Thermal structure of the Arctic summer mesosphere. *J. Geophys. Res.*, **104**, 9135–9149.

Matsuno, T., 1966: Quasi-geostrophic motions in the equatorial area. *J. Met. Soc. Jpn.*, **44**, 25–43.

Matsuno, T., 1970: Vertical propagation of stationary planetary waves in the winter Northern Hemisphere. *J. Atmos. Sci.*, **27**, 871–883.

McIntyre, M. E., 1980: Towards a Lagrangian-mean description of stratospheric circulations and chemical transports. *Phil. Trans. R. Soc. Lond.*, **A296**, 129–148.

McIntyre, M. E., 1981: On the "wave momentum" myth. *J. Fluid Mech.*, **106**, 331–347.

McIntyre, M. E., 1989: On dynamics and transport near the polar mesopause in summer. *J. Geophys. Res.*, **94**, 14617–14628 (Special Issue for the International Workshop on Noctilucent Clouds).

McIntyre, M. E., 1993: On the role of wave propagation and wave breaking in atmosphere–ocean dynamics (Sectional Lecture). In *Theoretical and Applied Mechanics 1992*, Proc. XVIII Int. Congr. Theor. Appl. Mech., Haifa (S. R. Bodner, J. Singer, A. Solan and Z. Hashin, Eds) Elsevier, Amsterdam & New York, 281–304.

McIntyre, M. E., 1994: The quasi-biennial oscillation (QBO): some points about the terrestrial QBO and the possibility of related phenomena in the solar interior. In *The Solar Engine and its Influence on the Terrestrial Atmosphere and Climate* (Vol. 25 of

NATO ASI Subseries I, Global Environmental Change) (E. Nesme-Ribes, Ed.), Springer-Verlag, Heidelberg, pp. 293–320.

McIntyre, M. E., 1998: Breaking waves and global-scale chemical transport in the Earth's atmosphere, with spinoffs for the Sun's interior. *Progr. Theor. Phys. (Kyoto) Suppl.*, **130**, 137–166 (invited review for 12th Nishinomiya Symposium in memory of Hideki Yukawa). See also an important Corrigendum, *Progr. Theor. Phys. (Kyoto)*, **101** (1999), 189.

McIntyre, M. E., 1999: How far have we come in understanding the dynamics of the middle atmosphere? Invited Symposium Lecture, *Proceedings 14th European Space Agency Symposium on European Rocket and Balloon Programmes and Related Research*, Potsdam, Germany, 31 May–3 June 1999, ESA SP-437 (B. Kaldeich-Schürmann, Ed.), ESA Publications Division, ESTEC, Noordwijk, The Netherlands.

McIntyre, M. E., 2000a: On global-scale atmospheric circulations. In *Perspectives in Fluid Dynamics: A Collective Introduction to Current Research* (G. K. Batchelor, H. K. Moffatt and M. G. Worster, Eds), Cambridge University Press, pp. 557–624.

McIntyre, M. E., 2000b: The Earth's middle atmosphere and the Sun's interior. In *Long and Short Term Variability of the Sun's History and Global Change* (Proceedings IUGG Symposium GA6.01 held in Birmingham, July 1999) (W. Schröder, Ed.), Interdivisional Commission on History, IAGA/IUGG, Newsletter **39**, pp. 13–38. Science Edition, D-28777 Bremen-Rönnebeck, Hechelstrasse 8, Germany.

McIntyre, M. E. and W. A. Norton, 1990: Dissipative wave-mean interactions and the transport of vorticity or potential vorticity. *J. Fluid Mech.*, **212**, 403–435 (G. K. Batchelor Festschrift Issue); Corrigendum, **220** (1990) 693. (A further, important correction is described in R. Mo, O. Bühler and M. E. McIntyre, 1998: Permeability of the stratospheric vortex edge: On the mean mass flux due to thermally dissipating, steady, non-breaking Rossby waves. *Q. J. R. Meteorol. Soc.*, **124**, 2129–2148.

McIntyre, M. E. and W. A. Norton, 2000: Potential-vorticity inversion on a hemisphere. *J. Atmos. Sci.*, **57**, 1214–1235. Corrigendum: *J. Atmos. Sci.*, **58**, 949.

McIntyre, M. E. and T. N. Palmer, 1984: The "surf zone" in the stratosphere. *J. Atmos. Terr. Phys.*, **46**, 825–849.

McIntyre, M. E. and T. N. Palmer, 1985: A note on the general concept of wave breaking for Rossby and gravity waves. *Pure Appl. Geophys.*, **123**, 964–975.

Mote, P. W. *et al.*, 1996: An atmospheric tape recorder: the imprint of tropical tropopause temperatures on stratospheric water vapor. *J. Geophys. Res.*, **101**, 3989–4006.

Mote, P. W. *et al.*, 1998: Vertical velocity, vertical diffusion, and dilution by midlatitude air in the tropical lower stratosphere. *J. Geophys. Res.*, **103**, 8651–8666.

Norton, W. A., 1994: Breaking Rossby waves in a model stratosphere diagnosed by a vortex-following coordinate system and a technique for advecting material contours. *J. Atmos. Sci.*, **51**, 654–673.

Plumb, R. A., 1977: The interaction of two internal waves with the mean flow: implications for the theory of the quasi-biennial oscillation. *J. Atmos. Sci.*, **34**, 1847–1858.

Plumb, R. A. and J. Eluszkiewicz, 1999: The Brewer–Dobson circulation: dynamics of the tropical upwelling. *J. Atmos. Sci.*, **56**, 868–890.

Plumb, R. A. and A. D. McEwan, 1978: The instability of a forced standing wave in a viscous stratified fluid: a laboratory analogue of the quasi-biennial oscillation. *J. Atmos. Sci.*, **35**, 1827–1839.

Robinson, W. A., 1988: Analysis of LIMS data by potential vorticity inversion. *J. Atmos. Sci.*, **45**, 2319–2342.

Rosenlof, K., 1996: Summer hemisphere differences in temperature and transport in the lower stratosphere. *J. Geophys. Res.*, **101**, 19129–19136.

Rosenlof, K. and J. R. Holton, 1993: Estimates of the stratospheric residual circulation using the downward control principle. *J. Geophys. Res.*, **98**, 10465–10479.

Rossby, C.-G., 1936: Dynamics of steady ocean currents in the light of experimental fluid mechanics. *Mass. Inst. Technol. Woods Hole Oc. Inst. Pap. Phys. Oceanogr. Meteorol.*, **5**(1), 1–43. See Eq. (75).

Rossby, C.-G., 1940: Planetary flow patterns in the atmosphere. *Q. J. R. Meteorol. Soc.*, **66**(Suppl.), 68–97.

Sankey, D., 1998: Dynamics of upwelling in the equatorial lower stratosphere. Ph.D. thesis, University of Cambridge.

Saravanan, R., 1990: A multiwave model of the quasi-biennial oscillation. *J. Atmos. Sci.*, **47**, 2465–2474.

Schröder, W., 1999: Were noctilucent clouds caused by the Krakatoa eruption? A case study of the research problems before 1885. *Bull. Am. Meteorol. Soc.*, **80**, 2081–2085.

Scott, R. K. and Haynes, P. H., 1998: Internal interannual variability of the extratropical stratospheric circulation: the low-latitude flywheel. *Q. J. R. Meteorol. Soc.*, **124**, 2149–2173.

Shine, K. P., 1989: Sources and sinks of zonal momentum in the middle atmosphere diagnosed using the diabatic circulation. *Q. J. Roy. Meteorol. Soc.*, **115**, 265–292.

Snodgrass, F. E. *et al.*, 1966: Propagation of ocean swell across the Pacific. *Phil. Trans. R. Soc. Lond.*, **A259**, 431–497.

Solomon, S., J. T. Kiehl, R. R. Garcia and W. L. Grose, 1986: Tracer transport by the diabatic circulation deduced from satellite observations. *J. Atmos. Sci.*, **43**, 1603–1617.

Sparling, L. S. *et al.*, 1997: Diabatic cross-isentropic dispersion in the lower stratosphere. *J. Geophys. Res.*, **102**, 25817–25829.

Spiegel, E. A. and J.-P. Zahn, 1992: The solar tacholine. *Astron. Astrophys.*, **265**, 106–114.

Taylor, G. I., 1915: Eddy motion in the atmosphere. *Phil. Trans. Soc. Lond.*, **A215**, 1–23.

Thomas, G. E. *et al.*, 1989: Relation between increasing methane and the presence of ice clouds at the mesopause. *Nature*, **338**, 490–492.

Thompson, M. J. *et al.*, 1996: Differential rotation and dynamics of the solar interior. *Science*, **272**, 1300–1305.

Thuburn, J. and M. E. McIntyre, 1997: Numerical advection schemes, cross-isentropic random walks, and correlations between chemical species. *J. Geophys. Res.*, **102**, 6775–6797.

Wallace, J. M. and J. R. Holton, 1968: A diagnostic numerical model of the quasi-biennial oscillation. *J. Atmos. Sci.*, **25**, 280–292.

Wallace, J. M. and V. E. Kousky, 1968: Observational evidence of Kelvin waves in the tropical stratosphere. *J. Atmos. Sci.*, **25**, 900–907.

Ward, W. E. and V. I. Fomichev, 1996: Consequences of CO_2–O collisional relaxation on the energy budget of the mesosphere and lower thermosphere. *Adv. Space Res.*, **17**(11), 127–137.

Yulaeva, E., J. R. Holton and J. M. Wallace, 1994: On the cause of the annual cycle in tropical lower stratospheric temperatures. *J. Atmos. Sci.*, **51**, 169–174.

Atmospheric Dynamics of the Outer Planets

Andrew P. Ingersoll

Division of Geological and Planetary Sciences, California Institute of Technology, Pasadena, CA, USA

The giant planets—Jupiter, Saturn, Uranus, and Neptune—are fluid objects. The winds are powered by absorbed sunlight, as on earth, and by internal heat left over from planetary formation. The main constituents of the atmospheres are hydrogen and helium. The clouds are made of ammonia, hydrogen sulphide, and water. All four giant planets are banded, with multiple zonal jet streams. Even Uranus, whose spin axis is tipped by 98° relative to the orbit axis, shows latitudinal banding and zonal jets. Equator-to-pole temperature differences are close to zero. Wind speeds are larger than on earth and do not decrease with distance from the sun. Although the power/area at Neptune is only 1/20 that at Jupiter, the winds at Neptune are three times stronger. Stable vortices like the Great Red Spot of Jupiter and similar spots on Neptune come in all size ranges and exhibit a variety of behaviours including merging, orbiting, filament ejection, and oscillating in both shape and position. At least at cloud-top levels, 90% of the long-lived vortices are anticyclonic and sit in anticyclonic shear zones. Features in the cyclonic zones tend to be chaotic, with lifetimes of several days or less. These mesoscale eddies tend to have lightning in them, which suggests that they get their energy from moist convection. The rate of conversion of eddy kinetic energy into kinetic energy of the zonal jets is more than 10% of the power/area radiated by Jupiter. This fraction is more than an order of magnitude larger than on earth. Several lines of evidence now indicate that the winds at cloud-top levels are the surface manifestation of deep-rooted motions that extend into the interior and are presumably driven by internal heat.

A. INTRODUCTION

The outer planets—Jupiter, Saturn, Uranus, and Neptune—are fluid objects. Their hydrogen–helium atmospheres merge imperceptibly with the liquid interiors. Because the interiors are warm, there is no phase change from gas to liquid. Internal heat left over from planetary formation stirs the atmospheres and maintains the temperature lapse rate close to the adiabatic value. With the exception of Uranus, which has only a small internal heat source, the total power convected up from the interior is comparable to that absorbed from the sun. The infrared heat flux—power per unit area radiated to space—decreases by a factor of 20 as one moves out from Jupiter to Neptune, yet Neptune's wind speeds are three times Jupiter's

In this chapter we will briefly review the discoveries of the Voyager era, which are summarized in Ingersoll (1990). Then we will review progress on modelling the long-lived anticyclonic vortices. Jupiter's Great Red Spot (GRS) is the most well-known example, but it is only the largest of a family whose members range in size from 20,000 km down to 1000 km, which is comparable to the radius of deformation.

We then review the discoveries of the Galileo mission. The probe took measurements down to the 20 bar level, well below the condensation level (5–6 bars) of the gases in Jupiter's atmosphere. Contrary to expectations, the composition was still varying with depth when the probe ceased communication. The probe entered an area of reduced cloud cover, perhaps resembling the subtropics on earth. Surprisingly, the associated downdraft extends well below cloud base.

The Galileo orbiter took images at 25 km resolution on both the day and night sides of the planet and thus was able to resolve features comparable in size to the atmospheric scale height. We will focus on the lightning observations, in which we were able to identify the sites of moist convection and link them to clouds visible in day-side images.

B. VOYAGER OBSERVATIONS OF WINDS AND TEMPERATURES

Figure 1 shows a composite of the four gas-giant planets. The relative sizes are accurate; Jupiter's equatorial radius is 71,492 km at the 1 bar level. Every

FIGURE 1 Colour composite of the four gas-giant planets. The relative sizes are accurate. The equatorial radius of Jupiter at the 1 bar level is 71,492 km (Ingersoll, 1990).

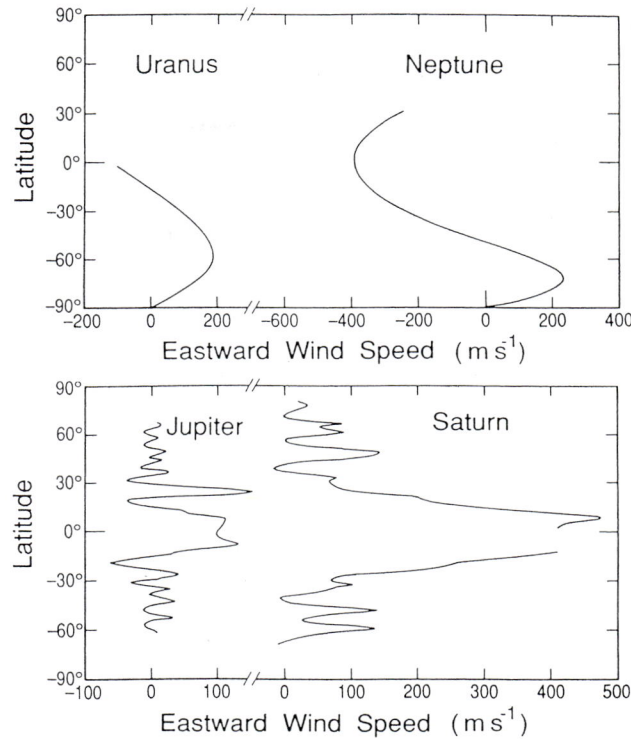

FIGURE 2 Zonal wind speed as a function of latitude. Note the scale change between top and bottom. Since the giant planets have no solid surfaces, the winds are measured relative to the magnetic field, which is generated in the interior and rotates with it (Ingersoll et al., 1995).

feature visible in the planets' atmospheres is a cloud. Colour is related to the chemical composition of the particles. Dynamics organizes the chemistry, which is affected by updrafts and downdrafts, lightning, and ultraviolet light. Nevertheless, details of this process and the exact composition of the colored compounds remain uncertain.

All four of the giant planets are banded. Even Uranus, whose clouds have extremely low contrast, exhibits faint bands that circle the planet on lines of constant latitude. Since the spin axis of Uranus is tilted 98° with respect of the orbit normal, the banding shows the dominant role that rotation plays in setting the patterns. The sun was almost directly overhead at the south pole at the time of the Voyager encounter.

Figure 2 shows the zonal wind speeds. Since the planets have no solid surfaces, the winds are measured relative to the magnetic field, which is tied to the interior and rotates with it. We use the term prograde and retrograde to describe gas that moves faster or slower than the internal rotation. On earth these winds would be called westerlies and easterlies, respectively. Notice the scale change between the bottom and top halves of Fig. 2. Despite being closer to the sun and having more power/area radiated to space, Jupiter has the weakest winds of any of the gas giant planets. Like the earth, Uranus and Neptune have retrograde winds at the equator, but Jupiter and Saturn have prograde equatorial winds.

The reasons for these differences are not known. One possibility is that the winds that we see are a historical accident. Although Jupiter's winds have remained constant for 100 years, it is possible that they change on longer timescales. On the other hand, there are differences in the compositions and equations of state of the interiors that could account for the differences in the wind profiles.

One explanation (Ingersoll et al., 1995) for why Jupiter has the weakest winds of the gas giants is that the dissipation of kinetic energy is greatest at Jupiter and decreases as one moves outward in the solar system. According to this idea, the power/area radiated to space is a measure of small-scale turbulence, which acts as a brake on the winds. Apparently as one moves outward in the solar system, the increase in the time constant for kinetic energy dissipation more than offsets the decrease in the generation rate of kinetic energy. The result is that the winds increase. This argument may also explain why the earth has the lowest wind speeds of any planet in the solar system: the earth radiates the most power/area and therefore has the most dissipative atmosphere.

Using cloud-tracked winds derived from Voyager images it was possible to estimate a Reynolds stress $\overline{\rho u'v'}$ at each latitude (Ingersoll *et al.*, 1981). This is the northward transport of eastward momentum by the eddies, which are departures of the wind from the zonal mean. We find that the momentum transport is into the jets. The transfer of eddy kinetic energy into zonal mean kinetic energy is positive. Although this transfer is only one term in the mechanical energy cycle, it is a useful diagnostic. Assuming the energy transfer involves mass/area over a pressure range of 2.5 bars, we find that the associated power/area is over 10% of the power/area radiated to space. On earth the corresponding ratio is less than 1%. Apparently the mechanical energy cycle is relatively more important on Jupiter than it is on earth when compared to the total energy radiated by the planet.

For Jupiter's zonal winds it is possible to compute the absolute vorticity and its gradient. The gradient is not monotonic, but changes sign at the latitudes of the westward jets. There the curvature of the zonal velocity profile $\bar{u}_{yy}$ is greater than β which is the gradient of planetary vorticity. The best estimates (Ingersoll *et al.*, 1981; Limaye, 1986) are that $\bar{u}_{yy}$ exceeds β by a factor of 2.5 to 3.0 at each of the westward jets. Ordinarily this would imply that the flow is unstable, since a non-monotonic absolute vorticity gradient violates the barotropic stability criterion. Numerical models (Maltrud and Vallis, 1991; Panetta, 1993; Cho and Polvani, 1996) that attempt to model the upscale energy transfer from eddies to zonal jets do not get this result: they find that the absolute vorticity gradient is always positive. Either the eddies are driving the zonal flow so hard that they can sustain the unstable profile, or else the flow extends into the interior where the barotropic stability criterion has a quantitatively different value (Ingersoll and Pollard, 1982).

Figure 3 shows the emitted infrared heat flux σT_e^4 vs. latitude for each of the giant planets. The effective temperature T_e is close to the physical temperature near the 400 mb level. The bumps and wiggles are associated with the zonal jets, and imply that the jets decay with height in the upper troposphere. On a planetary scale, the effective temperature is remarkably constant. The lack of an equator-to-pole temperature gradient may be due to convection in the interior (Ingersoll and Porco,

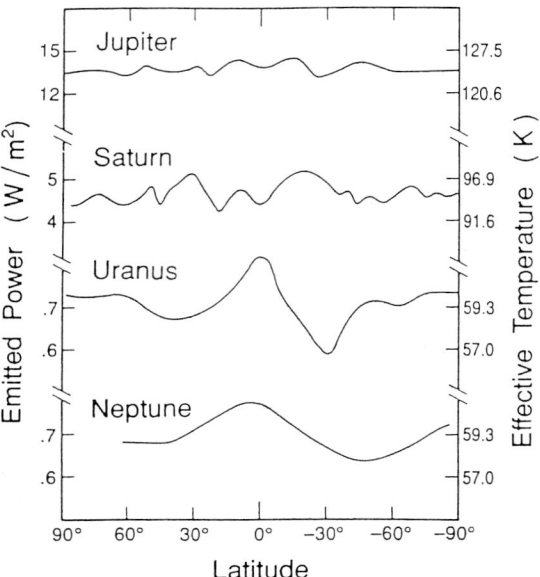

FIGURE 3 Emitted power and effective blackbody temperature as a function of latitude. Note the absence of appreciable equator-to-pole temperature differences. The largest differences are associated with the zonal jets and indicate that the zonal jets decay with altitude in the upper troposphere (Ingersoll *et al.*, 1995).

1978). The idea is that the high densities, great depth, and stirring by convection make the interior a better heat transport medium than the atmosphere. It is as if the earth's ocean–atmosphere system were getting half its energy from the sun and the other half at the bottom of the ocean. The oceans would be much more active, and would probably not allow appreciable temperature differences between equator and pole. On the giant planets, heat transport in the atmosphere is short-circuited by heat transport in the fluid interior.

C. MODELS OF VORTICES IN SHEAR

Perhaps the simplest model of a vortex in shear is the so-called Kida vortex (Moore and Saffman, 1971; Kida, 1981). One assumes inviscid, two-dimensional, nondivergent flow. Except for an isolated patch, the fluid has uniform vorticity. Within the patch the vorticity is uniform but different from the vorticity outside. For applications to the giant planets one assumes that the flow at infinity is a linear shear without a straining motion. A steady, stable configuration is when the anomalous patch has elliptical shape, with the long axis oriented parallel to the flow at infinity (east–west on the giant planets). The aspect ratio (ratio of long to short axes) depends on the ratio of the vorticity inside to that outside the patch. Finite-amplitude elliptical perturbations lead to stable oscillations in the aspect ratio and orientation. The relative phase and amplitude of the two oscillations are coupled, and depend on the amplitude of the initial perturbation.

Polvani *et al.* (1990) showed that the Kida vortex model does a good job of matching the observations of vortices on the giant planets. For the GRS and white ovals of Jupiter the vorticity inside and outside was measured by tracking small cloud features. The model accurately accounts for the average aspect ratio of the vortex (Ingersoll *et al.*, 1995). For the Great Dark Spot (GDS) of Neptune, which oscillates in aspect ratio and orientation, the model accurately accounts for the relative phase and relative amplitude of the two oscillations. The model does not explain the amplitude itself, which is a free parameter of the theory, nor does it account for the observed shedding of filaments.

It is remarkable that the Kida vortex model works as well as it does. It has no vertical structure, no gradient in the ambient vorticity or β effect, no forcing, and no dissipation. Polvani *et al.* argue that the radius of deformation must be large for the model to work. This means the stratification must be large or the layer must be deep, or both. LeBeau and Dowling (1998) have studied isolated vortices in a three-dimensional primitive equation model, and find Kida-type oscillations and a tendency toward filament formation. They reproduce other features of the GDS as well, including its drift toward the equator and ultimate dissolution into planetary-scale waves.

Achterberg and Ingersoll (1994) found Kida-type oscillations in a three-dimensional quasi-geostrophic model. They also found oscillations in latitude and longitude when the radius of deformation is large enough. The oscillations arise when the top and bottom halves of the vortex begin to orbit around a common vertical axis. Achterberg and Ingersoll propose that the latitude–longitude oscillation of Neptune's second dark spot could be due to this effect. They were not able to reproduce the large amplitude of the oscillation ($\pm 45°$ in longitude) because the two halves of the vortex tended to separate and drift off separately. Orbiting is common in vortex dynamics, but it usually involves two vortices at the same altitude.

These inviscid theories shed no light on what maintains the vortices or their oscillations against dissipation. Observations of mergers suggest that they feed on each other. Small vortices, 1000 km in diameter, approach the GRS from the east every few days (MacLow and Ingersoll, 1986). Sometimes they turn south before reaching the GRS and get caught up in the eastward current heading away from it. At other times they pass to the north of the GRS and either continue on to the west or get caught up in the GRS and are absorbed (the GRS is a southern anticyclone). Two historic mergers took place in 1999 when the three white ovals, which were then 60 years old, merged twice to form one white oval. Figure 4 shows two of the white ovals in 1998, as they were approaching each other (Vasavada *et al.*, 1998). The ovals are anticyclonic and are separated by an irregular yellow feature whose vorticity is cyclonic. There is also a white oval further south. The cyclonic patches seemed to keep the anticyclonic ovals from each other. For some reason, in 1998–99 the cyclonic patches weakened, and the ovals came together and merged. Mergers of smaller anticyclonic vortices were captured in the Voyager movies (MacLow and Ingersoll, 1986). Usually a filament is ejected shortly after the merger, presumably because the new larger vortex is unstable.

The ovals are easy to make in a numerical model and are remarkably stable. Initially they can be balanced or unbalanced. They can be rotated 90° and placed crosswise in the shear flow (Ingersoll and Cuong, 1981). Or they can be created from an initially unstable flow. Figure 5 shows an example from Dowling and Ingersoll (1989), in which the initial shear flow broke up into small eddies, which eventually merged with each other to produce a single large vortex. The final shear flow was weaker than the original one because energy went into the vortex.

D. GALILEO PROBE AND VERTICAL STRUCTURE

As one proceeds upward from the interior of each planet, the temperature falls to the point where the gases H_2O, H_2S, and NH_3 condense to form clouds. The pressure at cloud base depends on the abundance of the condensing species and the temperature. Figure 6 shows the temperature profiles and condensation levels for Jupiter and Saturn. The black dots are where the pressure is 0.01, 0.1, 1.0, and 10.0 bar, with $P = 0.1$ bar near the temperature minimum—the tropopause. Temperatures at Uranus and Neptune are so cold that CH_4 forms an additional cloud above the NH_3 cloud. On all the giant planets the level of emission to space is at 0.3–0.5 bar. Sunlight probably does not penetrate below the base of the water cloud.

For Jupiter the base of the water cloud is at a pressure of 5–6 bar. This estimate assumes that the interior of the planet is a mixture of elements resembling that on the sun, with all the oxygen bound up as water, all the nitrogen as ammonia, all the carbon as methane, and so on. In fact the Galileo probe results suggest that the elements heavier than helium are enriched relative

FIGURE 4 Colour image of three anticyclonic white ovals in 1997. The three ovals are in Jupiter's southern hemisphere. The two ovals at the top of the image were almost 60 years old when the image was taken. They later merged with each other and later with a third oval at the same latitude to form a single larger oval. The yellow pear-shaped region between the two ovals has cyclonic vorticity. Flow impinges on the region from the west, forming a white cloud, and then passes over the top and down the other side, exiting to the east on the south side of the rightmost oval (Vasavada *et al.*, 1998).

to hydrogen and helium by a factor of 3 relative to solar composition. Even with this enrichment, hydrogen and helium constitute 99% of the mixture by volume.

The Galileo probe reached the 20 bar level before it failed due to the high temperatures. This was deeper than it was designed to go. The assumption had been that the atmosphere below cloud base would be well-stirred (dry adiabatic) and chemically homogeneous. By penetrating below the base of the water cloud, the probe would be sampling the bulk composition of the planet, according to this assumption.

The atmosphere at the probe site turned out to be close to dry adiabatic, though on the stable side (Seiff *et al.*, 1998), but definitely not chemically homogeneous. Ammonia, which does not condense below the 1 bar level, was still increasing with depth at 7 bar (Folkner *et al.*, 1998). Hydrogen sulphide, which does not condense below the 2 bar level, was still increasing with depth at 12 bar (Niemann *et al.*, 1998). Water was still increasing with depth at 20 bar (Niemann *et al.*, 1998).

Part of the explanation is that the probe entered the atmosphere in an area of reduced cloud cover—one of the so-called 5 µm hot spots. Figure 7 shows an image of Jupiter at 5 µm with the location of the probe entry site (Orton *et al.*, 1996). The gaseous opacity is low at this wavelength, so the reduced cloud cover allows thermal radiation from the warm subcloud region to escape to space. Hence the area looks hot at 5 µm. Reduced cloud cover could occur if dry upper tropospheric air were forced downward, as in the subtropics on earth. The surprise is that the downdraft extends down to 20 bars, which is well below the cloud base and well below the depth of sunlight penetration.

The downdraft hypothesis has its own problems, as pointed out by Showman and Ingersoll (1998). First, the downward moving air must be less dense than the air it is displacing or else the two would mix. The displaced air is presumably typical of Jupiter's interior, and has the planet-wide abundances of ammonia, hydrogen sulphide, and water. In other words, the downdraft would not be dry if mixing had occurred. However, less dense air must be forced down, and the mechanism to do that is uncertain. Second, when mixing does occur, one would expect all the gaseous abundances to increase together. As we have noted, however, the ammonia

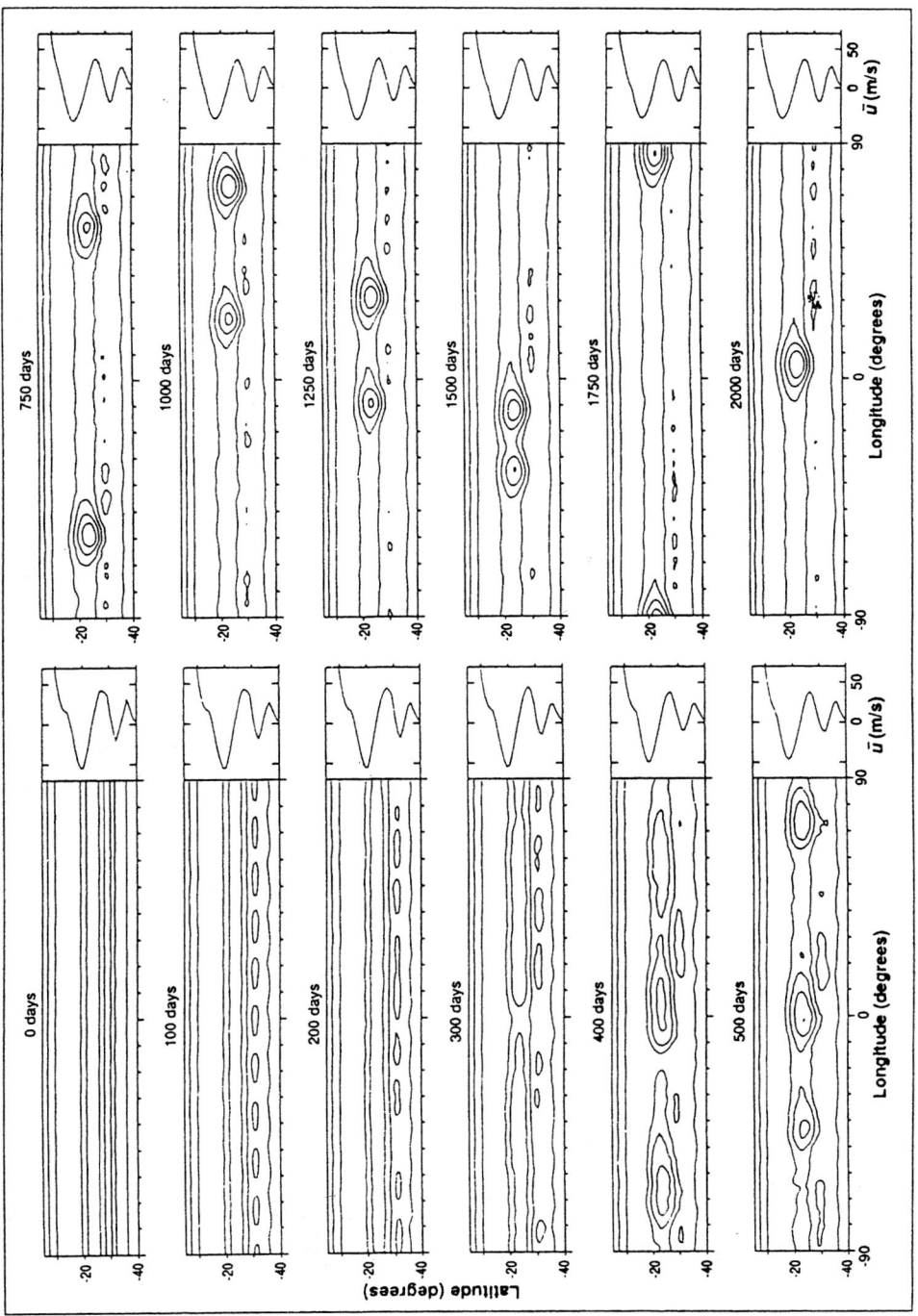

FIGURE 5 Genesis of vortices in an inviscid shallow water model. The initial state is constructed by adding a small perturbation to Jupiter's observed zonal velocity profile. The latitude of the integration is −5° to −40°. Latitudinally varying bottom topography is derived from observations of vortex tube stretching. The initial flow is unstable and breaks up into eddies, which merge to form a single large vortex at the latitude of the Great Red Spot (Dowling and Ingersoll, 1989).

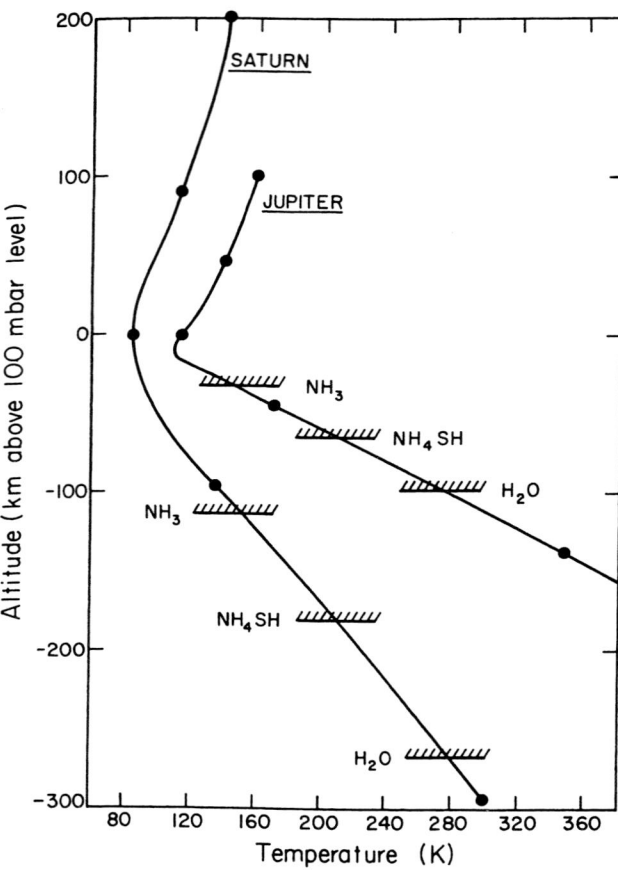

FIGURE 6 Temperature and cloud base as a function of altitude for Jupiter and Saturn. The large dots denote points where the pressure changes by a factor of 10, with the zero of altitude at the 100 mb level. Cloud base is computed for a solar composition atmosphere (elemental ratios H:O:S:N:C as on the sun). Increasing the abundance of a condensable gas by a factor of 3 lowers the altitude of cloud base by ~10 km.

increased before the H_2S, and the H_2S increased before the water. The third problem is that the probe detected large wind shear, i.e., the atmosphere below the 5 bar level was moving 80 m s^{-1} faster than the air at cloud top (Atkinson *et al.*, 1998). At this rate the top and bottom halves of the downdraft would be sheared apart in about 1 day, since the hot-spot diameter is 6000–8000 km.

Showman and Dowling (2000) have put forth a modified form of the downdraft hypothesis. They model the hot spot as part of a large-amplitude wave, in which the fluid approaches the structure from the west, is deflected downward, and then rises again as it exits to the east. Traversing the hot spot takes about 1 day. Since no mixing is involved, the layering associated with the three cloud layers (Fig. 6) is preserved. This explains why the ammonia increased before the H_2S and the H_2S increased before the water as the probe descended. The trick is to get a wave whose particle displacements cover a factor of 4 to 8 in pressure. Their numerical model almost succeeds in doing this.

The fact that the composition had not reached its final value at 20 bars complicates the interpretation of the wind measurements. Again, the assumption had been that the winds below cloud base would reflect the barotropic state of Jupiter's interior. Any zonal flow would extend downwards on cylinders according to the ideas of Poincaré, Taylor and Proudman. However, if the atmosphere is not adiabatic and the density is not horizontally homogeneous, then the flow is not barotropic. One then has the possibility of further wind shear at deeper levels. The basic observation is that the wind increased from ~100 m s^{-1} at cloud top to ~180 m s^{-1} at cloud base and then stayed roughly constant to 20 bars (Atkinson *et al.*, 1998). This certainly means that the winds are deep, i.e., below the depth of sunlight penetration. However, in view of the uncertainties surrounding the dynamics of hot spots, it is less clear that the deep winds are characteristic of Jupiter's interior or are necessarily driven by internal heat.

There are several complications to understanding the dynamical state of Jupiter's interior. First, at a pressure of a few megabars—at 80% of the planet's radius from the centre, molecular hydrogen liquid turns into a metallic liquid. This phase transition may involve a density discontinuity and possibly a velocity discontinuity. Second, even in the molecular hydrogen envelope the electrical conductivity is large enough to drag the magnetic field with it, so magnetohydrodynamic effects are important. Third, there may be a layer in the 1–40 k bar range where the atmosphere is not convecting and the principal mode of heat transport is radiation (Guillot *et al.*, 1994). A stably stratified layer such as this could decouple the winds in the interior from winds in the atmosphere. Gierasch (1999) and Allison (2000) discuss some of the possibilities.

E. GALILEO ORBITER OBSERVATIONS OF WATER AND LIGHTNING

In principle, instruments on the orbiter can measure the abundance of water at many places on the planet. These instruments have to "see down" to where the water is, since the water condenses out of the atmosphere ~100 km below the tops of the clouds (Fig. 6). This severely restricts the places that water is detectable. The Near Infrared Mapping Spectrometer (NIMS) found places where water was saturated at the highest temperatures reached (about 260 K), but the abundance of water below cloud base was not measured (Roos-Serote *et al.*, 2000). The Solid State Imaging

FIGURE 7 Thermal emission from Jupiter at a wavelength of 5 μm. The bright areas are holes in the clouds from which intense radiation from warmer levels below cloud top escapes to space. The Galileo probe entered on the south edge of a 5 μm hot spot similar to the one in the centre of the image just above the equator. (Ortiz *et al.* 1998)

(SSI) system detected clouds at the 4 bar level (Banfield *et al.*, 1998), where the only possible condensate is water, but again the abundance below cloud base was not measured. Further, the water cloud was visible only in a few locations, mostly in the clear spaces next to convective towers. Apparently on Jupiter, as on earth, the updrafts that produce high, thick clouds are bordered by downdrafts that produce unusually cloud-free areas. These afford the rare glimpses of the water cloud, which otherwise is obscured by the more uniform cloud cover of ammonia.

Lightning provides indirect evidence for water, if models based on terrestrial cloud electrification can be believed. These models (Gibbard *et al.*, 1995; Yair *et al.*, 1995) calculate the expected occurrence of lightning, for various assumptions about the composition of the atmosphere below cloud base. Water is the only substance present in sufficient abundance to produce lightning, assuming enrichment factors relative to solar composition no greater than 3. The minimum water abundance consistent with lightning corresponds to the solar ratio of oxygen to hydrogen.

The diameter of a lightning flash when it emerges from the upper cloud deck is a rough indicator of how deep it is. Models of photons diffusing up through a scattering layer imply that the diameter of the half power contour is 1.3 to 1.5 times the depth of the source. The observed diameters range from 90 to 160 km, implying that the light source ranges from 65 to 120 km below the tops of the clouds (Little *et al.*, 1999). This depth is greater than expected, and puts the lightning flashes below the base of the water cloud unless the water abundance is several times the solar value.

Lightning can only be seen on the night side, but apparently the storms last long enough so that the Galileo orbiter was able to follow them from day side to night side, a time interval of 2–3 h. Figure 8 shows one example. On the day side the lightning storms appear bright and white. They are isolated from each other and have diameters up to 1000 km. The clouds in the lightning storms are optically thick and higher than average. Similar clouds in the Voyager movies appear suddenly and grow to a diameter of 1000 km in 1 or 2 days. All of these properties are consistent with moist convection.

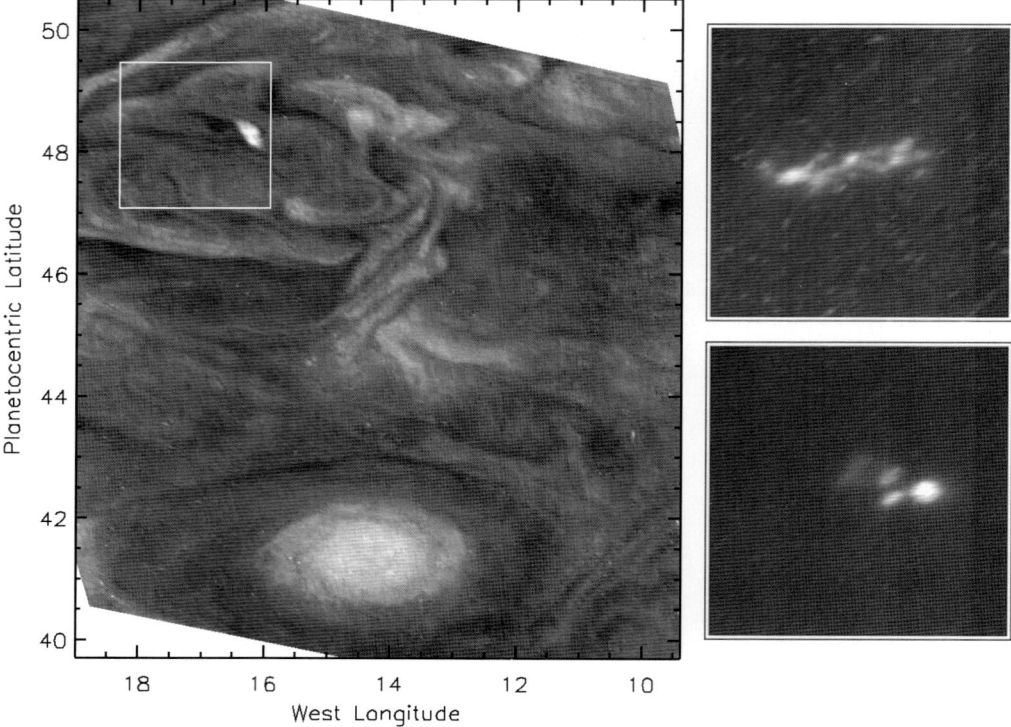

FIGURE 8 Images of a lightning storm on the day side and night side of Jupiter. The small white square in the upper left of the day-side image is shown on the right as it appeared on the night side ~2 h later. Two images of the night side, taken ~4 min apart, are shown. The size of the box in the night side images was enlarged by a factor of 2 relative to the box on the day-side image. Different camera settings of the two night-side images account for differences in signal-to-noise ratio (Little *et al.*, 1999).

The Galileo orbiter provided the confirmation that they do indeed have lightning in them (Little *et al.*, 1999; Gierasch *et al.*, 2000).

These bright, white, rapidly growing clouds are found predominantly in the cyclonic shear zones—those with a westward jet on the poleward side and an eastward jet on the equatorward side (Little *et al.*, 1999). These bands are also the sites of the 5 μm hot spots, which are areas of reduced cloud cover. Thus the cyclonic regions have a disturbed, chaotic appearance. The anticyclonic shear zones and the anticyclonic ovals like the GRS have a more uniform appearance. From the rate of divergence and the latent heat release, one can make a rough estimate of the vertical energy transfer in one of these moist convective structures (Gierasch *et al.*, 2000). Although they are spaced 10,000 km apart, collectively they carry a substantial fraction of the total internal energy generated by the planet.

One explanation (Ingersoll *et al.*, 2000) for the occurrence of moist convection in the cyclonic shear zones is that the eddy momentum transport induces upwelling in these zones. The argument parallels the discussion of zonally symmetrical circulations on earth. The convergence of eastward momentum into the eastward jets produces (is balanced by) a Coriolis torque associated with equatorward flow. The opposite is true of the westward jets. The cyclonic regions, which have an eastward jet on the equatorward side, therefore are sites of horizontal divergence. This is balanced by upwelling from below, which brings in moist air and induces moist convection.

The Voyager movies (MacLow and Ingersoll, 1986) and the Galileo lightning observations (Ingersoll, *et al.*, 2000) suggest that the small anticyclonic eddies, which regularly merge with the GRS and other large ovals, begin as moist convective structures. Their rapid divergence generates anticyclonic vorticity. They are either expelled from the cyclonic shear zones, or they are pulled apart. The latter produces the upgradient momentum transfer that pumps energy from eddies into the zonal jets. If they are expelled, the vortices usually merge with other anticyclonic ovals. Thus the moist convective structures are at the bottom of the food chain. They are the primary harvesters of energy, which cascades up to the larger scales by mergers and shearing action.

The next step is to add moist convection to a numerical model of the large-scale flow. The goal would be to see if moist convection behaves as it does in the images—originating in the cyclonic shear zones, developing anticyclonic vorticity, being pulled apart or merging with larger anticyclonic ovals, and so on. Perhaps such a model will explain why the zonal velocity profile violates the barotropic stability criterion. Perhaps it will explain where the GRS gets its energy. The interaction between moist convection and the large-scale dynamics is a difficult problem for earth. For Jupiter we have fewer *in situ* observations with which to test the model. But Jupiter is a photogenic planet, and the large number of time-lapse images will help.

References

Achterberg, R. K. and A. P. Ingersoll, 1994: Numerical simulation of baroclinic Jovian vortices. *J. Atmos. Sci.*, **51**, 541–562.

Allison, M., 2000: A similarity model for the windy Jovian thermocline. *Planet. Space Sci.*, **48**, 753–774.

Atkinson, D. H., J. B. Pollack and A. Seiff, 1998: The Galileo probe Doppler wind experiment: Measurement of the deep zonal winds of Jupiter. *J. Geophys. Res.*, **103**, 22911–22928.

Banfield, D., P. J. Gierasch, M. Bell, E. Ustinov, A. P. Ingersoll, A. R. Vasavada, R. A. West and M. J. S. Belton, 1998: Jupiter's cloud structure from Galileo imaging data. *Icarus*, **135**, 230–250.

Cho, J. Y.-K. and L. M. Polvani, 1996: The morphogenesis of bands and zonal winds in the atmospheres on the giant outer planets. *Science*, **273**, 335–337.

Dowling, T. E. and A. P. Ingersoll, 1989: Jupiter's Great Red Spot as a shallow water system. *J. Atmos. Sci.*, **46**, 3256–3278.

Folkner, W. M., R. Woo and S. Nandi, 1998: Ammonia abundance in Jupiter's atmosphere derived from the attenuation of the Galileo probe's radio signal. *J. Geophys. Res.*, **103**, 22847–22855.

Gibbard, S., E. H. Levy and J. I. Lunine, 1995: Generation of lightning in Jupiter's water cloud. *Nature*, **378**, 592–595.

Gierasch, P. J., 1999: Radiative-convective latitudinal gradients for Jupiter and Saturn models with a radiative zone. *Icarus*, **142**, 148–154.

Gierasch, P. J., A. P. Ingersoll, D. Banfield, S. P. Ewald, P. Helfenstein, A. Simon-Miller, A. Vasavada, H. H. Breneman, D. A. Senske and the Galileo Imaging Team, 2000: Observation of moist convection in Jupiter's atmosphere. *Nature*, **403**, 628–630.

Guillot, T., D. Gautier, G. Chabrier and B. Mosser, 1994: Are the giant planets fully convective? *Icarus*, **112**, 337–353.

Ingersoll, A. P., 1990: Atmospheric dynamics of the outer planets. *Science*, **248**, 308–315.

Ingersoll, A. P. and P. G. Cuong, 1981: Numerical model of long-lived Jovian vortices. *J. Atmos. Sci.*, **38**, 2067–2076.

Ingersoll, A. P. and D. Pollard, 1982: Motions in the interiors and atmospheres of Jupiter and Saturn: Scale analysis, anelastic equations, barotropic stability criterion. *Icarus*, **52**, 62–80.

Ingersoll, A. P. and C. C. Porco, 1978: Solar heating and internal heat flow in Jupiter. *Icarus*, **35**, 27–43.

Ingersoll, A. P., R. F. Beebe, J. L. Mitchell, G. S. Garneau, G. M. Yagi and J.-P. Muller, 1981: Interactions of eddies and mean zonal flow on Jupiter as inferred by Voyager 1 and 2 images. *J. Geophys. Res.*, **86**, 8733–8743.

Ingersoll, A. P., C. D. Barnet, R. F. Beebe, F. M. Flasar, D. P. Hinson, S. S. Limaye, L. A. Sromovsky and V. E. Suomi, 1995: Dynamic meteorology of Neptune. In *Neptune and Triton* (D. P. Cruikshank Ed.), University of Arizona Press, Tucson, pp. 613–682.

Ingersoll, A. P., P. J. Gierasch, D. Banfield, A. R. Vasavada and the Galileo Imaging Team, 2000: Moist convection as an energy source for the large-scale motions in Jupiter's atmosphere. *Nature*, **403**, 630–632.

Kida, S., 1981: Motion of an elliptic vortex in a uniform shear flow. *J. Phys. Soc. Japan*, **50**, 3517–3520.

LeBeau, R.-P. and T. E. Dowling, 1998: EPIC simulations of time-dependent, three-dimensional vortices with application to Neptune's Great Dark Spot. *Icarus*, **132**, 239–265.

Limaye, S. S., 1986: Jupiter: New estimates of the mean zonal flow at the cloud level. *Icarus*, **65**, 335–352.

Little, B., C. D. Anger, A. P. Ingersoll, A. R. Vasavada, D. A. Senske, H. H. Breneman, W. J. Borucki and the Galileo SSI Team, 1999: Galileo imaging of lightning on Jupiter. *Icarus*, **142**, 306–323.

MacLow, M.-M. and A. P. Ingersoll, 1986: Merging of vortices in the atmosphere of Jupiter: An analysis of Voyager images. *Icarus*, **65**, 353–369.

Maltrud, M. E. and G. K. Vallis, 1991: Energy spectra and coherent structures in forced two-dimensional and beta-plane turbulence. *J. Fluid Mech.*, **228**, 321–342.

Moore, D. W. and P. G. Saffman, 1971: Structure of a line vortex in an imposed strain. In *Aircraft Wake Turbulence and its Detection*. Plenum Press, New York, pp. 339–354.

Niemann, H. B., S. K. Atreya, G. R. Carignan, T. M. Donahue, J. A. Haberman, D. N. Harpold, R. E. Hartle, D. M. Hunten, W. T. Kasprzak, P. R. Mahaffy, T. C. Owen and S. H. Way, 1998: The composition of the Jovian atmosphere as determined by the Galileo probe mass spectrometer. *J. Geophys. Res.*, **103**, 22831–22845.

Ortiz, J. L., G. S. Orton, A. J. Friedson, S. T. Stewart, B. M. Fisher and J. R. Spencer, 1998: Evolution and persistence of 5 µm hot spots at the Galileo Probe entry latitude. *J. Geophys. Res.*, **103**, 23051–23069.

Orton, G. and 40 others, 1996: Earth-based observations of the Galileo probe entry site. *Science*, **272**, 839–840.

Panetta, R. L., 1993: Zonal jets in wide baroclinically unstable regions: Persistence and scale selection. *J. Atmos. Sci.*, **50**, 2073–2106.

Polvani, L. M., J. Wisdom, E. DeJong and A. P. Ingersoll, 1990: Simple dynamical models of Neptune's Great Dark Spot. *Science*, **249**, 1393–1398.

Roos-Serote M., A. R. Vasavada, L. Kamp, P. Drossart, P. Irwin, C. Nixon and R. W. Carlson, 2000: Proximate humid and dry regions in Jupiter's atmosphere indicate complex local meteorology. *Nature*, **406**, 158–160.

Seiff, A., D. B. Kirk, T. C. D. Knight, R. E. Young, J. D. Mihalov, L. A. Young, F. S. Milos, G. Schubert, R. C. Blanchard and D. Atkinson, 1998: Thermal structure of Jupiter's atmosphere near the edge of a 5-µm hot spot in the north equatorial belt. *J. Geophys. Res.*, **103**, 22857–22889.

Showman, A. P. and T. E. Dowling, 2000: Nonlinear simulations of Jupiter's 5-micron hot spots. *Science*, **289**, 1737–1740.

Showman, A. P. and A. P. Ingersoll, 1998: Convection without dissipation: Explaining the low water abundance at Jupiter. *Icarus*, **132**, 205–220.

Vasavada, A. R., A. P. Ingersoll, D. Banfield, M. Bell, P. J. Gierasch, M. J. S. Belton, G. S. Orton, K. P. Klaasen, E. DeJong, H. H. Breneman, T. J. Jones, J. M. Kaufman, K. P. Magee and D. A. Senske, 1998: Galileo imaging of Jupiter's atmosphere: The Great Red Spot, equatorial region, and white ovals. *Icarus*, **135**, 265–275.

Yair, Y., Z. Levin and S. Tzivion, 1995: Lightning generation in a Jovian thundercloud: Results from an axisymmetric numerical cloud model. *Icarus*, **115**, 421–434.

Palaeoclimate Studies at the Millennium
The Role of the Coupled System

Paul J. Valdes

Department of Meteorology, University of Reading, UK

The study of past climates represents an important test of our understanding of the processes of climate change, and as a test of the climate models used to predict future climate change. In this chapter, we will review the current state of knowledge of this subject, and discuss the exciting opportunities in the next decade. In particular, the past decade of research has shown us that a clear understanding of climate change requires models which include the atmosphere, ocean, cryosphere and biosphere. Such models have now been developed and need to be thoroughly tested against the geological record. The results will help us evaluate our confidence in the future climate predictions, and will allow us to address fundamental questions about earth history. The next decade of research promises to be an exciting and important time for palaeoclimate studies.

A. INTRODUCTION

From a palaeoclimate perspective, celebrating 150 years is a very short-term view. Palaeoclimatology mostly covers periods much older than this, indeed there is interest in understanding climates millions and even billions of years into the past. This chapter will briefly review some aspects of our current understanding of the past climate change, and will highlight some particularly exciting challenges for future research.

There are many reasons for studying past climates. Probably the most important motivation is related to concern about future climate change. It can be argued that a full understanding of past climate change is an essential step for confident predictions of future climate change. Moreover, quantifying past natural climate variability helps to give a perspective on the importance of future anthropogenic climate change, and is vital for better "fingerprinting" of the causes of climate changes during the last century.

In addition, past climates provide an essential test of models used to predict the future. This is particularly relevant for the latest generation of earth system models. These models now include representations of changes in the slow components of the climate system, such as the cryosphere, terrestrial vegetation, and carbon cycle. However, many of these components have already been significantly altered by mankind (in the case of vegetation, it has been perturbed for many centuries). Hence, the past is the only way to test the natural component of these models.

An additional role for paleoclimate studies is to help understand the evolution of life, and especially the evolution of humans. As well as being of great importance from a cultural perspective (in much the same way that cosmology contributes to our understanding of the evolution of the universe itself), it also helps in understanding the potential impact of climate changes on humans.

Finally, it is also worth noting that the studies of the distant past can aid in the exploration and exploitation of economically important minerals, such as oil. The formation of many minerals depend, in part, on past environmental conditions. Thus accurate models of the past climate conditions can be used to infer potential formation regions for some minerals.

B. PRE-QUATERNARY PALAEOCLIMATES

Palaeoclimate studies can be broadly divided into Quaternary (the last 2 million years) and pre-Quaternary. The latter periods are often somewhat neglected when it comes to testing climate models, though they are of much greater importance when considering economic minerals or understanding the full history of environmental change. For instance, for much of the last 250 million years the earth appears to have been essentially ice-free. Antarctica may have started to be glaciated around 37 million years ago, and it is thought that Greenland became glaciated approximately only 3–4 million years ago. Thus these pre-Quaternary climates clearly illustrate that our present climate conditions are not the "normal" condition for the planet, or even the typical conditions during the last few hundred million years.

Obviously, as we go further back into the past, the data becomes much more uncertain. This makes testing climate models more difficult compared to Quaternary climates, because model–data disagreements may be due to errors in the data, rather that errors in the models. However, the signal of climate change can also be very large, so that in some sense the "signal-to-noise" ratio may be as good for some pre-Quaternary periods as for the Quaternary.

One of the most challenging pre-Quaternary climate issues during the past decade has been the continental interior warm winter problem. Considerable geological evidence suggests that for much of the time between 100 and 55 million years ago (mid-Cretaceous to early Eocene) the climate was much warmer that present (e.g. Barron, 1983; Wing and Greenwood, 1996; Sellwood et al., 1994). Estimates of global warmth vary between 4 and 8°C. Low latitudes seem to have only changed very modestly, perhaps 1–2°C, whereas high latitudes were substantially warmer (as much as 30°C). There is no evidence for ice at either pole.

The precise causes of the global warmth are unclear. Atmospheric carbon dioxide concentrations are estimated to have been much higher during these times (Berner, 1991) but the precise amount is very uncertain (estimates vary between 2 and 16× pre-industrial concentrations). Also, the lack of permanent ice cover would have probably decreased the surface albedo and hence enhanced the warmth (sea levels were also high, with large parts of North America flooded, and this would also have had a tendency to decrease the surface albedo). The continents were also in a somewhat different position (see Fig. 1), although the main differences are that India is still attached to Antarctica, as is S. America. The latter is important because it prevents a circumpolar current developing, which has been argued to have been the cause of the cooling of Antarctica (Kennet, 1977).

Atmosphere-only climate model simulations of this period (e.g. Barron et al., 1993; Valdes et al., 1996) have shown that the models are generally in agreement with the global warmth, but they imply that ocean poleward heat transport may have had to be greater during this time. The models all show that the Northern Hemisphere continental interiors were very cold during winter (temperatures dropping to −30°C or colder) (see Fig. 1). The higher CO_2 and warmer oceans do warm the continents appreciably (compared to the modern), but the temperatures are still well below freezing.

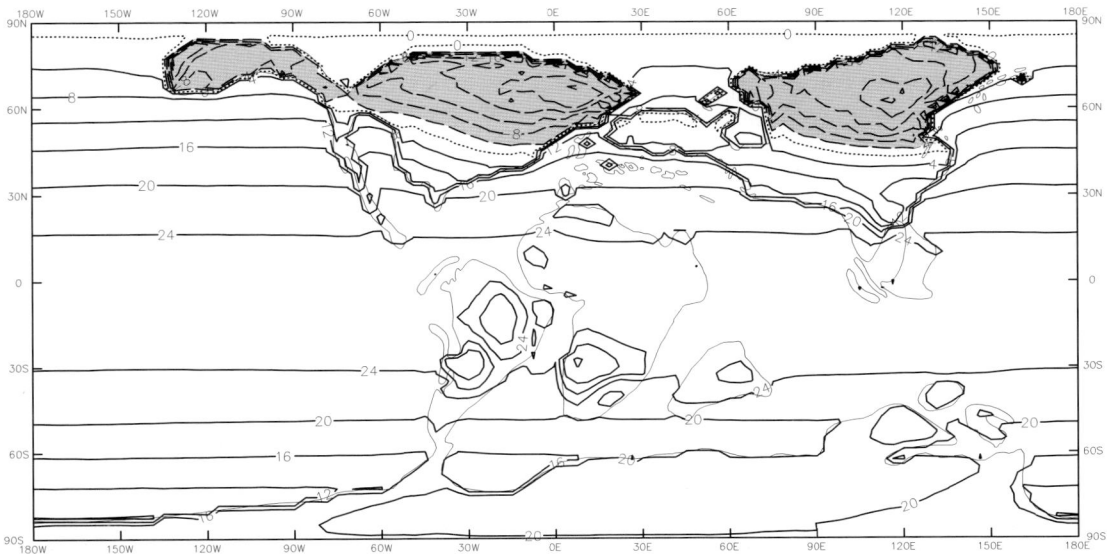

FIGURE 1 Hadley Centre climate model (HadAM3) simulation of the mid-Cretaceous (approximately 92 million years ago). The figure shows the December–January–February mean temperatures. The contour interval is 4°C. The shaded region indicates temperatures colder than −4°C. It can be seen that the continental interior temperatures are dropping well below zero.

However, estimates of climate based on plant fossils suggest that winters were nowhere near as extreme as the models suggest. These estimates, which are based on the morphological characteristics of the leaves rather than a nearest living relative approach, suggest that the winter temperatures in the interior of the continents of N. America and Eurasia barely dropped below zero. Thus there is a model–data disagreement of 30°C or more. Such a disagreement is much greater than any of the uncertainties in the methods and so represents a true failing of our understanding.

Although there have been various attempts to reconcile this disagreement, including questions about the accuracy of the data (Sloan and Barron, 1990), extra bodies of water (e.g. Sloan, 1994; Valdes et al., 1996, increased forest vegetation (e.g. OttoBliesner and Upchurch, 1997) or even enhanced polar stratospheric clouds (Sloan and Pollard, 1998), the basic problem has yet to be satisfactorily resolved. This remains a major challenge for the next decade.

An additional challenge for the climate models is to better represent the ocean circulation. To date, most climate models applied to the problem have either used prescribed sea surface temperatures (based on oxygen isotope analysis) or prescribed ocean heat transport. Thus such modelling studies have been unable to fully answer the question of why were these periods warm. In particular, there is an apparent paradox. Atmospheric models have suggested that in order to explain the correct latitudinal temperature gradient requires a significant increase in poleward ocean heat transport. However, the resulting weaker temperature gradients would suggest weaker winds and hence weaker forcing of the ocean circulation. We would therefore expect weaker poleward heat transport in the atmosphere and ocean. Reconciling this problem will require extensive use of coupled ocean–atmosphere models, which is only just beginning (e.g. Bice et al., 2000; Bush and Philander, 1997; Handoh et al., 1999).

C. QUATERNARY PALAEOCLIMATES

1. Observational Evidence

The last two million years, and especially the last 100,000 years, presents a demanding challenge for climate models of all forms. The richness of climate variability is at its best documented. During this period, climate was varying between cold, glacial times and warmer, interglacial periods. We are currently in a warm interglacial period. During the most extreme glacial times, ice sheets more than 2 km thick covered large areas of North America and Scandinavia (Peltier, 1994). The resulting sea-level drop was in excess of 120 m. European temperatures may have been more than 30°C colder than present (Peyron et al,. 1998).

At the start of the Quarternary, the variations of climate were dominated by oscillations on an approximately 41,000-year time period. However, at approximately 700,000 years ago the variability changed markedly. The amplitude increased and the period increased to approximately 100,000 years. In addition, throughout the Quarternary, there were oscillations of around 21,000 years. These three periods correspond to those calculated by Milankovitch (1941) for changes in the earth's orbit. More specifically the eccentricity of the orbit varies at 400,000-and 95,000-year periods, the obliquity (i.e. tilt) of the earth's axis varies on a 41,000-year period, and the time of the year when we are closest to the sun (called the precession of the equinoxes) varies on 19,000-and 23,000-year period. The closeness between the calculated periods and the observed periods has led to the suggestion that the changes in the earth's orbit causes the observed changes in climate. This is the so-called Milankovitch or orbital theory of climate variability, which is the dominant (but not the only theory) to explain the observed variations (for an alternative theory, see Muller and MacDonald, 1997).

Even if we accept the orbital theory, there are many aspects which require further research. The first point to note is that the changes in the earth's orbit result in large changes in the seasonal distribution of incoming solar radiation but relatively small changes in the annual mean. Hence we need to invoke climate feedbacks in order to explain why the annual mean climate changes because of these seasonal changes. The common explanation is that at times of low summer insolation, winter snows do not completely melt with the result that ice sheets start to build. This then changes the albedo and acts as a positive feedback which cools climate all year round.

The second problem is that a spectral analysis of the incoming solar radiation suggests that the dominant period is the precession one, at approximately 20,000 years. The changes in eccentricity (period near to 100,000 years) has relatively little spectral power, yet for the past 700,000 years the observed climate has been dominated by this period (Imbrie et al., 1993) Suggested explanations include the possibility of free, natural variability modes of the coupled climate system, or non-linear transitions between multiple equilibria (e.g. Saltzman et al., 1984; Paillard, 1988).

A third related problem is the question of why the observed variability changed at around 700,000 years ago. Prior to this, the observed variability was dominated by the 41,000-year period. Maasch and Saltzman (1990) suggested that it was related to a long-term cooling trend caused by a long-term decrease in atmospheric carbon dioxide. They were able to show, using a very simple three variable climate model (the three variables being atmospheric carbon dioxide, global ice volume, and global ocean temperature), that this long-term trend in atmospheric carbon dioxide could have changed the whole characteristic of variability.

Yet another challenge for our understanding is the observation from the ice cores from Greenland and Antarctica. The latter has now shown that atmospheric concentrations of carbon dioxide and methane have also varied by considerable amounts during the last 400,000 years (Petit *et al.*, 1999). During peak glacial times, these concentrations can drop to 190 ppmv and 350 ppbv, respectively. During peak interglacial periods, values were typically 280 ppmv (parts per million by volume) and 790 ppbv (parts per billion by volume) respectively (cf. 1990 values of 350 ppmv and 1700 ppbv). These large natural swings in radiatively active gases have yet to be fully explained, yet they have the potential for acting as important climate feedbacks. Recent work by Shackleton (2000) has suggested that the changes in carbon dioxide were in advance of changes in ice volume.

2. Modelling Studies

The above discussion clearly shows that a full understanding of this long-term climate variability will require a detailed knowledge of the whole earth's system, including the atmosphere, ocean, cryosphere and biosphere. All of these components have been shown to play an important role, and so must be included in any complete model of the system. A further challenge is that the time-scales involved in this problem will require models that can be run for many hundreds of millennia. Even given the likely rapid increase of computer power during the next decade, it will not be possible to achieve such intergrations with conventional GCMs.

This type of problem has led to a different type of climate model, which is now commonly called an Earth System Model of Intermediate Complexity (EMIC). Examples of this type of model include Gallee *et al.* (1992) and Claussen *et al.* (1999). The models work by representing the whole earth system, but with each component treated in a more simple way than in a GCM. For instance, Gallee *et al.* (1992) uses a 2-D atmospheric model with only latitude and height. There are no longitudinal variations, although the model does have a number of sectors representing the different continents. Similarly the ocean and cryosphere models are also relatively simple. When forced with known variations in the earth's orbital parameters and the observed variations in atmospheric carbon dioxide, the model is able to simulate the growth of the major Northern Hemisphere ice sheets. Analysis of the model results shows that the 100,000-year variation is related to the modulation of the precession cycle by the eccentricity variation, and is further amplified by the changes in carbon dioxide (Berger *et al.*, 1998).

Although these types of problem are very important, the use of simpler models means that we may be testing our understanding of the earth's system, but we are not directly testing the climate models used for predicting future climate change. For this, we must look at more recent problems and focus on specific aspects of the last 120,000 years. Within this period, there are major challenges for GCM type models.

a. The Glacial Inception Period

One of the long-standing challenges to GCMs has been the period around 115,000 years ago. According to orbital theory, this is the period during which there is less solar insolation in summer (see Fig. 2), which results in cooler summers and less melting of winter snow. This then starts the build-up of the great ice sheets. In theory a GCM simulation with orbital parameters and carbon dioxide concentrations modified to the appropriate values for 115,000 years ago should result in winter snows lasting through the summer. However, when this type of simulation was performed by a number of different climate models (e.g. Rind *et al.*, 1989; Royer *et al.*, 1983) the results were cooler summers (for an example, see Fig. 3, top), but no new areas in which winter snows lasted through the year. This suggests that these models were wrong, or we are missing some important process.

More recently, there have been a couple of GCMs which have been able to reproduce the required effect (e.g. Dong and Valdes, 1995; Syktus *et al.*, 1994) and another two models (deNoblet *et al.*, 1996b; Gallimore and Kutzbach, 1996) which have indicated that changes in high-latitude vegetation (not included in the other simulations) may be important. They suggested that the cooler summers resulted in southward migration of boreal forests and an expansion of tundra type vegetation. This is important because during winter, snow falls through the branches of boreal forest and thus the forest maintains a relatively low albedo despite high amounts of snow. By contrast, tundra is quickly covered by snow and thus has a high albedo. Thus an expansion of tundra acts as a positive feedback, enhancing the cooler conditions and helping shorten (or eliminate) snow-free conditions at high latitudes.

However, a note of caution should be added to all of these studies. Valdes and Glover (1999) have shown that a model's ability at simulating the cryosphere is

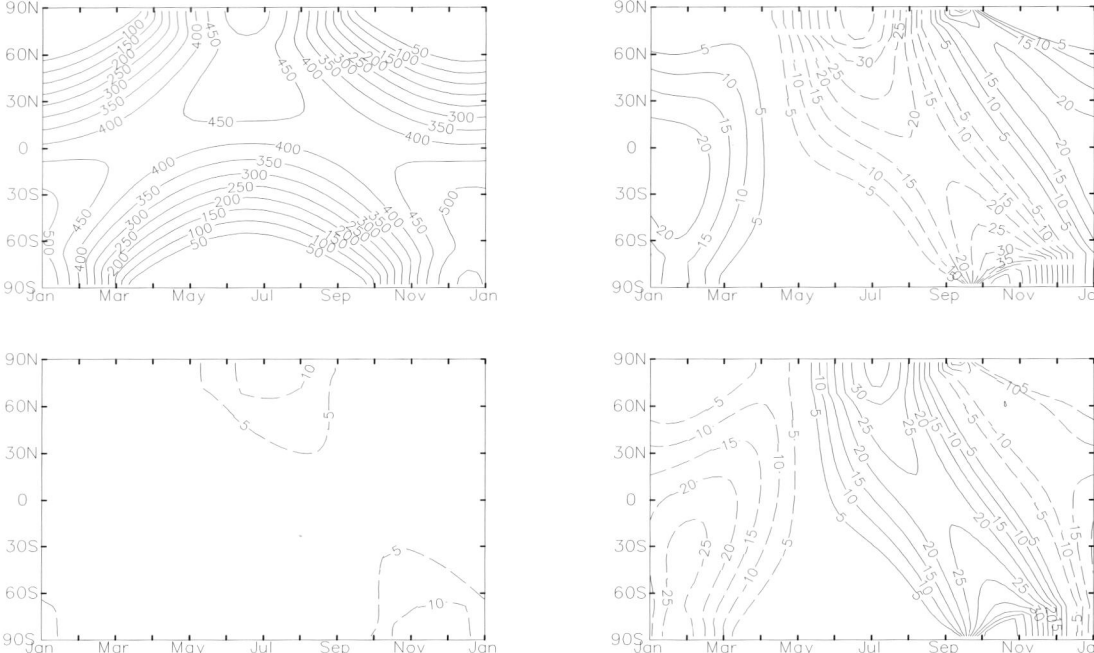

FIGURE 2 Changing patterns of incoming solar insolation, due to changes in the earth's orbit, for three key time periods. The upper-left panel shows the modern distribution, contour interval is 50 W m^{-2}. The upper-right panel shows the change from modern incoming solar radiation during glacial inception period (115,000 years ago), the lower-left panel for the Last Glacial Maximum (21,000 years ago), and the lower-right panel for the mid-Holocene (6000 years ago). The contour interval is 50 W m^{-2}. Note that the large changes in seasonal distribution (but small annual change), and the relatively small change at the Last Glacial Maximum.

strongly controlled by the model's ability at simulating the modern conditions. If the model has a warm summer bias in the modern control conditions, then it will require an extra large cooling to get the summer temperatures below zero during the conditions of 115,000 years ago, which is primarily required in order to have snow lasting through the summer. It is effectively a threshold-type behaviour. Similarly, a model with a cold bias in the modern climate will require an exceptionally small change to produce cool enough summers. Thus the success or failure of glacial inception simulations is being strongly controlled by the skill of the model in simulating modern, high-latitude summer temperatures. Models may have got the right result for the wrong reasons.

The above result also illustrates one of the challenges for the next decade. As we start to couple together different components of the earth system, we will find that small errors in the individual components may result in substantial errors in the coupled system. This has already been seen when atmosphere and ocean models were coupled and will become a growing problem when the biosphere and cryosphere are fully coupled to the other components.

b. The Last Glacial Maximum and Holocene Periods

Considering the more recent climates, further interesting challenges emerge. The last 21,000 years is an especially interesting period. It marks the transition from the coldest glacial conditions at 21,000 years ago (called the Last Glacial Maximum, or LGM), through a warm period between 13,000 and 15,000 years ago (called the Bolling–Allerod), a further cold period from about 11,000 to 12,500 years ago (called the Younger–Dryas), and eventually (at around 10,000 years ago) to the full interglacial conditions that we still enjoy today (called the Holocene). Carbon dioxide and methane concentrations also showed large, and rapid variations.

Many GCM studies of these periods have followed the snapshot style simulation in which the slow physical components (ice sheets, carbon dioxide, sea surface temperature or heat transport, etc.) are prescribed and an atmosphere-only model is integrated for a decade or two in order to compile a climatology of the period under consideration. Two key periods are the LGM itself (21,000 years ago), and the mid-Holocene period (6,000 years ago). Both periods have extensive data coverage available which can be compared to the model results, and numerous model simulations have been

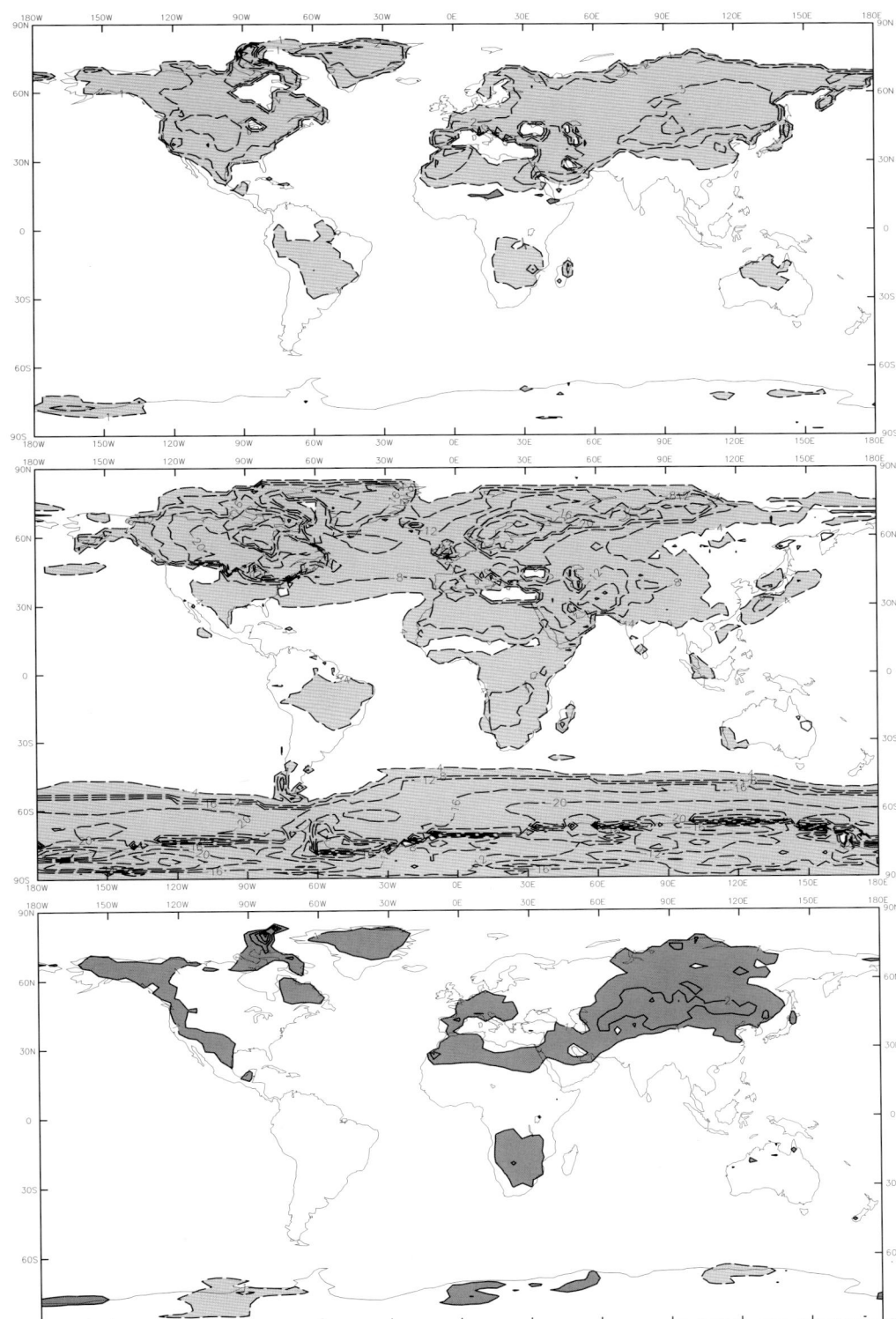

FIGURE 3 Changes in summer temperature (June–July–August mean) for three key periods during the last glacial cycle. Top panel shows the change between glacial inception periods (115,000 years ago) and the modern, the middle panel shows the change between the Last Glacial Maximum (21,000 years ago) and the modern, and the bottom panel shows the change between the mid-Holocene (6000 years ago) and the modern. For the glacial inception and the mid-Holocene figures, the contour interval is 1°C, and the light shading shows temperatures less than −1°C, and the darker shading shows temperatures greater than 1°C. For the LGM, the contour interval is 4°C, and the light shading shows regions where the temperature is less than −4°C. All results are from the Hadley Centre Climate Model (HadAM3).

performed, under the auspices of the Paleoclimate Model Intercomparison Project (PMIP) (Joussaume and Taylor, 1995).

PMIP complements the Atmospheric Model Intercomparison Project (AMIP) by investigating atmospheric model ability to simulate past climate change. The concept was to get a number of atmosphere-only models to perform identical simulations of the LGM and mid-Holocene. Comparing the different model predictions helps quantify uncertainty in the modelling process itself, and comparison of the model predictions with data helps to try and determine if any of the models produced a reasonable prediction.

Results for the LGM highlighted a number of important issues. All models predicted a cooling. This is perhaps not surprising given the boundary conditions, namely large ice sheets over North America and Scandinavia (Peltier, 1994) lower atmospheric carbon dioxide, sea surface temperatures based on either prescribed global reconstructions (CLIMAP, 1981) or using a simple slab ocean model with prescribed modern ocean heat transport. This latter assumption is probably a very poor choice, but we cannot observe past ocean heat transport and so there is no observationally based method for prescribing it. It was chosen because specifying ocean heat transport as the same as today's is a common assumption for future climate change scenarios.

Regional changes were more variable between the models. For instance, European winter temperatures were predicted to cool by 5–25°C, depending upon the climate model (for example, see Fig. 3 middle). Moreover, none of the models predicted sufficient cooling when compared with the data (Kageyama et al., 2000), mainly based on estimates from pollen. Does this show that models are fundamentally wrong? The answer is not simple. There are considerable uncertainties in the boundary conditions themselves. Pinot et al. (1999) have shown that CLIMAP may have over estimated the changes in the sea surface temperature in the North Atlantic, but this will generally make the data–model comparison even worse. In the future, work must focus on understanding this problem, and will entail running ocean and coupled ocean–atmosphere models (e.g. Bigg et al., 1998: Bush and Philander, 1998).

Another possible source of error is the data itself. As with most palaeo-data, the spatial coverage is not as good as for the modern. Much of the Western Europe pollen data originates from regions with relatively rich topographic structures, such as the Massif Central or the Pyrenees (Peyron et al., 1998). These orographic features are not resolved by the relatively coarse resolution of the GCMs. In addition, there was a small ice cap over the Alps but the GCMs again did not resolve this. Thus it is possible that at least some of the errors may be the result of the GCMs inadequately representing subgrid scale orography. A possible solution to this scale mismatch would be to make use of a limited area model, which would better represent the subgrid scale orography. These are now beginning to be used to better reconcile the scale of the data and the scale of the GCM (e.g. Hostetler et al., 2000).

The mid-Holocene simulations also presented a range of challenges for the models. The design of these simulations was just to include the orbital changes only (Fig. 2). No changes were made to sea surface temperature. The rationale behind this was that in general the observed changes of sea surface temperatures was less than the error bars of the method (typically 1 or 2°C). One of the most interesting results is related to the North African monsoon. The increased summer insolation results in warmer continents (for example, see Fig. 3, bottom), and a stronger monsoon (compared to the modern). All of the PMIP models predict such enhancement, but all of them underestimate the extent of the change. Pollen and lake level data all show a clear and substantial increase in precipitation (or precipitation–evaporation) during the mid-Holocene across most of North Africa. None of the models showed a sufficient latitudinal shift in the InterTropical Convergence Zone (Joussaume et al., 1999).

The causes for this model–data disagreement are still being debated. There are two leading theories. First, the neglect of any changes in the sea surface temperature is a serious limitation to the simulations. Hewitt and Mitchell (1998) and Kutzbach and Liu (1997) have shown that a coupled ocean–atmosphere model will produce an enhanced monsoon compared to the orbital-forcing-only simulation. However, even these simulations were insufficient to reproduce the observed changes.

Another weakness of the experimental design for the mid-Holocene PMIP simulations was that the land surface conditions were prescribed to be the same as today's. A number of simulations (e.g. Braconnot et al., 1999; de Noblet et al., 1996a; Kutzbach et al., 1996) have shown that soil and especially vegetation feedbacks can be an important aspect of the North African monsoon. Increased vegetation can lead to an enhanced monsoon, which then in turn encourages more vegetation. Moreover, Claussen (1994) has shown that this feedback process can result in multiple equilibria in today's conditions. If a model is initialized as vegetated, then it will stay vegetated. Claussen and Gayler (1997) have also shown that increased summer insolation during the mid-Holocene results in only the vegetated solution being stable. Claussen et al. (1999) have also suggested that this vegetated solution can rapidly collapse at around 5500 years ago, which is in reasonable agreement with the data.

The above work clearly shows that an accurate representation of the mid-Holocene will require running a fully coupled ocean–atmosphere–biosphere models (e.g. Braconnot et al., 1999). Indeed both PMIP simulations highlighted the need for running fully coupled models. However, the results also highlighted the problems associated with adding more components to the model. Even with atmosphere-only models, there was a large scatter in the results. It is unlikely that the scatter will diminish when fully coupled models are used. Indeed, the likelihood is that the scatter will become even larger.

D. SUBORBITAL TIMESCALE VARIABILITY

So far, we have only discussed long timescale variability. The palaeoclimate record can also give considerable information about shorter timescale variability. Ice core evidence, especially from Greenland, shows that there are many abrupt climate changes, such as Heinrich events (Bond et al., 1992) and Dansgaard–Oeshger cycles (GRIP Members, 1993: Taylor et al., 1993). There is also evidence for an approximate 1450-year oscillation (Bond et al., 1997). The former two appear to be related to large amounts of iceberg armadas entering the North Atlantic and modifying the ocean circulation. However, the processes involved in the 1450-year oscillation are less clear. The amplitude appears to be larger during glacial times, but it is still present in the Holocene. A satisfactory explanation for these oscillations is urgently required.

On even shorter timescales, there are some indicators of past El Niño events and the North Atlantic Oscillation (NAO). For instance, there is evidence that during the mid-Holocene, ENSO events were very different from their modern form, generally being much weaker (Rodbell et al., 1999). This represents an important test of our understanding and our ability to predict ENSO events. It is also very important as much of the recent warming trends are associated with enhanced ENSO events (Clement et al., 1999, Liu et al., 2000).

Similarly, changes in the NAO are of fundamental importance for European climate. High temporal resolution data from tree rings and other sources (e.g. Mann et al., 1999; Briffa, 2000; Cook et al., 1998) can help to better document the natural variability of the NAO. Moreover, this data can also be used to evaluate the climate model's ability to simulate natural variability (e.g. Jones et al., 1998; Osborn et al., 1999). This is a key aspect of methods to "fingerprint" anthropogenic climate change.

E. CONCLUSIONS

The above discussion has attempted to summarize some aspects of the past decade's research on palaeoclimate modelling, and to highlight a few of the many exciting opportunities for the next decade. A common theme running through much of the last decade's research on palaeoclimate modelling is that of experimental design. In general, this consists of specifying a number of aspects of the past climate, and then attempting to simulate the rest. Thus, the modelling studies have been unable to answer such questions as why was the LGM cold, because in large part the models were given the result. Thus when model–data discrepancies arose, there was always uncertainty about whether the internal aspects of the models were wrong (which would have important implications for future climate change simulations) or whether there were errors in the prescribed boundary conditions. Thus many results from climate models highlighted the processes needed to be included in a climate simulation, as opposed to truly testing our understanding of the models themselves.

The next decade will see a radical shift in this work. Climate models have now evolved into earth system models which include representations of the atmosphere, ocean, cryosphere and biosphere. These models are being used to predict the future and it is essential that they be fully tested. It will become increasingly possible to run models based on just a few fundamental external inputs, such as the orbital parameters and continental positions. Such inputs are generally well known, so the resulting comparisons between the model predictions and the observed data will be able to focus on either model data or incorrect interpretation of the observations. In addition, modelling will be much better able to answer fundamental questions about the reasons for past climate change.

A further exciting aspect of research over the next decade will be the increasing use of submodels which attempt to simulate the actual past climate observations. For instance, oxygen isotopes in ice cores and ocean sediments are extensively used for interpretation of climates but they are not a direct recorder of past temperatures. Rather than attempting to deduce climate from these isotopes, an alternative is to try and use a climate model to directly simulate the relevant ratios (e.g. see Joussaume and Jouzel, 1998; Schmidt, 1998). Such methods have the potential for greatly improving our quantitative comparisons of model outputs with observational data.

Another aspect to consider is the increasing computer power that is now becoming available. A decade ago, it was a major computing challenge to run an atmosphere-only GCM. Now it is possible to perform a similar simulation on a PC. It will soon become possible to run an earth system model for the whole transition from the Last Glacial Maximum to the modern. Such a simulation will be the ultimate test of our understanding of past climate change and the ultimate test of earth system models. The next decade promises to be an exciting and important time for palaeoclimate studies.

References

Barron, E. J., 1983: A warm, equable cretaceous—the nature of the problem. Earth Sci. Rev., **19**(4), 305–338.

Barron, E. J., 1989: Severe storms during earth history. Geo. Soc. Am. Bull., **101**(5), 601–612.

Barron, E. J., P. J. Fawcett, D. Pollard and S. Thompson, 1993: Model simulations of cretaceous climates—the role of geography and carbon dioxide. Phil. Trans. R. Soc. Lon. Se. B Bio. Sci., **341**(1297), 307–315.

Berger, A., M. F. Loutre and H. Gallee, 1998: Sensitivity of the LLN climate model to the astronomical and CO_2 forcings over the last 200 ky. Climate Dynam., **14**(9), 615–629.

Berner, R. A. 1991: A model for atmospheric CO_2 over phanerozoic time. Am. J. Sci., **291**, 339–376.

Bice, K. L., L. C. Sloan and E. J. Barron, 2000: Comparison of early Eocene isotopic paleotemperatures and three-dimensional OGCM temperature field: the potential for use of model-derived surface water $\delta^{18}O$. In Warm Climates in Earth History (B. T. Huber, K. .G. MacLeod and S. L. Wing, Ed.).

Bigg, G. R., M. R. Wadley, D. P. Stevens and J. A. Johnson, 1998: Simulations of two last glacial maximum ocean states. Paleoceanography, **13**(4), 304–351.

Bond, G., H. Heinrich, W. Broeker, L. Labeyrie, J. McManus, J. Andrews, S. Houn, R. Jantschik, S. Clasen, C. Simet, K. Tedesco, M. Klas, G. Bonani and S. Ivy, 1992: Evidence for massive discharge of icebergs into the North Atlantic ocean during the last glacial period. Nature, **360**, 245–249.

Bond, G. et al., 1997: A pervasive millennial-scale cycle in North Atlantic Holocene and glacial climates. Science, **278**(5341), 1257–1266.

Braconnot, P., S. Joussaume, O. Marti and N. deNoblet, 1999: Synergistic feedbacks from ocean and vegetation on the African monsoon response to mid-Holocene insolation. Geophys. Re. Lett., **26**(16), 2481–2484.

Briffa, K. R., 2000: Annual climate variability in the Holocene: interpreting the message of ancient trees. Quatern. Sci. Rev., **19**(1–5), 87–105.

Bush, A. B. G. and S. G. H. Philander, 1997: The late Cretaceous: Simulation with a coupled atmosphere-ocean general circulation model. Paleoceanography, **12**(3), 495–516.

Bush, A. B. G. and S. G. H. Philander, 1998: The role of ocean-atmosphere interactions in tropical cooling during the last glacial maximum. Science, **279**(5355), 1341–1344.

Claussen, M., 1994: On coupling global biome models with climate models. Climate Re., **4**, 203–221.

Claussen, M. and V. Gayler, 1997: The greening of the Sahara during the mid-Holocene: results of an interactive atmosphere-biome model. Glob. Eco. Biogeogr. Lett., **6**(5), 369–377.

Claussen, M. et al., 1999: Simulation of an abrupt change in Saharan vegetation in the mid-Holocene. Geophys. Re. Lett., **26**(14), 2037–2040.

Clement, A. C., R. Seager and M. A. Cane, 1999: Orbital controls on the El Niño/Southern Oscillation and the tropical climate. Paleoceanography, **14**(4), 441–456.

Climap, P. M., 1981: Seasonal Reconstructions of the Earth's surface at the Last Glacial Maximum. Geological Society of America Map and Chart Series, MC-36.

Cook, E. R., R. D. D'Arrigo and K. R. Briffa, 1998: A reconstruction of the North Atlantic Oscillation using tree-ring chronologies from North America and Europe. Holocene, **8**(1), 9–17.

deNoblet, N., P. Braconnot, S. Joussaume and V. Masson, 1996a: Sensitivity of simulated Asian and African summer monsoons to orbitally induced variations in insolation 126, 115 and 6kBP. Climate Dynam., **12**(9), 589–603.

deNoblet, N.I. et al., 1996b. Possible role of atmosphere-biosphere interactions in triggering the last glaciation. Geophys. Res. Let., **23**(22), 3191–3194.

Dong, B. W. and P. J. Valdes, 1995: Sensitivity studies of northern-hemisphere glaciation using an atmospheric general-circulation model. J. Climate. **8**(10), 2471–2496.

Gallee, H. et al., 1992: Simulation of the last glacial cycle by a coupled, sectorially averaged climate-ice sheet model. 2. Response to insolation and CO^2 variations. J. Geophys. Res. Atmos., **97**(D14), 15713–15740.

Gallimore, R. G., and J. E. Kutzbach, 1996: Role of orbitally induced changes in tundra area in the onset of glaciation. Nature, **381**(6582), 503–505.

GRIP Members, 1993: Climate instability during the last interglacial period in the GRIP ice core. Nature, **364**, 203–207.

Handoh, I. C., G. R. Bigg, E. J. W. Jones and M. Inoue, 1999: An ocean modeling study of the Cenomanian Atlantic: equatorial paleo-upwelling, organic-rich sediments and the consequences for a connection between the proto-North and South Atlantic. Geophys. Res. Lett., **26**(2), 223–226.

Hewitt, C. D. and J. F. B. Mitchell, 1998: A fully coupled GCM simulation of the climate of the mid-Holocene. Geophys. Res. Lett. **25**(3), 361–364

Hostetler, S. W., P. J. Bartlein, P. U. Clark, E. E. Small and A. M. Solomon, 2000: Simulated influences of Lake Agassiz on the climate of central North America 11,000 years ago. Nature, **405**(6784), 334–337.

Imbrie, J. et al., 1993: On the structure and origin of major glaciation cycles. 2. The 100,000-year cycle. Paleoceanography, **8**(6), 699–735.

Jones, P. D., K. R. Briffa, T. P. Barnett and S. F. B. Tett, 1998: High-resolution palaeoclimatic records for the last millennium: interpretation, integration and comparison with General Circulation Model control-run temperatures. Holocene, **8**(4), 455–471.

Joussaume, S. and J. Jouzel, 1988: Water isotope cycles and climate—investigations using atmospheric general-circulation models. Chemical Geol., **70**(1–2), 168.

Joussaume, S. and K. E. Taylor, 1995: Status of the Paleoclimate Modeling Intercomparison Project. Proceeding of the First International AMIP Scientific Conference, WCRP-92, Monterey, USA, pp. 425–430.

Joussaume, S. et al., 1999: Monsoon changes for 6000 years ago: Results of 18 simulations from the Paleoclimate Modeling Intercomparison Project (PMIP). Geophy. Res. Lett., **26**(7), 859–862.

Kageyama, M., O. Peyron, S. Pinot, P. Tarasov, J. Guiot, S.

Joussaume, G. Ramstein and PMIP participating groups, 2000: The Last Glacial Maximum climate over Europe and western Siberia: a PMIP comparison between models and data. Climate Dynam. In press.

Kennet, J. P., 1977: Cenozoic evolution of Antarctic glaciation, the circum-Antarctic ocean, and their impact on global paleoceanography. *J. Geophys. Res.*, **82**, 3843–3860.

Kutzbach, J., G. Bonan, J. Foley and S. P. Harrison, 1996: Vegetation and soil feedbacks on the response of the African monsoon to orbital forcing in the early to middle Holocene. *Nature*, **384**,(6610), 623–626.

Kutzbach, J. E. and Z. Liu, 1997: Response of the African monsoon to orbital forcing and ocean feedbacks in the middle Holocene. *Science*, **278**(5337), 440–443.

Liu, Z. Y., J. Kutzbach and L. X, Wu, 2000: Modeling climate shift of El Niño variability in the Holocene. *J. Geophys. Res. Lett.*, **27**(15), 2265–2268.

Maasch, K. A. and B. Saltzman, 1990: A low-order dynamic-model of global climatic variability over the full pleistocene. *J. Geophys. Res. Atmos.*, **95**(D2), 1955–1963.

Mann, M. E., R. S. Bradley and M. K. Hughes, 1999: Northern hemisphere temperatures during the past millennium: Inferences, uncertainties, and limitations. *Geophys. Res. Lett.* **26**(6), 759–762.

Milankovitch, M., 1941: Canon of insolation and the ice age problem (in Yugoslavian). *K. Serb. Acd. Beorg. Spec. Publ.* (English Translation by Israel Program for Scientific Translation, Jerusalem, 1969, 132.)

Muller, R. A. and G. J. MacDonald, 1997: Glacial cycles and astronomical forcing. *Science*, **277**(5323), 215–218.

Osborn, T. J., K. R. Briffa, S. F. B. Tett, P. D. Jones and R. M. Trigo, 1999: Evaluation of the North Atlantic Oscillation as simulated by a coupled climate model. *Climate Dynam.* **15**(9), 685–702.

OttoBliesner, B. L. and G. R. Upchurch, 1997: Vegetation-induced warming of high-latitude regions during the late Cretaceous period. *Nature*, **385**(6619), 804–807.

Paillard, D., 1998: The timing of Pleistocene glaciations from a simple multiple-state climate model. *Nature*, **391**(6665), 378–381.

Peltier, W. R., 1994: Ice-age paleotopography. *Science*, **265**(5169), 195–201.

Petit, J. R. *et al.*, 1999: Climate and atmospheric history of the past 420,000 years from the Vostok ice core, Antarctica. *Nature*, **399**(6735), 429–436.

Peyron, O. *et al.*, 1998: Climatic reconstruction in Europe for 18,000 yr B.P. from pollen data. *Quatern. Res.*, **49**(2), 183–196.

Pinot, S. *et al.*, 1999: Sensitivity of the European LGM climate to North Atlantic sea-surface temperature, *Geophys. Res. Lett.*, **26**(13), 1893–1896.

Rind, D., D. Peteet and G. Kukla, 1989: Can Milankovitch orbital variations initiate the growth of ice sheets in a general-circulation model. *J. Geophys. Res. Atmos.*, **94**(D10), 12851–12871.

Rodbell, D. T. *et al.*, 1999; An similar to 15,000-year record of El Niño-driven alluviation in southwestern Ecuador. *Science*, **283**(5401), 516–520.

Royer, J. F., M. Deque and P. Pestiaux, 1983: Orbital forcing of the inception of the Laurentide ice sheet. *Nature*, **304**(5921), 43–46.

Saltzman, B., A. R. Hansen and K. A. Maasch, 1984: The late quaternary glaciations as the response of a 3-component feedback-system to earth-orbital forcing. *J. Atmos. Sci.*, **41**(23), 3380–3389.

Schmidt, G. A., 1998: Oxygen-18 variations in a global ocean model. *Geophys. Res. Lett.*, **25**(8), 1201–1204.

Sellwood, B. W., G. D. Price and P. J. Valdes, 1994: Cooler estimates of Cretaceous temperatures. *Nature*, **370**(6489), 453–455.

Shackleton, N. J., 2000: The 100,000-year ice age cycle identified and found to lag temperature, carbon dioxide, and orbital eccentricity. *Science*, **289**(5486), 1897–1902.

Sloan, L. C., 1994: Equable climates during the early Eocene—significance of regional paleogeography for North-American climate. *Geology*, **22**(10), 881–884.

Sloan, L. C. and E. J. Barron, 1990: Equable climates during earth history. *Geology*, **18**(6), 489–492.

Sloan, L. C. and D. Pollard, 1998: Polar stratospheric clouds: A high latitude warming mechanism in an ancient greenhouse world. *Geophys. Res. Lett.*, **25**(18), 3517–3520.

Syktus, J., H. Gordon and J. Chappell, 1994: Sensitivity of a coupled atmosphere-dynamic upper ocean gcm to variations of CO_2, solar-constant, and orbital forcing. *Geophys. Res. Lett.*, **21**(15), 1599-1602.

Taylor, K. C., G. W. Lamorey, G. A. Doyle, R. B. Alley, P. M. Grootes, P. A. Mayewski, J. W. C. White and L. K. Barlow, 1993: The "flickering switch" of late Pleistocene climate change. *Nature*, **361**, 432–436.

Valdes, P. J., B. W. Sellwood and G. D. Price, 1996: The concept of Cretaceous equability. *Palaeoclimatol. Data Model.* **1**, 139–158.

Valdes, P. J. and R. W. Glover, 1999: Modelling the climate response to orbital forcing, *Phil. Trans. Roy. Soc.*, London, **357**, 1873–1890.

Wing, S. L. and D. R. Greenwood, 1996: Fossils and Fossil climate—the case for equable continental interiors in the Eocene. *Phil. Trans. R. Soc. Lon. Ser. B Biol. Sci.*, **341**(1297), 243–252.

Index

References in *italic* refer to Figures.

aerosols 124–5, 129–30
  biomass 171
  sulphur 156, 159, 252
  volcanic 125, 127–8, 146, 159, 252
aerosondes (robot aircraft) 22
albedo 69, 130, 165–6, *167*, 168, 252, 317, 319
Alpine Experiment (ALPEX) 34
ammonia 253, *254*, 309–10
Anaxagoras (Greek philosopher) 62
anomalies:
  forecast correlation 115
  high-pressure 150
  potential vorticity 35–7, *36*
  surface potential temperature 34–5
Arctic Oscillation (AO) 146
Aristotle (Greek philosopher) 62
Arrhenius, Svante 64, 152
atmosphere–ocean interchange 250–3, *251*
atmospheric:
  circulations global scale 283–302, 306–15
  composition 250
  forcing 185, 188–93
Atmospheric Model Intercomparison Project (AMIP) 25, 322

baroclinic:
  development 88
  instability 17–18, 20, 48–50, *51*, 64, 66, 308
barotropic models 112, 295–9
Bergen school 15–17, 83, 86, 101
Bergeron, T 86
Bernoulli function 35–7
billow turbulence 25, 223–4, *223*, *224*, 225
biogeochemical connections 249–57, 322–3
biospheric feedbacks 259, *260*
Bjerknes, Vilhelm 15–16, 17, 19, 64, 86, 106–8, *107*, 185, 195
boundary-layer:
  flow splitting 35, *35*, 37
  quasi-geostrophic flow 29–34, *31*, *32*
Buchan, Alexander 63
Buoyancy Effects in Fluids, J S Turner 221

carbon cycle 259–69
  biospheric feedbacks 272, *273–4*, 275–6
  climate coupled model 268–72, *270*
carbon dioxide concentrations 152, *153*, 156, 250–1, 257, 259–60, 317, 319
ceilometers 74
CFC (chlorofluorocarbons) 286
chaos theory 20
Charney, Jule G 64, 111–12, *111*, 113, 116
climate:
  sensitivity 123–4, 157

  UK 150, *151*
  variability 318–23
climate change 123, 133–41, 174–5
  anthropogenic 152–63
clouds:
  condensation nuclei (CCN) 252
  noctilucent (polar mesospheric) 284, 287, 295
  radiation feedback 181, 182, 252
Compendium of Meteorology 113
Conditional Instability of the Second Kind 63, 68
Convection, atmospheric 6
convective circulations:
  Betts-Miller 6, 7
  organized mesoscale 6
Cushing, D H 230
cyclones:
  extratropical 14–22, 63–4
    baroclinic instability 17–18
    cyclogenesis 6, 7, 18, 83, 86
    frontogenesis 20, 21, 66
    potential vorticity 19, *19*, 20, *20*, 22
    secondary cyclogenesis 21
    semi-geostrophic theory 20
    thermal theory 14–15, 17, 63

Deutscher Wetterdienst 84, 113
diffusive closures 6
dimethyl selenide (DMS) 252–3
dimethyl sulphide (DMS) 251–2
  CLAW hypothesis 256

Eady, E T 17, 18, 19, 64
eddy viscosity 287
Ekman transports 206, 212, 220–31, *221*, *227–9*
Electronic Computer Project 111
Electronic Numerical Integrator and Computer (ENIAC) 112, 113, 114
Ensemble Prediction System (EPS) 7, 8, *9*, 10–11, 12, 115, 194
ENSO (El Niño-Southern Oscillation) 199, 209–11, 266–8, *269*, 323
equations:
  Fokker-Planck 6, 7
  Liouville 3–5, *5*, 7
  Ricatti 5, *5*
equilibrium states:
  radiative-convective 67–9, *70*
  symmetrical 64–5
ERBE (Earth Radiation Budget Experiment) 166–71, 177
Espy, James Pollard 15, 63

FASTEX (Fronts and Atlantic Storm-Track Equipment) 90, 91, 96, *97*, 100, 101
Ferraro's law of isorotation 302

fisheries stock forecasting 230
Fitzroy, Vice Admiral Robert 15, 106
flows:
    near sea–surface laminar 220, 222, 224, 225, 230–1
    stratified 299
    unstratified shear 44–8, *45*
Fokker-Planck equation 6, 7
Forecast Factory, Lewis Richardson 117, *117*
forecasting:
    economic value 11, *11*, 81, 116
    numerical 3–12, 23–7
    probabilistic 8, 10–11, 12
forward tangent propagator 5
Fourier, Jean Baptiste-Joseph 64
Franklin, Benjamin 63
Friedman, R M 86
frontogenesis 20, 21, 66

Galileo observations 309–15
gas-giant planets 51, 54–5, *56*, 57–9, 60, 306–15
GEBA (Global Energy Balance Archive) 168, *170*, 171–2
Global Ocean Global Atmosphere integrations (GOGA) 148
gravity-wave:
    breaking 36, *36*, 39, 294–5
    propagation 29, 30, 32, *290*, 300
    solutions 111–12
Great Dark Spot (GDS), Neptune's 309
Great Red Spot (GRS), Jupiter's 309
greenhouse gases 69, 124, *125*, 126, 146, 152–7, 159, 162
grid reductions 24
gyroscopic pumping 292–4, *293*, 302*bis*

Hadley, George 63
Halley, Edmund 63
Hann, Julius 63–4
Helmholtz, H 15
High Resolution Limited Area Modelling 116
Hinkelmann, Karl–Heinz 113
hydrological cycle 165

Indian Ocean dipole 214–15, *216*, 217
InfraRed Interferometer (IRIS) 177–9, 181–2
initialization 110, 113
Institute for Advanced Studies 111
Interferometric Monitor for Greenhouse Gases (IMG) 178–82
International H2O Project (IHOP 2002) 80
IPCC (Intergovernmental Panel on Climate Change) 124, 125, 129, 153
iron particulates 254, *255*, 256, 257

Jupiter:
    atmosphere 306–9
    lightning 313–14, *314*
    water 312–15

Karman vortices 220, *221*, 222
Kelvin circulation 290, 291
Kleinschmidtz, Ernst 19
Komolgorov, A N 220

LGM (Last Glacial Maximum) 320, *321*, 322, 323
LIA (Little Ice Age) 140
lidars 73, 74, *75*, 77
lightning, Jupiter's atmosphere 312–15, *314*
Liouville equation 3–5, *5*, 7

lithium burning problem 302–3
Loomis, Elias 63*bis*
Lorenz (1963) attractor 4, *5*, 287, 302
LSVD (Lagged Single Value Decomposition) 187–8, 195

Madden-Julian Oscillation 69
magnetic diffusion layer 302
Margules, M 15, 64
MDI (Michelson Doppler Interferometer) 300
mesopause, summer polar 284–5, *285*
mesoscale: orography 29–40
Met. Office 113*bis*
Météo-France 84, 91, 97, 98, 100
Meteorologica, Aristotle 62
Michelson–Morley principle 291
Milankovich (orbital) theory 318, 319
model errors 5–7, 25–6, 318, 319–20, 322–3
models:
    atmospheric general circulation (AGCM) 144, 146, *147*, 148
    Atmospheric Model Intercomparison Project (AMIP) 25, 322
    baroclinic instability 17–18
    barotropic 113
    Bergen cyclone 16, *17*
    Canadian Mesoscale Compressible Community (MC2) 39
    CCM3 (Atlantic General Circulation) *147*, 148
    climate 23–4, 25–6, 126–9, 317–18, 319–20, 323
    coupled ocean-atmosphere 318–23
    dynamic global vegetation (TRIFFID) 262–3, *263*, 264, 265, 276
    Earth System Model of Intermediate Complexity (EMIC) 319
    European Centre for Medium–range Weather Forecasts (ECMWF) 6, *7*, 8, *9*, 10–11, 114–15
    Hadley Centre climate model (HadAM3) 167–9, 171–2, 260, 271, 276, *317*, *321*
    Hadley Centre general circulation model (HadAM2) 187, 194–5
    Hadley Centre ocean carbon cycle (HadOCC) 260–2, 265, 276
    High Resolution Limited Area Modelling (HIRLAM) 116
    Kida vortex in shear 308–9
    numerical 25, 26–7, 113
    ocean general circulation 243–4
    Palaeoclimate Model Intercomparison Project (PMIP) 322
    Swiss 39
moist convection, slantwise 314–15
momentum transport:
    eddy kinetic energy 308, 314
    wave-induced 287, 287–9, *289*, 291
monsoons:
    Indian Ocean;
        annual cycle 201–6
        surface heat balance 203–5, 212, *213*
        variability 198–201, 206–14
    North African 322
mountains, mesoscale 29–40
MWE (Medieval Warm Epoch) 140

NCEP (National Centres for Environmental Prediction) 113, 114
Near InfraRed Mapping Spectrometer (NIMS) 312
nitrogen oxides 253–4, *255*
non-linear dynamical systems 8
North Atlantic Oscillation (NAO) 143–51, *145*, 156, 185–6, 323
North Atlantic thermohaline circulation 154, *155*
numerical weather prediction 3–12, 23–7, 109–18
    quasi-geostrophic system 112, 113

ocean circulation:
    laminar 234–6

over North Atlantic 144, 186
　Stommel-Arons abyssal 238, *238*
　theorem 15–16
　turbulent 238–42, *239*, *240*, 243
ocean observations 233–45
Organization Météorologique Mondiale 84, 85
ozone:
　stratospheric 146, 162, 286
　tropospheric 159, 181, 250

Palaeoclimate Model Intercomparison Project 322
palaeoclimatology 137–8, 316–24
perturbations:
　dynamically constrained 8
　singular vector 8
planets, atmospheric dynamics 306–15
plankton multiplier mechanism 230–1
Plumb-McEwan experiment 289, *290*, 291, 295, 300
polar front meteorology 17
potential vorticity 19, *19*, 20, *20*, 48, *49*, 292, 299
　barotropic 55, *55*, 57, 59
precipitation, tropical 148–9, *150*
predictability:
　atmospheric 186–7
　potential 186, 194–6
prediction, ensemble 7, 8, *9*, 10–11, 12, 115, 194
Princeton Meteorology group 111, 112, 113
probability density function (pdf) 4, 5–7, 8
probability forecasts 8, 115
pyranometers 168, *170*

quasi-biennial oscillation (QBO) 289–91, *291*

radars 74, *77*, *78*, *79*
　Doppler 75–6, 78, *79*, 80, 291
　dual-polarization 74
radiation:
　balance 165–6, 168–72
　cloud absorption 171–2
　outgoing long-wave 166, 175–6, *176*, 177–82
　short-wave absorption *169*, 171–2
radiative:
　equilibrium temperature 286, 292, 294
　forcing 123–9, *125*, 130–1, 168
radio acoustic sounding system (RASS) 73
radiometers 73
reliability diagrams *10*
remote sensing 72–81
Reynolds numbers 220–1, 223, 283, 302, 308
Ricatti equation 5, *5*
Richardson, Lewis Fry 108–10, *108*, 112, 117, 220
Richardson numbers 221, 223, 224, 225, 231, 284, 301
Risk, weather-related 3
Rossby, Carl-Gustav 18–19, 111
Rossby waves 55, 60, 230, 292, 294–5, 295–6, 299, 300
Royal Meteorological Society 106, 112*bis*
RTE (radiative transfer equations) 226

satellite observations 174–82, 239, *244*, 286–7, *286*, 300
Shaw, Sir Napier 18
simulations, Eulerian *versus* semi–Lagrangian 25–6
SOHO (Solar and Heliospheric Observatory) 300
solar:
　insolation *320*
　interior magnetic field 299–302

radiation 165–72, 299–303, *301*
　variability 125, 303
Solberg, H 17, 87
solenoids 15
sonar 230
spectrometers:
　InfraRed Interferometer (IRIS) 177–9, 181–2
　Interferometric Monitor for Greenhouse Gases (IMG) 178–82
　Near InfraRed Mapping Spectrometer (NIMS) 312
　Total Ozone Mapping Spectrometer (TOMS) 81
SST (sea surface temperatures) 144, 146
　horizontal variance 230
　Indian Ocean 199, *200*, 201–2, *201*
　North Atlantic 185, 188–93
　ocean-to-atmosphere forcing 190–3
　tropical Atlantic 148–9
stellite measurements 73, 74
stochastic physics 6, 7, 8
Stuart solution 44, *45*, 46, *47*
surf zone 296, *298*, 299
Sutton, Sir Graham (Director, Met Office) 112
synoptic summary maps 84–8, *84*, *85*
　3- and 4-D 99–100
　ANASYG (analyse synoptique graphique) 85, 98, 101
　with dynamical graphics 88–90, 90–101, *91–7*, *99–100*, *102–4*
　PRESYG (prévision synoptique graphique) 85, 98, 99, 101
　reference synoptic trajectory 100–1
　standard French type *88*, *89*
　SYNERGIE forecasting 97, 98, *100*

tachocline (shear layer) 301, *301*, 302
targeted single vectors 22
Taylor caps 30, *31*, 34
Taylor, G I 221
TBO (tropospheric biennial oscillation) 207–8
television weather forecasts 81
temperatures:
　European instrumental records 137, *138*, 140, 174
　multiproxy averages 138–9, 140–1
　paleoclimatic 137–8, 317–24, *317*, *321*
　sea surface 144, 146, 148
　seasonal trends *135–6*, 318–23
　surface air 133–41, 158–62, *158*, *160*, 175
Thales (Greek philosopher) 62
thermocline, diurnal *224*, 225–6
topography, flow past 29–34
torsional oscillations 302, 303
Total Ozone Mapping Spectrometer (TOMS) 81
TRIFFID (dynamic global vegetation model) 262–3, *263*, 264, 265, 276
Tropical Atlantic Global Atmosphere integrations (TAGA) 148, *149*, *150*
Tropical Ocean Global Atmosphere integrations (TOGA) 148, *149*, *150*
tropical weather systems 66–9
turbulent flows 43–60, *298*
Tyndall, John 64

UARS (Upper Atmosphere Research Satellite) 286
UK Meteorological Office 8
uncertainty in forecasts 3–12
upscale cascades 43–60

volcanic aerosols 125, 127–8, 146, 159, 252
Von Neumann, John 110–11

vortices in shear 308–9, *310*, *311*
vorticity, absolute 295–7, *297*, 299, 308
Voyager observations 306–8, 313

wake generation 34, 36–7
    Alpine 37–9, *38*
water, Jupiter's atmosphere 312–15
wave-induced momentum transport 287, 287–9, 289, *289*, 291

Weather Prediction by Numerical Process, Lewis Richardson 108–9, 110, 111, 117
WMGG (well-mixed greenhouse gases) 69, 124, *125*, 126, 146, 152–7, 159, 162
World Meteorological Association 84, 85

yachts, drift prediction 226

International Geophysics Series

EDITED BY

RENATA DMOWSKA
Division of Engineering and Applied Science
Harvard University
Cambridge, MA 02138

JAMES R. HOLTON
Department of Atmospheric Sciences
University of Washington
Seattle, Washington

H. THOMAS ROSSBY
University of Rhode Island
Kingston, Rhode Island

Volume 1   BENO GUTENBERG. Physics of the Earth's Interior. 1959*

Volume 2   JOSEPH W. CHAMBERLAIN. Physics of the Aurora and Airglow. 1961*

Volume 3   S. K. RUNCORN (ed.) Continental Drift. 1962*

Volume 4   C. E. JUNGE. Air Chemistry and Radioactivity. 1963*

Volume 5   ROBERT G. FLEAGLE AND JOOST A. BUSINGER. An Introduction to Atmospheric Physics. 1963*

Volume 6   L. DUFOUR AND R. DEFAY. Thermodynamics of Clouds. 1963*

Volume 7   H. U. ROLL. Physics of the Marine Atmosphere. 1965*

Volume 8   RICHARD A. CRAIG. The Upper Atmosphere: Meteorology and Physics. 1965*

Volume 9   WILLIS L. WEBB. Structure of the Stratosphere and Mesosphere. 1966*

Volume 10   MICHELE CAPUTO. The Gravity Field of the Earth from Classical and Modern Methods. 1967*

Volume 11   S. MATSUSHITA AND WALLACE H. CAMPBELL (eds.) Physics of Geomagnetic Phenomena. (In two volumes.) 1967*

Volume 12   K. YA KONDRATYEV. Radiation in the Atmosphere. 1969*

Volume 13   E. PALMÉN AND C. W. NEWTON. Atmospheric Circulation Systems: Their Structure and Physical Interpretation. 1969*

Volume 14   HENRY RISHBETH AND OWEN K. GARRIOTT. Introduction to Ionospheric Physics. 1969*

Volume 15   C. S. RAMAGE. Monsoon Meteorology. 1971*

Volume 16   JAMES R. HOLTON. An Introduction to Dynamic Meteorology. 1972*

Volume 17   K. C. YEH AND C. H. LIU. Theory of Ionospheric Waves. 1972*

Volume 18   M. I. BUDYKO. Climate and Life. 1974*

Volume 19   MELVIN E. STERN. Ocean Circulation Physics. 1975*

Volume 20   J. A. JACOBS. The Earth's Core. 1975*

Volume 21   DAVID H. MILLER. Water at the Surface of the Earth: An Introduction to Ecosystem Hydrodynamics. 1977

* Out of Print

Volume 22   Joseph W. Chamberlain. Theory of Planetary Atmospheres: An Introduction to Their Physics and Chemistry. 1978*

Volume 23   James R. Holton. An Introduction to Dynamic Meteorology, Second Edition. 1979*

Volume 24   Arnett S. Dennis. Weather Modification by Cloud Seeding. 1980*

Volume 25   Robert G. Fleagle and Joost A. Businger. An Introduction to Atmospheric Physics, Second Edition. 1980

Volume 26   Kug-Nan Liou. An Introduction to Atmospheric Radiation. 1980*

Volume 27   David H. Miller. Energy at the Surface of the Earth: An Introduction to the Energetics of Ecosystems. 1981*

Volume 28   Helmut G. Landsberg. The Urban Climate. 1981

Volume 29   M. I. Budyko. The Earth's Climate: Past and Future. 1982*

Volume 30   Adrian E. Gill. Atmosphere-Ocean Dynamics. 1982

Volume 31   Paolo Lanzano. Deformations of an Elastic Earth. 1982*

Volume 32   Ronald T. Merrill and Michael W. McElhinny. The Earth's Magnetic Field. Its History, Origin, and Planetary Perspective. 1983*

Volume 33   John S. Lewis and Ronald G. Prinn. Planets and Their Atmospheres: Origin and Evolution. 1983

Volume 34   Rolf Meissner. The Continental Crust: A Geophysical Approach. 1986

Volume 35   M. U. Sagitov, B. Bodki, V. S. Nazarenko, and Kh. G. Tadzhidinov. Lunar Gravimetry. 1986*

Volume 36   Joseph W. Chamberlain and Donald M. Hunten. Theory of Planetary Atmospheres: An Introduction to Their Physics and Chemistry, Second Edition. 1987

Volume 37   J. A. Jacobs. The Earth's Core, Second Edition. 1987*

Volume 38   J. R. Apel. Principles of Ocean Physics. 1987

Volume 39   Martin A. Uman. The Lightning Discharge. 1987*

Volume 40   David G. Andrews, James R. Holton and Conway B. Leovy. Middle Atmosphere Dynamics. 1987

Volume 41   Peter Warneck. Chemistry of the Natural Atmosphere. 1988

Volume 42   S. Pal Arya. Introduction to Micrometeorology. 1988

Volume 43   Michael C. Kelley. The Earth's Ionosphere. 1989*

Volume 44   William R. Cotton and Richard A. Anthes. Storm and Cloud Dynamics. 1989

Volume 45   William Menke. Geophysical Data Analysis: Discrete Inverse Theory, Revised Edition. 1989

Volume 46   S. George Philander. El Niño, La Niña, and the Southern Oscillation. 1990

Volume 47   Robert A. Brown. Fluid Mechanics of the Atmosphere. 1991

Volume 48   James R. Holton. An Introduction to Dynamic Meteorology, Third Edition. 1992

Volume 49   Alexander A. Kaufman. Geophysical Field Theory and Method.

  Part A: Gravitational, Electric, and Magnetic Fields. 1992

  Part B: Electromagnetic Fields I. 1994

    Part C: Electromagnetic Fields II. 1994

Volume 50   Samuel S. Butcher, Gordon H. Orians, Robert J. Charlson, and Gordon V. Wolfe. Global Biogeochemical Cycles. 1992*

Volume 51   Brian Evans and Teng-Fong Wong. Fault Mechanics and Transport Properties of Rocks. 1992

Volume 52   Robert E. Huffman. Atmospheric Ultraviolet Remote Sensing. 1992

Volume 53   Robert A. Houze, Jr. Cloud Dynamics. 1993

Volume 54   Peter V. Hobbs. Aerosol-Cloud-Climate Interactions. 1993

Volume 55   S. J. Gibowicz and A. Kijko. An Introduction to Mining Seismology. 1993

Volume 56   Dennis L. Hartmann. Global Physical Climatology. 1994

Volume 57   Michael P. Ryan. Magmatic Systems. 1994

* Out of Print

Volume 58	THORNE LAY AND TERRY C. WALLACE. Modern Global Seismology. 1995
Volume 59	DANIEL S. WILKS. Statistical Methods in the Atmospheric Sciences. 1995
Volume 60	FREDERIK NEBEKER. Calculating the Weather. 1995
Volume 61	MURRY L. SALBY. Fundamentals of Atmospheric Physics. 1996
Volume 62	JAMES P. MCCALPIN. Paleoseismology. 1996
Volume 63	RONALD T. MERRILL, MICHAEL W. MCELHINNY, AND PHILLIP L. MCFADDEN. The Magnetic Field of the Earth: Paleomagnetism, the Core, and the Deep Mantle. 1996
Volume 64	NEIL D. OPDYKE AND JAMES E. T. CHANNELL. Magnetic Stratigraphy. 1996
Volume 65	JUDITH A. CURRY AND PETER J. WEBSTER. Thermodynamics of Atmospheres and Oceans. 1998
Volume 66	LAKSHMI H. KANTHA AND CAROL ANNE CLAYSON. Numerical Models of Oceans and Oceanic Processes. 2000
Volume 67	LAKSHMI H. KANTHA AND CAROL ANNE CLAYSON. Small Scale Processes in Geophysical Fluid Flows. 2000
Volume 68	RAYMOND S. BRADLEY. Paleoclimatology, Second Edition. 1999
Volume 69	LEE-LUENG FU AND ANNY CAZANAVE. Satellite Altimetry. 2000 (in press)
Volume 70	DAVID A. RANDALL. General Circulation Model Development. 2000 (in press)
Volume 71	PETER WARNECK. Chemistry of the Natural Atmosphere, Second Edition. 2000
Volume 72	MICHAEL C. JACOBSON, ROBERT J. CHARLSON, HENNING RODHE AND GORDON H. ORIANS. Earth System Science: From Biogeochemical Cycles to Global Change. 2000
Volume 73	MICHAEL W. MCELHINNY AND PHILLIP L. MCFADDEN. Paleomagnetism: Continents and Oceans. 2000
Volume 74	ANDREW E. DESSLER. The Physics and Chemistry of Stratospheric Ozone. 2000
Volume 75	BRUCE DOUGLAS, MICHEAL KEARNEY AND STEPHEN LEATHERMAN. Sea Level Rise: History and Consequences. 2000
Volume 76	ROMAN TEISSEYRE AND EUGENIUSZ MAJEWSKI. Earthquake Thermodynamics and Phase Transformations in the Earth's Interior. 2000
Volume 77	GEROLD SIEDLER, JOHN CHURCH AND JOHN GOULD. Ocean Circulation and Climate. 2001
Volume 78	ROGER PIELKE. Mesoscale. Meteorological Modeling, 2nd Edition. 2001
Volume 79	S. PAL ARYA. Introduction to Micrometeorology, 2nd Edition. 2001
Volume 80	SALTZMAN. Dynamical Paleoclimatology. 2001 NYP
Volume 81	LEE/KANAMORI/JENNINGS. International Handbook of Earthquake and Engineering Seismology. 2001 NYP
Volume 82	SHEPERD. Spectral Imaging of the Atmosphere. 2001 NYP

* Out of Print